Medical and Dental Guidance Notes (Second Edition)

A good practice guide on all aspects of ionising radiation protection in the clinical environment: IPEM Report 113

Online at: https://doi.org/10.1088/978-0-7503-2332-1

IPEM–IOP Series in Physics and Engineering in Medicine and Biology

About the Series

The series in Physics and Engineering in Medicine and Biology will allow the Institute of Physics and Engineering in Medicine (IPEM) to enhance its mission to 'advance physics and engineering applied to medicine and biology for the public good'.

It is focused on key areas including, but not limited to:

- clinical engineering
- diagnostic radiology
- informatics and computing
- magnetic resonance imaging
- nuclear medicine
- physiological measurement
- radiation protection
- radiotherapy
- rehabilitation engineering
- ultrasound and non-ionising radiation.

A number of IPEM–IOP titles are being published as part of the EUTEMPE Network Series for Medical Physics Experts.

A full list of titles published in this series can be found here: https://iopscience.iop.org/bookListInfo/physics-engineering-medicine-biology-series.

Medical and Dental Guidance Notes (Second Edition)

A good practice guide on all aspects of ionising radiation protection in the clinical environment: IPEM Report 113

Edited by

John Saunderson

Head of Radiation Protection and Radiology Physics, Hull University Teaching Hospitals NHS Trust, Hull, United Kingdom (Chair)

Mohamed Metwaly

Head of Dosimetry & Imaging, Radiotherapy Physics, United Lincolnshire Hospitals NHS Trust, Lincoln, United Kingdom

Contributors

William Mairs

Head of Imaging Physics and Radiation Protection, The Christie NHS Foundation Trust, Manchester, United Kingdom

Philip Mayles

Department of Physics, University of Liverpool, United Kingdom

Lisa Rowley

Head of Nuclear Medicine and Vascular Ultrasound, University Hospitals Coventry and Warwickshire NHS Trust, Coventry, United Kingdom

Mark Worrall

Head of Medical Physics, NHS Tayside, Ninewells Hospital, Dundee, United Kingdom

IOP Publishing, Bristol, UK

ISBN 978-0-7503-2332-1 (ebook)
ISBN 978-0-7503-2330-7 (print)
ISBN 978-0-7503-2333-8 (myPrint)
ISBN 978-0-7503-2331-4 (mobi)

DOI 10.1088/978-0-7503-2332-1

Version: 20240601

IOP ebooks

British Library Cataloguing-in-Publication Data: A catalogue record for this book is available from the British Library.

Published by IOP Publishing, wholly owned by The Institute of Physics, London

IOP Publishing, No.2 The Distillery, Glassfields, Avon Street, Bristol, BS2 0GR, UK

US Office: IOP Publishing, Inc., 190 North Independence Mall West, Suite 601, Philadelphia, PA 19106, USA

Contents

Preface

In 1921, the earliest predecessor to these Guidance Notes was published, when the newly formed British X-Ray and Radium Protection Committee released their '*Preliminary Report of the Committee*' (X-Ray and Radium Protection Committee 1921). That report set out to '*present knowledge in regard to equipment, ventilation and working conditions of X-ray and radium departments*'. That Committee of professional volunteers continued to publish revised guidance until the late 1940s when the UK government established a statutory Advisory Committee to take on this role. The '*Code of Practice for the Protection of Persons against Ionizing Radiations arising from Medical and Dental Use*' was subsequently published in 1957, and revised in 1964 and 1972 (Radioactive Substances Advisory Committee 1957, 1964, 1972). The final HM Stationery Office edition of the '*Guidance Notes for the Protection of Persons against Ionising Radiations arising from Medical and Dental Use*' was published in 1988 by the National Radiological Protection Board (NRPB *et al* 1988).

Following a change in regulations and government policy at the end of the twentieth century, the role of producing such guidance returned to the professional voluntary sector, and with the active assistance of the NRPB and regulators, the Institute of Physics and Engineering in Medicine (IPEM) published the first edition of '*Medical and Dental Guidance Notes: A Good Practice Guide on All Aspects of Ionising Radiation Protection in the Clinical Environment*' in 2002 (IPEM 2002).

This new 2024 edition of the IPEM Guidance Notes has been produced with the help of and in consultation with volunteers working in healthcare, representatives of UK regulators, and representatives of UK professional bodies. The working group consisted of senior NHS medical physicists representing the four IPEM Special Interest Groups (SIGs) for ionising radiations, namely:

- William Mairs for the Radiation Protection SIG;
- Philip Mayles for the Radiotherapy SIG;
- Lisa Rowley for the Nuclear Medicine SIG; and
- Mark Worrall for the Diagnostic Radiology SIG; with
- John Saunderson, who was chair of the Radiation Protection SIG at the time the project began, chaired the working group.

Mohamed Metwaly, editor-in-chief of the IPEM scientific report, oversaw the report's development with the chair of the working group and provided feedback, particularly during his final script editing, prepress work with the publisher, and final print check prior to publication.

These notes provide general guidance on good practice. They are not an attempt to repeat or interpret the legal requirements and advice contained in the legislation. Following the guidance is not mandatory but should suffice operationally to comply with the law. Other actions may be equally valid. Individuals who carry responsibilities under the legislation are advised to acquaint themselves appropriately with the legal requirements.

While the previous edition contained some very limited advice on non-ionising radiations, this edition exclusively addresses the issues surrounding the use of ionising radiations (e.g. radioactive substances, x-rays, protons, neutrons and electron beams) and does not attempt to offer advice on the safe use of lasers, MRI, ultrasound, ultraviolet, or other non-ionising radiations.

Acknowledgements

This latest version of the 'Guidance Notes' has been drafted by a working party of five nominees from the IPEM special interest groups with an interest in ionising radiation. A wide consultation process was followed to understand the needs and requirements of both professionals and regulators interested in the safe use of ionising radiation in medicine and dentistry. Before writing the first draft, invitations to request changes were sent out on the Medical–Physics–Engineering JISCmail list, and talks were given at several national meetings at which an invitation was made to delegates to send suggestions and requests for changes to the 2002 edition to the working party. The first complete draft was then shared with all the regulatory and professional bodies having an interest in its content, with an invitation to send comments on the draft to the working party. For some very specialist sections, particular experts were consulted.

So, as well as the five principal authors, a huge debt of gratitude is owed to the many kind and knowledgeable individuals and organisations who have generously contributed to and commented on this publication. The help given and received varied from a single comment to the drafting of whole sections. The Working Party would like to offer our hearty thanks to those listed below, with sincere apologies to any of those worthy of our deep gratitude who have inadvertently missed out.

Kathryn Adamson, Penny Allisy-Roberts, **AXREM**, Anna Barnes, Victoria Bassett-Smith, Alison Beaney, Michael Bernard, Helen Best, Simon Bishop, Mark Bradley, Andy Brennan, **British Institute of Radiology**, **British Nuclear Medicine Society**, John Byrne, Lindsey Cairns, Elly Castellano, Paul Charnock, Ian Chell, Conor Clancy, Philip Clewer, Thomas Cobley, Katrina Cockburn, Tim Coldwell, Pete Cole, Peter Colley, Duane Coombs, **CQC IR(ME)R Inspectorate**, Jilly Croasdale, Stephen Dainty, Lizzie Davies, Lindsey Devlin, John Dickson, David Dommett, Cliff Double, Ian Driver, Mike Dunn, David Eaton, Steve Ebdon-Jackson, Sue Edyvean, Mark Elliot, Sharon Ely, **Environment Agency**, Martyn Evans, Wil Evans, Richard Fernandez, Una Findlay, Louise Fraser, Karen Fuller, David Gallacher, Anne-Marie Gawronski, Daniel Gebretensae, Paul Giles, Tracy Gooding, Sarah-May Gould, David Grainger, Kate Griffiths, Andrew Gulson, David Hall, Graham Hall, Keith Hammond, Mark Hardy, Graham Hart, **Health and Safety Executive**, **Healthcare Improvement Scotland**, Peter Hiles, Paul Hinton, Keith Horner, Helen Hughes, Tony Hughes, **IPEM Special Interest Groups for Nuclear Medicine, Radiation Protection and Radiotherapy**, Andy Irwin, Lynda Johnson, Steven Jones, Peter Julyan, Mary Kelly, Yvonne Kinsella, Anastasios Konstantinidis, Ruth Lofts, Minder Louie, Jane MacKewn, Julian MacDonald, Daniel McGowan, Alastair McGown, Lynsey McKay, William Mairs, Anna Mayor, Laura Martin, Lucy Matthews, Philip Mayles, Mohamed

Metwaly, Thomas Melley, **MHRA**, Sofia Michopoulou, **Ministry of Defence**, Giles Morrison, Craig Moore, Rosemary Morton, Maria Murray, Michael Nettleton, **Office for Nuclear Regulation**, David Orr, Philip Orr, Nasreen Parker, Richard Peace, Debbie Peet, Sarah Peters, Lucy Pike, Jennifer Poveda, Amina Powell, Jennie Prince, **Public Health England**, Dick Putney, Stewart Redman, **Regulation and Quality Improvement Authority**, Jacqueline Roberts, Julie Robinson, Stewart Robinson, Andy Rogers, Nick Rowles, Lisa Rowley, John Saunderson, John Scott, **Scottish Environmental Protection Agency**, Joanne Shaw, Mary Simons, Mark Singleton, Ashley Smith, Tracy Soanes, **Society and College of Radiographers**, **Society for Radiological Protection**, Adam Stackhouse, Anne-Marie Stapleton, Joanne Stewart, Yvonne Sullivan, Lorna Sweetman, Catherine Taylor, Christie Theodorakou, Michael Thomas, Bill Thomson, Katharine Thomson, Jill Tipping, David Towey, **UK Health Security Agency**, **UK Radiopharmacy Group**, Aaron Vogl, Rachael Ward, Holly Warriner, Isabelle Watson, Michael Watt, Heather Williams, John Wilson, Richard Wilson, Tim Wood, Gail Woodhouse, Mark Worrall, Angela Wright.

And, of course, we also owe much gratitude to those giants upon whose shoulders we have stood; those who wrote the previous codes of practice and guidance notes that have steered us in our careers and upon which this publication is founded.

Editor biographies

John Saunderson

John Saunderson is a consultant medical physicist, leading the Radiation Protection and Radiology Physics service operating from Hull, UK, where he has provided support and advice to radiotherapy, nuclear medicine, radiology, dental radiology, and other radiation users for over 35 years. He is an accredited Radiation Protection Adviser, Radioactive Waste Adviser, Medical Physics Expert, Clinical Scientist and Consultant Scientist. John is a Fellow of the Institute of Physics and Engineering in Medicine. He has previously chaired the IPEM Radiation Protection Special Interest Group, and is currently an Association of Clinical Scientists' assessor, and external adviser for the IPEM Clinical Scientist Guided Training Scheme. He has peer-reviewed papers for the journals *Medical Physics, the Journal of Radiological Protection*, and *Physics in Medicine and Biology.*

Mohamed Metwaly

Mohamed Metwaly, PhD, FIPEM, is a lead consultant clinical scientist and registered medical physics expert (MPE) in the RPA2000 record (UK). He is the head of the Dosimetry and Imaging Quality Assurance Service (radiotherapy physics) at the United Lincolnshire Hospitals NHS Trust. He is the editor-in-chief of the IPEM Report Series. Since 2018, he has been a medical physics expert at the Health Research Authority (HRA), which reviews and approves ionisation radiation exposure for research and clinical trials. He joined the UK Accreditation Service's (UKAS) technical evaluation team for BS 70000 in 2018. Mohamed was appointed as the representative to the European Federation of Organisations for Medical Physics (EFOMP) in 2021. He chaired the publication and communication committee of the EFOMP in the same year for one term (2 years). In 2022, he joined the professional matter committee in the EFOMP, the International Organisation for Medical Physics (IOMP) publication and communication committee and the RPA2000 assessors' team for MPE certification.

William Mairs

William Mairs, BSc DipIPEM, PGDip, MSc MIPEM, is Group Leader for Imaging Physics and Radiation Protection at Christie Medical Physics and Engineering (CMPE). After gaining a Masters at Surrey University when starting a career in medical physics, Will undertook the Institute of Physics and Engineering in Medicine (IPEM) Part 1 and 2 training scheme to achieve HCPC registration as a Clinical Scientist. Training was undertaken at Hull and East Yorkshire Hospitals then CMPE, Manchester. Will has worked towards and attained a Radiation Protection Adviser (RPA) certificate and is recognised as a Medical Physics Expert, with an interest in Mammography Physics. He has a Postgraduate Diploma in Leadership and Management of Health and Social Care undertaken at the Universities of Bradford and Salford. He is now a Consultant Clinical Scientist and leads the HSE-recognised RPA body for CMPE. Through his work with the IPEM Radiation Protection Special Interest Group, Will has been part of the working party charged with updating the Medical and Dental Guidance Notes on behalf of IPEM and has led on chapters 1 (IRR17) and chapter 2 (IRMER17), as well as a number of appendices. Will is taking the lead in expanding the imaging science workforce for the NW Imaging Networks to ensure sustainable support for imaging services now and in the coming **years**.

Philip Mayles

Philip Mayles was Head of Physics until 2014 and then Radiation Protection Adviser and Radioactive Waste Adviser at the Clatterbridge Cancer Centre until he finally retired in 2019 after 50 years in the NHS. During this time, he advised on the building of two new radiotherapy facilities. He was an accredited RPA until 2022 and an accredited RWA until 2021. He contributed to both editions of the IPEM's Report 75 on the design of radiotherapy treatment facilities. He was responsible for the radiation protection section in the second edition of the Handbook of Radiation Physics, which he co-edited and which was published in 2022. He was a member of the Department of Health Medical Exposures Working Group for the 2013 revision of the EU Medical Exposures Directive. As a visiting professor at the University of Liverpool, he helped establish the MSc programme for STP and he continues to teach on this course. He is a fellow of the IPEM.

Lisa Rowley

Lisa Rowley, DClinSci, MIPEM, is Head of Nuclear Medicine and Vascular Ultrasound at the University Hospitals of Coventry and Warwickshire NHS Trust. She is an accredited Radiation Protection Adviser, Radioactive Waste Adviser and Medical Expert. She was also a Dangerous Goods Safety Adviser when first joining the working party for the medical and dental guidance notes update, although this has since lapsed. Lisa has worked in nuclear medicine at the Christie, Northampton, Leicester, and Oxford, was a corresponding member of the IPEM nuclear medicine special interest group during the update, and has led on the nuclear medicine chapters 10 to 20 and Appendix 21.

Mark Worrall

Mark Worrall, PhD, is the Head of Medical Physics at NHS Tayside, based at Ninewells Hospital, Dundee. His specialisation is diagnostic radiology and radiation protection, and he has provided support to the radiology, dental, nuclear medicine, and radiotherapy departments for 20 years. He is an accredited Radiation Protection Adviser, Radioactive Waste Adviser and Medical Physics Expert. He has previously chaired the IPEM diagnostic radiology special interest group and was the founder of the digital radiography user group. He is an assessor for RPA 2000, the Academy for Healthcare Science and the Association of Clinical Scientists. He is an external adviser for the IPEM Clinical Scientist Guided Training Scheme.

IPEM working group

Report editorial team
William Mairs, Head of Imaging Physics and Radiation Protection, The Christie NHS Foundation Trust, Manchester, UK

Philip Mayles, Department of Physics, University of Liverpool, UK

Lisa Rowley, Head of Nuclear Medicine and Vascular Ultrasound, University Hospitals Coventry and Warwickshire NHS Trust, Coventry, UK

John Saunderson, Head of Radiation Protection and Radiology Physics, Hull University Teaching Hospitals NHS Trust, Hull, UK

Mark Worrall, Head of Medical Physics, NHS Tayside, Ninewells Hospital, Dundee, UK

Report editor
John Saunderson, Head of Radiation Protection and Radiology Physics, Hull University Teaching Hospitals NHS Trust, Hull, UK

Editor in chief of the IPEM scientific reports
Mohamed Metwaly, Head of Dosimetry & Imaging, Radiotherapy Physics, United Lincolnshire Hospitals NHS Trust, Lincoln, UK

List of contributors

William Mairs
Head of Imaging Physics and Radiation Protection, The Christie NHS Foundation Trust, Manchester, UK

Philip Mayles
Department of Physics, University of Liverpool, UK

Lisa Rowley
Head of Nuclear Medicine and Vascular Ultrasound, University Hospitals Coventry and Warwickshire NHS Trust, Coventry, UK

John Saunderson
Head of Radiation Protection and Radiology Physics, Hull University Teaching Hospitals NHS Trust, Hull, UK (Chair)

Mark Worrall
Head of Medical Physics, NHS Tayside, Ninewells Hospital, Dundee, UK

Introduction

These medical and dental guidance notes (MDGN) have been prepared for those who use ionising radiation in medical and dental practice and in allied research involving human subjects. They apply wherever humans are irradiated for diagnostic, therapeutic, research, or other medical or dental purposes, or where *in vitro* medical tests are conducted. This includes all private and National Health Service (NHS) practice in hospitals, medical schools, clinics, mobile units, laboratories, surgeries and consulting rooms, including medical departments in industry and prisons. They apply also to ancillary activities such as the maintenance, testing, and calibration of equipment and the storage and disposal of radioactive substances where these are carried out in the above premises. The guidance given is for the protection of employed and self-employed persons, apprentices and students, patients and their friends and relatives who are acting as carers and comforters, volunteers in research projects, and members of the public.

These guidance notes are a guide to good radiation protection practice for the use of ionising radiation in medicine and dentistry. They include additional information and practical advice related to some, but not all, of the requirements of the Ionising Radiations Regulations (IRR 2017, IRR(NI) 2017) and its Approved Code of Practice (ACoP) and non-statutory guidance (HSE 2018) and the requirements of the Ionising Radiation (Medical Exposure) Regulations (IR(ME)R) and supporting guidance (IR(ME)R 2017, DHSC 2018, IR(ME)R(NI) 2018). It is essential, therefore, that employers and those who advise them are fully aware of these requirements and those of other relevant acts and regulations, in particular radioactive substances regulations (RSA 1993, CDG 2009, CDG(NI) 2010, RSA(A)R(NI) 2011, EPR 2016, EA(S)R 2018, RS(MoE)R(NI) 2018, REPPIR 2019, REPPIR(NI) 2019).

Throughout these notes, the words 'must', 'required', 'should', 'advised' and 'recommended' are used. This is intended to provide general guidance on good practice and is not an interpretation of the legal requirements. When in doubt, the appropriate regulations should be consulted.

All persons (e.g. employers, scientific and technical staff, medical and dental staff) whose work directly concerns the use of ionising radiation in medical or dental practice should consult the relevant sections of these guidance notes for practical advice on radiation protection.

In chapters 1 and 2, the foundations of the regulatory requirements are given forth, and in subsequent chapters, the requirements are created for specific applications. Safe procedures for using diagnostic x-rays for medical exposures and interventional radiology are discussed in chapters 3 and 4, whereas chapters 5 and 6 focus on dental departments at hospitals and universities, as well as the necessary equipment for them. Further guidance on dental radiology in other settings has been published by Public Health England and Faculty of General Dental Practitioners (PHE and FGDP 2020). Particulars of radiotherapy, including both external beam and brachytherapy, are discussed in chapters 7–9, while those of nuclear medicine are covered in chapters 10–15. Note that this second edition includes an additional chapter on Positron Emission Tomography (chapter 11). Therefore, the numbering of subsequent chapters differs from that of the 2002 edition. For those working in research and teaching establishments, the Association of University Radiation Protection Officers (AURPO) has produced AURPO Guidance Notes on Working with Ionising Radiations in Research and Teaching (AURPO 2019).

Chapters 16–20 are of interest to all those working with radioactive sources, whether in the field of nuclear medicine or in the field of brachytherapy. There is guidance concerning radioactive patients leaving the administering department, guidance on storage, transport, and disposal of radioactive materials, and guidance on contingency planning and emergency procedures for radioactive substance activity.

This edition has removed some appendices from the previous edition that are no longer relevant here and added a number of additional appendices that, it is hoped, will provide further practical help to readers in applying the regulations and safety requirements to their departments. The Appendices provide detailed information that is referred to in the text. A list of acronyms used in the MDGN is given in appendix 24. The number of appendices has increased from 21 in the 2002 edition to 24 in this second edition. Please note that the numbering of some of the appendices has changed.

Advice should be sought from the relevant Health Department, regulator, or authority with regard to any future developments in the usage of ionising radiation that are not covered by these medical and dental guidance notes.

References

AURPO 2019 *Guidance Notes on Working with Ionising Radiation in Research and Teaching* (UK: Association of University Radiation Protection Officers)

CDG 2009 *The Carriage of Dangerous Goods and Use of Transportable Pressure Equipment Regulations 2009* (https://legislation.gov.uk/uksi/2009/1348) (accessed 31 January 2023)

CDG(NI) 2010 *The Carriage of Dangerous Goods and Use of Transportable Pressure Equipment Regulations (Northern Ireland) 2010* (https://legislation.gov.uk/nisr/2010/160) (accessed 31 January 2023)

DHSC 2018 Guidance to the Ionising Radiation (Medical Exposure) Regulations *2017, Gov.UK Website* (Department of Health and Social Care) (https://gov.uk/government/publications/ionising-radiation-medical-exposure-regulations-2017-guidance) (accessed 31 January 2023)

EA(S)R 2018 The Environmental Authorisations (Scotland) Regulations 2018 (http://legislation.gov.uk/ssi/2018/219) (accessed 31 January 2023)

EPR 2016 *Environmental Permitting (England and Wales) Regulations* (https://legislation.gov.uk/uksi/2016/1154) (accessed 31 January 2023)

HSE 2018 *L121 work with Ionising Radiation Ionising Radiations Regulations 2017—Approved Code of Practice and guidance* 2nd edn (Norwich: Health and Safety Executive) https://hse.gov.uk/pubns/books/l121.htm (accessed 31 January 2023)

IPEM 2002 *Medical and Dental Guidance Notes. A Good Practice Guide on All Aspects of Ionising Radiation Protection in the Clinical Environment* ed P Allisy-Roberts, A Brennan, H Porter, M Rose and A Workman (York: Institute of Physics and Engineering in Medicine)

IR(ME)R 2017 *The Ionising Radiation (Medical Exposure) Regulations* (www.legislation.gov.uk/uksi/2017/1322) (accessed 31 January 2023)

IR(ME)R(NI) 2018 *The Ionising Radiation (Medical Exposure) Regulations (Northern Ireland)* (https://legislation.gov.uk/nisr/2018/17) (accessed 31 January 2023)

IRR 2017 *The Ionising Radiations Regulations.* Great Britain (https://legislation.gov.uk/uksi/2017/1075) (accessed 31 January 2023)

IRR(NI) 2017 *The Ionising Radiations Regulations (Northern Ireland)* (http://legislation.gov.uk/nisr/2017/229) (accessed 31 January 2023)

NRPB, HSE, Department of Health and Social Security, Department of Health and Social Services (Northern Ireland), Scottish Home and Health Department and Welsh Office 1988 *Guidance Notes for the Protection of Persons against Ionising Radiations Arising from Medical and Dental Use* (Chilton: National Radiological Protection Board)

PHE and FGDP 2020 *Guidance Notes for Dental Practitioners on the Safe Use of X-ray Equipment (2nd Edition)* 2nd edn (London: Public Health England, Faculty of General Dental Practice (UK)) https://rqia.org.uk/RQIA/files/44/449bdd1c-ccb0–4322-b0df-616a0de88fe4.pdf (accessed 31 January 2023)

Radioactive Substances Advisory Committee 1957 *Code of Practice for the Protection of Persons Exposed to Ionizing Radiations* 1st edn (London: Her Majesty's Stationary Office)

Radioactive Substances Advisory Committee 1964 *Code of Practice for the Protection of Persons Against Ionizing Radiations Arising from Medical and Dental Use* 2nd edn (London: Her Majesty's Stationary Office) https://iiif.wellcomecollection.org/pdf/b32179534 (accessed 23 March 2023)

Radioactive Substances Advisory Committee 1972 *Code of Practice for the Protection of Persons against Ionising Radiations Arising from Medical and Dental Use* 3rd edn (London: Her Majesty's Stationary Office) https://ia801008.us.archive.org/1/items/b3222011x/b3222011x.pdf (accessed 23 March 2023)

REPPIR 2019 *The Radiation (Emergency Preparedness and Public Information) Regulations 2019* (http://legislation.gov.uk/uksi/2019/703) (accessed 31 January 2023)

REPPIR(NI) 2019 *The Radiation (Emergency Preparedness and Public Information) Regulations (Northern Ireland)* (http://legislation.gov.uk/nisr/2019/185) (accessed 31 January 2023)

RSA 1993 *Radioactive Substances Act 1993* (http://legislation.gov.uk/ukpga/1993/12) (accessed 11 February 2023)

RSA(A)R(NI) 2011 *The Radioactive Substances Act 1993 (Amendment) Regulations (Northern Ireland)* (https://legislation.gov.uk/nisr/2011/290) (accessed 31 January 2023)

RS(MoE)R(NI) 2018 *The Radioactive Substances (Modification of Enactments) Regulations (Northern Ireland)* (Northern Ireland) https://legislation.gov.uk/nisr/2018/116 (accessed 23 March 2023)

X-Ray and Radium Protection Committee 1921 X-ray and radium protection *J. Roengen Soc.* **17** 97–144 (accessed 23 March 2023)

IOP Publishing

Medical and Dental Guidance Notes (Second Edition)
A good practice guide on all aspects of ionising radiation protection in the clinical environment:
IPEM Report 113
John Saunderson, Mohamed Metwaly, William Mairs, Philip Mayles,
Lisa Rowley and Mark Worrall

Chapter 1

General measures for radiation protection

Scope

1.1 Chapter 1 contains guidance on the organisational arrangements and general measures for radiation protection for staff and members of the public within a hospital or healthcare environment. It aims to set out a coherent introduction to various topics and is largely based on the Ionising Radiations Regulations 2017 (IRR 2017) and the Ionising Radiations Regulations (Northern Ireland) 2017 (IRR(NI) 2017), subsequently referred to as IRR17 unless a previous version is referenced. It also contains some measures for the radiation protection of patients, but this is covered in more depth in chapter 2, *Radiation protection of persons undergoing medical exposures*, along with protection for 'carers and comforters'. The guidance in this and the following chapters should be read in conjunction with *L121, Work with Ionising Radiation* (HSE 2018a), taking into account the legal status of the Regulations, the Approved Code of Practice and Health and Safety Executive (HSE) guidance in that publication, along with non-statutory guidance (e.g. HSE *Information Sheets*, as available) and professional advice. Where comprehensive guidance on a topic exists, this is indicated in the text.

The employer

1.2 The legal responsibility for compliance with IRR17 lies with the Employer, who may be a company, an NHS organisation, a visiting contractor, or a self-employed person such as a dentist or partner in a group practice. The role of the Employer is clearly identified in IRR17, as are the different duties of the Employer separately defined under the Ionising Radiation (Medical Exposure) Regulations (IR(ME)R 2017, IR(ME)(A)R 2018, IR(ME)R(NI) 2018), subsequently referred to as IR(ME)R unless a previous version is referenced (see paragraphs 2.8–2.14. *The Employer* in chapter 2).

1.3 The Employer should demonstrate a clear commitment to optimisation of exposure through a structured approach to the operational management of radiation protection. The range of Employers covered by this document is wide, and each will have a different management structure. To avoid undue repetition, the core of the text assumes a structure typical of a large hospital. In some instances, guidance specific to the smaller user is given. However, where reference is made to posts that may not exist in their organisation, such as head of a department or line manager, readers are encouraged to look at the functions and responsibilities assigned to that post and pose the question: 'Is it relevant to our work with radiation and who should undertake that role?'

1.4 At times, it may not be obvious who the Employer is. For example, private companies may use the equipment and staff at a working NHS facility to carry out procedures on private patients. Alternatively, buildings and equipment may be owned by a privately funded initiative (PFI) but used by an NHS Employer. Contracts and procedures should make clear where the responsibilities lie so there is no ambiguity, and these should be reviewed periodically.

1.5 Where work with ionising radiation by one Employer could lead to exposure of another Employer's employees or when employees have multiple Employers, regulation 16 of IRR17 stipulates that the various Employers must cooperate with each other to ensure they each meet the requirements of the regulations (see paragraphs 1.127–1.152 *Cooperation between Employers and others*).

Responsibilities

Employers

1.6 Under regulation 9(1) of IRR17, the Employer has overall responsibility for restricting exposure to ionising radiation and will need to cooperate with any other Employers (regulation 16) whose employees are affected by their work with ionising radiation. All Employers must make sure that the cumulative exposure of their employees from work activity over the year does not exceed a relevant dose limit (Regulation 12 along with Paragraphs 212 and 219 of L121 (HSE 2018a)). For employees that are not normally exposed to ionising radiation, the Employer should take steps to ensure doses are unlikely to exceed public dose limits in the course of their work (see paragraph 86 of L121 and paragraphs 1.127–1.152 *Cooperation between Employers and others* below).

1.7 The Employer may delegate tasks and allocate functions required by the regulations to suitably trained individuals, but cannot delegate responsibility. The Employer's responsibilities are summarised in appendix 1 *Roles and*

Responsibilities of the Employer using Ionising Radiation. Failure to meet the requirements may result in an Employer, such as a company, being prosecuted as a distinct legal personality. However, there are powers to proceed against, amongst others, employees and directors in addition to, or instead of, the Employer where warranted. Consideration would be given to their role if an offence was committed with their consent or attributable to neglect on their part (HSE no date a).

1.8 The commitment of senior management to restrict exposure to ionising radiation should be clearly demonstrated in a written radiation safety policy (section 2(3) of the *Health and Safety at Work etc Act* (HASWA 1974)). The safety policy should clearly identify those with responsibility, and the scope of that responsibility (see appendix 2, *Ionising radiation protection and medical exposures policy*). Heads of departments and line managers should be involved in the implementation of radiation protection, as with other health and safety (H&S) requirements.

1.9 Employers should not become divorced from radiation protection matters through an overdependence upon Radiation Protection Advisers (RPAs) and Radiation Protection Supervisors (RPSs). For instance, it is not the function of an RPA acting in that capacity to actively manage radiation protection, nor is it their function to act on the Employer's behalf. Similarly, it is not the function of an RPS to carry out investigations on the Employer's behalf nor is it their function to actively manage dosimetry. If a person who is an RPS is doing any of these tasks, they are doing them as an employee, not in their role as an RPS.

Employees

1.10 Employees have responsibilities that are concerned with limiting their own and others' radiation exposure. They are required to use, return to storage, and report damage to any Personal Protective Equipment (PPE) provided (see paragraphs 1.72–1.78 on *PPE*).

1.11 If they have cause to believe that they or some other person have received an overexposure or that a radioactive source has been mismanaged (e.g. lost or stolen), they must immediately notify their Employer of that belief.

1.12 Employees must also comply with the Employer's reasonable requests regarding radiation dose monitoring or assessment (see paragraphs 1.152–1.165 on *Personal Monitoring for External Radiation* and paragraphs 1.175–1.179 on *Personal Monitoring for Internal Radiation*) and medical surveillance (see paragraphs 1.187–1.198 on *Classified Workers*), if applicable. Paragraph 620 of L121 (HSE 2018a) makes clear that employees must wear, take reasonable care of, and return for processing any personal dosemeters provided by their Employer or they may be committing an offence under section 7—General duties of employees at work—of the *Health and Safety at Work etc Act* (HASWA 1974). Radiation protection culture in the medical sector is not always at the high level expected and achieved in other radiation industries, as demonstrated by the number of late or unreturned personal dosemeters (Cole *et al* 2014). Compliance with dosimetry requirements will likely be of interest at regulatory inspection, and poor compliance makes it difficult to prove that radiation protection measures are effective.

Manufacturers and installers

1.13 There are duties imposed on manufacturers, etc, of articles for use in work with ionising radiation to ensure that such articles are designed and constructed so as to restrict, so far as is reasonably practicable (SFAIRP), the extent to which employees and other persons are likely to be exposed to ionising radiation. The manufacturer's responsibility in regard to restriction of exposure does not extend to those undergoing medical exposure, which falls under the *Ionising Radiation (Medical Exposure) Regulations* (IR(ME)R 2017, IR(ME)R(NI) 2018), except for the assessment of those safety features and warning devices that protect patients from unintended exposures as part of a critical examination. The critical examination is the responsibility of the installer (see paragraphs 1.245–1.259 on *Critical Examination*).

1.14 Manufacturers and installers also have relevant responsibilities under the Medical Device Regulations (MDR 2002). Those manufacturing 'in-house' devices should also consider if their device must comply with these regulations. Current arrangements and guidance can be found on the UK Government website (MHRA 2022).

1.15 The Medical Device Regulations stipulate that a device must meet the essential requirements as set out in Annex I of the relevant Medical Device Directive (European Commission 1993, 1998, MHRA 2022). The requirements are classified as, for example, general requirements, protection against radiation, and information supplied by the manufacturer. A number of examples from Annex I follow that are likely to be sources of information to be used as part of the risk assessment process (see paragraphs 1.44–1.54 *Risk Assessments*).

- Clause 2. The solutions adopted by the manufacturer for the design and construction of the devices must conform to safety principles, taking account of the generally acknowledged state of the art. In selecting the most appropriate solutions, the manufacturer must apply the following principles in the following order:
  - eliminate or reduce risks as far as possible (inherently safe design and construction).
  - where appropriate, take adequate protection measures, including alarms if necessary, in relation to risks that cannot be eliminated.
  - inform users of the residual risks due to any shortcomings of the protection measures adopted.

- Clause 11.4.1: The operating instructions for devices emitting radiation must give detailed information as to the nature of the emitted radiation, means of protecting the patient and the user and ways of avoiding misuse and eliminating the risks inherent in installation.
- Clause 13.1: Each device must be accompanied by the information needed to use it safely and to identify the manufacturer, taking account of the training and knowledge of the potential users.
- Clause 13.6. Where appropriate, the instructions for use must contain the following particulars: (d) all the information needed to verify whether the device is properly installed and can operate correctly and safely, plus details of the nature and frequency of the maintenance and calibration needed to ensure that the devices operate properly and safely at all times;

The self-employed

1.16 Those who are self-employed have duties under IRR17 in the capacity of both the Employer and the employee.

Everyone

1.17 It is the responsibility of every person, employee or otherwise, not to misuse or interfere with sources of ionising radiation.

Culture

1.18 A strong radiation protection culture should permeate the organisation if radiation exposures are to be reduced to levels considered as low as reasonably practicable (ALARP). Despite the documentation and procedures in place, it is the safety culture that impacts the reality of behaviours. An organisation with a good radiation protection culture will have leaders who demonstrate commitment to safety, a questioning workforce that is willing to challenge unsafe activities, and a system whereby everyone is personally responsible for safety. For more on this subject and guiding principles on how to assess and establish a radiation protection culture, see the following references (INPO 2013, Cole *et al* 2014, IRPA 2014, Chapple *et al* 2017).

RSC

1.19 Employers are recommended to establish a Radiation Safety Committee(RSC) as part of the framework for the management of radiation protection (HSE 2015a). This will assist in reviewing the implementation of advice and should have a good liaison with the health and safety committee. An additional function of the RSC might be to oversee medical exposures, as explained in chapter 2, paragraphs 2.60–2.63 on *Governance*.

1.20 Many radiation protection decisions are taken routinely by appropriately trained staff in the departments undertaking work with ionising radiation. An RSC might be consulted about new, unusual, or difficult decisions relating to:

- establishing a programme of testing and maintenance of engineering controls etc;
- measures that affect health and safety, such as the provision of PPE;
- the content of local rules;
- dosimetry management and review of data;
- setting formal investigation levels;
- investigation of overexposures or those that will result in a special entry in a classified workers dose record.

The committee is also likely to act as a forum to discuss and provide feedback on local errors and near misses.

1.21 The composition and function of the RSC will be a local matter but should include senior management representatives, the medical director, clinical directors who use radiation, the health and safety manager, Radiation Protection Supervisors, Radiation Protection Advisers, Radioactive Waste Advisers if appointed as a requirement of radioactive substances legislation (see paragraph 19.9 and appendix 21, *Role of the RWA*), Medical Physics Experts (MPEs) under medical exposure regulations (IR(ME)R 2017, IR(ME) R(NI) 2018) (unless in a separate Medical Exposures Committee (MEC)], staff, trade union, and safety representatives. In very large establishments, there could be separate RSCs with appropriate communication arrangements at the managerial and operational levels for:

- X-ray work (e.g. for radiology, cardiology, accident and emergency, theatre screening and any service that does not fall under the supervision of a radiation specialist department e.g. consultant led use of mini c-arms outside the radiation protection framework of radiology);
- radionuclides (e.g. for laboratories, Nuclear Medicine department, radionuclide dispensary);
- radiotherapy (external beam, brachytherapy and other sealed sources).

Communication

1.22 There must be effective communication for an organisation to have a good health and safety culture and successful implementation of its radiation protection policy. Information will have to be provided to employees and, when appropriate, to other Employers and their employees (see paragraphs 1.127–1.152 on *Cooperation between Employers and others*). There will be a need to consult and communicate with other organisations or individuals as circumstances demand.

1.23 The HSE provides guidance on the establishment of effective systems of internal communication within a health and safety framework in *Managing for Health and Safety* (HSE 2013b). Communication of radiation protection policy should be supported with the reasons for doing activities in a particular way, evidence of standards and the monitoring process, examples of good and bad practice and lessons learned from incidents. Ways of communicating this information include leading by example, face-to-face discussions, presentations, and through written material such as policy statements or posters. All those involved must have an opportunity to contribute. It is a legal requirement to consult employees about health and safety issues that affect them, either directly or via a representative (SRSCR 1977, SRSCR(NI) 1979, HSCER 1996, HSCER(NI) 1996). Participation should encourage ownership and drive the need for individuals to take personal responsibility for safety.

1.24 The Employer must establish adequate arrangements for communication regarding staff safety, supervision, and radiation monitoring. The adequacy of these arrangements should be reviewed on a regular basis. Examples of what the Employer must communicate internally are:

- Information, instruction and training to restrict exposure (see paragraphs 1.92–1.103), including systems of work where appropriate (see paragraphs 1.116–1.118 on *Local Rules*).
- That pregnant employees and those planning to breastfeed should inform the Employer in writing and may be subject to restrictions as described in *Guidelines for Expectant or Breastfeeding Mothers* (HSE 2015b).
- Arrangements for personal dosimetry and provision of results including notifying staff of the intention to designate them as classified.

1.25 Other bodies or individuals that may have to be communicated with are listed below, with more details elsewhere in these guidance notes:

- Regulators for obtaining certificates, permits, etc and for reporting notifiable incidents.
- Qualified experts and services, such as RPA, RWA, MPE, Appointed Doctor, Dangerous Goods Safety Adviser (DGSA) or Approved Dosimetry Service (ADS).
- Members of the public who work with radiation which may affect them directly.

Radon

1.26 Radon is a colourless, odourless radioactive gas that is formed by the radioactive decay of elements that occur naturally in rocks and soils. It is present in the air around us, both indoors and outdoors. Near-surface rocks and soils are the main sources from which radon can migrate into overlying buildings or any enclosed space. Although the indoor radon levels are dependent on the local geology, they vary greatly with time and from one building to the next, owing to factors such as the building structure, use, heating, and ventilation. High radon levels occur in both homes and workplaces, which can lead to very significant exposure for the occupants. Radon levels in basements and rooms on the ground floor are likely to be higher than those on higher floors, as they are closer to the source. As radon is a public health issue (as well as a workplace issue), PHE has provided information online at www.ukradon.org. Information on radon in workplaces can be found on the HSE and HSENI websites (HSE no date d, HSENI 2019).

1.27 Employers have a legal responsibility to assess risks to their employees (and others as appropriate) and mitigate the hazard (regulation 3(1)(a) of the *Management of Health and Safety at Work Regulations* (MHSWR 1999, MHSWR(NI) 2000). Radon should be considered among these risks. This applies to all Employers, even those who may not obviously be considered to be carrying out work with ionising radiation. The responsibility lies with the Employer of the staff, who may be different from the owner of the property where work takes place. Employers should document their radon risk assessment.

1.28 Maps have been produced from a combination of radon measurements in more than 500,000 homes and the underlying geology to identify areas where at least 1% of homes have high radon levels. These are called Radon Affected Areas. The maps are available as an indicative atlas and a definitive dataset. The indicative maps can be accessed for free for the various countries in the UK on the UKradon website (UKHSA no date b).

1.29 Employers and homeowners can check if their premises are located in a Radon Affected Area more accurately, for a very small fee, using the definitive data found at the UKradon website (UKHSA no date a). However, the only fully reliable method of assessing the risk is to take radon measurements. Surveys can be carried out by leaving detectors in rooms of interest for a number of months. Employers are strongly encouraged to recommend home testing to their employees who may live in a Radon Affected Area. PHE offers a home testing kit for a charge, and this comes with guidance on detector placement (UKHSA no date a). PHE offers a separate radon measurement service for Employers, with guidelines for radon monitoring in workplaces (UKHSA no date a). Other 'validated laboratories' may be able to provide a service and details are on the Ukradon website (PHE 2020).

1.30 As basements in all areas are at increased risk of high radon levels, the HSE guidance (HSE no date d) is that any basement area that is occupied for more than 1 h per week should be monitored.

1.31 IRR17 comes into effect where work is carried out in an atmosphere containing radon-222 gas at an annual average activity concentration in the air exceeding 300 Bq m^{-3}. Employers are required to take action to restrict resulting exposures and 'notify' the HSE (see paragraph 1.39, *Notification*). Remedial measures to reduce radon levels may include permanent building modifications, such as installing a radon sump or a positive ventilation system. If engineered systems are fitted to reduce levels, then procedures must be in place to ensure that they remain operational and switched on. The system needs to be maintained by regular physical checks, such as ensuring that fans are still running. In addition, radon measurements (e.g. annually) will provide reassurance that the whole system is working effectively.

1.32 If initial radon levels were significantly lower than 300 Bq m^{-3}, the HSE advises reviewing the risk assessments and re-monitoring at least every ten years. Re-monitoring should be undertaken more frequently if the initial radon levels were closer to 300 Bq m^{-3} and after significant changes in building structure (including extensions), usage, heating, or ventilation.

1.33 Hospital estates departments typically manage radon risk on behalf of an Employer although the responsibility remains with the Employer (see paragraphs 1.6–1.9 on *Employers and responsibility*). Consideration should be given to the need to report on radon to the organisation's RSC. Employers who do not have an estates department, such as dental practices, will have to consider their premises themselves and take all necessary actions.

Notification, registration and consent

1.34 IRR17 adopted a graded approach (risk-based) to regulatory control of work involving radiation. This means that, depending on the work being carried out, Employers may need to apply to the HSE or HSENI to notify them of the work, register a practice or get consent for specified practices. A practice is a human activity that can increase the exposure of individuals to radiation from a radiation source and is managed as a planned exposure situation. The HSE and HSENI strongly advise that an RPA is consulted about this process and have provided a lot of information on their websites and the methods for applying (HSE no date f, HSENI no date a). There are also flow charts to aid decision-making on pages 20–21 of L121 (HSE 2018b) and these are free to access.

1.35 The graded approach applies to all Employers working with radioactive material (artificial and naturally occurring radionuclides), radiation generators (such as x-ray devices), and who carry out work in an atmosphere containing radon-222 gas above an annual average concentration of 300 Bq m^{-3} (see paragraphs 1.26–1.33 on *Radon*). However, certain dose rates and work practices are exempt from notification, and these are detailed in Schedule 1 of IRR17. For example, there is no need to notify or register an apparatus containing a sealed source that is of a type approved by the HSE, if using electrical devices operating at a potential difference of $< 30kV_p$ or if using radionuclides with concentrations and quantities below the specified values in Schedule 7 of IRR17.

1.36 The decision to notify or register work with radioactive material depends on the quantity and concentration of the radionuclides as well as the radionuclide itself. Schedule 7 of IRR17 details these specific values, along with the exemption levels and levels above which a notification of an occurrence should take place. Part 3 of Schedule 7 explains how to calculate exemption values where more than one radionuclide is involved.

1.37 An application to the HSE or HSENI must be submitted online by the Employer or an authorised employee before starting new work. The application cannot be submitted by a third party. If more than one risk category is required for the work performed, one application will address the different tiers. A single application covers all the sites operated by an Employer. As part of the submission, an Employer will be asked questions about the organisation, employees and its activities. The higher the risk category applied for, the more information requested. A new application will be required any time there is a significant change to the information (e.g. a change of Employer's address or the work is no longer under the Employer's control) or if the work ceases and will not be performed again by that Employer.

1.38 Note that:

- an x-ray unit on a trailer, such as a mobile CT or mammography service, is considered a portable source by the HSE.
  a. the HSE have conveyed a view on what is considered 'industrial radiography'.
  b. If an Employer is looking for a foreign body, x-raying a painting or archaeological finds, or x-raying for security or customs purposes, these are not to be considered industrial radiography.
  c. Testing shielding around a medical room, or radiography of lead aprons (for example) searching for cracks is not industrial radiography for the purposes of applications made under the graded approach.
  d. If x-raying a pipe weld for a plumber, that is industrial radiography.

Notification

1.39 This is the lowest risk category. It applies to any work that is not exempt (from notification) and that does not require registration or consent. In practice that means work:

- with < 1000 kg of radioactive material containing artificial or naturally occurring radionuclides that are processed for their radioactive, fissile or fertile properties
  - If the concentration and quantity are as directed by Schedule 7 Part 1 of IRR17;
- with < 1000 kg of radioactive material containing naturally occurring radionuclides that are not processed for their radioactive, fissile or fertile properties
  - If the concentration and quantity are as directed by Schedule 7 Part 2 of IRR17;
- in an atmosphere containing radon gas above an annual average concentration of 300 Bq m^{-3} (HSE 2018b, paragraph 31)

Registration

1.40 This is the medium risk category. All practices must be registered, except those that are exempt, require notification, or require consent. In practice this means work:

- with radiation generators that are not a specified practice requiring consent.
- with radioactive material containing artificial or naturally occurring radionuclides that are processed for their radioactive, fissile or fertile properties.
- If the concentration and quantity are as directed by Schedule 7 Part 1 of IRR17. Note there is a different level for registration once there is ⩾ 1000 kg of material.
- with radioactive material containing naturally occurring radionuclides that are not processed for their radioactive, fissile or fertile properties
  - If the concentration and quantity are as directed by Schedule 7 Part 2 of IRR17. Note there is a different level for registration once there is ⩾ 1000 kg of material.

1.41 In the registration process, an Employer is asked to confirm that they have various regulatory requirements in place, such as risk assessments (see paragraphs 1.44 to 1.54 *Risk assessments*), an appointed RPA (see paragraphs 1.88 to 1.91), and local rules for designated areas (see paragraphs 1.116 to 1.118 *Local rules*).

1.42 If patients are returned to a care facility after a nuclear medicine procedure (diagnostic or therapeutic), that facility must register to work with artificial radionuclides. They must also have in place all relevant documents, such as a risk assessment, and appoint an RPA. The HSE acknowledges that an application may be made after the event in this special case. It is recommended that the registration remains in place unless the care facility closes, i.e. do not inform the HSE that the care facility has ceased dealing with radioactive patients after each event, as it may happen again in the future. It might be appropriate to:

- distinguish between direct care of a patient in a managed home, which would be considered work with radiation, and a patient who is returned to a dwelling where they are significantly independent (supported/assisted living), for example, where there is no direct care.
- appoint the RPA from the hospital that undertakes the nuclear medicine procedures for this purpose, as they will be able to advise about the patient and their radioactivity levels (note that there may be a fee associated with the appointment, and in any case, it should be confirmed that valid insurance is in place). It is certainly expected that there is cooperation between Employers and the facility that undertook the procedure, and should provide as much information as appropriate to allow patient management with radiation protection in mind. It is not required that the nuclear medicine facility check to see that the advice is followed.

Consent

1.43 This is the highest risk category and applies to specified practices as set out in Regulation 7 of IRR17. These are:

(a) the deliberate administration of radioactive substances to persons and, in so far as the radiation protection of persons is concerned, animals for the purpose of medical or veterinary diagnosis, treatment, or research (e.g. nuclear medicine and brachytherapy). (Note that protection of the patient in the deliberate administration of radioactive substances to persons for medical purposes is addressed within IR(ME)R which is covered in chapter 2);

(b) the exploitation and closure of uranium mines;

(c) the deliberate addition of radioactive substances in the production or manufacture of consumer products or other products, including medicinal products—e.g. operation of a radiopharmacy;

(d) the operation of an accelerator emitting radiation with an energy higher than 1 MeV (except when operated as part of a practice within sub-paragraph (e) or (f) below and except an electron microscope);

(e) industrial radiography;

(f) industrial irradiation;

(g) any practice involving a high-activity sealed source (other than one within sub-paragraph (e) or (f) above);

(h) the operation, decommissioning or closure of any facility for the long-term storage or disposal of radioactive waste (including facilities managing radioactive waste for this purpose) but not any such facility situated on a site licensed under section 1 of the Nuclear Installations Act 1965;

(i) practices discharging significant amounts of radioactive material with airborne or liquid effluent into the environment, where 'significant' means a single discharge is expected to exceed a quantity in column 5 of Part 1 of Schedule 7 of IRR17. Ga-67 is not listed in IRR17, and HSE has stated that 100 GBq or more of Ga-67 is a significant amount (HSE no date j).

Risk assessment

1.44 Employers who intend to undertake an activity involving work with ionising radiation must make a 'suitable and sufficient' assessment of the associated radiation risk to employees and any other person exposed (excluding those undergoing a medical or non-medical exposure falling within the IR(ME)R; see chapter 2, paragraphs 2.1–2.3, *Scope*) prior to the commencement of the work. A risk assessment should be made for each installation and all existing activities. The risk assessment must be reviewed when new techniques are introduced to an activity that may affect radiation exposure. The Employer should seek advice from an RPA on the format and content of a risk assessment that identifies the control measures that need to be in place to ensure that exposures are ALARP. The assessment should consider how exposures can be restricted both during routine operations and, importantly, in the event of foreseeable radiation accidents (where the aim is to prevent or limit the consequences of such an accident).

1.45 An additional general risk assessment that considers electrical, mechanical, manual handling, chemical, and biological hazards, etc for that practice may also be needed to satisfy the requirements of the *Management of Health and Safety at Work Regulations* (MHSWR 1999, MHSWR(NI) 2000). Use the following general stages to guide your thoughts:
(a) identify the hazards;
(b) decide who might be harmed and how;
(c) evaluate the risks and decide whether existing precautions are adequate or need to be improved;
(d) record the findings;
(e) review the assessment and revise if necessary.

1.46 It is best practice to document the risk assessment, but it is strictly a legal requirement only where there are more than five employees. Documentation facilitates a review of the risk, the associated control measures and actions required, and their communication with the relevant staff. The HSE has provided guidance on the development of risk assessments in *INDG163: Risk Assessment. A brief guide to controlling risks in the workplace* (HSE 2014a) and a basic template on their website (HSE no date i). Although an Employer may wish to keep all their risk assessments in a similar format, typically, a standard risk assessment template will not be appropriate to satisfy IRR17. Each of the detailed requirements for a radiation risk assessment is set out in paragraphs 70 and 71 of *L121, Work with Ionising Radiation. Ionising Radiations Regulations 2017—Approved Code of Practice and Guidance* (HSE 2018b)—must be addressed for an assessment to be suitable and sufficient. These include but are not limited to, specific, detailed information about:
- dose rates;
- projected doses to staff and the public in normal situations and any possible accident situations. This may require dose calculations (see appendix 4, *Examples of dose calculations for risk assessments*) in advance of starting the activity, which will be supported by monitoring results once the activity is underway;
- whether staff will be designated as classified and why;
- who will be provided with personal dosimetry and how frequently?
- appropriate personal protective equipment and who will wear it;
- training needs (including refresher training);
- audit arrangements to ensure the risk assessment content is being upheld.

1.47 An example of a pro-forma approach to risk assessment is given in appendix 5, *Radiation risk assessment pro-forma*. The complexity of a risk assessment should be commensurate with the level of risk. There can be hazards that pose no risk, and it is not expected that all risks will be eliminated where they exist. A suitably adapted radiation risk assessment can be used for an area where very similar activities are undertaken, such as in a suite of general radiography rooms.

1.48 Be careful not to introduce other radiological and conventional risks associated with alternative techniques to satisfy both IRR17 and Regulation 3 of the *Management of Health and Safety at Work Regulations* (MHSWR 1999, MHSWR(NI) 2000). For example, some control methods for restricting exposure to ionising radiation by using distance and shielding might pose unacceptable risks of falls or back strain (paragraph 66 of L121 (HSE 2018b)).

1.49 There are responsibilities on manufacturers and installers (see paragraphs 1.13–1.15) regarding the design and construction of devices that require the elimination or reduction of risks as far as possible. Where risks are not eliminated, adequate protection measures, including alarms if necessary, are to be utilised, and users are to be informed of the residual risks due to any shortcomings of the protection measures adopted. Therefore, the Instructions for Use (IFU) supplied by the manufacturer or installer should be used to inform a risk assessment.

1.50 It may be necessary to consult safety representatives or established committees about the introduction of new measures that affect health and safety (regulation 4(a) of the *Safety Representatives and Safety Committees Regulations* (SRSCR 1977, SRSCR(NI) 1979)). Where there is no appointed safety representative, the Employer must consult the employees (*The Health and Safety (Consultation with Employees) Regulations* (HSCER 1996, HSCER (NI) 1996). The HSE gives guidance on this legal requirement in *INDG232—Consulting employees on health and safety: a brief guide to the law* (HSE 2013a). Consulting (or directly involving) those working in an area can help identify hazards and groups of people who are not present under typical conditions (such as visitors and contractors), and they can suggest control measures that an Employer may not have considered.

1.51 Do everything that is ‘reasonably practicable’ to protect people. This requires balancing the level of risk against the measures needed to control it in terms of money, time, and effort. Where there are several actions necessary, prioritise and tackle the most serious risk first.

1.52 An Employer should consider and document possible accident situations, their likelihood, and their potential severity. Where there is a radiation risk to employees or other persons from an identifiable radiation accident, the Employer must take all reasonably practicable steps to prevent the accident, limit the consequences if an accident does occur, and provide employees with sufficient information, instruction, training, and equipment necessary to restrict their exposure to ionising radiation. Typical risk likelihood descriptions are ‘rare’, ‘unlikely’, ‘possible’, ‘likely’, and ‘almost certain’. These could be linked with the estimated (or evidence-based) frequency of occurrence over years or decades, e.g., they could occur once or more in a year, in ten years, or in a working lifetime. Those accidents that are unlikely in a working life would not need to be included in the projected annual dose of individuals nor contribute to their potential designation as a classified workers. The possible dose that could be received in reasonably foreseeable accident situations should be recorded and fed into the projected annual dose estimate. The risk assessment details could be brief for those situations identified as rare and unlikely while focusing on robust contingency plans for those that are reasonably foreseeable (perhaps those described as possible, likely, and almost certain). (See paragraphs 1.82 to 1.87 on contingency plans.) An RPA should be able to help with dose calculations and evidence of expected doses in various facilities. Dose calculations performed when an incident takes place can be used to update a risk assessment with hindsight.

1.53 There are specific requirements under IRR17 concerning dose limitation for pregnant or breastfeeding staff and for young or inexperienced staff. These should be addressed in the risk assessment. The HSE has provided guidance on workers that have particular requirements in *The Health and Safety Toolbox: How to Control Risks at Work* (HSE 2014c).

1.54 Risk assessments should be kept under review to ensure they remain suitable and sufficient. This is a requirement of Regulation 3(3) of the MHSWR (MHSWR 1999, MHSWR(NI) 2000)] and must take place if there is reason to believe it is no longer suitable or if there has been a significant change in work practices, including as a result of staff turnover. The Employer should decide on the frequency of review based on the nature of the work, the degree of risk, and the extent of any likely change in the work activity (paragraph 78 of L121 (HSE 2018b)). They should likely be reviewed at least every three years, although this does not necessarily mean re-issued, as long as there is documented evidence of the review.

Dose limits

1.55 Dose limitation is one of the fundamental principles of radiation protection, along with justification and optimisation (ICRP 2007b). IRR17 Regulation 12 requires an Employer to limit the dose to employees and other specified classes of persons. This does not include those undergoing medical or non-medical exposure (see chapter 2, paragraphs 2.1–2.3, *Scope*), which falls within the IR (ME)R. Schedule 3 details the limits on the effective dose and equivalent dose (for the lens of the eye, the skin, and the extremities) for:

- Employees and trainees of 18 years of age or above;
- Trainees aged under 18 years;
- Other persons.

1.56 The most relevant dose limits are summarised in table 1.1. In addition to these, the effective dose limit for other persons (not being a carer or comforter) who may be exposed to radiation resulting from the medical exposure of another is 5 mSv in any period of five consecutive calendar years, and this may be

Table 1.1. General dose limits per calendar year as set out in Part 1 of Schedule 3 of the IRR17.

Class of person	Effective dose	Lens of eye	Extremities or any 1 cm^2 of skin
Employees and trainees of 18 years of age or above	20 mSv	20 mSv	500 mSv
Trainees aged under 18 years	6 mSv	15 mSv	150 mSv
Other persons	1 mSv	15 mSv	50 mSv

particularly relevant for some nuclear medicine therapies. There are also some higher limits for employees that the HSE or HSENI may permit in very exceptional circumstances.

1.57 Regulation 9(1) requires the restriction of exposure, and a radiation risk assessment should detail the measures required to ensure exposures are as low as reasonably practicable (ALARP). Dose limits are not considered a target dose for optimisation and in the majority of healthcare exposure scenarios, annual occupational doses are a small fraction of the limits.

1.58 If an individual is exposed to ionising radiation to the extent that they exceed the relevant dose limit, it is called 'overexposure' (see paragraphs 1.221–1.23 *Radiation Incidents*).

1.59 Where overexposure of an employee takes place, regulation 27 requires dose limitation for that employee for the remainder of the dose limitation period (calendar year).

Dose constraints

1.60 Dose constraints can be used to help restrict the exposure of individuals by ensuring that appropriate consideration is given to radiation protection measures. They can be based on experience or recommendations from professional bodies and are appropriate:

- for members of the public who may be affected by radiation work activities, not least because they may be exposed to more than one source. The HSE has recommended that the dose constraint on optimisation for a single new source should not exceed 0.3 mSv $year^{-1}$ for a critical group likely to receive the highest average dose from the work (paragraph 155 of L121 (HSE 2018b)).
- for those exposed occupationally where individual doses from a single source are likely to be a significant fraction of a dose limit. An example of developing an occupational dose constraint for interventional cardiologists is given by Mairs (Mairs 2016).

1.61 Constraints define a level of dose that ought to be achievable in well-managed practices and should help to filter out options for radiation protection that could lead to unreasonably high levels of individual dose, even though the collective dose for the workforce as a whole is optimised. They are used at the planning stage and should be documented in the risk assessment process. It should be noted that dose constraints are neither investigation levels nor necessarily routinely acceptable levels of exposure. The ALARP principle must always be followed for individual exposures.

1.62 Dose constraints for outside workers should involve cooperation between Employers.

1.63 It is essential, as well as a legal requirement (see paragraphs 1.10–1.12 *Employees responsibilities*), that employees follow the protection measures specified for a practice in order to limit their own radiation exposure and that of others, such that an appropriate dose constraint can be met.

1.64 Although it is not referred to as a dose constraint, in relation to a pregnant employee, Regulation 9(6)(a) requires that the equivalent dose to a foetus is ALARP and unlikely to exceed 1 mSv from when an Employer is notified of a pregnancy for the remainder of the pregnancy.

Formal investigation levels and action on dosimetry results

1.65 Staff can receive unintended exposures as a result of poor radiation protection practice and may be exposed over a period of time rather than in a one-off event. Staff can also receive higher than expected doses through increased workloads, even in facilities with well managed radiation protection.

1.66 As part of ensuring doses are kept as low as reasonably practicable (ALARP), IRR17 requires formal investigation levels to be established locally by the Employer and recorded in the local rules. The purpose of investigation levels is to trigger a review of working conditions where staff have exceeded the specified level of dose so that an assessment can be made about the adequacy of the control measures in place. Therefore, it is likely that the Employer will select different investigation levels for different groups of staff or in different areas of work where projected doses vary.

1.67 The Employer must carry out a formal investigation in consultation with the RPA if a member of staff exceeds an effective dose of 15 mSv in a calendar year (IRR 2017, IRR(NI) 2017). However, it is recommended that investigation levels be set at a lower effective dose, between 1 mSv and 4 mSv, depending on the type of work. HSE guidance (paragraph 183 of L121 (HSE 2018b) sets out details that should be covered as part of the investigation. The report of the investigation should be kept for at least two years.

1.68 Formal investigation levels strictly only apply to effective doses, but it is not uncommon that investigation levels are also set locally in terms of equivalent doses for other positions on the body that are monitored.

1.69 It is not considered ALARP to only consider staff doses once they exceed an investigation level. Therefore, it is recommended that Employers establish action levels based on expected doses in each monitoring period, which include effective doses and equivalent doses for the extremities and eye, where appropriate. It is appropriate that an Employer take action to reduce exposure before an investigation level is exceeded if it is suspected that doses for an individual will exceed that level.

1.70 It is advisable to discuss a dose that exceeds the action level with that member of staff. As health and safety law is in place to protect individuals, it will be appropriate to consider if an action level, as applied to a group, should be adapted for individual circumstances (be aware that a part-time member of staff may have a larger workload than a full-time member of staff, and adjustment of an action level may be more appropriate if based on workload, e.g. dose per Dose Area Product rather than whole time equivalent). There should be established communication arrangements between the dosimetry service, the Employer and the RPS to flag these doses in a timely manner.

1.71 See paragraphs 1.233 to 1.240 for information on how to perform an investigation.

Personal protective equipment

1.72 Personal Protective Equipment (PPE) is used to restrict exposure to employees (or other persons) when other control measures, such as engineering controls, design features, and systems of work, are not sufficient. This is usually the case when someone is positioned close to the radiation source as a requirement of the task they are undertaking. The provision of PPE is included in the risk assessment process. For example:

- when working in interventional radiology, a clinician is required beside the patient and x-ray tube during the procedure. As a result, the risk assessment should establish if it is necessary and reasonably practicable to wear a lead apron, thyroid collar, leg protectors, and lead glasses to limit occupational exposure.
- palpation of a patient close to the radiation field in diagnostic examinations, such as barium enemas, can make use of protective lead gloves.
- manipulation of radioactive material in nuclear medicine or radiotherapy may require a person to use their hands very close to the radioactive source. Gloves will likely be appropriate to ensure radioactive material does not contaminate the operator.

1.73 It is the responsibility of the Employer to provide adequate (sufficient protection, in line with the risk assessment) and suitable (correct match for the task and the person) PPE, to provide adequate storage facilities for it, and to take all reasonable steps to ensure it is used correctly. This includes the provision of training on its use. It is the responsibility of employees to use the PPE, return it to storage, and report if they suspect it is defective. The

Employer's quality assurance programme should include assessments of PPE to ensure its ongoing suitability.

1.74 When assessing the need for PPE, also consider if there is an additional risk as a result of the PPE itself. If a member of staff cannot support the weight of a lead apron due to musculoskeletal problems, another approach may be required, such as a different style or type of apron.

1.75 HSE guidance on the Personal Protective Equipment at Work Regulations (PPER 1992, PPER(NI) 1993) provides advice about the selection and use of PPE, which is equally applicable to protection against ionising radiation. A brief overview of PPE is also available from HSE (HSE no date h).

1.76 IRR17 Regulation 10 requires that PPE comply with any applicable provision of the Personal Protective Equipment Regulations 2002 (*The Personal Protective Equipment Regulations* 2002 *(revoked)* (2002)). Subsequent legislation (*The Personal Protective Equipment (Enforcement) Regulations* 2018 (2018)) and guidance (the Office for Product Safety and Standards 2022) have amended the requirements for placing PPE on the UK market.

1.77 The British Institute of Radiology has published clear recommendations on PPE for diagnostic x-ray use (BIR 2016a), posters (on 'wearing your PPE' and 'caring for your PPE') and free videos (on 'screening PPE', 'wear and fit' and 'care and storage') available from its website (BIR 2016b),

1.78 Specific advice on modalities can be found in sections of the relevant chapters.

Young persons and trainees

1.79 A 'young person' is defined as anyone less than 18 years of age (MHSWR 1999, MHSWR(NI) 2000), and a 'child' is anyone who has not yet reached the official minimum school leaving age (MSLA), which they reach in the school year in which they turn 16 (HSE 2013d). A young person who is not a child can carry out work with radiation if the work is necessary for their training, the person is properly supervised by a competent person, and the risks are reduced to the lowest level, so far as is reasonably practicable (HSE no date g). It is a local management decision as to whether allowing young people to observe x-ray procedures is consistent with restricting exposure to ionising radiation so far as is reasonably practicable: however, Judith Hackitt, former HSE Chair, has made a very supportive statement in regards to offering work experience to young people (Hackitt 2013). Information is provided for schools/colleges, work experience organisers, the Employer, parents/carers/guardians and for students/learners on the HSE website (HSE no date e).

1.80 Under IRR17, a 'trainee means a person aged 16 years or over (including a student) who is undergoing instruction or training that involves operations that would, in the case of an employee, involve work with ionising radiation'. For trainees aged under 18 years (but above the MSLA), the dose limits in IRR17

Schedule 3, paragraphs 3 and 4, apply. Anyone under the age of 16 is not specifically defined in IRR17, and as such, they fall under the category of 'other person' which has dose limits as set out in IRR17 Schedule 3 paragraphs 5 and 7. For trainees aged over 18 years and for employees, the dose limits in IRR17 Schedule 3 Paragraphs 1 and 2 apply.

1.81 A young person (which can include a trainee or student on work placement or experience) is subject to the requirements of IRR17 which include the following:

- A risk assessment, if applicable and if not already in place, for someone visiting or working in the department, taking into account the age (and maturity) of the person involved. The risk assessment should:
  - include the estimation of likely doses—based on previous staff doses or measurements, if applicable. Cooperation between Employers may be required to account for total annual exposure, depending on the employment/placement circumstances of the individual (see paragraphs 1.127–1.152 *Cooperation between employers*).
  - consider whether personal dosimetry is appropriate. Where the duration of the work is brief and not suited to the typical wear period of a passive dosemeter (e.g. usually 1–3 months), the use of an Electronic Personal Dosimeter (EPD) will be more appropriate.
  - address control measures that should be proportionate to the risk, taking into consideration the inexperience of young people and their lack of awareness of potential or existing risks.
  - not be considered in isolation from other health and safety considerations. For example, the use of a lead apron to restrict exposure may lead to a risk of back strain.
- Where applicable, the local rules should address the arrangements for young people. For short visits, the written arrangements for visitors should normally be adequate, but this can be expanded depending on local practices. For longer periods of attendance, a separate section in the local rules should be considered.
- Restriction of exposure must be ensured and doses minimised (IRR17 regulations 9 and 12). Trainees and young people should not be designated as classified workers.
- Adequate information, instruction, and training must be provided, appropriate for the length of the visit or employment (IRR17 regulation 15). This may include ensuring that the radiation safety policy, local rules, and any other relevant documentation (such as work instructions) are read and that they understand the requirements contained in these documents, especially the contingency plans.
- They must follow the duties of employees (IRR17 regulation 35), unless under supervision as visitors.

Contingency plans

1.82 Where a risk assessment shows that a radiation accident (defined in IRR17 as where immediate action is required to prevent or reduce exposure and also known as a 'significant event') is reasonably foreseeable, the Employer must put a contingency plan (i.e. immediate action) in place to restrict exposure and ensure the health and safety of anyone who may be involved in the accident, so far as is reasonably practicable. An accident is one that has or could result in significant exposure (that significantly exceeds normal planned exposures). The plan may include employees, members of the public, emergency services, etc as required and should address the incident, clean up, and any associated recovery. Paragraph 238 of L121 (HSE 2018b) makes clear that the actions should be proportionate to the risk and magnitude of the exposure. It goes on to give examples of contingency plans and confirms that small, contained spillages of radioactive material (and other incidents that could not result in exposures of concern) are not considered radiation accidents. The Employer should consult an RPA in regard to the contingency plans.

1.83 It may be useful to consider actions to take as a result of a radiation incident separately from 'contingency plans' associated with radiation accidents. Describe the actions in sufficient detail to allow staff to follow them easily. The key message is 'make it safe, then follow up'. Include arrangements for sending personal dosemeters to be read and actions to reassure staff about their exposure, where relevant.

1.84 L121 (HSE 2018b) gives a lot of guidance on contingency plans. A few key points are listed below:

- The plans must be recorded in the local rules (as a summary or reference).
- Training must be provided on the implementation of the plans. Rehearsals of the arrangements should be carried out at specified intervals, where appropriate, and as guided by L121 paragraph 246.
- Some accident scenarios (e.g. fires) will require prior consultation with the emergency services so they know the role they will play and how they will control the exposure of their employees.

1.85 If a contingency plan (or even a small part of it) has to be put into action:

- The cause of those circumstances is to be analysed so that recurrence can be mitigated.
- A record of the analysis must be kept for at least two years from the date it was made.
- Any exposure that occurred due to the accident must be recorded on the relevant dose record.

1.86 When considering the actions that are to be taken in the contingency plan (or systems of work when an incident is not considered a significant event), make sure that the impact of the action is fully addressed and seek out advice from those with more experience if required. For example, in a diagnostic x-ray facility, it might not be appropriate to hit the emergency stop button when something goes wrong, and experience shows that this feature is rarely used as a result of clinical practice. If a fault is suspected but the system is not emitting radiation, hitting the emergency stop button may clear the fault/error log on the system. If a clinician's hands accidently stray into the x-ray beam during an interventional CT procedure, an instruction can be made over the intercom to move them (this is a balance of the patient's clinical needs versus the high occupational exposure in the beam). Hitting the emergency stop button is likely to be appropriate when the radiation is continuously on, where exposure is not intended. The risk assessment and training should ensure appropriate actions are planned and understood by the staff.

1.87 For more on contingency plans, see chapter 20 (*Contingency planning and emergency procedures for radioactive substances*).

Radiation protection adviser

1.88 Employers who use ionising radiation and are required to notify, register, or gain consent from the HSE (see paragraphs 1.34–1.43, *Notification, registration, and consent*) will need to appoint a suitably qualified Radiation Protection Adviser (RPA), in writing. An RPA will not be required where the only work carried out is below the exemption levels contained in Schedule 1 of IRR17.

1.89 The RPA can be an individual or a group of individuals (i.e. a HSE recognised RPA body) with recognised certification and knowledge and experience appropriate to the scope of their advice. Certification is granted (by RPA2000, www.rpa2000.org.uk) based on core competence to give advice on compliance with the IRR17. It does not demonstrate an individual is suitable to advise an Employer on all situations where radiation is used—typically, an individual gains competence in a particular field. If the RPA is unable to fulfil an advisory function due to a lack of competence, a lack of information, or a lack of facilities, professional codes of conduct require that it be brought to the attention of the Employer. An Employer should ensure that an RPA is suitable for the practice they are seeking advice on. An Employer might consult more than one RPA, for example, where multiple modalities are to be utilised, to ensure the depth and breadth of competence is available.

1.90 The RPA should visit the facility they advise and should regularly review radiation protection matters with the RPS (and the relevant clinical lead or manager, if appropriate). The frequency of reviews should be determined by the Employer (though the task is usually delegated to the manager of the service) in consultation with the RPA. The frequency should account for the extent of the hazards involved in any change in practices or professional guidance and should be identified in the contractual arrangement with the RPA. It is suggested that each substantial department be reviewed at least annually. Small departments under the managerial control of larger departments but which have their own RPS may be reviewed less frequently. Dental practices, where risks are smaller and more easily managed, could be reviewed at least every three years (this may not extend to those who have the CBCT modality, if it is considered a greater risk). Contact, other than site visits, can be made more frequently as required through various forms of communication and meetings, such as a RSC. All aspects of radiation protection should be reviewed at least every three years by the Employer.

1.91 Appendix 3, *The role of the radiation protection adviser*, summarises the matters that legally require consultation with an RPA and indicates other matters where consultation with an RPA will be helpful for the Employer. The RPA should, where appropriate, liaise with other expert advisers such as the Medical Physics Expert (MPE) (see chapter 2, paragraphs 2.56–2.59), *Radioactive Waste Adviser (RWA)* (see paragraphs 19.9, 19.10, 19.20, 19.28, 19.32, and 19.49, and appendix 21, *Role of the RWA*), or Dangerous Goods Safety Adviser (see paragraphs 18.67, 18.68, and 18.159).

Information, instruction and training

1.92 Employers must provide training to all employees working with ionising radiation, including management, in order to restrict exposure. All employees should be informed of their responsibilities and cooperate with their Employer and any health and safety training provided. A Radiation Protection Adviser (RPA) should be consulted on training needs.

1.93 The regulations differentiate between those who are ‘engaged in work with’ ionising radiation and those who are ‘directly concerned with the work’ with ionising radiation. Those engaged in work with radiation need radiation protection training plus sufficient information and instruction so they know the:

- health risks of radiation exposure;
- general radiation protection procedures and precautions to take;
- specific radiation protection procedures and precautions to take in connection with the work they may be assigned;
- the importance of complying with the regulations.

1.94 Those directly concerned with the work are to be provided with adequate information to ensure their health and safety, so far as is reasonably practicable.

1.95 Those in training must be under appropriate supervision until they have been signed off as competent.

1.96 Specific training is essential where staff are required to:

- follow systems of work (including how to control access or what measures to take to enact contingency plans for reasonably foreseeable accidents);
- wear PPE (and dosimetry equipment);

- work in the controlled area of another Employer as an outside worker (see paragraphs 1.199–1.220 *Outside Workers*);
- work with high activity sealed sources (HASS);
- become a Radiation Protection Supervisor (RPS). The HSE provide criteria for this (HSE 2000a).
- make entries in radiation passbooks (see paragraphs 1.199–1.220 *Outside workers*);
- monitor radiation levels for designated areas;
- consult an RPA or other expert.

1.97 Individuals who may become pregnant or start breastfeeding and are engaged in work with ionising radiation should be informed about the possible risks arising from ionising radiation to the foetus and to a nursing infant. The HSE has published guidance on such risks and control measures (HSE 2013c, 2015b). They should also be informed of the importance of notifying the Employer about becoming pregnant or breastfeeding (if working with unsealed sources), in writing, as soon as possible.

1.98 Staff in professions that receive radiation protection training as part of their initial qualification will still need to be trained in local arrangements when they start working at a particular location. Radiation protection training should be on an induction checklist, where appropriate. IRR17 also requires that refresher training be undertaken at appropriate intervals. This could be through a combination of continual professional development, reading the local rules, and mandatory training after a specified period. The regulations do not state a frequency for updating training, only that staff are to be adequately trained. A risk-based approach could be used to consider the frequency of refresher training to ensure that knowledge and awareness are maintained. Those who have responsibilities in accident situations but otherwise have little involvement with radiation protection should have more frequent training, e.g. annual. Those who regularly work with ionising radiation but have low-risk duties could be re-trained every three years, for example. If new equipment or working practices are introduced, then staff will require further training. Certain professions have radiation protection training included in their continuing professional development cycles. Where this is deemed satisfactory (based on content and frequency), it may be relied on for general radiation protection training and could be supplemented with local-specific training.

1.99 Employers should ensure that the training provided meets the intended aims and check the adequacy of the training (Paragraph 269 L121 (HSE 2018b)). Where training issues are identified through, for example, incidents or audits, action must be taken to improve the standard of training methods/materials and to address any shortcomings of the staff in question.

1.100 The duty to provide adequate information, instruction, and training is extended to all Employers, not just those who are undertaking work with ionising radiation. As such, there may have to be cooperation between

Employers and an exchange of information (see paragraphs 1.127–1.152 *Cooperation between employers and others*) to ensure all employees are informed of the risks and precautions to take.

1.101 Training records must be kept and be available to demonstrate employers have met their obligations.

1.102 It is important not to overlook staff who may not be working with radiation but who work in the same building, e.g. porters, cleaners, estates personnel, admin staff, etc. They must have sufficient information to ensure their safety. A basic provision of radiation protection training might include how to recognise safety and warning signs (such as controlled area signs) and enough information about where staff could seek help if required. For certain staff groups, it might also need to be supplemented with specific information (maybe even systems of work) to enable them to do their role, e.g. when can domestic staff enter x-ray equipment rooms to clean or when should they not remove potential radioactive waste from a nuclear medicine facility? Local rules intended to legally restrict exposure in designated areas have to contain specific information and may be inappropriate for certain staff groups. Systems of work can be shorter than the local rules and more targeted, so the information is easier to understand and follow.

1.103 It is essential to establish which employer is responsible for the provision of training in regard to outside workers (see paragraphs 1.199–1.220 *Outside workers*).

Controlled areas and supervised areas

1.104 IRR17 Regulation 17 requires areas where persons may be exposed to elevated levels of ionising radiation or where there is a significant risk of the spread of contamination to be designated as ‘controlled areas’ or ‘supervised areas’. See appendix 6, *Designation of controlled and supervised areas*, for further guidance on when an area should be designated. In general terms, inside controlled areas, special procedures need to be followed (i.e. systems of work within the local rules) to restrict the possibility of significant exposures, and inside supervised areas, the conditions need to be kept under review. Undesignated areas require no special measures for members of the public or workers.

1.105 In the medical and dental settings, radiation sources are commonly used within specially designed rooms (see specific chapters for guidance on facility/shielding design in that modality and paragraphs 1.60–1.64 *Dose constraints*). It is accepted practice in some modalities to designate the whole room as a controlled area limited by physical barriers such as the walls, doors, ceilings, and floors (although the designation may be temporary; see below) in keeping with regulation 19(2) of IRR17, which requires physical demarcation of a controlled area where reasonably practicable (or delineation by some other means). Some areas are also delineated by an operator area, a visualisation screen or the end of a maze/chicane without a physical barrier such as a door. Staff will work under the system of their local rules when in these areas. Doses and/or dose rates may not be at the levels requiring designation under the regulations in all parts of the room, but this is a pragmatic means to identify the area with any signs, lights, and engineering controls, as appropriate, required at the entrance(s).

1.106 Rooms will be designated as controlled areas when a source of radiation (either radioactive substances or electronically generated radiation such as x-rays or proton or electron beams) is present or in use, which could lead to significant internal or external radiation exposure. Other areas where there is a significant risk of radionuclide contamination outside the working area will need to be designated as controlled areas. Gamma camera rooms used for injection are expected to be controlled areas. SPECT/CT rooms will generally be controlled when the equipment is on because of the potential for x-ray exposure from the CT component. The radiation risk assessment will be used to inform the decision.

1.107 Where x-ray units are switched off at the mains power supply or put into standby mode at times (e.g. overnight or on weekends), some sites choose to designate rooms as controlled only when the equipment is powered up and capable of making an exposure. This enables cleaners, maintenance workers, and others to safely access the room outside of working hours without having to work under a written system of work.

1.108 For some uses of x-ray equipment, e.g. intraoral or DEXA equipment, it may not always be reasonably practicable to designate the whole room as a controlled area. In such cases, the controlled area might be designated as the area between the x-ray tube and the patient. The operator of the equipment prevents any member of staff from entering the controlled area, either physically or by observation. When the x-ray set is not in use (i.e. when switched off or incapable of making an unauthorised exposure, e.g. through password control), the temporary controlled area no longer exists, and staff may enter the area without working under the systems of work in the local rules. Suitable and sufficient warning signs are required for controlled areas, but HSE acknowledges that this may not be practicable where the controlled area cannot be physically demarcated (L121 paragraph 365 (HSE 2018b)) and verbal warnings can be used instead. Any warning signs and lights in use must reflect the true status of the area designation at that point in time.

1.109 Where the limit of a controlled area is a barrier such as a wall, door, ceiling, or floor, the barrier will normally be designed to be of a sufficient thickness of a suitable material to limit the exposure outside the room so that adjoining areas do not need to be designated as controlled areas. The exception to this is where access is rarely required and access by persons can be controlled, such as on the roofs of linear accelerators, which are sometimes designed with less shielding in the roof to reduce construction costs.

1.110 The use of mobile x-ray equipment will require the designation of a temporary controlled area during exposures. The designation of a controlled area is generally limited to a radius around the tube and/or the patient, except for the primary beam, where a solid barrier such as a floor or wall constructed of a sufficient thickness of a suitable material is required to provide sufficient protection to those beyond that barrier. Mobile fluoroscopic devices usually include material that satisfactorily attenuates the primary beam, so only scatter needs to be considered.

1.111 Regulation 20 of IRR17 requires that designated areas have radiation levels adequately monitored (see paragraphs 1.1–1.1 *Area monitoring*).

1.112 There should be a documented radiation risk assessment for each controlled area, clearly identifying the control measures and actions required to restrict exposure. These are the basis of the local rules. The workload and particular procedures in each room should be defined and regularly reviewed to ensure that the assumptions made remain valid. Changes in practice (e.g. new techniques, new equipment) will generally require a reassessment of the designation of the surrounding area. The 'instantaneous dose rate' averaged over one minute (IDR) and 'time average dose rate' averaged over an eight-hour working day (TADR) values (typically calculated by the RPA) should be included, together with the results of any area monitoring measurements carried out.

1.113 Additional guidance for designation on the basis of radionuclides dispersed in a human body is dealt with in paragraphs 9.14, 9.48, 10.29, and 10.36 (also L121 Paragraphs 297 and 312 (HSE 2018b)). Designation is likely in most brachytherapy work and some radionuclide therapy work, such as radioiodine therapy for ablation of the thyroid.

1.114 Regulation 19(1)(b) of IRR17 requires suitable and sufficient signs displayed in suitable positions warning that the (designated) area has been so designated and indicating the nature of the radiation sources, e.g. x-ray or 'unsealed sources' and the risks arising from such sources e.g. external gamma-radiation, cloud beta-radiation, inhalation, or ingestion (see appendix 7, *Warning signs and notices*, for more details on warning signs). The signage should incorporate a trefoil in a black triangle with a yellow background. It should also include whether or not an entry is permitted, along with any conditions and sufficient information to enable employees to take appropriate action before entering the area (e.g. to wear appropriate personal protective equipment).

1.115 An illuminated warning light at the room entrance (preferably at eye level) may accompany the warning notice to indicate when access is strictly forbidden e.g. during radiotherapy when the beam is 'on' and during diagnostic x-ray exposures if entry is directly into an unprotected area of the room or into the control area if the room is designated to include this area. The light normally incorporates appropriate wording depending on the conditions and actions necessary. Such lights are a 'warning device' and are expected when sources can cause significant exposure in a very short period of time (L121 paragraph 126 (HSE 2018b)). L121 paragraph 126(b) sets out expectations on signals associated with exposure conditions, e.g. capable of exposure, prepping to expose and exposing. Guidance in L121 paragraph 127 says that for most x-ray generators it should be reasonably practicable to have automatic warning devices of this nature.

Local rules

1.116 Written local rules are required for controlled areas to restrict exposure to ionising radiation and to control exposures in the event of a radiation accident (IRR17 regulations 13 and 18; (IRR 2017, IRR(NI) 2017). If arrangements are needed in supervised areas to restrict exposures or prevent accidents, e.g. where unsealed sources are being used in a pathology laboratory, local rules are also likely to be appropriate. A summary of the required contents of local rules is given in table 1.2.

1.117 Local rules do not need to contain detailed protocols of working practices, to which reference should be made, but should contain at least the information listed in paragraph 336 (essential contents) and a brief summary of or reference to the general arrangements listed in paragraph 337 (optional

Table 1.2. Summary of the contents of local rules.

Essential contents	Optional contents	
Identification/description of designated areas.	Management/supervision responsibilities for radiation protection.	Significant findings of the risk assessment.
Name(s) (and contact details) of the RPS(s).	Testing/maintenance of engineering controls and safety features.	Programme to review whether doses are ALARP.
Arrangements for restricting access.	Radiation/contamination monitoring.	Programme to review local rules.
Dose investigation level.	Testing of monitoring equipment.	Procedures for initiating investigations.
Summary of working instructions, including written arrangements for non-classified persons. Reference to detailed work instructions if relevant.	Personal dosimetry arrangements.	Procedures for ensuring staff have information, instruction and training.
Contingency arrangements.	Arrangements for pregnant and breastfeeding staff.	Procedures for contact and consultation with the RPAs.

contents) of the Health and Safety Executive 2018 (HSE 2018b). A diagram or a clear description of the designated areas should be provided.

1.118 The local rules should be brief and relevant, kept up to date, and reviewed preferably annually but at least three-yearly. An Employer should balance the level of detail against the ability of individuals to understand and take action on the content. Employees should read the local rules for their work area and sign an undertaking to that effect at induction and after any amendments have been made to the local rules. Local rules can be electronic, but readily available paper copies may be needed for the RPS or staff working in the area, especially where the work is carried out away from the Employer's base location (L121 paragraph 334 (HSE 2018b)). Local rules must be available at or near the area concerned, and it may therefore be appropriate to display specific instructions in a particular work area (L121 paragraph 343 (HSE 2018b)).

Radiation protection supervisor

1.119 For work carried out in any area subject to local rules, the Employer must appoint one or more RPSs to assist in securing compliance with IRR17 in respect of work carried out in those areas. However, it should be noted that the legal responsibility for ensuring compliance remains with the Employer.

1.120 With regard to suitability for appointment, the prospective RPS will typically be a full-time employee of the Employer who understands the requirements of IRR17 and the role they are expected to fulfil, works in the area often enough to monitor local practice and is in a sufficiently senior position to supervise and ensure compliance with the arrangements set out in the local rules. It is good practice to confirm their role and their understanding in writing. The Employer should consider the need to consult the RPA with any queries about RPS selection and appointment.

1.121 The RPS should be given appropriate training (HSE 2000a) to inform them on radiation protection principles and procedures, the requirements of IRR17, and the arrangements in the local rules to enable them to supervise the work safely. The extent of the training and refresher training should be appropriate to the nature of the duties, and Employers are advised to consult their RPA.

1.122 It should be clear which aspects of an individual's role are legally required of the RPS under IRR17 and which are additional tasks that may be delegated to that person through their job description or agreed job functions. The RPS's role is to ensure compliance with the local rules. They are not responsible for management functions within their RPS role.

1.123 Depending on the nature of the practice, the tasks of the RPS in assisting the Employer to comply with the regulations may include the following [article 84 of *Council Directive* 2013/59/*Euratom of* 5 *December* 2013 *laying down basic safety standards for protection against the dangers arising from exposure to ionising radiation, etc* (European Commission 2014)]:

(a) ensuring that the work with radiation is carried out in accordance with the requirements of any specified procedure or local rules (note that in line with paragraph 1.122, ensuring compliance with local rules is not optional);
(b) supervising implementation of the programme for workplace monitoring;
(c) maintaining adequate records of all radiation sources;
(d) carrying out periodic assessments of the condition of the relevant safety and warning systems;
(e) supervising implementation of the personal monitoring programme;
(f) supervising implementation of the health surveillance programme;
(g) providing new workers with an appropriate introduction to local rules and procedures;
(h) giving advice and comments on work plans (processes, systems of work and use of resources e.g. staff);
(i) establishing work plans;
(j) providing reports to the local management;
(k) participating in the arrangements for prevention, preparedness and response for emergency exposure situations;
(l) providing information and training for exposed workers;
(m) liaising with the RPA.

1.124 The Employer must provide sufficient resources and managerial support for the RPS to perform their role effectively (HSE 2000a). In addition, the *Management of Health and Safety at Work Regulations* (MHSWR 1999, MHSWR(NI) 2000) state that the Employer must ensure adequate time is made available for the appointed person to fulfil their duties. An RPS experiencing difficulties in meeting the requirements of their role should report immediately to the head of the department, who, in consultation with the RPA if necessary, should decide what action is to be taken and should implement appropriate arrangements.

1.125 The name and contact details of the RPS must be included in the local rules associated with the areas they will supervise. The Employer may also display these in the area where the work is undertaken.

1.126 In deciding how many RPSs are required, the Employer will need to consider the type and complexity of the work, the number of different locations to be covered, staff rotation, shift work, and absences due to sickness, training, and holidays. HSE generally expects an RPS to be available while work with ionising radiation is being undertaken. This is particularly important if the contingency plans require action by an RPS. Where more than one RPS covers an area of work, all have equal responsibility to fulfil the role of ensuring compliance with IRR17 (i.e., there is no 'deputy RPS' defined under

IRR17). The Employer may wish to consult the RPA on sufficient coverage by RPSs.

Cooperation between employers and others

1.127 There should be cooperation and an appropriate exchange of information between Employers. When employees of one Employer work on the premises of another Employer, it is necessary to coordinate the measures required to comply with statutory duties, inform each other of the risks to employees and the control measures in place, and allocate responsibilities for aspects of management. Under IRR17, this applies where work with ionising radiation undertaken by one Employer is likely to lead to the exposure of an employee of another Employer. This includes a need to share information when work will take place in a building where radon levels exceed 300 Bq m^{-3} and, therefore, where IRR17 applies (see paragraphs 1.26–1.33 *Radon*).

1.128 In some organisations, there may be several different Employers involved in various aspects of work within a radiation facility. It can be challenging to understand the terminology used and how the regulations apply to the range of individuals that may be exposed to radiation. A flow chart to help has been provided in appendix 8, *Identifying persons exposed to ionising radiation.*

1.129 There are a number of scenarios that will result in the need for cooperation between Employers to ensure the safety of staff and others and compliance with IRR17. These can be broadly grouped into the following categories A, B, and C:

A. Employers sharing the same workplace

1.130 When employees of a number of different Employers work at a particular location alongside each other (not all necessarily working in designated areas).

- A particular radiation source could be associated with the facility or it could be brought on site by another Employer, such as when using an americium source to test/confirm the attenuation of a newly built x-ray room.
  - Includes the need to address those working in radon atmospheres > 300 Bq m^{-3}. The Employer controlling the premises should notify and cooperate with other Employers working in the building (paragraph 284 L121(HSE 2018b)).
- A designated area, if it exists, could be under the control of the facility or another Employer. Laboratory spaces may host both clinical and university research employees working in the same controlled areas. Consideration is needed regarding whose policies and procedures are being followed to avoid confusion and ambiguity.
  - Includes the need to address Outside Workers (see paragraphs 1.199–1.220 *Outside workers*). Outside workers may be classified or

non-classified radiation workers and are those individuals who carry out work or services on behalf of their Employer at the site of another Employer. Examples include agency staff or MPEs from a regional service provider working alongside local staff in their 'customer's' designated areas.

1.131 When employees of a number of Employers are working on the same site, it is important that protection is optimised and exposures are ALARP, no matter who created the exposure pathway or who has control of it. The training and protection in place for another Employer's staff should be as good as those in place for the staff of the Employer undertaking work with ionising radiation. The precise allocation of responsibility should be agreed upon between Employers in advance of the work (preferably when contracts are being drawn up) and communicated in writing to the employees. These responsibilities and agreements should be documented and reviewed as appropriate.

1.132 When addressing which Employer has responsibility for particular aspects of protection, consider which Employer is under regulatory control for the particular source of radiation and which Employer has specialist knowledge on how to manage radiation protection. It would be prudent to consult with an RPA on arrangements. It is the Employer in charge of the source who must ensure protection under ALARP, and the Employer of any staff working with or alongside that source must be in total agreement with the policies and procedures that are in place.

1.133 Those with practising privileges can be considered employees of, for example, private healthcare facilities for health and safety purposes, and they are not considered outside workers. Even though they are technically self-employed for employment purposes, they may not be acting as their own health and safety Employer. There is still a need for cooperation between Employers, i.e. the facility Employer and the individual with practising privileges.

1.134 The Employer must establish adequate communication arrangements for informing one another when employees are pregnant or breastfeeding, as appropriate.

1.135 Whether the sharing of space is temporary or permanent, there are legal duties for all involved Employers to cooperate with each other in line with IRR17 Regulation 16.

1.136 All involved Employers must exchange general information on the following (L121 paragraph 282 (HSE 2018b)):

- Health and safety risks.
- Measures taken to control the risks and ensure safe systems of work.
- Fire precautions, including information to enable outside workers to identify the person nominated to implement evacuation procedures within premises where the work is taking place.

1.137 Compliance with regulation 12 of the Management Regulations (MHSWR 1999, MHSWR(NI) 2000) should assist Employers in complying with regulation 16 of IRR17 (HSE 2018b). Health and safety information that should be shared includes the following:

- The source, magnitude and type of radiological risk.
- Details relating to controlled areas that could be used by either Employer.
- Relevant arrangements to prevent or mitigate the consequences of a radiation accident.
- Contingency plans to be used in the event of a radiation accident.

B. Patients undergoing diagnostic or therapeutic procedures

1.138 Employers undertaking work with ionising radiation may expose another's employees at a different location as a result of their activities. For example, when nuclear medicine patients are discharged into the care of another Employer such as a nursing home or a prison (see chapter 16 *Patients leaving hospital after administration of radioactive substances*).

C. Staff with multiple employers—sharing dose information

1.139 Individuals may have multiple Employers simultaneously, including self-employment for some, and in this situation, the Employers must cooperate to ensure that occupational doses are restricted across employments (see paragraphs 1.166–1.174 *Dosimetry systems across multiple employers* for more). It is important to note that agencies supplying staff to work with radiation are Employers under IRR17 and therefore must meet their responsibilities under the regulations. In the case of classified staff with multiple Employers, cooperation between Employers is required to decide which Employer will undertake the classification process for the employee and to ensure there is only one statutory dose record per employee. A report of that individual's total annual dose must be provided to the Central Index of Dose Information (CIDI) via the Approved Dosimetry Service (ADS).

1.140 The sharing of dose data and dose history between all relevant groups is required under IRR17 in order for Employers to make informed decisions on employee roles and activities. Cooperation between approved dosimetry services, HSE-appointed doctors, occupational health services, and medical physics departments, should ensure that employees can be adequately monitored and classified where necessary (note that classification itself is the sole responsibility of the Employer).

1.141 Where more than one Employer works with radioactive substances on the same site, suitable cooperation arrangements must be in place and adhered to in order to satisfy the requirements of relevant radioactive substance legislation.

Cooperation between employers policy

1.142 Employers should set out their approach to cooperation between Employers in a policy. This should identify the information that is required about visiting staff, how they will be managed from a radiation protection perspective, what information will be shared with their Employer and what level of detail is required in a contract, if any. It should also state who is allowed to communicate and collaborate on the Employer's behalf. Guidance on what should be included can be found in L121 (HSE 2018b). An aide memoire for developing a policy to cover cooperation between Employers, using a risk-based approach, is given in appendix 9. This is not considered exhaustive, and an RPA should be consulted on local arrangements. If relevant, it should be expanded to include sending radioactive patients to nursing homes, etc and managing carers and comforters (see paragraphs 2.156–2.168, *Carers and comforters*) which are currently not included.

Risk-based cooperation and radiation protection management

1.143 The level of detail involved in the cooperation is to be commensurate with the risk. It is not practical or necessary to let every Employer who has staff working at a facility know that there are radiation activities taking place. The regulations specify 'where it is likely to give rise to exposure'. Where visiting employees do not enter designated areas as part of their role, there should be no need to exchange information about radiation risks with their Employer. A basic radiation protection induction, including how to recognise designated area warning signs and knowledge of access restrictions, may be sufficient. Warning signs are in place to control access where induction training is not delivered. This is the method used to deter members of the public from entering designated areas, and this should be sufficient for unsupervised workers who are on site but are not working with ionising radiation.

1.144 Visiting workers who are to enter designated areas to perform a service will be outside workers (see paragraphs 1.199–1.220 *Outside workers* for more on this), unless the designated area is 'handed over' (see paragraphs 1.147–1.151 *Area handover to another employer*). An example of this could be an outsourced cleaner who has to enter controlled areas to clean. It is likely that they will not be exposed to radiation while performing their duties. Where the exposure risks are minimal and the employee is considered an 'other person' for the purposes of radiation protection, they will need some instruction (very basic but relevant training), and a copy of the risk assessment for the activities should be shared with their Employer. This should indicate their likely dose, including any foreseeable accidental exposures. It should also state any personal dosimetry arrangements or alternative means of demonstrating restriction of dose. This approach should be appropriate for those visiting workers who need to know when areas have been undesignated to allow their access, e.g. an x-ray room where the generator has been turned off for the night to allow cleaning.

1.145 Occupationally exposed individuals of another Employer who are exposed as a result of supporting a patient are outside workers (e.g. social workers or care assistants). They are not carers and comforters (see chapter 2, paragraphs 2.156–2.168 *Carers and comforters*). Although they may only visit a facility once, it is possible that individuals may be exposed frequently at different facilities, and there is a need to track the dose they receive. The facility where the exposure takes place should provide information describing the likely dose received, any risks, and how they are managed. This could be in the form of a standard letter or piece of text that should be passed to the support worker themselves and to their Employer (probably via the employee). A copy of the radiation risk assessment can be shared or made available on request. Their Employer should be alerted to the fact that their employees are being exposed to radiation during the course of their work, and this will allow for a risk assessment to be drawn up. Paragraphs 212 and 219 of L121 (HSE 2018b) state that all Employers, whether or not they are involved in work with ionising radiation, must make sure that the cumulative exposure of their employees over the year does not exceed a relevant dose limit (although ALARP is the aim).

1.146 Those visiting workers who are not performing a service for the radiation facility but who are required to enter designated areas are not outside workers (for example, prison officers escorting a prisoner patient in need of a scan or an external regulator inspecting the premises). For such persons, there is still a requirement to cooperate with their Employer, and the radiation facility should share a copy of the risk assessment in the same way as for the care worker described in paragraph 1.145.

Area handover to another employer

1.147 It is common practice (and preferable) in some situations to temporarily hand over a facility's controlled area to the control of another Employer where the first Employer's risk assessments and supervision do not extend to the activities that may be carried out by the other Employer. This is often used for service engineers carrying out maintenance testing of equipment and for medical physicists performing QA tests, etc. The use of a handover form removes the requirement for outside worker arrangements in the area, although the sharing of some information will still be required.

1.148 A widely used handover form is available on the AXREM website (AXREM 2018). This particular form has a dual function (a) to transfer responsibility for the controlled area (area handover), and (b) to pass information on the state of the x-ray set (equipment handover). A good system requires all visiting service engineers and maintenance staff to report to a central point in order to receive the handover form and inform the department of their presence. The handover form should be signed and handed back to the same point once the work has been completed.

1.149 It is preferable to hand over the controlled area in person, but in situations where this is not possible, such as on a mammography van that is not constantly manned or 'out of hours' maintenance on equipment, this can be done by agreement between the Employers. An equipment handover is still needed, and this may be achieved by leaving the paperwork filled in as much as possible for the incoming employee. Regular audits must be conducted to ensure compliance and governance in such situations.

1.150 The person/organisation taking responsibility for the controlled area will work under their risk assessments and local rules. Assumptions will be made about the suitability of the protection measures in place for routine work, such as the shielding of boundary walls. Communication at handover should be used to identify any specific control measures required for a particular controlled area (that a visitor will not be aware of) and ensure there are no conflicting arrangements with those typically in place for the facility. This will ensure local employees can continue to follow the standard arrangements (such as access restrictions) and not be put at risk. A guidance document (AXREM 2018) recommends that signs be placed on the entrances to the designated areas to say who is in control of the area, along with a copy of the local rules, so that others know the systems currently in place.

1.151 A potentially complicated example of shared workspace management and area handover is where various Employers are responsible for controlled areas at different times during the critical examination of new installations and the installation and testing of new or modified equipment, particularly where the site is a new-build and has not yet been handed over to the purchasing Employer. It is important to have clarity and agree on responsibilities as appropriate to the work. The Employer in charge of the controlled area will need to consider who they may expose as a consequence of their work.

This could include medical physics staff, building contractors, or sub-contractors who are directly involved or those who are working in adjacent areas.

Cooperation of the employee

1.152 Employers can only meet certain specific duties under IRR17 with the cooperation of employees and outside workers (L121 paragraph 621 (HSE 2018b)). Employers are therefore required to ensure that their workers are adequately trained and understand their role in the radiation protection programme. IRR17 Regulation 35 sets out employee and outside worker duties, which are supported by guidance (HSE 2018b). Employers should check that employees are complying with the required control measures and challenge and correct unsafe behaviour (L121 Paragraph 621). Where Employers have instructed employees to wear personal dosimetry, employees have a legal duty under section 7 of the *Health and Safety at Work etc Act* (HASWA 1974) to do so.

Personal monitoring for external radiation

1.153 To demonstrate that doses are ALARP, non-classified persons may be subject to personal dosimetry. This would be determined by risk assessment, supported by the radiation policy on dosimetry, and could be optional content in the local rules. L121, paragraph 381 (HSE 2018b) says that it is advisable for employees who spend a significant amount of their time working in controlled areas to be provided with personal dosimetry. Personal monitoring that is performed should be carried out by an Approved Dosimetry Service (ADS) or by a service that conforms to similar satisfactory standards. Classified staff must be monitored using approved dosimetry.

1.154 Where exposure is from external sources, personal dosimetry could be performed by means of one or more passive dosemeters (i.e. those that must be processed to provide a dose reading, i.e. not an instant readout) worn on an appropriate body part. Typical personal dosemeter positions are the trunk or chest to indicate 'whole body' dose, the collar to indicate both whole body and eye dose, at the eye, the foot or leg and the fingers/wrist. Dosemeters must be appropriate for the task, and the advice of an RPA should be sought when establishing monitoring programmes.

1.155 The length of time for which a dosemeter will be allocated will depend on the manufacturer's or supplier's recommendations, the doses likely to be received during the period, and the probability of an accidental exposure. Each dosemeter should be returned promptly after use for a valid dose assessment and replaced with a new one. Most employees who have personal dosimetry are monitored with 1 or 2 month wear periods. However, periods ranging from 2 weeks to 3 months can be appropriate in certain circumstances, e.g. 2 weeks could be appropriate in a high-risk area with high dose rates. If someone loses a dosemeter or starts employment within a typical wear period, they should be issued a spare until they have a replacement or their own dosemeter.

1.156 A direct-reading or 'active' device (instantaneous readout of dose/dose rate) may be worn if an immediate indication of the dose received is necessary. This can be in addition to or instead of passive dosemeters, noting that classified staff must have approved dosimetry at all times. A direct-reading electronic dosemeter would be appropriate for classified staff where an indication of dose is required in the short term, e.g. a classified outside worker needs an entry in their radiation passbook after working in another Employer's controlled area. Again, ensure that the dosemeter chosen is suitable for the radiation field it is to measure, as some electronic personal dosimeters are not suitable in pulsed radiation fields, such as those experienced in interventional radiology.

1.157 Where there is a significant risk of an accidental exposure, a direct-reading dosemeter (one that has an alarm would be an advantage) should be worn by classified staff in addition to their normal personal monitoring, so that there is an early indication of the accidental exposure and then appropriate measures can be taken to ensure doses are ALARP. Such situations might include a sealed source replacement, a stuck brachytherapy source, nuclear medicine therapy administrations, or any contingency arrangements previously identified in risk assessments. For non-classified staff involved in such situations, a direct-reading dosemeter may be worn instead of a passive device. L121, paragraph 138 (HSE 2018b) states that all employees working with high-dose-rate sealed sources should wear a dosemeter that gives an audible alarm when high dose rates are detected. Records of accidental exposures exceeding the 6 mSv effective dose, 15 mSv eye dose, and 150 mSv skin or extremity dose must be kept (see appendix 1, *Record keeping*).

1.158 Pregnant classified (and occasionally non-classified) workers could also wear a direct-reading dosemeter in addition to the normal personal monitor. This will enable them to monitor their body dose on a daily and monthly basis to identify those procedures that result in higher doses. Appropriate measures should then be taken to ensure doses are ALARP and that the equivalent dose to the foetus is unlikely to exceed 1 mSv from when the Employer is notified of the pregnancy until the end of the pregnancy (regulation 9(6)(a) of IRR17). A risk assessment should help identify those non-classified workers for whom these arrangements are appropriate (HSE 2015b).

1.159 Where it is unlikely that an individual performing a particular function in a radiation facility will receive a radiation dose, it is possible to monitor whoever performs that function with a 'job badge'. An example of this could be where a number of individuals carry out the role of a scrub nurse in a surgical theatre. Each individual will wear the monitor while they are doing the job, and the reading on the dosemeter indicates the worst-case scenario, which is that a single person receives all the exposure during the wear period. Job badges are only appropriate when used to confirm that doses are negligible (i.e. below the threshold of the dosemeter) and where there is no identified risk of a radiation accident in the risk assessment. Where this is not the case, individuals may need to be monitored in line with current guidance. This is a slightly different scenario from monitoring an individual (when they are in a controlled area) who is assumed to receive the same exposure as a number of individuals (see L121 paragraph 382 (HSE 2018b) for more).

1.160 A risk assessment should establish if monitoring will be routine or periodic. It is recommended that an initial assessment of occupational exposure be made for any new facility or when a new technique is introduced. If routine monitoring will not take place, then periodic monitoring should take place to ensure the risk assessment is still valid. A significant effective dose is considered to be 1 mSv $year^{-1}$ (L121 paragraphs 452 and 453 (HSE 2018b)), and routine body monitoring is appropriate for those individuals likely to receive this level of effective dose through their work (including possible accident situations). The 2002 edition of the Medical and Dental Guidance Notes (IPEM 2002) and previous Government Guidance Notes (NRPB *et al* 1988) recommended routine monitoring of the extremities and eyes when doses were likely to exceed 1/10 of the relevant dose limit. However, IPEM has published specific guidance on personal dosimetry for personnel working in healthcare in the UK (Martin *et al* 2018). The newer guidance puts forward a higher level of exposure at which individuals would routinely be monitored for eye dose and, in certain situations, a higher level of exposure at which foot/leg doses would be assessed and attention given to the protection in place, as well as guidance on where and when monitors (body, finger, and eye) should be worn for different radiology, nuclear medicine, and radiotherapy procedures.

1.161 The 2018 IPEM publication (Martin *et al* 2018) is a comprehensive document providing recommendations for monitoring in the majority of healthcare environments and setting out an approach for Employers to assess dose levels and determine monitoring requirements based on established rules for new activities (part of the risk assessment process). Dose levels from the literature are provided, which can be used to estimate exposure in different clinical scenarios, and the document addresses the fact that the eye dose limit for those occupationally exposed (20 mSv for those employees 18 years of age and over) is now more likely to be approached or even exceeded than the whole body effective dose limit (also 20 mSv $year^{-1}$). It also covers topics such as establishing the most exposed area of the body when radiation fields are not uniform, the interventional radiology 'double-dosemeter' approach (one near the eye or at the collar above the lead apron and one underneath the apron), how to establish alteration factors when it is not reasonable to monitor at certain locations on the body (e.g. a dosemeter on the fingertip could adversely affect tactile function) or when it is necessary to account for protection worn (e.g. lead glasses), and when those factors can and cannot be used to alter a dose record. For example, alteration factors may be used to estimate the exposure of non-classified staff because they do not legally have to be monitored by approved dosimetry. However, for classified staff, it is not appropriate to alter a measured dose unless the ADS has received approval of the method from the HSE.

1.162 L121 paragraph 620 (HSE 2018b) makes clear that employees must wear, take reasonable care of and return for processing any personal dosemeters provided by their Employer or they may be committing an offence under section 7 of Health and Satiety at Work Act (HASWA 1974), which states:

'It shall be the duty of every employee while at work—

(a) to take reasonable care for the health and safety of himself and of other persons who may be affected by his acts or omissions at work; and

(b) as regards any duty or requirement imposed on his Employer or any other person by or under any of the relevant statutory provisions, to co-operate with him so far as is necessary to enable that duty or requirement to be performed or complied with'.

1.163 Care should be taken to prevent the dosemeter, while not being worn, from being exposed inadvertently to ionising radiation or subject to other conditions, e.g. heat or moisture, which could affect the assessment of doses.

1.164 A common reason given for not wearing a dosemeter that is provided by an Employer is sterility. The dosemeter supplier should be able to provide information with regard to sterilisation methods that will not adversely affect the dosemeter readout process. It is suggested that infection control staff be asked for their advice (where they are available) to help reach a satisfactory arrangement that can be used to encourage those who may be concerned about this issue.

1.165 There are various requirements for record retention with regard to personal monitoring (see appendix 10, *Record keeping*). Staff should have access to their dosimetry results as this will encourage a good radiation protection culture, but it is also necessary that staff are aware of their dose and can share this with any other Employers that they may be working for or when they are working in the designated area of another Employer (see paragraphs 1.127–1.152 *Cooperation between employers* and paragraphs 1.199–1.220 *Outside workers*). Employers should provide training on accessing and sharing dose records as appropriate. An Employer may have to share their employees' doses with another Employer for the same reason. A contract of employment gives an Employer consent to share this information as it is a legal requirement, but consideration of data protection legislation is also essential. Only the necessary information should be used, e.g. the name of the employee and the associated dose may be sufficient rather than including a national insurance number and the names of other employees documented on the same dose report.

1.166 A system for managing personal dosimetry will be effective when it addresses the following points (Rogers *et al* 2017):

- Written arrangements describe the system which should include performance indices.

- Training records are available to demonstrate that staff have been trained appropriately. This will include how to use dosemeters, their responsibilities, and what to do if something goes wrong.
- There is robust management of issuing and returning of dosemeters.
- Dosemeter results are promptly reviewed and actions are taken when investigation levels are breached (see paragraphs 1.65–1.71 *Formal investigation levels and action on dosimetry results*).
- Management supervises compliance and takes action when it is not achieved.
- A record of staff doses is available when it is required and shared as appropriate.

Dosimetry systems across multiple employers

1.167 There is a need to manage the occupational exposure of staff across employments to ensure that dose limits are upheld. This requires cooperation between Employers (see paragraphs 1.127–1.152) and a management system capable of supporting non-standardised inputs and approaches to monitoring individuals. Employers will have to gather information about their employees, communicate with other Employers, make decisions based on likely annual doses, perform appropriate dosimetry, and collate and share results. All the information and decisions should be recorded, with due respect for information governance.

1.168 These decisions will include deciding which staff are monitored, who supplies the dosemeters, who receives the results, and who may share them with others. It will also include when the results are shared (e.g. periodically or only in the event of an unexpected dose), who takes action on results that exceed investigation levels or indicate that doses are not ALARP (and how the investigation will be shared), who classifies certain staff, and who reports an overexposure. This has been a requirement of the regulations for many years and is not new in that regard; however, it is challenging in practice. Guidance has been published (Rogers *et al* 2017, Martin *et al* 2018), which represents a means of compliance and should help standardise management systems in the UK, thus making cooperation between Employers more straightforward. This aspect of the regulations will be of interest at the regulatory inspection.

1.169 Rogers *et al* (2017) set out a risk-based approach to managing dosimetry across organisational boundaries as well as giving general advice on dosimetry management systems. It should be noted that health and safety do not allow for generalisations. Frameworks can be provided to guide decisions, but the outcome must be that employees are protected as individuals.

1.170 It should be set out in a policy that employees have a responsibility to tell their Employer if they are working with ionising radiation or working in a facility where they may be exposed to ionising radiation for another Employer.

Employers should also ask their staff periodically. This is particularly important when staff may be classified.

1.171 The collection of the required information will be best facilitated through a pro-forma that includes employee-specific information, details (including contact details) of other Employers, the type of work carried out, and details of personal monitoring. If the employee has their current dose record for the other employments that they can share, this will be advantageous and will determine the urgency of making contact with the other Employers or highlight if exposure must be limited until accurate details are established. By providing this information, the employee is giving their consent to their information being shared, but in any case, a contract of employment is sufficient to allow the sharing of information in this way (also consider the appropriate data protection regulations).

1.172 Contact should then be made with the other Employers. This could be done initially through a group email, or in cases of increased urgency, a phone call could be made. Example text is provided below:

- A member of staff in my employment has stated that they also work under your employment; therefore, we must cooperate under IRR17 to ensure we can meet the requirements of the regulations. Attached is our approach to managing dosimetry, which is in line with best practice. Can you please agree to share dose information as appropriate and provide the most suitable contact details for ongoing communications where specifics will be established?

1.173 Once contact has been made, a copy of the appropriate radiation risk assessment and the dosimetry policy will be key sources of information for decision-making. Using the information available, an estimate should be made of the employee's likely total annual exposure and decisions made by management in line with the published guidance.

1.174 If a worker exceeds the dose limit, the Employer at the particular point when the dose limit was exceeded is deemed to be responsible (L121, paragraph 286 (HSE 2018b)).

1.175 See also paragraphs 1.213–1.220 *Classified outside workers* and paragraphs 1.216–1.220 on *Non-classified persons*.

Personal monitoring for internal radiation

1.176 The dose limits for staff specified in IRR17 apply to the total dose received from all sources of radiation, excluding natural background radiation at normal levels and an individual's own medical exposure (IRR regulation 2(6) (IRR 2017, IRR(NI) 2017)). Additionally, regulation 20(6) requires the Employer to measure and record any potentially significant dose received by classified workers, including that due to internal radiation. L121, paragraph 453 (HSE 2018b), clarifies that an individual dose component is only significant if it is likely to deliver doses > 1 mSv in a calendar year.

1.177 The radiation doses delivered to the vast majority of healthcare workers will come from external radiation; however, doses received from internal radiation must also be considered whenever unsealed radionuclides are involved. The requirement for internal radiation dose monitoring should be decided at the risk assessment stage. If it is decided that classified staff require internal dose monitoring, then the Employer must ensure that the dosimetry service they engage in has been approved specifically for this purpose (Regulation 22 (2)). If internal dose monitoring is required for staff working in a particular area, then it should also be borne in mind that outside workers may also need to be monitored, depending on their duties. Ideally, this will need to be discussed by the RPAs of the two Employers (see paragraphs 1.127–1.152 *Cooperation between employers*).

1.178 Methods of monitoring for internal radiation can involve the analysis of inhaled air samples, the analysis of biological samples taken from staff, or in-vivo measurements of internally deposited radionuclides using external detector systems. These methods have been briefly reviewed elsewhere (Martin *et al* 2018).

1.179 Due to the physical properties and quantities of radionuclides handled in medical facilities, it is highly unlikely that any healthcare staff will require routine internal dose monitoring. However, it may be necessary to consider methods of internal dose assessment in scenarios involving accidental releases of radionuclides. For incidents involving gamma-emitting radionuclides, dose estimation can be performed by taking extensive readings on the surface of the body using calibrated detectors. For alpha and beta emitters, biological specimens may need to be collected and counted. This should be considered as part of the risk assessment.

1.180 Although routine internal dose monitoring of individuals may not be justified, there may still be a requirement for adequate monitoring of surface contamination levels under IRR17 regulation 20(6) (IRR 2017, IRR(NI) 2017) for those areas designated as controlled due to the risk of radioactive contamination; see paragraphs 1.180–1.182 on area monitoring.

Area monitoring

1.181 Regulation 20 of IRR17 requires that designated areas (i.e. controlled or supervised) have their radiation levels adequately monitored, and L121 (HSE 2018b) says this should be both inside and outside the boundary of the designated area. This is in addition to personal monitoring of staff doses. Area monitoring, including contamination monitoring (see chapter 10, *Diagnostic use of open radioactive substances*), should be carried out at appropriate intervals to check the dose, dose rate, and surface or airborne contamination levels to which persons are exposed and to detect inadequacies in controls or systems. The programme of monitoring (on which an Employer should consult an RPA) will be based on the nature and extent of the risks resulting from exposure to ionising radiation and should make use of suitable equipment that meets testing requirements (performed by or under the supervision of a Qualified Person—see paragraphs 1.183–1.186).

1.182 The designation of each radiation area should be confirmed periodically by reviewing the working conditions (e.g. workload, techniques, occupancy) and occasional radiation measurements. This should be performed at the commissioning time, repeated at a frequency determined by the risk assessment, and when the situation or practice changes significantly.

1.183 Area monitoring might include measurement of the instantaneous dose rate around linear accelerators and CT scanners, or a longer environmental measurement that might use thermoluminescent dosemeters (TLDs) or optically stimulated luminescence (OSL) technology. Passive neutron dosimeters can be used to confirm low neutron dose levels around linear accelerators and cyclotron bunkers. Measurements carried out inside controlled areas should demonstrate that staff doses are ALARP.

Qualified person

1.184 It is the responsibility of the Employer to ensure that all monitoring equipment has a current, appropriate, and traceable calibration. An inventory of equipment will help with the management of the equipment and indicate when testing is due. The Employer should appoint a Qualified Person or Persons (QP) to be responsible for performing the examination, testing, calibration, and maintenance of all equipment used for monitoring radiation in and around designated areas, in accordance with the requirements of measurement good practice guides, e.g. NPL Guide 14 *The Examination, Testing, and Calibration of Portable Radiation Protection Instruments* (NPL 2014). Monitoring equipment should normally be tested and thoroughly examined at least once a year, and a risk assessment can be used to indicate whether more frequent testing is required (L121 paragraph 409 (HSE 2018b)).

1.185 The QP role under IRR17 is different from that of the RPA (see paragraphs 1.88–1.91). The RPA should be consulted on the implementation of requirements for designated areas, including the type and extent of a monitoring programme. The RPA should also be consulted on the calibration (e.g. energy and dose/dose rate response) and regular checking (to ensure the monitor is operational on a daily basis e.g. battery checks, zeroing, and response) of monitoring equipment to ensure it is serviceable and correctly used. The RPA does not actually have to perform the tasks. The QP performs or supervises the tasks.

1.186 A QP does not need to be appointed in writing by the Employer but it is important that the QP is fully aware of the type of radiation to be measured and the type of designated area. The QP should possess the necessary expertise in instrumentation, theory, and practice to calibrate and maintain the type of instrument being tested, or supervise others to undertake the tasks. The QP role will likely be performed by an organisation with individuals who specialise in this field, but there is no reason that Employers cannot do their own examinations and tests, provided they have the facilities and a QP with the necessary (and documented) expertise. Proper tests will require radiation sources with accurately determined activity or the use of equipment with calibrations traceable to a national standard. Details of the local arrangements for the examination and testing of radiation monitoring equipment could be included in the local rules (L121 paragraph 337 (HSE 2018b)), including contact details for the QP.

1.187 Any records of testing regimes must be authorised, e.g., signed off by a suitably qualified person. The name and contact details of that person should be stated (L121 paragraph 426 (HSE 2018b)). A record of the results should be kept for at least two years.

Classified workers

1.188 Employees who are likely to receive a dose in excess of 3/10 of any relevant dose limit (except that for the eye, where 15 mSv in a year is the classification level) as a result of the work undertaken, including the possibility of accidents that are likely to occur (e.g. those that are reasonably foreseeable, as identified and documented in the radiation risk assessment), must be classified (see paragraph 1.52 in the risk assessment section for more on this). The results of the risk assessment should be used to identify those persons. Employees under 18 years of age cannot be designated as classified.

1.189 Classification on the basis of potential extremity doses may be expected for (but not limited to) those engaged in the following types of work:

- therapeutic interventional radiology and cardiology procedures;
- endoscopy procedures using x-rays;
- radiopharmaceutical or radionuclide preparation;
- diagnostic and therapeutic radiopharmaceutical administrations;
- source preparation, insertion or removal in brachytherapy procedures; or
- intravascular and intraoperative radiotherapy procedures.

1.190 As the dose limit to the lens of the eye has been reduced to 20 mSv $year^{-1}$ it is likely that some of these staff groups, such as interventional radiologists and interventional cardiologists, will have to be given close attention in regard to their radiation protection (ceiling suspended shields and/or lead glasses, for example) and considered for classification on the basis of eye dose.

1.191 Classified employees are subject to medical surveillance and more stringent requirements for the assessment and recording of their doses. However, classification is not an alternative to restricting exposure so far as is reasonably practicable. An RPA will be able to provide assistance in determining who should be classified. If a member of staff is taken on for a short-term contract, the following process must still be adhered to.

1.192 If it is shown by risk assessment or as a result of personal monitoring that a member of staff will have to be designated as classified, the Employer should take the following steps. In practice, it is likely that the Employer will delegate this duty, for example, to a radiology manager or the chair of the RSC.

- Write to the employee to let them know they are to be classified and inform them of their responsibility to cooperate with the Employer under IRR17 regulation 35 (HSE 2018b)—comply with monitoring and present for a medical (during their working hours and at the cost of the Employer). The classified person must be given training, information and instruction to allow them to adhere to the administrative arrangements surrounding their monitoring and records.
- Arrange for the employee to be assessed as fit to work with ionising radiation by a relevant doctor (appointed doctor or 'employment medical adviser' appointed under section 56 of the *Health and Safety at Work etc Act* (HASWA 1974) BEFORE they are classified. An appointed doctor is a registered medical practitioner who is appointed in writing by the HSE or HSENI (HSE 2018c). It could take at least one month for this to be arranged and carried out. The employee should take their dose record to the medical. The current lists of appointed doctors are available on the HSE and HSENI websites (HSE 2022, HSENI 2022).
- This exercise should be performed in conjunction with occupational health. The assessment by the relevant doctor will form part of a 'health record' for the employee which should contain the particulars listed in Schedule 6 of the regulations IRR17 (HSE 2018b). This record must be maintained by the Employer until the member of staff reaches 75 years old, or in any case, at least 30 years from when the last entry was made. While the employee is classified, an entry is required in the health record every 12 months (±1 month) or sooner if stipulated by the relevant doctor. If they are a 'Classified Outside Worker' with a Radiation Passbook (see paragraphs 1.199–1.220 *Outside workers*) they should take that along to the medical assessment. If a new employee was classified at a previous Employer's facility, it is sufficient for the new Employer to check that they have been certified fit to work within the last 12 months, unless the work is significantly different.
- Arrange to have the radiation dose received by the employee systematically monitored by an Approved Dosimetry Service (ADS). This includes the issue and return of dosemeters for assessing doses from external radiation and the making and maintaining of dose records. The ADS should provide

summaries of dose records, and the Employer should keep these for at least two years. The type of dosemeter(s) worn should be suitable to assess all doses (whole body dose, extremity dose, or eye dose) that are likely to be significant (L121 paragraphs 452 and 453 (HSE 2018b)). If necessary, neutron doses or committed doses from internal radiation should also be assessed by appropriate means.

1.193 A significant dose is described in L121 (HSE 2018b) as any component of an effective dose that may exceed 1 mSv year^{-1}. It is not made clear what a significant eye or extremity dose is, so this must be based on an individual assessment, such as through a period of appropriate monitoring. The potential dose from radiation accidents must also be considered. Once a likely level of exposure is established, it will be down to professional judgement as to what is considered significant; however, IPEM has published specific guidance on personal dosimetry for personnel working in healthcare in the UK (Martin *et al* 2018). Recommendations are given on when to routinely monitor staff dose (see also paragraphs 1.152–1.165 *Personal monitoring for external radiation*), and these recommendations are as applicable to classified staff as they are to non-classified staff. The publication also recommends the routine use of the 'double-dosemeter' approach to body monitoring for classified staff working with x-rays.

1.194 Consideration should be given to the level at which routine monitoring is recommended for non-classified staff (see paragraphs 1.152–1.165 *Personal monitoring or external radiation*), as it would not be appropriate to monitor non-classified staff while neglecting classified staff.

1.195 If a dosemeter belonging to a classified person is lost or damaged, the Employer should carry out an investigation to determine the dose received during the period and give it to the ADS to record. If there is insufficient information available to make an estimate of the dose received, the Employer should have the ADS record a 'notional dose'. A notional dose is the proportion of the total annual dose limit for the relevant period.

1.196 If an Employer suspects that the classified worker received a dose that was either significantly higher or lower than what would normally be recorded for a dosemeter wear period, they should investigate and provide the ADS with an estimated dose, which is recorded as a 'special entry'. The investigation should be kept for two years, and the outcome can be appealed by the employee within three months of being notified.

1.197 The person remains classified until the end of a calendar year unless they are to be taken off those work duties as a result of the medical assessment or due to moving to a role where doses are not expected to be significant. When declassifying the employee or ending employment, a 'termination record' (which details doses recorded during a period of classification) will be provided by the ADS to the HSE/HSENI and the Employer. The Employer should make this available to the employee.

1.198 When the classified person has more than one Employer or works as a self-employed person on occasion, each Employer has responsibilities under regulation 22 of IRR17. This means that each Employer should make separate arrangements for dosimetry with an appropriate ADS, although in practice the same ADS may be used. In addition, each Employer will need to establish the total dose received by the employee to demonstrate that a dose limit has not been exceeded or whether an investigation level has been reached (see paragraphs 1.127–1.152 *Cooperation between employers* and paragraphs 1.166–1.174 *Dosimetry systems across multiple employers*).

1.199 When entering the controlled area of another Employer while still acting in the employ of the original Employer, the classified person will be a classified outside worker, which comes with other legal requirements (see paragraphs 1.213–1.220 *Classified outside workers*).

Outside workers

1.200 This section should be read in conjunction with paragraphs 1.125–1.152 on *Cooperation between employers.*

1.201 The term 'outside worker' encompasses both classified and non-classified outside workers.

- A non-classified outside worker is a person, who is not a classified worker, who carries out services in the controlled or supervised area of an Employer other than that of his or her own Employer.
- A classified outside worker carries out services in the controlled area of another Employer.

1.202 'Services' implies a benefit to the Employer in control of the area. Any employee of another Employer who is not carrying out a service is therefore considered an 'other person' as specified in IRR17. Appendix 8 has a flow chart that can help identify those being exposed to radiation and match it with the terminology used in the regulations. Outside workers will include visiting applications specialists demonstrating or operating radiation equipment within an organisation, agency staff such as radiographers, or possibly contracted medical physics teams performing QA on equipment (unless the area has previously been formally handed over; see paragraphs 1.152–1.151 *Area handover to another employer*).

1.203 Outside workers are of concern within the regulations because, without careful management, they may be at particular risk as they may be unfamiliar with local safety arrangements, procedures, and policies in place to reduce hazards and risks for employees.

1.204 A useful document covering outside workers, including the practicalities of putting the requirements into practice, has been published by the IAEA (IAEA 2015). Of interest are examples of checklists that can be used to develop contractual arrangements and a flow chart for assessing the competence of contracted workers. An aide memoire to help an Employer develop a policy to cover cooperation between Employers (including managing outside workers) using a risk-based approach is given in appendix 9. This is not meant to be exhaustive, and an RPA should be consulted on local arrangements.

1.205 A risk assessment (e.g. the radiation risk assessment or a standalone risk assessment) should cover the arrangements for outside workers to ensure doses are ALARP. The assessment should consider (in particular) predicted doses, the level of supervision required, previous training and experience, appropriate personal dosimetry, and an individual's dose history. It is recommended that there be specific instructions for outside workers in the local rules (a subset or an appendix would be appropriate). These should include the need to share information with the Employer (likely through the RPS or facility manager, as appropriate) regarding dose history, especially when an individual's dose is approaching or exceeding the investigation level detailed in said local rules.

Entry into designated areas

1.206 There are a number of additional requirements for classified outside workers (see paragraphs 1.213–1.215) but other than the applicable dose limit, there is little difference in the management of radiation protection for non-classified outside workers and other persons entering a designated area. The purpose is to ensure doses are ALARP. Before allowing entry to the designated area, the Employer must ensure:

- arrangements are in place for entry as recorded in the local rules.
- appropriate training (general and specific) has been provided or the individual is under appropriate supervision. This may be straightforward if a department manager is assessing the training of an agency radiographer but it might not be as obvious where they are assessing the training of an external specialist such as an engineer. This is a good reason to use the handover form as described in paragraphs 1.127–1.152 *Cooperation between employers and others.*
- doses are restricted (e.g., by means of engineering controls, systems of work, PPE etc) and this is demonstrated through monitoring or other means.
- where the 'other person' is an employee entering the designated area as a result of their work, there must also be cooperation between Employers (see paragraphs 1.127–1.152).

Outside workers—cooperation between employers

1.207 Decisions on how and when information will be shared between Employers should ideally be made when contracts are established, along with the actions that will be taken when, for example, an individual exceeds a formal investigation level or in the event of a breach of the regulations or a radiation event, e.g. a significant contamination incident or a serious unauthorised exposure. If there is a breach of the regulations surrounding outside workers and it is detailed in a contract that one particular Employer is taking responsibility for that aspect of compliance, it will be that Employer who is liable (regulation 37(6) of IRR17 (HSE 2018b)).

1.208 In order to make decisions on the correct worker to send, the Employers of outside workers should be informed of any specific risks, likely doses that will be received, and any training needs prior to the deployment of the outside worker. L121, paragraph 293 (HSE 2018b) lists the information that should be provided to the Employer of an outside worker as the following:

- The detail of the actual work to be carried out.
- The type of likely radiation exposure.
- An estimate of the total dose likely to arise from the work.
- The work procedures that will be required to keep doses ALARP (including use of any special protective equipment).
- The risks associated with the work and the precautions to be taken.
- Any local restrictions that will be applied.
- The local rules that apply (including emergency arrangements and contingency plans).
- RPS names and contact details.
- Any relevant dose constraints and associated local investigation levels.
- If the work will take place outside of normal working hours, the Employers should exchange information on emergency contacts.

1.209 The level of detail involved in the information exchange should be commensurate with the risk. A copy of the risk assessment and the local rules, if written with sufficient detail, should address all of these requirements. Alternatively, a simple template summarising the relevant information will be satisfactory.

1.210 For those working with ionising radiation, the host organisation also requires certain information on the proposed outside worker, such as dose history, training levels, and any special requirements. This information should be shared through prior communication with the other Employer. However, it will also be appropriate to have a pre-work 'induction' meeting between the host Employer and the outside worker, in the same way as would be expected for new workers on a building site, for example. This should occur before the first work session and periodically, as appropriate, to review the current information and radiation protection arrangements. The incoming outside worker should report on arrival to an appropriate contact (local decision), such as the RPS or local manager, who will (prior to allowing entry to the controlled area):

- gather/confirm the details of the worker, including fitness for work for classified outside workers;
- identify the appropriate dose limits (from IRR17) and check dose history (workers must have access to their dose record);
- review monitoring arrangements and provide additional monitors where necessary;
- explore their training and ensure any shortfalls are addressed (general training and specific training in regards to the work they will undertake and use of any relevant PPE). This will include the provision of local rules which are to be read and understood, where appropriate;
- share details of the risks and likely doses that will be received;
- provide PPE.

1.211 A number of checklists that could be used to help gather information from certain categories of visiting employees are given as examples in appendix 11, *Pre-work checklist for outside workers and trainees*. They should be developed and expanded to reflect local practice and may be referred to in local rules or radiation protection management policies as part of the systems of work for allowing entry to designated areas, for example. The pre-work checklists will facilitate compliance with the regulations, but it is hoped that a simpler method will present itself.

1.212 The pre-work checklist is a pragmatic way to satisfy an Employer that they have assessed and upheld their duties towards an outside worker. Healthcare environments are not always predictable, and prior cooperation between Employers may not always be possible. Indeed, an Employer can only take reasonable steps to cooperate with another Employer who may not be fulfilling their duties to cooperate in return. The Employer of the facility must therefore either prevent the outside worker from working or ensure the correct information is available before making a risk-based decision about the suitability of the work taking place. The necessary information will likely come from the outside worker themselves in this case, so employees must be well aware of their dose history. If suitable evidence of appropriate training cannot be provided, the Employer must provide the training or ensure direct supervision of the work by a competent person.

1.213 The HSE is unlikely to look kindly on Employers that are not cooperating with other Employers and facilitating the radiation protection arrangements of their employees who are outside workers. Even when Employers have robust arrangements, employees must comply with the systems of work and facilitate the radiation protection arrangements by sharing training records so that additional training needs can be identified and sharing radiation dose records so that investigation levels (see paragraphs 1.65–1.71) can be upheld, etc. IRR17 regulation 35 sets out employee and outside worker duties (see paragraphs 1.10–1.12 on responsibilities) which are supported by guidance (HSE 2018b). It is best practice for outside workers to understand their current dose history and training situation and be able to share this with other Employers to ensure work can take place when planned.

Practical examples putting the theory into practice

Example 1

A prison officer has accompanied a prisoner for a radiation procedure and must remain handcuffed to the individual at all times. The prison officer must therefore be present in the controlled area when the exposure takes place. They are not carrying out a service for the radiation facility, and therefore they are not outside workers. They are also not a 'carer and comforter' (see chapter 2, paragraphs 2.155–2.167, *Carers and comforters*) as defined in the *Ionising Radiation (Medical Exposure) Regulations* (IR(ME)R 2017, IR(ME)R(NI) 2018) because they are working. Therefore, they must be treated as an 'other person', but the facility making the exposure must cooperate with the Employer of the prison guard to the extent necessary to satisfy the regulations (i.e. provide a risk assessment and an estimate of the likely dose in line with regulation 16 of IRR17).

Example 2

A student is on rotation through a radiotherapy department for a number of months and will work in several controlled areas. They are not employed by the hospital where they are training. They are not considered outside workers as they are not providing a service to the hospital because they are receiving instruction and training and are under direct supervision. More importantly, under IRR17, the student is considered an 'employee' of the person whose undertaking is providing the instruction or training for the purposes of the regulations (see the definition of 'trainee' in IRR17 and Regulations 2(2)(b) and 3 (IRR 2017, IRR(NI) 2017)) and is therefore not an outside worker. Although students are 'employees', they should be managed in a similar way to outside workers in terms of gathering information about their training, dosimetry arrangements, etc. Note that IRR17 requires different dose limits for those under the age of 18. It would be useful to have a pre-work meeting at the start of the student's rotation (and periodically as appropriate). A pre-work checklist is given as an example in appendix 11, *Pre-work checklist for outside workers and trainees.*

Example 3

A cardiologist is undertaking work with limited supervision in the later years of their specialist training to gain experience in performing cases. They rotate through a number of different Employers' cardiology departments but are not directly employed by them. Given they are performing a service and are receiving limited instruction at this stage, they could be considered an outside worker. It is recommended that consideration be given to the role individuals are undertaking to identify their designation under the regulations and at which point they cease to be trainees/students (this should be recorded in their scope of practice and the IR(ME)R entitlement procedure and should be competence-based). In some respects, it is less important whether the cardiologist is classified as an outside worker or not, as long as the requirements of Regulation 16 (*Co-operation between Employers)* of IRR17 are being complied with. It is expected that some organisations will accept honorary contracts as a means of limiting their outside worker numbers to a more manageable size; however, an honorary contract will not excuse Employers from compliance with Regulation 16.

Example 4

Staff from an outsourced cleaning service must enter the controlled area of a radiation facility to carry out domestic chores while it is still designated as such (e.g. in some cases in nuclear medicine). This is a benefit to the Employer in charge of the controlled area, and the cleaner is considered a (non-classified) outside worker. There will be many examples of non-classified outside workers across organisations. The majority of non-classified outside workers in the healthcare sector will receive negligible doses. If they are not considered to be working with ionising radiation, they must be subject to dose restriction as if they were an 'other person' in line with L121, paragraph 86 (HSE 2018b).

If the facility has areas that are temporarily designated as controlled areas when x-rays are in use, when the area is no longer controlled e.g. the risk of exposure has been removed and the warning light is off, the cleaner can enter without restriction and is not an outside worker under IRR17. Similarly, in areas such as nuclear medicine, where decontamination of areas including patient toilets is carried out prior to the cleaner's arrival, the cleaner is no longer entering a supervised or controlled area and can perform duties without being subject to the outside worker arrangements. In cases such as these, it would be advisable to have a reversible designation sign that can indicate which areas are no longer designated. In some cases, it may not be possible to completely decontaminate an area (such as the bathroom of an isolation suite used for radioiodine therapies), and in such cases, the contracted domestic staff should be considered to be outside workers. However, regardless of their status and whether they are contracted with or employed by the organisation, cleaning and domestic staff should receive appropriate training and instruction before entering any areas that use or store radiation or radioactive materials.

Example 5

A medical physicist attends the x-ray room of a local healthcare facility to which they provide contracted services. The physicist receives an appropriate handover form prior to carrying out the routine QA, and the area then falls under the control of the physicist's Employer, meaning they are not an outside worker. The physicist then calls in a radiographer from the organisation to help with the exposures. The radiographer then becomes an outside worker (the example also works if an engineer has to come into the room to make adjustments to a system). It is therefore required that the medical physicist's local rules detail the arrangements for outside workers and how they will manage this situation. On a separate occasion, the physicist attends alongside the local staff to observe procedures and give advice as an MPE. In this capacity, they are an outside worker.

Example 6
Consider a care worker escorting and supporting a patient for an x-ray as a working example of a non-classified outside worker. Proportionate 'training' could be addressed by instructions from a radiographer (direct supervision), and a copy of the risk assessment should be made available to the outside worker and their Employer. It may be appropriate to provide a generic cover letter explaining that the outside worker has received a dose during the course of their employment and that their Employer should risk assess and track the total exposure of all their staff. It is the responsibility of all Employers to ensure IRR17 is upheld, not just those undertaking work with ionising radiation.

Classified outside workers

1.214 An example of a classified outside worker is a classified interventional radiologist who works in the controlled areas of other healthcare organisations while under the employment of a particular organisation due to a contract to provide services in that way.

1.215 Classified outside workers must be issued with an HSE-approved radiation passbook. This is issued by the Approved Dosimetry Service on request. Details of the required content of the passbook are given in Schedule 5 of IRR17. Radiation passbooks have a unique serial number and cannot be shared or transferred between individuals. Entries in a radiation passbook should only be made by those who have had suitable and sufficient training and are authorised to do so. In practice, this is likely to be an RPS or manager, along with employees of the Approved Dosimetry Service.

1.216 Classified outside workers are subject to dose assessment and dose estimation.

- The Employer of the classified outside worker must arrange for the appropriate dosimetry assessments to be provided by an ADS (because they are, by definition, classified workers). The classified outside worker must wear personal dosemeters provided by their Employer at all times. However, for practical reasons, it may be sensible to issue dosemeters to classified outside workers for each of the locations where they work, provided the management of this is agreed upon between the various Employers and all monitors are provided by an ADS. As with all classified staff, dosemeters should be issued for the assessment of all significant components of dose, i.e. effective dose, skin dose, or dose to the lens of the eye.
- In addition to the dosemeters provided by the classified outside worker's Employer, the Employer in control of the controlled area in which the classified outside worker is working must have facilities for estimating the radiation dose accrued by the worker while in the controlled area, and this

estimate must be entered in the passbook. HSE guidance from 2018 (see the copyright page of L121 (HSE 2018b) for an explanation of the legal status of 'Approved Code of Practice' and guidance) says that estimation 'should' be made using a suitable personal dosemeter (or air sampler in the case of internal radiation) (paragraph 383 of L121 (HSE 2018b), where the previous edition from 2000 formerly stated that this was the 'ideal' method of estimation (HSE 2000b). It is not necessary that this additional dose monitoring equipment be provided by an ADS, provided the original Employer's approved dosemeters are in use to allow dose assessment. It should, however, be appropriate for its intended use, have a traceable calibration, and come under the supervisory control of the RPS and Qualified Person (see paragraphs 1.183–1.186). It is preferable to have a quick dose estimate that tends to overstate the dose than a more accurate estimate that cannot be made for some days or even weeks (L121 Paragraph 383 (HSE 2018b)).

Non-classified persons

1.217 If a worker is unlikely to receive an effective and/or equivalent dose in excess of 3/10 of any relevant dose limit (15 mSv for the eye), there is usually no need for them to be classified. This will apply to many people working in controlled areas.

1.218 Where non-classified persons (including employees, non-classified outside workers, and any other person) enter a controlled area, the Employer must put in place arrangements to restrict their exposure so that they do not exceed those levels set out in Regulation 19(4) of IRR17, i.e. the classification level for employees and non-classified outside workers or the relevant dose limits for any other persons. It is important to note here that Employers should take particular steps to restrict the exposure of any employees who would normally not be exposed to ionising radiation in the course of their work. They should in fact be treated as members of the public, and control measures should be put in place to make it unlikely that they would receive a dose greater than the dose limit for any 'other persons' (L121 paragraph 86 (HSE 2018b)).

1.219 To demonstrate that doses are restricted, non-classified persons may be subject to personal dose monitoring. This would be determined by risk assessment, supported by the radiation policy on dosimetry, and details may be recorded in the local rules for clarity (optional content—see table 1.2). L121, paragraph 381 (HSE 2018b), says that it is advisable for employees who spend a significant amount of their time working in controlled areas to be provided with personal dosimetry. Personal monitoring that is performed should be carried out by an ADS or by a service that conforms to similar satisfactory standards.

1.220 Where personal monitoring is not provided, it is possible to determine individual doses through knowledge of dose rates and time spent in the controlled area where these are known to be stable and/or constantly below a particular value (L121 paragraph 380 (HSE 2018b)). It is also possible to monitor a representative person (or worst-case person) and use their dose to indicate the dose to others who will have a similar exposure.

1.221 Paragraphs 1.127–1.152 on cooperation between Employers, paragraphs 1.167–1.175 on dosimetry systems across Employers and paragraphs 1.199–1.220 on outside workers may also be of interest where employees are working on more than one Employer's premises.

Radiation incidents

1.222 A radiation 'incident' is not defined in IRR17, and unless an Employer has defined the term in a local policy, it is often used arbitrarily to describe an adverse event with an undefined impact. The word fails to adequately describe the full complement of actions to be taken in response to an event, although individuals who have received sufficient basic radiation protection or general health and safety training should understand the need to report, investigate, and learn from adverse events.

1.223 The term incident is typically used to describe an event where an individual receives a dose that is higher than expected (although in the case of a medical exposure, it can include an underexposure; see chapter 2) or where there is a loss of control of radioactive substances. In general, any significant increase in exposure from that received under normal, well managed radiation protection conditions (i.e. optimised) can be seen as an unexpected exposure and should be investigated sufficiently so lessons may be learned. There may not always be a dosemeter to indicate that someone has received an unexpected exposure. An investigation may have to take place on the grounds of non-standard practice. See paragraphs 1.233–1.24 on *Performing an investigation.* The level of the dose received and the person who is exposed impact the action to be taken at that stage (seek advice from an RPA/RWA/MPE as appropriate). Near misses that become apparent should also be investigated. A record should be made, and it is important to look for trends or common themes that, if identified, may lead to the prevention of future incidents (see also appendix 12, *Incidents—learning*).

1.224 In addition to internal reporting systems, incidents may have to be reported to the authorities. There are subtleties about when external reporting is required, and it is recommended that professional advice be sought without delay, e.g. from an RPA/MPE/RWA/DGSA, so that where necessary, incidents can be notified immediately to the authorities, and where this is not necessary, incidents are addressed appropriately without unnecessary notification.

1.225 IRR17 does define the terms 'overexposure' and 'radiation accident'.

- An *overexposure* is when the exposure of someone to whom a dose limit applies, exceeds that limit.

- A *radiation accident* is a 'significant event' which requires action to be taken immediately to reduce the level of exposure.
- All hazards with the potential to cause a radiation accident should be identified in the risk assessment process (see paragraphs 1.44–1.54 on risk assessment).
- Contingency plans are covered in paragraphs 1.82–1.87 and chapter 20 *Contingency planning and emergency procedures for radioactive substances.*
- IRR17, Regulation 13, (HSE 2018b) requires analyses of the circumstances when it is necessary for some or all of the arrangements in the contingency plan to be carried out to determine so far as is reasonably practicable, the measures required to prevent a recurrence. A record of the analysis must be kept for two years.
- IRR17, Regulation 24, sets out the requirements for dosimetry for accidents or other occurrences that take place that are likely to result in a person receiving an effective dose of ionising radiation greater than 6 mSv or an equivalent dose greater than 15 mSv for the lens of an eye or greater than 150 mSv for the skin or the extremities. The Employer must notify the HSE or HSENI of the result of the dose assessment as soon as possible.

1.226 Any suspected overexposure, i.e. an exposure in excess of any dose limit, must be investigated immediately and, if it cannot be ruled out, must be notified as soon as practicable to the HSE or HSENI, the Employer of the individual concerned, a relevant doctor (an appointed doctor or employment medical adviser), and the individual concerned. Reports of immediate investigations must be kept for two years, and if the overexposure is confirmed, the report must be kept for at least 30 years, or until the person is 75 years old. It would be worth noting that these time periods for document storage are required by IRR17, and there may be other demands set out in information governance documents such as the Records Management Code of Practice (NHSX 2021). The requirement for investigation and reporting includes suspected over-exposures for all employees and for members of the public who are exposed in excess of the public dose limit.

1.227 The term 'incident' is also used when talking about notifiable events (such as losing radioactive material) or when equipment faults may have implications for health and safety.

1.228 The HSE and HSENI have guidance on ionising radiation notification on their websites (HSE no date b, HSENI no date b).

1.229 Notifiable events during the transport of radioactive material must be reported as soon as is reasonably practicable by the driver to the emergency services, the consignor, the carrier, and the relevant local authority (which may be identified by the emergency services attending the scene) (ONR 2022). Further notifications by the consignor(s), appointed DGSA, and/or RPA may be required to the ONR (if the incident takes place in Great Britain) or HSENI (if it takes place in Northern Ireland).

1.230 Notifications in respect of matters related to radioactive materials, e.g. suspected loss of a source, or breach of limits, are made to the relevant regulators: HSE/HSENI and the relevant environmental agencies.

1.231 The Employer should establish a radiation incident reporting procedure providing for compliance with the legal duty to report incidents of specified kinds to the appropriate authority. The procedure should follow relevant guidance and make clear the respective roles of those involved in investigating and reporting where the Employer delegated tasks (not responsibility).

1.232 It should be noted that IRR17 regulations 8, 9, 12, 17, 18, 19, 24, 26, 32(1), and 35(1) do not apply in relation to persons undergoing medical exposures, although the ALARP principle still applies.

1.233 Accidental and unintended medical exposures (including foetal exposure or exposure of a breastfeeding infant as a result of a medical exposure and exposures of carers and comforters) are addressed in chapter 2.

Performing an investigation

1.234 Any suspected unintended exposure should be investigated to a level appropriate to the risk, i.e. what level of dose is involved and how many people might be affected etc. Investigations should be carried out in a methodical way to ensure quality information is gathered, collated, and analysed. An investigation should help identify weaknesses so that measures can be taken to prevent their reoccurrence. An appropriate review of the evidence should lead to firm conclusions and actions to be taken. An RPA should be consulted.

1.235 Investigations (with varying scope and purposes) are specifically required under IRR17 when:

- staff exposure exceeds a formal dose investigation level. This is to bring about a review of the working conditions.
- a classified worker has lost, damaged, or destroyed their dosemeter. This is to provide either an estimate of the dose for the dosimetry record or grounds to enter a notional dose.
- it is believed that the dose recorded in the dosimetry record of a classified worker is not accurate. To find out if this is true and provide an estimate for the record.
- it is believed that a radioactive substance has been released or spilt, lost or stolen etc. To establish if this is true and to allow notification to the authorities.

1.236 In addition, the regulations require an initial, immediate investigation and a detailed investigation (where appropriate) if it is suspected that there has been an overexposure, i.e. someone has exceeded an effective or equivalent dose limit. Through the immediate investigation, the Employer must be able to demonstrate (beyond a reasonable doubt) that the overexposure did not occur; otherwise, a detailed investigation must be performed. Therefore, it is important that those wearing dosimetry notify the Employer about lost, damaged, or accidentally exposed dosemeters before they are read by the dosimetry service (Martin *et al* 2018).

1.237 The extent of the investigation will depend on the difficulty in establishing the facts and the magnitude of the doses involved. A root cause analysis may prove useful. The detailed investigation should establish the following:

- What occurred, where, when, and why the incident occurred, and who was involved.
- The dose received by those involved.
- The actions to prevent reoccurrence.

1.238 Investigation is the Employer's responsibility, though the task is often delegated (through a radiation protection or dosimetry policy) to the local manager or RPS. The right people must be involved, i.e. those who have firsthand information. At the detailed investigation stage, senior staff should at least be aware of the investigation, if not actively involved. It may also be appropriate to seek the views of relevant safety representatives and/or established safety committees, i.e. Health and Safety, Risk, or RSCs.

1.239 In the case of personal dosimetry results, it must be shown beyond a reasonable doubt that a dose received is not real in the event of a higher than expected reading. Potential sources of invalid dosemeter readings include wearing the dosemeter in the wrong location, wearing over/under lead protection inappropriately, wearing someone else's dosemeter, or storing it in a location where it could be exposed to radiation. If dosemeters have been worn and stored correctly, it should be established if there was an increase in workload or complexity of case mix, new procedures, or modified equipment settings (did these require an increased patient dose or closer proximity to the patient) associated with the wear period(s) in question. If no cause can be established, the worker should be observed by the RPS and/or RPA during routine practice.

1.240 An investigation might include evidence about the following:

- the work routine of the individual and immediate work colleagues;
- details of any radiation monitors in use in the areas concerned during the period under investigation to identify any deterioration in physical control measures;
- involvement of the individual in any known incidents in which they may have received an unusual exposure;
- assessed or estimated doses over the last few years compared to those of work colleagues undertaking similar work;

- results of any special radiation survey in the areas concerned (e.g. as part of a reconstruction advised by the RPA);
- evidence from the RPS, the individual concerned or work colleagues about non-adherence to local rules or deficiencies in local rules;
- deficiencies in the information, instruction or training received and general competence for the work undertaken;
- other possible explanations for a suspected unintended exposure (e.g. there is evidence that the employee has worn their dosemeter while receiving a medical exposure, or crossing radiation backscatter inspection machines/gates).

1.241 All relevant information should be documented in a format that can be easily distributed. The regulators may request a copy of the detailed investigation for reportable incidents. The key facts concerning the incident should be recorded, including the information and calculations used to estimate doses as well as the recommendations to avoid recurrence. The report should be signed and dated by the person who prepared it. It is suggested that a record also be made of the actions taken to implement the report's recommendations. Disseminating learning outcomes to staff is essential.

Incidents—shared learning

1.242 Paragraphs 1.233–1.240 on performing an investigation talk about the need to investigate when things go wrong and, where appropriate, there is a legal requirement to report to external bodies such as the HSE. As well as the learning that should take place within an organisation as a result of the investigation and analysis of incidents (and near misses), it can help to share those occurrences with a wider audience so that measures can be taken to prevent a repeat elsewhere. This is voluntary reporting rather than mandatory reporting. To help Employers decide how best to share incidents, appendix 12, *Incidents—shared learning*, introduces a number of pathways relevant in the UK.

Control of radioactive substances and articles

1.243 The holding, use and disposal of radioactive substances are governed in the UK by radioactive substances regulations, the principal ones (as amended) being:

- In England and Wales, the *Environmental Permitting Regulations* (EPR 2016).
- In Northern Ireland, the *Radioactive Substances Act* 1993 *(Amendment) Regulations (Northern Ireland)* 2011 (RSA(A)R(NI) 2011).
- In Scotland, the *Environmental Authorisations (Scotland) Regulations* (EA(S)R 2018).

More guidance on complying with the requirements of these regulations can be found in chapter 18, *Keeping, accounting for, and moving radioactive substances (including transport)*.

1.244 The transport of radioactive substances is principally governed (as amended) by:

- In Great Britain, the Carriage of Dangerous Goods and Use of Transportable Pressure Equipment Regulations (CDG 2009).
- In Northern Ireland, the Carriage of Dangerous Goods and Use of Transportable Pressure Equipment Regulations (Northern Ireland) (CDG(NI) 2010).
- For waste to outside the UK, the Transfrontier Shipment of Radioactive Waste and Spent Fuel Regulations (UK Parliament 2008).

More guidance on complying with the requirements of these regulations can be found in chapter 18, *Keeping, accounting for, and moving radioactive substances (including transport)*.

1.245 IRR17 also lays out requirements for the control of radioactive substances and articles in Regulations 28–34. (IRR17 Regulation 33 is now omitted as stated in Schedule 4(3)(4) of IR(ME)R17). It should be noted that IRR17 applies to transporting radioactive substances as this is considered working with ionising radiation (Paragraph 25 of L121 (HSE 2018b)). Refer to:

- Chapter 18 for *Keeping, accounting for and moving radioactive substances (including transport)*.
- Paragraphs 1.245–1.259 on *Critical examinations*.
- Chapter 2 for *Equipment used for medical exposures etc*.
- Paragraphs 1.6–1.17 on *Responsibilities*.

Critical examination ('Crit Ex')

1.246 Detailed guidance on critical examinations has been previously published by HSE in PM77 (HSE 2006) (withdrawn by HSE but still useful), IPEM Report 107 (IPEM 2012) for x-ray generating equipment used in diagnostic radiology, and IPEM Report 75 (IPEM 2017) for radiotherapy treatment facilities.

1.247 Where an article e.g. equipment, machinery, or appliances (or components of these) for use in work with ionising radiation (HASWA 1974, IRR 2017, IRR (NI) 2017) is installed or erected, a critical examination must be carried out where there are radiation protection implications associated with incorrect installation. This applies to the exposure of anyone (employees, members of the public, and patients). The purpose of the critical examination is to demonstrate to the purchaser that the designed safety features and warning devices operate correctly, that there is sufficient protection for persons from exposure to ionising radiation, and that the equipment is safe to use in normal circumstances. It is the responsibility of the equipment installer to ensure that a critical examination is performed where necessary, although undertaking the examination may be contracted out. An RPA should be consulted on the extent of testing and the adequacy of the test results (with regard to safety features and whether the level of protection is sufficient).

1.248 The installer and the purchasing Employer should cooperate, exchange information, and establish the arrangements for the critical examination in a written agreement. This will include who will carry out the critical examination (e.g. service engineer or the purchasing Employer's medical physics staff) and which RPA will be consulted (appointed by the installer or the Employer).

1.249 The results of the critical examination and details on how it was undertaken should be made available to the Employer's RPA if they are not undertaking it.

1.250 A critical examination will not only be appropriate following the installation or relocation of equipment but also following major service/repair work where incorrect reassembly of equipment could have radiation safety implications. For example, in the case of the following modification or replacement of:

(a) an x-ray tube insert;
(b) an automatic exposure control (AEC) device;
(c) a klystron on a linear accelerator;
(d) an ion chamber on a dose calibrator;
(e) sources in brachytherapy or intravascular radiotherapy units;
(f) interlocks designed to prevent or terminate exposure;
(g) hot cells, etc for radiopharmaceutical preparation;
(h) installed hand and foot monitors or air monitors for detecting radioactive substances.

1.251 Following major service/repair work, appropriate routine performance tests should also be performed by the Employer's representative, the results of which should be checked against manufacturer data where available and the baseline values of the commissioning tests prior to returning equipment to clinical use. Note that these tests legally fall under *The Ionising Radiation (Medical Exposure) Regulations* (IR(ME)R 2017, IR(ME)R(NI) 2018) (see chapter 2, paragraphs 2.88–2.94, *Acceptance, commissioning and routine testing*) and are not part of the critical exam.

1.252 The Employer must not allow any equipment to be used for medical exposures (including ancillary equipment that can control or influence the extent of medical exposures) unless and until the results of the critical examination are satisfactory (IRR17 requirement) and the equipment has been accepted and commissioned to specification (IR(ME)R requirement), as a check of the initial integrity of the equipment. The critical examination should not be confused with the recommended standards for acceptance testing, commissioning tests, or routine performance tests (e.g. IPEM

Reports: 91 (IPEM 2005), 94 (IPEM 2007), and 111 (IPEM 2015)), some of which may be undertaken at the same time as the critical examination following an installation. The life cycle of x-ray equipment is shown in appendix 13.

1.253 Critical examinations should be carried out on articles containing radioactive substances (e.g. a hot cell for F-18 production for both the F-18 vial and any sealed source to check the calibrator) and on x-ray generating equipment, and include associated components intended to restrict exposure. For example, exposure interlocks, back-up timer, light-to-x-ray beam alignment, and shielding from leakage radiation can all be included in a critical examination. The examination should include the protection provided for persons undergoing medical exposures (and non-medical exposures using medical radiological equipment, see chapter 2, paragraphs 2.1–2.3, *Scope*), as well as the adequacy of protection for staff and members of the public.

1.254 Other devices and features (e.g. walls, shields, warning lights, and signals) that are not necessarily components of the article itself should be included in the critical examination by the appropriate installer if these devices are intended to restrict exposure and are likely to be affected by the way the article has been erected or installed. For example, if a supplier instals equipment and connects it to the door interlocks and warning signals that are already in place, the supplier would have to check that the interlocks and warning signals operated correctly as part of the critical examination. On the other hand, if a hospital instals interlocks and warning signals in a room that already contains installed equipment, the hospital must include the checks as part of its critical examination. IPEM Report 107 (IPEM 2012) gives further clarification on this when, in section 4.5, it states that a lead screen installed as part of a mammography operator console should be included in a critical exam, but shielding of a room would not usually be included. However, section 1.4 states that consideration should be given to the safety of persons in the vicinity while undertaking the critical examination; therefore, cooperation between Employers is required to ensure that room shielding is satisfactory before the critical examination commences. This could be done through the confirmation of the shielding specification, and RPA advice may be required. Whether included in the critical examination or not, engineering controls provided to restrict exposure under IRR regulation 9(2)(a) must be maintained and examined/tested at suitable intervals under IRR regulation 11(1) (HSE 2018b).

1.255 A critical examination is not formally required following the delivery of mobile equipment (HSE 2006) such as an x-ray unit or specimen cabinet, as these are usually delivered fully assembled and have not been installed or erected. It is expected that the manufacturer has undertaken tests equivalent to a critical examination at the factory or wherever the assembly of the equipment took place, and a copy of the report should be provided with the equipment or be requested by the equipment purchaser if it is not. It would be prudent to seek assurance from an RPA on the adequacy of the testing undertaken and the results provided by the manufacturer. Similarly, a mobile dental x-ray unit does not require a critical examination, but a wall-mounted dental x-ray unit does. Hand-held mobile dental x-ray equipment is not 'installed' and therefore not subject to the formal requirement for a critical examination. However, the radiation protection afforded to the operator by the tubehead shielding and other design features is crucially important with hand-held equipment, and certain features must be examined prior to commencing clinical use (see Dental Guidance Notes for more (PHE and FGDP 2020)). It should be noted that in these cases, where a local critical examination by the installer is not a requirement, the functioning of safety features should still be checked by the Employer operating the equipment prior to first use and acceptance testing, and commissioning will still be required. The replacement of an x-ray tube insert in any mobile x-ray unit does require a critical examination.

1.256 Table 1.3 indicates radiation safety features and warning devices that should be included for consideration in a critical examination. It is the duty of the installer of the particular feature or device to undergo a critical examination. (Note that the list is not exhaustive.)

Table 1.3. A selection of items that need a critical examination by the appropriate installer. Read in conjunction with paragraph 1.253 above.

Interlocks	Warning systems	Safety design features	Barriers
Door interlocks. Emergency off button. Beam off/disable button. Microswitch interlocks. Alignment and filter interlocks. Last man out systems. Image receptor sensor interlocks.	Warning signals. Entry warning signs. Beam on indication. Fluoroscopy duration alarms. Tube/beam selection indications e.g. energy, dose rate, tube (where more than one). Unambiguous labelling. Environmental dose rate monitors. Exit installed contamination monitors.	Exposure termination. Leakage radiation. Beam filtration and collimators. Position of exposure switch. Protection of exposure switch against accidental activation. Fluoroscopy features (termination after fixed duration, maximum skin dose rate limitation). AEC (dose rate, termination, back-up timer, chamber/mode selection). Ventilation from radiopharmacy hot cells	Entrance doors/maze design. Primary and secondary barriers (position and adequacy for adjacent areas). Protective cubicle and mobile screens. Heating, ventilation, air-conditioning and cable penetrations. Provision/adequacy of personal protection (drapes, aprons, shields).

1.257 There is often confusion surrounding the influence new software/firmware will have on safety systems or exposure to ionising radiation. It is therefore essential to ensure appropriate cooperation (information sharing) between the manufacturer, the installer, and the Employer regarding any proposed change. The Installer should have (in conjunction with the appointed RPA and manufacturer) previously considered the need for a critical examination and should be able to convey this decision (and reasoning) to the Employer.

1.258 Software that affects dose and/or dose rate would require a critical examination. On the other hand, software dealing with patient demographics or data handling would not. Once again, it is the responsibility of the installer to determine whether or not a critical examination is required and to ensure that one is undertaken.

1.259 Accompanying software release notes should be available and sent to the Employer in advance of any planned upgrade/change that identify what changes will be made. This will also enable the local MPE and/or RPA to consider the planned changes and advise on action prior to use. Records of software changes should be made and kept by the Employer in accordance with local handover arrangements.

1.260 The associated risk assessment should be reviewed and updated following the critical examination. This should include consideration of occupancy and supervision of adjacent areas, as well as planned changes in the intended workload.

Radiation emergencies and REPPIR

1.261 Emergency exposure situations are unexpected situations that may require the implementation of urgent protective actions and sometimes longer-term protective actions (ICRP 2007a). International Commission on Radiological Protection Publication 109 (ICRP 2009) gives recommendations for applying the ICRP system to emergency exposure situations. The approach taken is one of constrained optimisation, in which a reference level is set (in the range of 20–100 mSv effective dose) and residual doses are evaluated after protective measures have been implemented. If they are above the reference level, protective measures are improved. If they are below the reference level, detailed optimisation is undertaken. The ICRP recommends that plans be prepared for all types of emergency exposure situations, indicating the responsibilities of different agencies, methods of communication and organisation and a framework for guided decision-making.

1.262 The Euratom Basic Safety Standard (European Union 2014) defines an 'emergency' as a non-routine situation or event involving a radiation source that necessitates prompt action to mitigate serious adverse consequences for human health and safety, quality of life, property, or the environment, or a hazard that could give rise to such serious adverse consequences. Articles 69 and 70 lay down, respectively, the emergency response plan to be implemented in the case of a radiation emergency and the information to be provided to members of the public likely to be affected. Articles 97 and 98 set out the responsibilities of member states and competent authorities, and other requirements for regulatory control in terms of establishing an emergency management system and emergency preparedness.

1.263 Emergency preparedness and response requirements are implemented in the UK through the Radiation (Emergency Preparedness and Public Information) Regulations (REPPIR 2019, REPPIR (NI) 2019). These lay down requirements where dispersible sources above levels set in Schedule 1 of those regulations are used (e.g. 300,000 GBq Tc-99m) which could lead to a dose to persons off-site exceeding 1 millisievert in a year. There is an associated Approved Code of Practice and guidance (HSE and ONR 2019). See chapter 20. *Contingency planning and emergency radioactive procedures for radioactive substances*, and an overview on the HSE and HSENI websites (HSE no date c) for more information.

1.264 For those engaged in practices involving a high-activity sealed source, training should address the emergency procedures that must be followed in the event of any foreseeable accident that could result in damage to the source (s) e.g. a fire, or in the event of loss, theft, or unauthorised use of HASS (paragraph 278 of HSE 2018a, 2018b).

1.265 It is worth noting a number of national resources at this point:

- National arrangements for incidents involving radioactivity (NAIR) to assist the emergency service where no formal contingency plan exists (See information provided by Public Health England (PHE 2021).
- National Poisons Information Service (NPIS 2022). Healthcare professionals can obtain specialist help if they suspect or know that someone is suffering from the effects of radiation.
- Accident and Emergency departments were provided with RAM GENE radiation monitors in around 2005 and basic procedures for their use for the assessment of potentially radiologically contaminated patients (HPA 2011, PHE 2012).

References

AXREM 2018 *Radiation Controlled Area and Equipment Handover Form* (UK: AXREM) www.axrem.org.uk/resource/ensuring-compliance-with-radiation-safety-regulations/ (accessed 31 January 2023)

BIR 2016a *Personal Protective Equipment for Diagnostic X-ray Use* ed P Hiles, H Hughes, D Arthur and C Martin (London: British Institute of Radiology) https://birpublications.org/page/ppe (accessed 25 February 2022)

BIR 2016b *Personal Protective Equipment for Diagnsotic X-ray Use, BIR Website* (London) https://birpublications.org/page/ppe (accessed 23 November 2022)

CDG 2009 *The Carriage of Dangerous Goods and Use of Transportable Pressure Equipment Regulations 2009* https://legislation.gov.uk/uksi/2009/1348 (accessed 31 January 2023)

CDG(NI) 2010 *The Carriage of Dangerous Goods and Use of Transportable Pressure Equipment Regulations (Northern Ireland)* 2010 https://legislation.gov.uk/nisr/2010/160 (accessed 31 January 2023)

Chapple C L, Bradley A, Murray M, Orr P, Reay J, Riley P, Rogers A, Sandhu N and Thurston J 2017 Radiation safety culture in the UK medical sector: a top to bottom strategy *Radiat. Prot. Dosim.* **173** 80–6

Cole P *et al* 2014 Developing the radiation protection safety culture in the UK *J. Radiol. Protect.* **34** 469–84

EA(S)R 2018 *The Environmental Authorisations (Scotland) Regulations 2018* http://legislation.gov.uk/ssi/2018/219 (accessed 31 January 2023)

EPR 2016 *Environmental Permitting (England and Wales) Regulations* https://legislation.gov.uk/uksi/2016/1154 (accessed 31 January 2023)

European Commission 1993 Council Directive 93/42/EEC of 14 June 1993 concerning medical devices *Off. J. L169* (EU) https://eur-lex.europa.eu/legal-content/EN/TXT/?uri=CELEX%3A31993L0042 (accessed 31 January 2023)

European Commission 1998 *Directive 98/79/EC of the European Parliament and of the Council of 27 October 1998 on* In Vitro *Diagnostic Medical Devices* https://eur-lex.europa.eu/legal-content/EN/TXT/PDF/?uri=CELEX:31998L0079&from=EN (accessed 31 January 2023)

European Commission 2014 1.123, and repealing Directives 89/618/E89/618/Euratom, 90/641/Euratom, 96/29/Euratom, 97/43/E *Off. J. Eur. Commun. L13* (December 2003) 1–73 https://eur-lex.europa.eu/LexUriServ/LexUriServ.do?uri=OJ:L:2014:013:0001:0073:EN:PDF (accessed 31 January 2023)

European Union 2014 *Council Directive 2013/59/Euratom of 5 December 2013 Laying Down Basic Safety Standards for Protection Against the Dangers Arising from Exposure to Ionising Radiation* https://eur-lex.europa.eu/legal-content/EN/TXT/?uri=OJ:L:2014:013:TOC (accessed 31 January 2023)

Hackitt J 2013 *Work Experience* https://na.eventscloud.com/file_uploads/693898fcef56976813277a8ed47cf662_workexperience.pdf (accessed 31 January 2023)

HASWA 1974 *Health and Safety at Work etc Act 1974* https://legislation.gov.uk/ukpga/1974/37 (accessed 8 February 2023)

HPA 2011 *HPA-CRCE-014: Internal Radioactive Contamination: Guidance on Screening People* (UK: Health Protection Agency) https://gov.uk/government/publications/internal-radioactive-contamination-guidance-on-screening-people (accessed 31 January 2023)

HSCER 1996 *The Health and Safety (Consultation with Employees) Regulations 1996* (Great Britain) https://legislation.gov.uk/uksi/1996/1513 (accessed 31 January 2023)

HSCER(NI) 1996 *The Health and Safety (Consultation with Employees) Regulations (Northern Ireland) 1996* (Northern Ireland) https://legislation.gov.uk/nisr/1996/511 (accessed 31 January 2023)

HSE 2000a *HSE Information Sheet. Radiation Protection Supervisors, IRIS6* https://hse.gov.uk/pubns/irp6.pdf (accessed 31 January 2023)

HSE 2000b *L121 Work with Ionising Radiation. Approved Code of Practice and Practical Guidance on the Ionising Radiations Regulations 1999* 1st edn (Sudbury: Health and Safety Executive)

HSE 2006 *Guidance Note PM77 (Third Edition): Equipment Used in Connection with Medical Exposure* 3rd edn (Sudbury: Health and Safety Executive) https://hse.gov.uk/pubns/guidance/pm77.pdf (accessed 31 January 2023)

HSE 2013a *Consulting Employees on Health and Safety: A Brief Guide to the Law* 2nd edn (Health and Safety Executive) https://hse.gov.uk/pubns/indg232.htm (accessed 31 January 2023)

HSE 2013b *HSG65 Managing for Health and Safety* 3rd (Norwich: Managing for Health and Safety) https://hse.gov.uk/pubns/books/hsg65.htm (accessed 31 January 2023)

HSE 2013c *INDG334: New and Expectant Mothers Who Work. A Brief Guide to Your Health and Safety* https://hse.gov.uk/pubns/indg334.pdf (accessed 31 January 2023)

HSE 2013d *INDG364: Young People and Work Experience: A Brief Guide to Health and Safety for Employers. Rev 1* (Norwich: Health and Safety Executive) https://stopslaverynetwork.org/wp-content/uploads/2018/01/HSE-Guide-Young-People.pdf (accessed 31 January 2023)

HSE 2014a *INDG163: Risk Assessment. A Brief Guide to Controlling Risks in the Workplace* 4th edn (Norwich: Health and Safety Executive) https://www.egx.net/content/dam/sitebuilder/rna/egx/2021/_docs/health-and-safety---space-only-exhibitors/Risk Assessment Guide.pdf.coredownload.281187949.pdf (accessed 31 January 2023)

HSE 2014c *The Health and Safety Toolbox: How to Control Risks at Work* (Norwich: Health and Safety Executive) https://www.recruiteasy.co.uk/downloads/resources/Health-&-Safety-Toolbox-How-to-control-risks-at-work.pdf (accessed 31 January 2023)

HSE 2015a *HSG263 Involving Your Workforce in Health and Safety: Guidance for all Workplaces* (Norwich: Health and Safety Executive) https://hse.gov.uk/pubns/books/hsg263.htm (accessed 31 January 2023)

HSE 2015b *INDG334 Working Safely with Ionising Radiation. Guidelines for Expectant or Breastfeeding Mothers* http://hse.gov.uk/pubns/indg334.htm (accessed 31 January 2023)

HSE 2018a *MS33 Guidance for Appointed Doctors on the Ionising Radiations Regulations 2017* 1st (Health and Safety Executive) https://hse.gov.uk/pubns/ms33.htm (accessed 31 January 2023)

HSE 2018b L121 *Work with Ionising Radiation Ionising Radiations Regulations 2017—Approved Code of Practice and guidance* 2nd (Norwich: Health and Safety Executive) https://hse.gov.uk/pubns/books/l121.htm (accessed 31 January 2023)

HSE 2022 *Appointed Doctors, HSE Website* (Norwhich: Health and Safety Executive) https://hse.gov.uk/doctors/ (accessed 5 December 2022)

HSE no date a *HSE—Identifying the Accused, HSE Website* https://hse.gov.uk/enforce/enforcementguidesc/identifying/ (accessed 23 November 2022)

HSE no date b *Ionising radiation: protecting workers and others, HSE Website* (Norwich: Health and Safety Executive) https://www.hse.gov.uk/radiation/ionising/index.htm (Accessed: 14 February 2024)

HSE no date c *Radiation (Emergency Preparedness and Public Information) Regulations 2019 (REPPIR), HSE Website* (Norwich: Health and Safety Executive) https://hse.gov.uk/radiation/ionising/reppir.htm (accessed 5 December 2022)

HSE no date d *Radon in the Workplace, HSE Website* (Health and Safety Executive) https://hse.gov.uk/radiation/ionising/radon.htm (accessed 23 November 2022)

HSE no date e *Work Experience, HSE Website* https://hse.gov.uk/youngpeople/workexperience/ (accessed 25 February 2022) https://www.hse.gov.uk/young-workers/employer/work-experience.htm

HSE no date f *Working with Ionising Radiation: Notify, Register or Get Consent, HSE Website* www.hse.gov.uk/radiation/ionising/notification-process.htm (accessed 23 November 2022)

HSE no date g *Young Workers: Guidance for Employers, HSE Website* (UK: Health and Safety Executive) https://hse.gov.uk/young-workers/index.htm (accessed 23 November 2022)

HSE no date h *Personal Protective Equipment (PPE) at Work, HSE Website* (Health and Safety Executive) https://hse.gov.uk/ppe (accessed 21 April 2023)

HSE no date i *Managing risks and risk assessment at work - 3. Risk assessment template and examples, HSE website* (Norwich: Health and Safety Executive) https://www.hse.gov.uk/simple-health-safety/risk/risk-assessment-template-and-examples.htm (accessed 14 February 2024

HSE no date j *Work with ionising radiation that requires consent, HSE website* (Norwich: Health and Safety Executive) https://www.hse.gov.uk/radiation/ionising/consent.htm (accessed 14 February 2024)

HSE and ONR 2019 L126 The Radiation (Emergency Preparedness and Public Information) Regulations *2019—Approved Code of Practice and guidance* 2nd edn (London, UK: TSO) https://onr.org.uk/documents/2020/reppir-2019-acop.pdf (accessed 31 January 2023)

HSENI 2019 *Radon Information Leaflet* (Health and Safety Executive for Northern Ireland) https://hseni.gov.uk/sites/hseni.gov.uk/files/radon-information-leaflet_0.pdf (accessed 31 January 2023)

HSENI 2022 *Appointed Doctors under the Ionising Radiation Regulations (NI) 2017, HSENI website* (Belfast: Health and Safety Executive for Northern Ireland) https://hseni.gov.uk/appointed-doctors-under-ionising-radiation-regulations-ni-2017 (accessed 5 December 2022)

HSENI no date a *Overview - Notification, registration or consent for work with ionising radiation, HSENI Website* (Belfast: Health and Safety Executive for Northern Ireland) https://www.hseni.gov.uk/overview-notification-registration-or-consent-work-ionising-radiation (Accessed: 14 February 2024)

HSENI no date b *Reporting incidents to HSENI, HSENI Website* (Health and Safety Executive for Northern Ireland) https://www.hseni.gov.uk/reporting-incidents-hseni%0A (accessed 14 February 2024)

IAEA 2015 *Safety Reports Series No. 84 Radiation Protection of Itinerant Workers, Safety Reports Series* (Vienna, Austria: International Atomic Energy Agency) https://iaea.org/publications/10788/radiation-protection-of-itinerant-workers (accessed 31 January 2023)

ICRP 2007a Scope of radiological protection control measures. ICRP Publication 104 *Ann. ICRP* **37** 104 https://www.icrp.org/publication.asp?id=ICRP Publication 104 (accessed 31 January 2023)

ICRP 2007b The 2007 Recommendations of the International Commission on Radiological Protection. ICRP Publication 103 *Ann. ICRP* **37** 103 https://icrp.org/publication.asp?id=ICRP Publication 103 (accessed 31 January 2023) ed S V S Mattsson

ICRP 2009 Application of the Commission's Recommendations for the Protection of People in Emergency Exposure Situations. ICRP Publication 109 *Ann. ICRP* **39** 109 https://www.icrp.org/publication.asp?id=ICRP Publication 109 (accessed 31 January 2023)

INPO 2013 *Traits of a Healthy Nuclear Safety Culture* (Atlanta, GA: Institute of Nuclear Power Operations) https://nrc.gov/docs/ML1303/ML13031A707.pdf (accessed 31 January 2023)

IPEM 2002 *Medical and Dental Guidance Notes A Good Practice Guide on all Aspects of Ionising Radiation Protection in the Clinical Environment* (UK: Institute of Physics and Engineering in Medicine)

IPEM 2005 *Report 91: Recommended Standards for the Routine Performance Testing of Diagnostic X-ray Imaging Systems* (York: Institute of Physics and Engineering in Medicine)

IPEM 2007 *Report 94: Acceptance Testing and Commissioning of Linear Accelerators* ed M Kirby, C Hall, S Ryde and York (York: Institute of Physics and Engineering in Medicine)

IPEM 2012 *Report 107: The Critical Examination of X-ray Generating Equipment in Diagnostic Radiology* (York: Institute of Physics and Engineering in Medicine)

IPEM 2015 *Report 111: Quality Control of Gamma Cameras and Nuclear Medicine Computer Systems* (York: Institute of Physics and Engineering in Medicine)

IPEM 2017 *Report 75: Design and Shielding of Radiotherapy Treatment Facilities* ed P Horton and D Easton (Bristol, UK: Institute of Physics) 2nd edn

IR(ME)(A)R 2018 *The Ionising Radiation (Medical Exposure) (Amendment) Regulations* https://legislation.gov.uk/uksi/2018/121 (accessed 31 January 2023)

IR(ME)R 2017 *The Ionising Radiation (Medical Exposure) Regulations* www.legislation.gov.uk/uksi/2017/1322 (accessed 31 January 2023)

IR(ME)R(NI) 2018 *The Ionising Radiation (Medical Exposure) Regulations (Northern Ireland)* https://legislation.gov.uk/nisr/2018/17 (accessed 31 January 2023)

IRPA 2014 *IRPA Guiding Principles for Establishing a Radiation Protection Culture* (International Radiation Protection Association) www.irpa.net/page.asp?id=179

IRR 2017 *The Ionising Radiations Regulations* (Great Britain) https://legislation.gov.uk/uksi/2017/1075 (accessed 31 January 2023).

IRR(NI) 2017 *The Ionising Radiations Regulations (Northern Ireland)* http://legislation.gov.uk/nisr/2017/229 (accessed 31 January 2023).

Mairs W D A 2016 Occupational dose constraints for the lens of the eye for interventional radiologists and interventional cardiologists in the UK *Br. J. Radiol.* **89** 1062

Martin J C, Temperton D H D, Hughes A and Jupp T 2018 *Guidance on the Personal Monitoring Requirements for Personnel Working in Healthcare* (UK: IOP Publishing Ltd)

MDR 2002 *The Medical Devices Regulations 2002* http://legislation.gov.uk/uksi/2002/618/pdfs/uksi_20020618_en.pdf (accessed 31 January 2023)

MHRA 2022 *Guidance: Regulating Medical Devices in the UK, GOV.UK Website* (Medicines and Healthcare Products Regulatory Agency) https://gov.uk/guidance/regulating-medical-devices-in-the-uk (accessed 23 November 2022)

MHSWR 1999 *The Management of Health and Safety at Work Regulations* www.legislation.gov.uk/uksi/1999/3242/regulation/3 (accessed 31 January 2023)

MHSWR(NI) 2000 *Management of Health and Safety at Work Regulations (Northern Ireland) 2000* http://legislation.gov.uk/nisr/2000/388 (accessed 31 January 2023)

NHSX 2021 *Records Management Code of Practice 2021 A Guide to the Management of Health and Care Records*. AUGUST 202, *Nhs.* AUGUST 202. NHSX https://nhsx.nhs.uk/information-governance/guidance/records-management-code/ (accessed 31 January 2023)

NPIS 2022 *TOXBASE, National Poisons Information Service Website* www.toxbase.org (accessed 5 December 2022)

NPL 2014 *Good Practice Guide No.14: The Examination, Testing and Calibration of Portable Radiation Protection Instruments, NPL Good Practice Guide* ed C Lee and P Burgess (Teddington: National Physical Laboratory) https://npl.co.uk/special-pages/guides/gpg14_portable (accessed 31 January 2023)

NRPB, HSE, Department of Health and Social Security, Department of Health and Social Services (Norhtern Ireland), Scottish Home and Health Department and Wlesh Office 1988 *Guidance Notes for the Protection of Persons against Ionising Radiations Arising from Medical and Dental Use* (Chilton: National Radiological Protection Board)

Office for Product Safety & Standards 2022 *Guidance for Businesses on Regulation 2016/425 and the Personal Protective Equipment (Enforcement) Regulations 2018*, *GOV.UK Website* https://gov.uk/government/publications/personal-protective-equipment-enforcement-regulations-2018 (accessed 5 December 2022)

ONR 2022 *Transport of Radioactive Material, Office for Nuclear Regulation Website* https://onr.org.uk/transport/index.htm (accessed 30 June 2022)

PHE 2012 *Guidance: Rapid Screening for Internal Radioactive Contamination: Training Resource, GOV.UK Website* (Public Health England) https://gov.uk/guidance/rapid-screening-for-internal-radioactive-contamination-training-resource (accessed 5 December 2022)

PHE 2020 *PHE Validation Scheme for Laboratories* http://ukradon.org/cms/assets/gfx/content/resource_3462cs9edda0fd4d.pdf (accessed 31 January 2023)

PHE 2021 *Guidance: National Arrangements for Incidents Involving Radioactivity (NAIR), GOV.UK Website* https://gov.uk/guidance/national-arrangements-for-incidents-involving-radioactivity-nair (accessed 22 July 2022)

PHE and FGDP 2020 *Guidance Notes for Dental Practitioners on the Safe Use of X-ray Equipment* (London: Public Health England, Faculty of General Dental Practice (UK)) 2nd edn https://www.rqia.org.uk/RQIA/files/44/449bdd1c-ccb0-4322-b0df-616a0de88fe4.pdf (accessed 31 January 2023)

PPER 1992 *The Personal Protective Equipment at Work Regulations 1992* https://legislation.gov.uk/uksi/1992/2966 (accessed 31 January 2023).

PPER(NI) 1993 *Personal Protective Equipment at Work Regulations (Northern Ireland) 1993* (Northern Ireland) https://legislation.gov.uk/nisr/1993/20 (accessed 31 January 2023)

REPPIR 2019 *The Radiation (Emergency Preparedness and Public Information) Regulations 2019* http://legislation.gov.uk/uksi/2019/703 (accessed 31 January 2023)

REPPIR(NI) 2019 *The Radiation (Emergency Preparedness and Public Information) Regulations (Northern Ireland)* http://legislation.gov.uk/nisr/2019/185 (accessed 31 January 2023)

Rogers A, Chapple C L, Murray M, Platton D and Saunderson J 2017 UK guidance on the management of personal dosimetry systems for healthcare staff working at multiple organizations *Br. J. Radiol.* **90** 1079

RSA(A)R(NI) 2011 *The Radioactive Substances Act 1993 (Amendment) Regulations (Northern Ireland)* https://legislation.gov.uk/nisr/2011/290 (accessed 31 January 2023)

SRSCR 1977 *The Safety Representatives and Safety Committees Regulations 1977* (Great Britain) https://legislation.gov.uk/uksi/1977/500/ (accessed 31 January 2023)

SRSCR(NI) 1979 *The Safety Representatives and Safety Committees Regulations (Northern Ireland) 1979* (Northern Ireland) https://legislation.gov.uk/nisr/1979/437 (accessed 31 January 2023)

The Personal Protective Equipment (Enforcement) Regulations 2018 https://legislation.gov.uk/uksi/2018/390 (accessed 31 January 2023)

The Personal Protective Equipment Regulations 2002 (Revoked) 2002 https://legislation.gov.uk/uksi/2002/1144 (accessed 31 January 2023)

UKHSA no date a *Radon in the Workplace, Ukradon Website* https://ukradon.org/information/workplace (accessed 23 February 2022)

UKHSA no date b *UK Maps of Radon, UKHSA Website* https://ukradon.org/information/ukmaps (accessed 23 November 2022)

UK Parliament 2008 *The Transfrontier Shipment of Radioactive Waste and Spent Fuel Regulations* https://legislation.gov.uk/uksi/2008/3087 (accessed 31 January 2023)

IOP Publishing

Medical and Dental Guidance Notes (Second Edition)
A good practice guide on all aspects of ionising radiation protection in the clinical environment:
IPEM Report 113
John Saunderson, Mohamed Metwaly, William Mairs, Philip Mayles, Lisa Rowley and Mark Worrall

Chapter 2

Radiation protection of persons undergoing exposures

Scope

2.1 This chapter gives guidance on the roles and responsibilities of all duty holders under the Ionising Radiation (Medical Exposure) Regulations 2017 (IR(ME)R 2017) as amended (IR(ME)(A)R 2018) and the Ionising Radiation (Medical Exposure) Regulations 2018 in Northern Ireland (IR(ME)R(NI) 2018), subsequently referred to as IR(ME)R unless a previous version is referenced. It includes other professional and general guidance on the radiation protection framework for justification (individual exposures rather than justification of practices, which is covered in the Justification of Practices Involving Ionising Radiation Regulations 2004 (JoPIIRR 2004) as amended (JoPIIRR 2018) and optimisation of all medical exposures (and non-medical exposures using medical radiological equipment) to ensure the adequate protection of patients or other individuals being exposed, e.g. carers and comforters (chapter 1 covers occupational exposures and exposures of members of the public). This chapter aims to introduce topics and signpost key references that should be read in conjunction with the Regulations and the Department of Health and Social Care Guidance to the Ionising Radiation (Medical Exposure) Regulations 2017 (DHSC 2018a).

2.2 As stated in Regulation 3, IR(ME)R applies to exposure to ionising radiation of:
   (a) patients as part of their own medical diagnosis or treatment;
   (b) individuals as part of a health screening programme;
   (c) patients or other persons voluntarily participating in medical or biomedical, diagnostic or therapeutic, research programmes;
   (d) carers and comforters;

(e) asymptomatic individuals;
(f) individuals undergoing non-medical imaging using medical radiological equipment.

Exposures (a)–(e) are considered 'medical exposures'.

2.3 There are excellent publications discussing the implementation of IR(ME)R in radiotherapy (Radiotherapy Board 2020) and in diagnostic and interventional radiology (BIR *et al* 2020). There is also specific guidance for breast screening services (NHSBSP 2023).

General principles of patient protection

2.4 Exposure to ionising radiation carries a risk of stochastic effects such as cancer and, depending on the level of the dose delivered to particular body tissues, also a risk of harmful tissue reactions (ICRP 2007). Irradiation during pregnancy may also involve a risk to the fetus, and the administration of radioactive substances to a breastfeeding patient may also involve a risk to the infant. It is important, therefore, that only those medical exposures that are justified are undertaken.

2.5 Justification of types of practice involving medical and non-medical imaging exposures is covered by the JoPIIRR (JoPIIRR 2004) as amended (JoPIIRR 2018). Existing relevant justified practice examples (BEIS 2019) include:

- use of ionising radiation for medical diagnosis and treatment, including occupational health and other health screening, and medical and biomedical research;
- manufacture of radioactive sources, substances and radiopharmaceuticals;
- non-destructive testing using radioactive sources, substances and radiation generators for radiography;
- use of ionising radiation for teaching;
- use of ionising radiation to support National Measurement Systems and calibration sources in the testing of equipment;
- transport of radioactive material by road in accordance with ADR (ADR no date).

2.6 New practices must be justified prior to their first use, following the process outlined in JoPIIRR. In addition, IR(ME)R requires that each exposure be justified on an individual basis. This requires the Practitioner to consider the purpose of the exposure, whether the benefits of the exposure (to the individual and/or society) outweigh the risks, and whether alternative techniques involving less or no exposure to radiation might produce the required diagnosis or therapeutic effect, e.g. by using non-ionising radiation. If a type of practice involving medical exposure (this does not apply to non-medical imaging) is not justified in general, a specific individual exposure of this type could be justified in special circumstances, to be evaluated on a case-by-case basis. Hence, medical Practitioners may deviate from utilising only types of practices that have previously been justified where this would be beneficial to the health of a particular patient (BEIS 2019). This is accounted for in Regulation 21 of the 2004 Regulations, as amended by Regulation 16 of the 2018 Regulations.

2.7 Doses arising from exposures to which IR(ME)R applies (except radiotherapeutic procedures) must be kept as low as reasonably practicable (ALARP), consistent with the intended purpose. This involves selecting equipment and methods that will optimise exposure. In relation to all radiotherapeutic exposures, IR(ME)R Practitioners must ensure that the exposures of target volumes are individually planned, taking into account that doses to non-target volumes and tissues must be ALARP consistent with the intended radiotherapeutic purpose.

The employer

2.8 The 'Employer' is defined in IR(ME)R as any person who, in the course of a trade, business, or other undertaking, carries out (other than as an employee) or engages others to carry out those exposures described in Regulation 3 (i.e. medical exposures and non-medical exposures using medical radiological equipment) or practical aspects of a given radiological installation. In practice, where the Employer is not an individual (such as a private dentist, for example), the Employer would be the Trust or organisation, i.e. the legal entity, and not the chief executive (who is also an employee of the legal entity).

2.9 The primary responsibility for providing a framework for radiation protection covering these exposures lies with the Employer through the provision of the Employer's written procedures (hereafter referred to as Employer's procedures —see paragraphs 2.71–2.72 for more). The Employer should be identified and documented within the IR(ME)R framework. Appendix 14 gives consideration to the Employer's procedures in Schedule 2 of the IR(ME)R in more detail.

2.10 The legal duties of the Employer (see also appendix 1, *Roles and responsibilities of the employer*) are clearly identified in IR(ME)R and are consistent with clinical governance. DHSC guidance on IR(ME)R (DHSC 2018a) states that the legal duties of the Employer cannot be delegated; however, it is worth mentioning that in practice, the Employer is not usually best placed to undertake the task of writing procedures etc so tasks can be delegated through governance processes and committees, which are clearly set out in overarching policies. In circumstances where more than one Employer may exist for certain aspects of an exposure, cooperation between Employers is recommended to agree on the roles and responsibilities of each party. In practice, the designated Employer for each aspect of the exposure would usually be the person best placed to undertake the functions and responsibilities required in the Regulations.

2.11 The Employer's procedures should be sufficiently robust to ensure adequate cooperation and agreement between such persons as may be separately responsible for the justification (IR(ME)R Practitioner role), authorisation

(Practitioner or Operator role), and optimisation (Practitioner and Operator role) of medical exposures. There should also be adequate cooperation and agreement between the Referrer and the IR(ME)R Practitioner. If duty holders believe that the Employer's procedures are inadequate, they have a professional responsibility to bring this to the attention of the Employer, preferably in writing.

2.12 When IR(ME)R Practitioners and Operators from different Employers are entitled and involved in an exposure, their respective Employers must provide evidence of 'adequate training' to each other. DHSC IR(ME)R 2017 guidance (DHSC 2018a) recommends that it be agreed (for example, in the contract) that those engaged are adequately trained and are undertaking continuing education and training.

2.13 The Employer must have quality assurance (QA) programmes for all procedures and protocols which would include document control, frequency of review, etc (in addition to those required for equipment performance—see paragraphs 2.79–2.84).

2.14 Even in those types of practices where the Employer is also the IR(ME)R Practitioner, and indeed perhaps also functions as the Operator, e.g. a general dental practitioner, the appropriate Employer's procedures must be established and followed.

The referrer

2.15 IR(ME)R defines the Referrer as a '*registered health care professional who is entitled, in accordance with the Employer's procedures, to refer individuals for exposure to a Practitioner.*' The Referrer must belong to a profession regulated by a body, as detailed in section 25(3) of the National Health Service Reform and Health Care Professions Act 2002 (National Health Service Reform and Health Care Professions Act 2002 2002). In Northern Ireland, the regulations also allow for Referrers who are medical practitioners registered with the Medical Council of Ireland.

2.16 The Referrer must be entitled to act in this capacity by the Employer, and the scope of entitlement should also be specified. Under IR(ME)R, the Referrer has the responsibility for providing sufficient medical data to allow the Practitioner to justify the exposure. They should therefore have sufficient competence in history-taking, assessment, and decision-making to gather suitable and relevant data. It will be beneficial if the Referrer can check imaging history, either through electronic systems or a review of patient notes (to the extent that they have access to these), and have a discussion with the patient in regards to images that may have been undertaken in the independent sector, for example, before making a request to help ensure that the referral is needed (this will be backed up with further checks along the referral pathway, such as by the IR(ME)R Practitioner when considering justification and/or by an Operator immediately in advance of the exposure).

2.17 Training for Referrers (see also paragraphs 2.109–2.120 *Training*) is not mandatory under IR(ME)R but it is highly desirable that they understand referral practices for the specific organisations to which they refer. The Referrer should have developed their understanding of IR(ME)R through training and experience, including awareness of radiation risks, how to make, amend, or cancel a referral, and how to access the clinical evaluation. The appropriate level of training should be decided in discussion between the clinical imaging and medical physics departments and documented in the entitlement procedure. The Referrer should also engage in CPD appropriate to their professional scope of practice (not a requirement of IR(ME)R).

2.18 The Employer must establish referral guidelines and make these available to the Referrers. In most cases, these will be based on national and professional guidance and agreed upon with those health professionals involved in the exposures. The Royal College of Radiologists (RCR) has published evidence-based guidelines for clinical imaging services, focusing on those most relevant to primary and emergency care as well as some guidance on the exposure of asymptomatic individuals (RCR 2017). 'iRefer' does not include all necessary referral guidelines, for example, it does not include all those used in cardiology or interventional radiology. Dental referral guidelines are also available (Faculty of General Dental Practice 2018). Adopting these guidelines is not mandatory, and others are available. An organisation can develop its own guidelines, for example, for specialist orthopaedic services within a children's hospital. Referral guidelines are required for concomitant exposures in radiotherapy that are included as an adjunct to treatment, where these are not included within the radiotherapy protocol or where separate justification is required.

2.19 The Referrer requesting the examination or treatment does so within their entitled scope of practice, in a manner that conforms to the Employer's procedures and referral guidelines. The request should be clear, complete, and legible, with all mandatory fields completed. The following information should be provided:

- unique patient identification;
- all relevant previous diagnostic information;
- sufficient details of the clinical question being asked (or request for imaging where there is no anticipated clinical benefit, i.e. non-medical imaging) to allow the IR(ME)R Practitioner to justify the exposure or an Operator to authorise the exposure against authorisation guidelines produced by the IR (ME)R Practitioner;
- laterality of the body part to be imaged, e.g. left or right knee, where appropriate;

- timing of the imaging/treatment where referral is made in advance, e.g. if it is follow-up imaging required in a number of months;
- if applicable, documentation of inquiries into the possibility of pregnancy and, if a nuclear medicine procedure is envisaged, whether the individual is breastfeeding;
- a signature uniquely identifying the Referrer (which may be electronic). The Employer's procedures should be written in such a way that pre-signed blank request cards by a Referrer do not comply with the Employer's procedures with respect to referrals. Such practices should be identified through audit, and then action must be taken to correct the individuals involved. This could include removing entitlements until the individual is competent. Any entries on the request card made by others (e.g. patient identity or clinical details) should be checked and initiated by the Referrer prior to signing the card.

2.20 In addition, the Referrer may indicate the examination that is thought to be appropriate, for example, following discussion at a Multidisciplinary Team (MDT) meeting.

2.21 With the increased use of electronic requesting, there is a risk that the wrong patient is selected for an examination, for example, when drop-down clinic or ward lists are used to search for patients rather than searches with unique identifiers. It is recommended that the Referrer be vigilant and be able to check, 'Is this the correct patient, modality, and examination?' The use of the 'pause and check' framework for Referrers (SCoR 2017a) should facilitate a final check of the right patient and the right test at the right time. An additional problem is when people fail to log-out and their details and privileges are used to make referrals by someone other than themselves. The Employer should have details of their expectations in regard to this very serious issue, methods to ensure appropriate training and support, and an agreed disciplinary approach stated in their procedures for clarity and fairness. For more on electronic referrals, please see guidance from the RCR (RCR 2022).

2.22 Regulation 6(2) of IR(ME)R states that the Employer must take steps to ensure that any Employer's procedures are complied with by the Referrer, Practitioner, and Operator. This can be more difficult in the case of Referrers who are outside the managerial control of the Employer, e.g. general practitioners referring to a hospital. A good way to demonstrate compliance is through audits and feedback. For example, the quality and suitability of referrals should be audited on a regular basis and action taken to address issues that could adversely affect patient care. There should be a local procedure for returning poor-quality referrals to the Referrer.

2.23 The NHS and the independent sector have evolved, extending the roles of many non-medical healthcare professionals. As a result, clinical imaging departments have seen a rise in diagnostic imaging requests from non-medical Referrers (e.g. radiographers, chiropractors, physiotherapists, osteopaths, and nurses), which is recognised as an important factor in improving patient care. A collaboration of professional organisations has developed guidance to address the functions of non-medical Referrers (BIR *et al* 2020), and the BIR released a position statement on training and governance (BIR 2022). Some expanded examples of non-medical Referrers and their scopes of practice are:

- Nurse Practitioners in emergency departments referring for extremity x-rays, or following the specific care pathways for stroke or head trauma.
- In nuclear medicine, some radiographers may be able to refer for x-rays to complement nuclear medicine procedures.
- Radiographers may be able to refer for x-rays pre-MRI scanning, e.g. to exclude metal fragments in the orbits.

2.24 One way to ensure that the requirements for non-medical Referrers are fulfilled is via a training and approval process, an approach taken by some departments in the UK. The information supplied in an application allows the Employer to assess an individual's knowledge and experience as relevant to their scope of practice. Training can be provided in local radiological practices. The process also ensures that all non-medical Referrers have a sufficient level of clinical competence, as evidenced by the endorsement of the application by the clinical lead.

2.25 An up-to-date list of non-medical Referrers, including the range of referrals that may be made by each individual (or professional group, if they are all appropriately trained and entitled), should be included in the Employer's IR(ME)R procedure for identifying duty holders (or referenced standalone register). It is important that this list be readily available to the clinical imaging department.

2.26 The cancellation of a referral that is no longer required is an important process that should be documented in the Employer's procedures and may require a direct phone call as soon as possible (CQC 2014). The sole use of electronic systems for cancellations is rarely effective due to interface issues (CQC 2018a).

The practitioner (and justification)

2.27 The Practitioner is defined in IR(ME)R as a 'registered health care professional…who is entitled, in accordance with the Employer's procedures, to take responsibility for an individual exposure.'

2.28 The primary responsibility of the Practitioner is to ensure that the exposure is justified in accordance with the Employer's procedures, considering the benefits and risks of the exposure, the individual's clinical information (including imaging history), and any alternative techniques involving a lower, or no, radiation dose.

2.29 The Justification of Practices Involving Ionising Radiation Regulations 2004 (JoPIIRR 2004) as amended (JoPIIRR 2018) are in place to ensure that practices are accepted for wide use based on their economic, social, or other benefits in relation to the health detriment they may cause. Guidance on these regulations (BEIS 2019) explains this higher level of justification and gives a list of justified practices, such as the 'use of ionising radiation in radiography, fluoroscopy, computed tomography, and in-vivo and in-vitro nuclear medicine'. Once the practice is justified, it is the Practitioner's role to ensure that the particular procedure or exam is appropriate for the individual to whom they wish to apply the technique.

2.30 A useful paper (Malone *et al* 2012) sets out that justification can best be facilitated by:

- awareness (of risk, for both the individual and the medical professionals involved);
- appropriateness (of the procedure);
- audit (of justification).

2.31 There are a number of categories of exposure that feature in the justification process and the optimisation process, to which the Practitioner and/or Operator must pay particular attention. The table below (table 2.1) highlights these and who has the responsibility to fulfil the requirement.

2.32 The legal responsibility for justification always remains with the IR(ME)R Practitioner. However, authorising that the exposure has been justified is a separate function, the responsibility for which can rest with the IR(ME)R

Table 2.1. Exposures to which special/particular attention should be given during justification and optimisation.

Justification (Practitioner responsibility)	Optimisation (Practitioner and Operator)
IR(ME)R does not explicitly mention the medical exposure of children as a special category within justification; however, justification based on an assessment of 'individual detriment' already considers the higher risks associated with exposing children. Health screening programmes. Carers and comforters. Asymptomatic individuals. Urgency of exposure where pregnancy cannot be excluded (particularly if abdominal and pelvic regions are involved). Urgency of exposure where the individual is breastfeeding and the exposure involves administration of radioactive substances.	Medical exposures of children. Health screening programmes. Medical exposures involving high doses for the individual being exposed. Exposure where pregnancy cannot be excluded (particularly if abdominal and pelvic regions are involved). Where the individual is breastfeeding and the exposure involves administration of radioactive substances.

Practitioner or a suitably trained and entitled IR(ME)R Operator who may also authorise defined exposures against authorisation guidelines produced by the IR(ME)R Practitioner (see paragraphs 2.133–2.136 *Authorisation*).

2.33 The Practitioner (along with the Operator) must ensure that doses arising from exposures to which IR(ME)R applies (except radiotherapeutic procedures) are kept ALARP, consistent with the intended purpose. In relation to all radiotherapeutic exposures, IR(ME)R Practitioners must ensure that target volume exposures are individually planned and their delivery appropriately verified, taking into account that doses to non-target volumes and tissues must be ALARP, consistent with the intended radiotherapeutic purpose.

2.34 For a discussion about asymptomatic individuals and recommendations on the justification of CT scanning for Individual Health Assessments (assessment of asymptomatic individuals who may be at risk of a disease but are not assessed as part of a screening programme), please see the reference (Department of Health 2014).

Clinical radiology

2.35 The IR(ME)R Practitioner for diagnostic x-ray procedures is, in most cases, a clinical radiologist, but for specifically identified procedures, it may be an adequately trained and entitled radiographer, orthopaedic surgeon, cardiologist, A&E consultant, or other registered healthcare professional, as identified in the Employer's procedures and agreed upon with the other health professionals involved in the exposures.

2.36 The Society and College of Radiographers has written guidance (SCoR 2018b) on diagnostic radiographers who may be entitled to act as IR(ME)R Practitioners once they have completed adequate training and practical assessment relevant to their scope of practice as detailed in Schedule 3 of IR(ME)R. The document contains a useful tool for assessing this competence, and the model could be applied to other professionals.

2.37 For many standard diagnostic x-ray examinations, a suitably entitled Operator may be identified as responsible for authorisations. They do this by confirming the applicability of the justification on the basis of the authorisation guidelines issued by the IR(ME)R Practitioner against the medical data included in the referral for an individual patient.

Nuclear medicine

2.38 The Administration of Radioactive Substances Advisory Committee (ARSAC) issues licences on behalf of the Secretary of State that authorise the administration of radioactive substances. The Employer needs a licence for each radiological installation at which radioactive substances are to be administered. The IR(ME)R Practitioner also needs a licence in order to justify an exposure involving the administration of radioactive substances (a licence for each procedure carried out by the Practitioner and also for each research trial). For example, for radioiodine treatment of thyrotoxicosis, the IR(ME)R Practitioner may be a nuclear medicine consultant, an oncologist, or perhaps an endocrinologist. Although the IR(ME)R Practitioner for radionuclide procedures is always the licence holder, it is possible for authorisation of routine diagnostic exposures to be carried out by an entitled Operator, such as a physicist, radiographer or technologist, following authorisation guidelines issued by the licenced IR(ME)R Practitioner. See paragraphs 2.133–2.136 on *Authorisation*, for more on this.

Radiotherapy

2.39 The process of radiotherapy treatment may involve both treatment exposure and additional concomitant imaging. Concomitant exposures include those necessary for planning the treatment as well as verification exposures at the time of treatment associated with confirmation of the patient setup. As these concomitant exposures would normally not be justified except as an adjunct to the therapeutic dose, it is common practice for a fixed number of exposures associated with treatment to be justified as a whole (within the treatment or imaging protocol). Any additional exposures over and above the protocol would need to be justified separately. The concomitant exposures usually involve the irradiation of non-target tissues, and the principle of optimisation requires that the irradiation of such healthy tissues be restricted to that necessary for the radiotherapeutic purpose, as such irradiation can lead to the formation of secondary cancers. In order to be in a position to consider the justification of the exposures, the Practitioner needs to have information available on the expected doses and tolerances for the organs at risk.

2.40 Large doses of radiation can cause both tissue reactions in healthy tissues as well as stochastic effects. Consequently, the necessary Practitioner training requires an in-depth understanding of the interpretation of the diagnostic data, the indications for radiotherapy, the side effects of radiotherapy, obtaining informed consent (RCR 2014), and alternative treatments not involving radiation (as required under IR(ME)R regulation 11(2)(d)). The Practitioner will therefore normally be on the GMC's specialist register for clinical oncology (RCR 2014). It is advisable that a clinical oncologist be part of the associated MDT when radiation is being used therapeutically. In order to act as Practitioner for therapeutic doses of radionuclides, the Practitioner must hold a license for the relevant procedures, as described under the Nuclear Medicine section above. This will also apply to therapies involving sealed sources, i.e. brachytherapy.

2.41 Entitlement of Practitioners is the responsibility of the Employer. Some Employers have entitled other staff, including associate specialist doctors (formerly known as Speciality and Associated Speciality Grade doctors), oncologists who have completed FRCR Part I, and some radiographers, to act as Practitioners for particular indications, such as some palliative treatments. In all instances, the scope of practice must be clearly defined, and there must be evidence of appropriate additional training and competency assessment. (Often, the justification of palliative radiotherapy is a more complex medical decision than some radical therapy). It is also important that the Employer's procedures clarify whether the duty holder's entitlement is as an Operator, to authorise an exposure following authorisation guidelines issued by the Practitioner, or to act as a Practitioner taking responsibility for the justification of the whole treatment.

2.42 A treatment planning exposure would not be justified if it was not a precursor to treatment. The IR(ME)R inspectors for England annual report for 2017–18 (CQC 2018a) records instances where treatment planning exposures have taken place pending the results of diagnostic examinations that have subsequently shown radiotherapy to be inappropriate. Radiographers and other registered health professionals may also be entitled to act as Practitioners in deciding whether concomitant exposures in addition to those defined in the protocol are justified. However, as the clinical oncologist has overall responsibility for the therapy treatment, it may be more appropriate for the oncologists to create procedures defining the circumstances in which it is justifiable to carry out additional exposures so that the other professionals can then authorise following these guidelines.

2.43 Whichever approach is adopted, it is important that departmentally agreed estimates of dose from all concomitant exposures be recorded in the relevant patient record; this is particularly important for children and young adults. The IR(ME)R Practitioner for the treatment exposures may need to review the patient's treatment management if the dose from concomitant exposures significantly exceeds that expected when the original justification was undertaken. This may also require notification to the regulator.

2.44 The document prepared jointly by the Royal College of Radiologists, the Institute of Physics and Engineering in Medicine, and the Society and College of Radiographers contains helpful advice on the justification and entitlement of Practitioners (Radiotherapy Board 2020).

The operator

2.45 An Operator is defined as any person who is entitled, in accordance with the Employer's procedures, to carry out practical aspects of exposures, including those to whom practical aspects have been allocated, medical physics experts (MPEs), and, except where they do so under the direct supervision of a person who is adequately trained, persons participating in practical aspects as part of practical training.

2.46 The range of functions covered by the term 'practical aspects' is broad and directly influences the extent of an individual's radiation exposure. They include:

- patient identification;
- checking the pregnancy status of the individual in accordance with the Employer's procedure;
- the physical conduct of the exposure (including optimisation);
- handling and use of radiological equipment;
- authorising an individual exposure following guidelines issued by a Practitioner;
- assessment of technical and physical parameters e.g. dose, calibration or maintenance;
- preparation and administration of radiopharmaceuticals;
- administration of intravenous contrast agents (an extravasated contrast agent in CT may lead to repeat exposure);
- computer planning and calculation of monitor units to be delivered in radiotherapy;
- clinical evaluation and reporting of images.

2.47 It is possible that several Operators will carry out different functions associated with an individual exposure. All Operators must be identified as specified in the Employer's procedures.

2.48 Useful 'pause and check' tools (The Society and College of Radiographers 2017b, 2019a, 2019b, 2019c) have been developed to help reduce accidental or unintended exposures associated with 'Operator errors' in both clinical imaging and radiotherapy. The tools are a concise summary of all the key information required to make sure the right individual gets the right exposure at the right time. They can be effective in reducing the number of radiation incidents (CQC 2014); however, the Care Quality Commission (CQC) continues to receive notifications of unintended exposures and over-exposures that could have been prevented with a simple 'stop moment' (CQC 2018a). They urge Employers to clarify and reinforce the responsibilities of all IR(ME)R duty holders and staff, and to remind them not to become complacent. Speed and efficiency should not come at the expense of vital safety checks for patients.

2.49 It should be apparent which Operator undertook the exposure through documentation, e.g. a real or electronic signature. Any additional information required (by the Employer's procedures) to facilitate a retrospective estimation of the effective dose to the patient (or patient dose audit) should be recorded. This may not be required where standard examination protocols have been followed or where digital systems record details automatically. For example, consideration should be given to the need to record patient weight to facilitate dose audits for paediatric patients.

2.50 An individual may be entitled to be both a Practitioner and an Operator, for example, if the Practitioner has control of the footswitch during fluoroscopic procedures. In this example, there are two different Operators involved in undertaking the exposure—the clinician who depresses the footswitch and the radiographer who changes the factors (e.g. pulse rates) and adjusts the collimation. Operators must use their professional judgement to optimise the exposure for each patient, for example, by taking the appropriate radiographic views with the appropriate exposure factors.

2.51 There are a number of categories of exposure that feature in the optimisation process to which the (Practitioner and) Operator must pay particular attention (see table 2.1).

2.52 The use of 'mini C-arm' fluoroscopy systems outside the radiation protection framework of a radiology department is not uncommon now. It should be noted that those using the equipment must meet all the requirements of the regulations (although the Employer will still be responsible for the procedures and protocols, etc) and it is therefore better to work with radiology and use their expertise and experience where possible. The arrangements for justification should be clear and understood, e.g. who justifies the exposures in theatre. The IR(ME)R Operator must be adequately trained and entitled (see paragraphs 2.109–2.120 on *Training*). This training must meet the requirements of IR(ME)R Schedule 3 and include theoretical knowledge and practical experience. This should be assessed before an Operator is left unsupervised to work on practical aspects. Clinical evaluation (paragraphs—2.189–2.194) and equipment quality assurance (paragraphs 2.73–2.94) should not be overlooked.

2.53 Third party service engineers will not normally be considered IR(ME)R Operators (except where they have a contractual arrangement with the Employer to perform additional commissioning tests post-servicing to demonstrate and confirm the acceptance of the equipment for clinical purposes). Although QA and maintenance are 'practical aspects, it is the job of the provider's duty holders, e.g. the MPE and/or radiographer acting as Operator to assess if the equipment is fit for clinical use after maintenance, etc, and to escalate to an MDT as appropriate for a decision when the equipment is not within expected tolerances, for example, to confirm if an artefact may impact image quality such that it would prevent a return to clinical use.

2.54 Medical physics staff who undertake equipment testing should be identified in the Employer's procedures (by name or staff group) as Operators. It would be wise for the local staff to re-boot a system or at least check the default factors are set correctly before using a system on patients after medical physics checks because the range of conditions assessed during advanced (often referred to as 'Level B') quality control testing means that it is very likely the setup has been altered from the standard protocol used for patients, e.g. pulse rate, dose rate, image processing, etc. This process should be documented in a local handover procedure for returning equipment to clinical use.

2.55 In some cases, e.g. a dentist, clinical oncologist, cardiologist, or orthopaedic surgeon may act as the Referrer, IR(ME)R Practitioner, and Operator. In this case, the individual must comply with all the duties placed on these duty holders.

The medical physics expert (MPE)

2.56 The MPE is defined as an individual or a group of individuals having the knowledge, training, and experience to act or give advice on matters relating to radiation physics applied to exposure, whose competence in this respect is recognised by the Secretary of State for Health or the Department of Health in Northern Ireland. The Department of Health and Social Care has appointed the certification body RPA2000 as the assessing body for this purpose (Department of Health and Social Care 2018). A list of current certificate holders can be seen on the RPA2000 website (RPA2000 2023)

2.57 There is a legal requirement for the Employer to appoint an MPE (or MPEs) suitable for the types of exposures carried out by the Employer.

2.58 The Employer needs to ensure that the MPE(s) shall:

(a) be closely involved in radiotherapy practices other than standardised nuclear medicine therapy administrations. This is achievable if the MPE is contracted full-time to the Employer and available during operational hours;

(b) be involved in diagnostic nuclear medicine practices and standardised nuclear medicine therapy administrations. This is achievable if the MPE is contracted full-time or contracted to work sessions and be contactable at other times (see chapters 10 and 13);

(c) be involved in interventional radiology and CT. This is achievable if the MPE is contracted full-time or contracted to work sessions and is contactable at other times;

(d) be involved as appropriate for consultation on optimisation in other practices. This is achievable if the MPE is available under contract and the areas for consultation are specified in the contract;

(e) give advice on medical radiological equipment, dosimetry, and quality assurance.

2.59 The role of the MPE is described in IR(ME)R Regulation 14. Note that, as the MPE is expected to act as well as give advice, the actions of the MPE are classified as Operator actions. The MPE should, where appropriate, liaise with the Radiation Protection Adviser (RPA) (see chapter 1) and Radioactive Waste Adviser (RWA) (see chapter 19, *Accumulation and disposal of radioactive waste*). The role of the MPE is described in more detail in appendix 15.

Governance

2.60 In large hospitals, it will be helpful if the Employer establishes a governance committee e.g. a Medical Exposures Committee (MEC), to review the Employer's procedures, particularly regarding justification and optimisation. Once the framework is established, these procedures should be reviewed periodically (for example, the CQC suggests that at least every three years in diagnostic radiology and every two years for nuclear medicine and radiotherapy might be appropriate) and in all cases where there has been a significant change in practice/equipment etc. Such a committee should facilitate the development of policy through a multidisciplinary team to allow for its timely implementation.

2.61 The responsibility for establishing procedures, however, remains with the Employer. To reinforce this, the MEC should be part of the Clinical Governance management arrangements with a reporting route to the Chief Executive or Medical Director. In smaller hospitals, this committee may be part of the Radiation Safety Committee (RSC) with relevant representation. The MEC can undertake other useful roles. For example, it could liaise closely with the Research Ethics Committee (REC) for research proposals that include medical exposures and review diagnostic reference levels.

2.62 There should be strong ties and effective communication between this group and modality-specific work taking place, as laid out in the section on Optimisation through a multidisciplinary team, paragraphs 2.137–2.142.

2.63 If the regulators inspect a facility against the requirements of IR(ME)R they will examine the governance framework and test if it is robust. For example, it is expected that there be senior management buy-in (ideally through appropriate attendance at the meetings) and that there be representation from all services undertaking exposures in order for a committee to be quorate. Meeting minutes will be reviewed to ensure that actions are completed in a timely fashion and that escalation is effective when required.

Licencing requirements under IR(ME)R

2.64 On 6th February 2018, the enactment of IR(ME)R revoked the Medicines (Administration of Radioactive Substances) Regulations 1978 (MARS 1978) in England, Wales, Scotland, and Northern Ireland. The issuing of certificates to individual doctors under MARS78 has now been replaced by a licencing system under IR(ME)R for Employers and Practitioners.

2.65 For any radiological installation wishing to administer radioactive substances for the purposes of diagnosis, treatment, or research, IR(ME)R requires that both the Employer and Practitioner hold a licence issued by the relevant Licensing Authority. The licences will list the radioactive medicinal products that can be administered and the names of the licenced Practitioners.

2.66 As a transitional arrangement, any certificate issued to an individual under MARS78 that was valid on the Feb 6th 2018 will be considered to licence both the Practitioner and the Employer listed for the scope of that certificate until its expiration.

2.67 The UK Health Security Agency (UKHSA) will process and issue licences on behalf of the Licencing Authorities and is advised by the Administration of Radioactive Substances Advisory Committee (ARSAC) on the suitability of applicants. Application forms and guidance on the application process are available at www.gov.uk/arsac.

Employer licences

2.68 The Employer (as defined under IR(ME)R) must hold a separate licence for each installation on which the administration of radioactive substances takes place. Employer licences are usually issued for a period of five years but may be revoked or varied at any time by the Licencing Authority. Up-to-date guidance on when and how the Employer will need to amend their licence or notify the Licencing Authority of material changes can be found on the ARSAC website (www.gov.uk/arsac) and in the most current version of the ARSAC Notes for Guidance, currently referenced (ARSAC 2023). Note that under the Ionising Radiations Regulations (IRR 1999), 'consent' is also required from HSE or HSE(NI) (see chapter 1, paragraph 1.43).

Practitioner licences

2.69 Any individual who is clinically responsible for the justification of the administration of radioactive substances to humans must apply for a Practitioner Licence. A Practitioner Licence is not specific to a radiological installation; however, safeguards will need to be in place for individuals who may work across multiple sites.

2.70 Licences are usually issued for five years but may be revoked or varied at any time by the Licencing Authority. Further information is available in the ARSAC Notes for Guidance (ARSAC 2023).

Employer's procedures

2.71 Regulation 6 of the IR(ME)R requires the Employer to establish a framework of general procedures, protocols, and quality assurance programmes under which professionals can practice. The framework of procedures must cover the matters set out in Schedule 2 as a minimum and should be considered controlled documents. Employer's procedures must also include provision for the carrying out of 'clinical audits' as appropriate (see paragraphs 2.195–2.200). It is recommended that the Employer seek advice from professional colleagues from relevant specialities when establishing the procedures (DHSC 2018a), and a multidisciplinary approach to writing them will be most effective.

2.72 Appendix 14 gives consideration to the Employer's procedures in Schedule 2 of IR(ME)R in more detail.

Medical radiological equipment (used for medical and non-medical exposures)

2.72 As of 6th February 2018 the radiation protection framework associated with medical radiological equipment falls under IR(ME)R, except for the critical examination (see chapter 1, paragraphs 1.246–1.260, *Critical examinations*). IR(ME)R defines:

- 'equipment' as that which delivers ionising radiation to a person undergoing exposure or which directly controls or influences the extent of such exposure;
- 'medical radiological' as pertaining to radiodiagnostic and radiotherapeutic procedures, and interventional radiology or other medical uses of ionising radiation for planning, guiding, and verification purposes.

2.73 IR(ME)R applies to both medical and non-medical exposures with medical radiological equipment and sets out responsibilities with regard to the equipment detailed below.

2.74 The Employer (who has control over any equipment) is required under IR (ME)R to

- Regulation 15(1):
  - implement and maintain a quality assurance programme in respect of that equipment, which must, as a minimum, permit:
    - the assessment of the dose of ionising radiation that a person may be exposed to from an exposure to which the Regulations apply, by way of the ordinary operation of that equipment;
    - the administered activity to be verified.
  - draw up, keep up-to-date and preserve at each radiological installation an inventory of equipment at that installation and, when so requested, provide it to the relevant enforcing authority (see *Equipment inventory* section, paragraphs 2.99–2.102).
- Regulation 15(3) (see paragraphs 2.88–2.94): undertake adequate:
  - testing of any equipment before it is first used for a medical or radiological purpose;
  - performance testing at regular intervals;
  - performance testing following a maintenance procedure that is capable of affecting the equipment's performance.
- Regulation 15(6):
  - put in place any measures necessary to improve inadequate or defective performance of equipment;
  - specify acceptable performance criteria for equipment;
  - specify what corrective action is necessary when, following the application of any criteria specified, equipment is ascertained to be defective; such corrective action may include taking the equipment out of service.

- Schedule 2(d): have written procedures to ensure that quality assurance programmes in respect of written procedures, written protocols, and equipment are followed.

2.76 The Operator is required under IR(ME)R to:

- Regulation 12(3): select equipment and methods to ensure that each exposure is ALARP consistent with the intended diagnostic or therapeutic purpose with regard to QA, assessment and evaluation of dose or administered activity, and adherence to DRLs (see *Diagnostic reference levels* section, paragraphs 2.146–2.154) where appropriate.

2.77 The MPE is required under IR(ME)R to:

- Regulation 14(2): give advice on:
  - dosimetry and quality assurance matters relating to radiation protection concerning exposures;
  - physical measurements for the evaluation of dose delivered;
  - medical radiologic equipment.
- Regulation 14(3): contribute to:
  - the definition and performance of quality assurance of the equipment;
  - acceptance testing of equipment;
  - the preparation of technical specifications for equipment and installation design;
  - the surveillance of the medical radiological installations;
  - the selection of equipment required to perform radiation protection measurements.

2.78 It is evident that the Employer's procedures must allow for appropriate selection of equipment, continuous QA, optimisation, and ultimately disposal at an appropriate time.

2.79 The QA programme should detail the planned and systematic actions necessary to provide adequate confidence that equipment will meet defined standards (through testing before first use and throughout the life of the equipment against agreed performance criteria as well as planned preventative maintenance) and have regard for the manufacturer's Instructions For Use (IFU). The procedures must also state the actions to be taken when equipment does not meet the standards.

2.80 The MPE should contribute to the definition and performance of quality assurance for the equipment. The QA programme should include all equipment used in connection with the exposure (i.e. including ancillary equipment). The extent of the QA programme depends on the nature and range of equipment in use and should make it clear who has the responsibility for organising, performing, and acting on various specific functions. The programme should be reviewed periodically to ensure it is adequate, and it is advisable to audit compliance. A document from the HSE (HSE 2006) gives detailed guidance on QA programmes and how to tailor them to an Employer's equipment inventory. Although the document was written when equipment issues fell under the Ionising Radiations Regulations (IRR 1999), it is still relevant (although it may be removed from the HSE website at a later date).

2.81 The optimisation of exposures should be at the forefront of the minds of the Employer, Practitioner, Operator and MPE. In equipment terms, this will start when specifying and purchasing the equipment. Arrangements should be in place to ensure that the most appropriate equipment is selected for each exposure and that equipment is not used for procedures for which it is not suitable.

2.82 A QA programme must be consistent with the manufacturer's recommendations and national/professional standards as appropriate, e.g. IPEM 2005, 2007, 2015, 2018. It should include the provision of appropriate test equipment and its calibration and maintenance. Measurements should be taken at suitable intervals to assess representative patient doses for those undergoing exposure. This might be performed, for example, by assessing the dose administered to PMMA phantoms during quality control testing. This is different from the requirement to establish and use DRLs (see the *Diagnostic reference levels* section, paragraphs 2.146–2.154). See paragraphs 2.88–2.94 on *Acceptance, commissioning and routine testing* for more information on tests before first use and during the lifetime of the equipment.

2.83 A QA programme should include procedures for dealing with and recording equipment faults. This should address the following:

- When equipment should be taken out of service. This should be clear and give authority to those who will have to do it. For example, if a radiographer has to remove the only CT scanner from service when on night shift they must have confidence they will be supported.
- When the equipment manufacturer and/or the MPE/RPA should be contacted for advice.
- What tests should be performed to investigate the fault? What should the outcome be? And who can make decisions on its fitness for use? This may refer to the main routine tests within the QA programme or detail others, such as investigative phantom studies. Keep any test/phantom images on picture archiving and communication systems (PACS) as evidence of the investigation.
- What tests are to be done before return to use and the required outcome.
- Who reviews the fault log and at what frequency? Fault logs contain information about the suitability of the equipment for its intended use, and a review of faults over a period of time can provide information that is missed when focused on individual fault events.
- If there appears to have been an incident, then staff should refer to the incident policy.

2.84 It is necessary to maintain all equipment, ideally in accordance with the manufacturer's specifications and on a regular basis. The manufacturer and/or installer should provide the Employer with adequate information about the proper use, testing, and maintenance of the equipment under the Ionising Radiation Regulations 32(2)(c) (IRR 2017, IRR(NI) 2017). The Employer should consult their MPE on the specification of the maintenance contract, the adequacy of the maintenance arrangements, the information supplied at each service and the information provided to the Employer, the handover arrangements (also involve the RPA as this includes both radiation protection control and equipment quality), and the procedures for the acceptance of equipment back into clinical use following maintenance and repair. These equipment maintenance arrangements should be reviewed on at least a three-yearly basis or when the equipment or maintenance contract is reviewed/changed.

2.85 Ensure these arrangements are applied equally to equipment that falls outside of the management control of larger imaging departments, such as radiology. For example, a mini C-arm fluoroscopy system used in orthopaedics or dual-energy x-ray absorptiometry (DXA) imaging in endocrinology must have a robust radiation protection framework, including equipment QA. Regulatory inspectors, such as the CQC, are keen to inspect these areas when on-site and will take enforcement action where necessary.

2.86 The Employer should establish and implement an appropriate equipment replacement policy with regard to the MPE advice. When justifying keeping equipment in use, consider its performance, the magnitude of patient doses delivered, the frequency of use and number of patients affected, the patient demographic, the availability of parts, and the cost of replacement, although this will not justify patients being exposed to non-typical risks (HSE 2006).

2.87 Any Employer who disposes of equipment for future use by another person assumes responsibility and liability as a supplier. Disposal to a dealer may not necessarily negate this responsibility. *Managing Medical Devices* (MHRA 2021) contains useful guidance on selling/donating for reuse, decommissioning, and disposal of medical devices. Also, see paragraphs 2.103–2.106 for more on *Equipment disposal.*

Acceptance, commissioning and routine testing

2.88 New or second-hand equipment used for medical and non-medical exposures (including equipment on loan or being demonstrated) should be safe and appropriate for use and should be accompanied by the relevant operating instructions, the original source activity calibration certificate (if applicable), and safety information.

2.89 For the *critical examination*, it is the responsibility of the installer to ensure that the safety features and warning devices operate correctly and that there is sufficient protection from radiation exposure. A critical examination must be undertaken after installation and before clinical use. Documentation and an adequate demonstration of the radiation safety features must be provided. See chapter 1, paragraphs 1.246–1.260 for more on critical examinations.

2.90 In *acceptance testing*, the onus is on the installer to demonstrate to the purchaser's representative that all the equipment specified in the purchase contract has been supplied and that it meets the specifications laid down in that contract or in the manufacturer's documentation for the equipment. Acceptance testing is described in Medicines and Healthcare Products Regulatory Agency (MHRA) guidance (MHRA 2021). IR(ME)R states that the MPE must contribute to acceptance testing of equipment.

2.91 *Commissioning tests* are carried out by the purchaser's representative to ensure that the equipment is ready for clinical use and to establish baseline values against which the results of subsequent routine performance tests can be compared. Commissioning tests is an integral part of optimisation and as such, the relevant MPE should be involved directly or through advice on the appropriate tests and performance criteria. In diagnostic radiology, these tests are usually performed jointly by both the installer and the purchaser's representative, as this facilitates corrective action and optimisation of the system. In radiotherapy, commissioning tests are performed by radiotherapy physics staff (or a third party contracted by the Employer to undertake all or part of the commissioning) over an extended period and may include the commissioning and integrity of the equipment data transferred to treatment planning systems. In nuclear medicine, some testing of the imaging systems is performed between the installer and the purchaser's representative, then many more are carried out by local staff. Calibrators are checked by local staff, and fume cupboards/hot cells are commissioned by outside contractors. With increased uptake of multimodality imaging and treatment systems, MPEs will have to work together to ensure the optimisation of the full system (see also the section on optimisation through a multidisciplinary team, paragraphs 2.137–2.142).

2.92 *Routine performance tests* are undertaken at the frequency specified in the QA programme or after maintenance or repair to establish that the equipment continues to perform satisfactorily. Routine performance testing involves frequent, potentially quicker, or less complex testing (often called Level A tests) carried out by radiographers or assistant practitioners (SCoR 2019h), for example, and less frequent, more time-consuming tests (often called Level B tests), which may require special expertise and instrumentation (IPEM 2005) which are typically carried out by medical physicists. It is very important that Level A tests are performed and that the results are scrutinised, as these tests can identify when equipment is on its way out of tolerance before it actually breaches remedial or suspension levels. Specific results should be recorded rather than just a pass/fail outcome. The results should be recorded independently of the equipment, so as not to rely on the computer of the equipment to store the data as the only record.

2.93 The Employer must establish and maintain appropriate arrangements for the performance of equipment testing before first use, routinely, and following a maintenance procedure that is capable of affecting the equipment's performance (under IR(ME)R regulation 15(3)) (IR(ME)R 2017, IR(ME)R(NI) 2018), and ensure that there are adequate resources to undertake the tests. A number of current guidance documents that recommend tests and tolerances are referenced in the section on medical radiological equipment (see paragraph 2.82).

2.94 Although many of these tests may be combined in a single (acceptance) survey at installation that tests the initial integrity of the equipment, their purpose should remain distinct, and their components should be clearly identified (preferably documented in protocols), as should the identity and role of the installer, purchaser, and MPE.

Equipment on loan

2.95 The use of loan equipment for medical exposures and non-medical exposures with medical radiological equipment is subject to the general requirements of the Regulations as set out in this document (see chapter 1, paragraphs 1.246–1.259 on *Critical examinations*; paragraphs 2.73–2.87 on *Medical Radiological Equipment (used for medical and non-medical exposures)* and paragraphs 2.88–2.94 on *Acceptance, commissioning and routine testing* for more information). An example of when loan equipment is used is when trialling a fluoroscopy mobile in order to evaluate the system before considering a purchase or when one is used as a temporary arrangement to allow procedures to take place while a manufacturer investigates an Employer's faulty system.

2.96 The arrangements (outlined below) for the Employer and the supplier must be in place before the equipment may be used clinically. An appropriate medical equipment loan or indemnity agreement should be signed unless the supplier is another NHS organisation that performs exposures: in which case an indemnity agreement may not be required. Such agreements ensure there is adequate protection for the supplier (e.g. if their equipment is damaged) and the recipient who will use the equipment (e.g. the supplier has valid insurance covering their responsibilities). There is more on this topic and the DHSC Master Indemnity Agreement on the DHSC website: https://www.gov.uk/government/publications/master-indemnity-agreement-mia

2.97 The Employer should:

(a) consult their RPA, relevant RPS, relevant MPE and device manager prior to accepting the loan of medical equipment;

(b) agree an appropriate medical equipment loan or indemnity agreement with the supplier as necessary;

(c) perform appropriate risk assessments (including, where appropriate, an assessment of the location of the equipment) leading to the development of appropriate systems of work (see chapter 1 paragraphs 1.44–1.54, *Risk assessments* and 1.116–1.118, *Local rules*);

(d) ensure that the handover documentation includes:
  (i) operating instructions;
  (ii) critical examination (as appropriate) and other safety information;
  (iii) relevant safety information (e.g. electrical, mechanical, laser, Control of Substances Hazardous to Health (COSHH 2002));
  (iv) source activity calibration (when relevant); and
  (v) an agreed protocol for clinical use;

(e) confirm that appropriate arrangements are in place for the preventative maintenance and inspection of the equipment as specified in any medical equipment loan or indemnity agreement;

(f) confirm that commissioning tests have been performed and that the equipment has been included in a QA programme;

(g) make appropriate arrangements (usually with the supplier) for training of the employees who will use the equipment (Operators), which is activity and equipment specific, prior to the equipment being put into clinical use, clearly identifying the individuals involved and the scope of this training under IR(ME)R;

(h) document all the appropriate IR(ME)R training (see *Training* paragraphs 2.109–2.120);

(i) include details of the Employer's radiological equipment inventory; and

(j) cease using the equipment at the end of the agreed loan term.

2.98 The supplier of equipment on loan should:

(a) agree and comply with the terms of the medical equipment loan or indemnity agreement;

(b) provide handover documentation as specified above;

(c) ensure a critical examination is performed, as appropriate;

(d) confirm that the equipment is performing to specification;

(e) provide preventative maintenance and inspection as specified in the medical equipment loan or indemnity agreement;

(f) specify and agree on the scope of the training needed to operate the equipment under IR(ME)R;

(g) provide appropriate training as agreed (activity and equipment specific) under IR(ME)R to the identified employees;

(h) supervise an agreed number of clinical cases; and

(i) remove the equipment at the end of the agreed term and reinstate the premises where appropriate.

Equipment inventory

2.99 IR(ME)R Regulations 15(1) and 15(2) (IR(ME)R 2017, IR(ME)R(NI) 2018) require the Employer to keep an equipment inventory, which should list equipment that delivers ionising radiation to individuals undergoing exposures and also equipment that directly controls, or influences, the extent of the exposure. For example:

- In nuclear medicine, the inventory should include imaging equipment and dose calibrators.
- In diagnostic radiology, each x-ray installation, mobile system, CR reader, DXA scanner etc.
- In radiotherapy, each linear accelerator, onboard imaging system (if it has its own serial number), CT scanner, etc (see chapter 8, paragraph 8.5).

2.100 The inventory should include, at a minimum:

(a) type of equipment;
(b) name of manufacturer;
(c) model number;
(d) serial number or other unique identifiers;
(e) year of manufacture;
(f) year of installation (note that this is not necessarily the same as the manufacture date and often generic asset registers do not have both dates as standard fields).

2.101 It is recommended that the following be included too:

(g) location, and

(h) service agent.

2.102 This inventory does not need to be distinct from the mandatory medical device inventory (see MHRA guidance on managing medical devices (MHRA 2021), which should already encompass all radiation equipment).

Equipment disposal

2.103 There is guidance on decommissioning and disposal of devices in reference (MHRA 2021). Update the relevant inventory when equipment is disposed of.

2.104 HSE guidance (which may be removed from the HSE website at a future date) states that an Employer should consider the destination of redundant equipment; e.g. if the equipment is supplied to a veterinary practice for use in animal radiology, the Employer has responsibilities under Regulation 32 (IRR 2017). Equipment that is going to be scrapped should have the mains lead removed or the circuit boards removed and separated from the generator. Any radioactive sources must be removed from redundant equipment, disposed of via an authorised route, and properly accounted for. If it is not reasonably practicable to remove sources from the equipment, then the equipment itself must be disposed of via an authorised route (HSE 2006). For example, in nuclear medicine, old liquid scintillation counters may have calibration sources within (often not that obvious), or PET/CT scanners may contain sealed sources, i.e. point sources, for QA purposes. Megavoltage accelerators, such as linacs in radiotherapy and cyclotrons in radiopharmaceutical production, may have significantly activated internal components, which must be carefully considered with regard to safe handling, transport, and disposal, and appropriate action taken to ensure compliance with all relevant radiation regulations. You should consult an RWA (see chapter 19, *Accumulation and disposal of radioactive waste*).

2.105 The route for disposing of x-ray equipment depends on its construction. Where x-ray equipment is being replaced, it is easiest to include the removal and disposal of the old equipment in the purchase contract for the new. Where the equipment requiring disposal is not being replaced, there are several points to consider. Older equipment may contain hazardous metals, such as beryllium. If the equipment has cooling oil, this may contain toxic substances. It is important that the owner is aware of what is in the equipment; they should consult the documentation that came with the equipment or contact the manufacturer for clarity. With this information, the requirements of the relevant hazardous waste regulations (*The Special Waste Amendment (Scotland) Regulations* 2004, *The Hazardous Waste (England and Wales) Regulations* 2005, *The Hazardous Waste Regulations (Northern Ireland)* 2005) and waste electrical regulations (WEEE 2013) must be observed.

2.106 Data protection requirements must be complied with, e.g. removal of patient data.

Written protocols

2.107 Written protocols are required under IR(ME)R. They need to be recorded for every standard examination and procedure on each piece of equipment, yet allow latitude for professional input. In some modalities, they may be relatively simple, such as an exposure chart in an x-ray room. In others, much more detail and variation might be required, such as in external beam radiotherapy, where planning, verification, and treatment protocols all need to be considered. Non-medical imaging using medical equipment also needs to have written protocols. The protocols should fall under a programme of quality assurance.

2.108 Where exposure factors are pre-programmed into a unit, these must be backed up and recorded elsewhere for use in the event of equipment failure.

Training

2.109 IR(ME)R Regulation 17 states that Practitioners and Operators must not perform a medical exposure or any practical aspect for which they have not been adequately trained. An Employer must take steps to ensure every Practitioner and Operator engaged by them complies with this regulation and that they undertake continuing education and training after initial qualification. This continuing training should address developments in the field, new equipment, new techniques, and the relevant radiation protection requirements.

2.110 Trainees can participate in practical aspects under the supervision of a member of staff who is adequately trained. The DHSC guidance (DHSC 2018a) states that the nature and extent of the supervision will depend on the task and the level of pre-existing training and experience. For example, radiographers or specialist registrars will become more autonomous as they develop competence, and their supervisor will document their development. This gradual autonomy can be recorded in the entitlement procedure. The Society and College of Radiographers (SCoR) advise that student radiographers and trainee assistant practitioners should not be entitled as Operators under IR(ME)R, while acknowledging that if an Employer is satisfied that evidence of an appropriate assessment and an up-to-date training record is held and maintained by the relevant clinical imaging or radiotherapy services departments, it is legally possible for a student to be entitled as an IR(ME)R Operator with a restricted scope of practice (SCoR 2019f). Further professional body advice can be found in the IR(ME)R Implications for Clinical Practice reports (BIR *et al* 2020, Radiotherapy Board 2020).

2.111 Adequate training must satisfy the requirements of IR(ME)R Schedule 3. This schedule details the range of theoretical knowledge and practical experience that Practitioners and Operators should be trained against. The individuals should cover the relevant topics to enable them to complete their role well; this is a local decision, and an Employer has the scope to decide what is adequate (likely supported with advice from specialists such as an MPE and clinical leads). In some cases, professional bodies or Royal Colleges will define a syllabus that meets the needs of IR(ME)R training, and an Employer can rely on that syllabus. For example, those with a Fellowship in the Royal College of Radiologists or those with Cardiologist status will have completed IR(ME)R Practitioner and Operator training, both theoretical and practical. This should be supplemented with local training on the specific equipment in place in a department.

2.112 A certificate issued by a competent institution or person will provide sufficient proof that adequate training was provided to the person to whom the certificate has been issued. The training records must show the date(s) on which training was completed and the nature of the training.

2.113 Note that there is a difference between training and competence, and that although training may be provided, without an assessment of some sort to demonstrate that the knowledge and experience have been absorbed, the adequacy of the individual's training cannot be assumed.

2.114 The Employer must keep up-to-date training records for all Practitioners and Operators separately from general personnel records, make them available for inspection, and keep them for periods consistent with current guidelines on the retention of records (see appendix 10, *Retention of records*).

2.115 Adequate training (i.e. IR(ME)R Schedule 3 training) is not required for Referrers under IR(ME)R. However, it is good practice for the Employer to ensure that, at a minimum, they understand the referral guidelines in use (including the benefits and risks of exposure) and the local means of referral. It is a requirement that Employers ensure their procedures are followed by Referrers (and Practitioners and Operators) and this can be challenging if they are not within the Employer's organisation. A good way to demonstrate compliance is through audits and feedback.

2.116 Non-medical Referrers (see *Referrer* section, paragraphs 2.23–2.25) may require a higher level of training and competence assessment by an Employer before entitlement, compared to medical Referrers who have a certificate supporting their medical competence. A useful reference is BIR (2022). Non-medical Referrers are to be adequately trained to take a patient history and provide the appropriate referral information to the IR(ME)R Practitioner. They must also understand some relevant specifics of medical exposures, such as:

(i) the IR(ME)R regulations and their role as a Referrer;
(ii) the benefits and risks of medical exposures;
(iii) their scope of practice;
(iv) the referral process and any relevant Employers procedures / standard operating procedures;
(v) appropriate examinations for typical cases that they would be requesting.

2.117 There should be adequate communication of training records between Employers when employees of one Employer work in the establishment of another Employer. DHSC guidance (DHSC 2018a) gives the example that agency staff training records should be maintained by the agency and be made available on request to the entitled Employer. Another example is if one organisation routinely refers to another organisation using non-medical referrals (e.g. for musculoskeletal imaging), it may be appropriate to have the referring organisation provide adequate training through a contractual agreement.

2.118 The use of imaging equipment outside the protection framework of a typical radiation department (e.g. use of mini C-arm fluoroscopy systems outside the control of a radiology department), must still be supported with the full complement of radiation protection activities, policies, and procedures. These standalone services are likely to need enhanced support, through training and auditing, to ensure they comply with the regulations. It is worth considering if, for example, a limited period of training for an orthopaedic surgeon on a particular piece of equipment is comparable with that of a qualified radiographer who took at least three years to train in a wide scope of practice. The

Regulations do allow for IR(ME)R Operators to be trained to use equipment safely and to optimise exposure; however, they must be adequately trained in all relevant aspects of Schedule 3. Where Operators are given training to meet a limited scope of practice, such as for orthopaedic surgeons, practical training and signoff could include a teaching session with the applications specialist for the equipment and direct supervision by an experienced radiographer for a minimum number of cases (note that the Employer defines the adequate training requirements and competence must be demonstrated before someone is entitled). Peer supervision by other competent Operators may also be an option for this practical experience and signoff. Periodic audits of practice and refresher training should be provided.

2.119 There is limited current guidance on IR(ME)R update training frequencies. An Employer can use a risk-based approach with fresher training (in addition to continuing education and training) likely to be appropriate on a one-to-five-yearly cycle. At least every five years is in line with the professional revalidation cycle for cardiologists (British Cardiovascular Society 2014), dentists (PHE and FGDP 2020), and RPAs (RPA2000 2014), for example. The Employer should make these decisions, with advice from suitable experts (including an MPE), and record the decision in the entitlement procedure (or reference a training procedure). Consideration should be given to the level of dose associated with the procedures the individual is involved with, the level of their initial training, how frequently they are involved in the work (e.g. the Operator uses the equipment every day or they use it once a year), and if they teach the subject (which indicates a high level of understanding), for example. The method of delivery should also be considered and may include face to face training for higher risk activities and printed material or an online presentation for lower risk activities.

2.120 Some practical training ideas:

- Run interactive sessions addressing the Employer's procedures so that staff can ask questions and discuss issues or concerns. This is useful for new procedures and to ensure they remain appropriate over time.
- Give lectures/classroom teaching on the various general topics covered in Schedule 3 periodically so staff can refresh their knowledge. A test at the end for each individual will assess if the knowledge has been absorbed.
- Arrange specialist training with application specialists for specific equipment to enable staff to make the best use of the dose saving features.
- Involve clinical leads, MPEs, advanced practitioners, and academics, as appropriate, making use of their advanced knowledge and experience.
- Re-reading of procedures should take place at least every two years to ensure staff remain familiar with their duties.

Communicating radiation risk or 'benefit and risk'

2.121 IR(ME)R requires that the risks associated with exposures be communicated to individuals, as well as the benefits (which are also very important) in certain situations. In particular,

- Regulation 6(8) requires the Employer to take measures to raise awareness of the effects of ionising radiation among individuals capable of child-bearing or breastfeeding.
  - This measure should not be restricted to inquiries by Operators prior to exposure and might include measures such as signs in waiting rooms or relevant information in patient appointment letters (DHSC 2018a). Related guidance for Operators can be found in (SCoR 2019g).
- Regulation 12(4) requires the Employer's procedures to provide that, individuals voluntarily taking part in medical or biomedical research programmes are informed in advance about the risks of the exposure.
- Regulation 12(6&7) requires the Employer's procedures to provide that, where appropriate, written instructions and information are provided to patients (or an appropriate person), being administered radioactive substances, setting out the risks associated with ionising radiation and how they can restrict exposure of others.
- Schedule 2(i) requires an Employer's procedure setting out that wherever practicable, and prior to an exposure taking place, the individual to be exposed or their representative is to be provided with adequate information relating to the benefits and risks associated with the radiation dose from the exposure. Related guidance for Operators can be found in (SCoR 2019a)
  - The amount of information and the method of delivery should be commensurate with the risk. Posters, leaflets, and information provided by the professionals involved (and as set out in the Employer's procedures) will be useful in fulfilling this requirement. It is recommended that information be simple, succinct, and qualitative (rather than quantitative) and, where possible, include the clinical context. For example, 'We believe the risks of the procedures are small, and the procedure will help us address your suspected clinical condition through diagnosis and treatment, where appropriate' (DHSC 2018a). There are examples of posters developed at SCIN (SCIN 2019) and CIB (CIB 2019).

2.122 Appendix 16 provides some theory on risk communication, and appendix 17 gives an example of basic x-ray benefits and risk information that could be shared in a leaflet or on a healthcare facility's website. Higher radiation risk procedures, such as some procedures in interventional radiology, may require a more individual approach to information sharing and consent (see section below), although consent is not addressed in IR(ME)R (other than in regulation 12(6) in regard to patients undergoing treatment or diagnosis with radioactive substances).

Consent

2.123 There is a general legal and ethical principle that informed consent must be obtained before undertaking an examination or treatment of a patient (National Health Service 2019, GMC 2020). To give valid consent, the person needs to understand the nature and purpose of the procedure, including the benefits and risks, so that they are included in the decision making aspect of their treatment. In deciding what information should be shared, it is important to note that social and legal thinking now favours the 'reasonable patient', where the emphasis is on what the average prudent patient would want to know about potential risks and treatment options rather than what the average prudent doctor feels they should tell the patient (Malone *et al* 2012, Royal College of Radiologists 2012). Healthcare organisations should have procedures for consent that have regard for the law and consider special circumstances such as capacity issues.

2.124 Although IR(ME)R does not address informed consent, IR(ME)R Schedule 2 (i) requires the Employer's procedures to ensure that adequate information on benefits and risks (see paragraphs 2.121–2.122 on risk communication) is provided for exposures wherever practicable and prior to the exposure taking place. Practice guidance for the diagnostic imaging and radiotherapy workforce is available at (SCoR 2018a).

2.125 It is important that the appropriate professionals (the Referrer, Practitioner, or Operator) have sufficient knowledge of the levels of dose and risk associated with procedures, as that will impact their ability to impart the information to patients. MPEs must contribute to the training of Practitioners and other staff in relevant aspects of radiation protection, which should help advance such knowledge. Guidance such as the RCR iRefer system (RCR 2017), which includes risk data, can help proliferate awareness of risk (Malone *et al* 2012).

2.126 Consent can be implied, given orally, or provided in writing. The level of formality should increase with the risks involved (GMC 2008). Prior to accepting consent, it should be considered whether the wanted and needed information has been adequately provided. It is active consent that is more important than a signature on a form where the process was not followed correctly. Guidance (Department of Health 2009) states that '*the validity of consent does not depend on the form in which it is given*'. Written consent merely serves as evidence of consent: if the elements of voluntariness, appropriate information, and capacity have not been satisfied, a signature on a form will not make the consent valid.

2.127 When putting this into practice for an exposure falling within the IR(ME)R, the level of risk communication and formality of consent will vary depending on the level of exposure involved. Low risk procedures can be managed with the provision of information in the form of leaflets or notices, time to consider the information in a suitable environment, and the opportunity to ask questions. Historically, for those procedures, consent has been presumed if the patient shows up for their appointment and complies with inquiries regarding patient identification, etc. For higher dose procedures, the need for consent becomes more explicit. Forms and written consent, including details of the radiation risk, are recommended for high-dose procedures (Malone *et al* 2012) and those recognised to be more complex or carrying a more serious risk of complication (Royal College of Radiologists 2012).

2.128 There is currently no consensus on what is classified as a low or high dose exposure for the purposes of consent or on when implied or explicit (verbal or written) consent is required. The RCR standards on this topic (Royal College of Radiologists 2012) were deliberately left generic, giving the authority to the care provider to decide what is appropriate, taking the individual patient into account. At a 'group of experts' consultation on justification held by the International Atomic Energy Agency in 2008, low level exposures were considered 'arbitrarily' as those below 1 mSv effective dose (Malone *et al* 2012). A paper published on the practicalities of justification (Malone *et al* 2012) gives an example of the high dose category where forms should be used during the consent process for exposures with an effective dose of $\geqslant$ 10 mSv. Another approach to this decision could be based on risk rather than dose. It would be worth considering this topic locally, perhaps as part of an RSC and/or MEC, taking the practicalities of providing information and obtaining consent into account.

2.129 The RCR (Royal College of Radiologists 2012) recommends that explicit consent be obtained from patients where their images are to be used for teaching and training, following the 'anonymise or ask' principle.

Pregnancy and breastfeeding

2.130 IR(ME)R regulation 11(1)(f) states that a person must not carry out an exposure unless, in the case of an individual of childbearing potential, the person has enquired whether that individual is pregnant or breastfeeding, if relevant. Pregnancy and breastfeeding do not preclude an individual from exposure, but the information does feed into the justification process, and Regulation 11(3)(d) sets out that the urgency of the exposure must be considered where:

- pregnancy cannot be excluded, in particular, if abdominal and pelvic regions are involved, taking into account the exposure of the person concerned and any unborn child; and
- an individual who is breastfeeding undergoes an exposure involving the administration of radioactive substances, taking into account the exposure of both the individual and the child.

2.131 The Employer's procedures are to provide for making enquiries of individuals of childbearing potential to establish whether the individual is or may be pregnant or breastfeeding (see paragraph 2.71–2.72 on Employer's procedures). The procedure should facilitate a diverse gender spectrum (SCoR 2019g). For example, consider a transgender patient who recently identified as male; their NHS notes have been updated to reflect this, but they still have childbearing potential. The Employer's procedure must identify and address the risk to that individual. A suitable sign in a patient waiting area may help communicate that there are risks associated with radiation exposure during pregnancy. Guidance to support inclusive practice in regard to pregnancy status enquiries is available (SoR 2021).

2.132 See also the following publications to inform and reduce the risk of radiation exposure to those who may be pregnant or breastfeeding and their offspring:

- Protection of pregnant patients during diagnostic medical exposures to ionising radiation (HPA, RCR and CoR 2009).
- Notes for Guidance on the Clinical Administration of Radiopharmaceuticals and Use of Sealed Radioactive Sources (ARSAC 2023)
- IR(ME)R: Implications for clinical practice in imaging, interventional radiology and diagnostic nuclear medicine (BIR *et al* 2020) (In particular the flowchart in appendix 7 and the main body text that supports it in regard to high/low dose procedures and the 10-day and 28-day rules).
- IR(ME)R: Implications for clinical practice in radiotherapy guidance from the Radiotherapy Board (Radiotherapy Board 2020)
- The impact of IR(ME)R 2017 IR(ME)R (NI) 2018 on pregnancy checking procedures (SCoR 2019g).

Authorisation

2.133 Regulation 11(1) of the IR(ME)R states that '*a person must not carry out an exposure unless (b) it has been justified by the Practitioner…and (c) it has been authorised by the Practitioner or…the Operator …*'. This illustrates the separate nature of justification and authorisation.

2.134 Authorisation is the means of demonstrating that justification has taken place. The method of authorisation should be decided locally and specified in the Employer's procedures. The person authorising must be identifiable, and this is often recorded as a signature on the request card or electronically through the Radiology Information System (RIS) or radiotherapy record and verify system (RVS). It should be noted that the person who authorises an exposure does not have to be the person who carries out the exposure. For example, a senior radiographer may authorise but another radiographer may take the radiographs; in nuclear medicine, a physicist may authorise, but a medical technologist may administer the radionuclide. In cases where the Referrer, Practitioner and/or Operator are the same person (e.g. dentists), justification and authorisation must still take place and should be documented.

2.135 Regulation 11(5) makes it possible for a Practitioner to produce guidelines under which an Operator can authorise an exposure. This is often called 'authorising under protocol' or under 'authorisation guidelines'. In this situation, the Practitioner has still made the decision that justified the exposure and therefore retains legal responsibility, despite not having considered the individual request. It is the Operator who considers the request, compares the clinical information in the referral to the guidelines, and authorises the exposure as justified if the guidelines are met. It is important that the guidelines issued by the Practitioner are controlled so that only the current version is available to Operators and that an individual Practitioner can be identified for all exposures authorised under the guidelines. If an Operator authorises an exposure that falls outside the guidelines laid down, they will be responsible for that exposure, and unless they have been entitled to justify said exposure, they will be working outside their scope of practice and entitlement.

2.136 Each member of staff involved in an exposure should be aware of the role they are performing and their scope of practice. Entitling individuals to act as Practitioners and Operators is a local decision, and Employers may take different approaches. Many radiographers act as IR(ME)R Practitioners for general radiography examinations and theatre procedures but act as Operators who authorise CT exposures under protocol. See paragraph 2.36 for more on the entitlements of Radiographers as Practitioners.

Optimisation through a multidisciplinary team

2.137 Optimisation of exposures is best achieved using a multidisciplinary team approach. The membership of the MDT will vary depending on the complexity of the examination or treatment, and of the equipment used but as a minimum, it should consist of:

- One or more representatives from the group making a diagnosis or intervention using the images, and for radiotherapy, representation from those planning the treatment (e.g. MPE). For most diagnostic examinations, this will be a radiologist, but dentists, cardiologists, oncologists, and surgeons should attend for the examinations in which they are involved.
- One or more representatives from the group operating the equipment during the examination or treatment. For most diagnostic examinations this will be a radiographer, but dentists, cardiologists and surgeons should attend where local practice has them acting as the primary Operator during the procedure.
- An MPE, for whom the examination is within their scope of practice. Note that an individual could fulfil more than one role on the MDT, where appropriate.

2.138 Consideration should also be given to including:

- A representative of the manufacturer where the equipment is more complex, such as interventional or CT units. This representative will be better acquainted with the behaviour of the unit under all conditions and with the optimal operation of any dose reduction capabilities the unit has. They can also assist with matters relating to image processing.
- Any specialist involved with the patient cohort or the specific procedures under discussion, e.g. a paediatric specialist or nurses who may be providing close patient care.

2.139 All members of the MDT (which has been referred to as the 'Image Optimisation Team' by the DHSC) should be appropriately trained, command sufficient authority and seniority, and have details of their role, including available resources and management support, in writing from their Employer (Department of Health 2016).

2.140 It is for those who will be making a diagnosis or intervention to specify the level of image quality they require. Achieving this level of image quality with the lowest dose is then the goal towards which all members of the MDT should work. Different staff groups will be involved in different clinical protocols as relevant to their field and experience. The optimisation work should be a continuous, iterative process that will likely take some time to produce the desired outcome; therefore, long term buy-in is required. Patient considerations are as relevant as technical considerations when it comes to achieving an optimised protocol.

2.141 Where relevant, the MDT should also consider the optimal positioning of, and protection given to, anyone acting as a carer and comforter during the examination.

2.142 For more on this, see recommendation 7 and the response in reference (Department of Health 2016).

Dose constraints under IR(ME)R

2.143 IR(ME)R defines a dose constraint as a restriction on the prospective doses of individuals that may result from a given radiation source. Dose constraints can be used to help restrict exposure by ensuring that appropriate consideration is given to radiation protection measures. They can be based on experience or recommendations from professional bodies and should be used for:

(i) medical or biomedical research exposures of individuals for whom no direct medical benefit is expected from the exposure. Such research should be subject to a dose constraint based on the total dose from all radiodiagnostic procedures included in the protocol (DHSC 2018b). For studies involving healthy volunteers, it is recommended that the total annual dose be considered and not exceed an effective dose of 10 mSv and that participants are all over 50 years old. Where this is not the case, the

Clinical Radiation Expert should ensure that specific justification is provided to address this (HRA *et al* 2020a, 2020b).
(ii) the protection of carers and comforters (see paragraphs—2.156–2.168)

2.144 IR(ME)R Regulation 6(6) states that a dose constraint must be established by the Employer in terms of individual effective or equivalent doses over a defined appropriate time period. The Employer's procedures should provide the dose constraints as required.

2.145 Note that a dose constraint is different from a dose limit. A constraint is used as a tool in optimisation, and if it is exceeded, it should trigger a review of the radiation protection measures in place. That alone may not be reportable to the regulators; however, local procedures should make clear what action will be taken if a constraint is exceeded. Exceeding a dose limit, as set out in chapter 1 on IRR, is an 'overexposure' and is reportable to the HSE.

Diagnostic reference levels

2.146 Diagnostic reference levels (DRLs) consistent with the appropriate diagnostic image quality should be established by the Employer for standard-sized patients undergoing:
(a) standard radiological investigations;
(b) routinely undertaken and well defined interventional procedures;
(c) exposures as part of a health screening programme;
(d) asymptomatic exposures;
(e) non-medical imaging using medical radiological equipment; and
(f) nuclear medicine investigations.

2.147 Radiotherapy planning CT scans are not considered diagnostic scans, and therefore the use of the term DRLs is not appropriate. However, the use of dose reference levels is a useful method of demonstrating that dose optimisation has taken place (UKHSA 2022).

2.148 DRLs should be expressed in easily recordable quantities that have a direct relevance to patient dose (screening time, milliampere seconds (mAs), dose–area product (DAP), dose-length product (DLP), radionuclide activity, etc) and should be set locally with due regard to regional, national, or European data.

2.149 Where the capacity exists for the analysis of all examinations undertaken—either by downloading data recorded in a Radiology Information System or directly from a Dose Management System—an MPE should be consulted on an appropriate approach to analysing the data, which could include data that is incorrect or that has been wrongly attributed to a single examination when it represents the cumulative dose from many. The MPE must also consider that the sample will include patients of all sizes and that weight information may not be available.

2.150 The MPE in conjunction with the MEC/RSC should review DRLs periodically (protocol audit frequency based on risk, ensuring that all are subject to review every three years for each modality) or when changes are made to equipment or procedures. Patient dosimetry, imaging performance, and QC measurements should be examined to identify the potential for dose and performance optimisation and a report should be made for the Practitioners. This should be a component of the dose reduction strategy and the imaging equipment replacement policy.

2.151 DRLs are designed to monitor the overall performance of a specified examination or procedure on a given system or set of systems and should not be applied to individual patients. The exposure of an individual patient should be based on that individual's clinical condition, physique, and diagnostic needs.

2.152 However, in specific incidences, for example, in interventional procedures (where screening time and dose are known to be variable between patients), knowledge of the DRL for the procedure can help the clinician undertaking the procedure identify when the exposure is unusually high. Depending on how the DRL is used locally (as set out in the Employer's procedures), the DRL (or more likely a multiple of it) may be used as a warning or notification level during the procedure to raise the clinician's awareness of the dose. This might prompt a less experienced clinician to seek assistance to complete the case, for example. In such cases, the operator should record the dose quantity and any contributing clinical factors (e.g. patient size, complexity etc). This will help identify if DRLs are being consistently exceeded.

2.153 If DRLs are consistently exceeded for standard-sized patients, there should be an investigation, followed by corrective action. If DRLs are never exceeded, they should still be reviewed and possibly revised downward. If locally set DRLs are not relevant to local practice they will have little value as an optimisation tool.

2.154 The latest guidance on significant accidental and unintended exposures under IR(ME)R (CQC 2023) requires that in interventional radiology and cardiology when a procedure results in unintended or unpredicted observable deterministic effects, this will be notifiable to the authorities even when there has been no procedural failure.

Information and instructions for nuclear medicine patients

2.155 Patients undergoing treatment or diagnosis with radioactive substances must be appropriately informed about the risks and the precautions to be taken to protect those around them prior to leaving the radiological installation where the exposure was carried out. The Employer's procedures must provide for this. Although the primary responsibility lies with the Employer to ensure that this is performed, the task may be delegated to an Operator as identified in the Employer's procedures. Generic risk assessments, drawn up in conjunction with the MPE and RPA as appropriate, should be documented and available to support this advice. Specific risk assessments will be needed in non-routine circumstances. Written instructions and information should be reviewed at least every three years (see chapter 16 *Patients leaving hospital after administration of radioactive substances* for more information).

Carers and comforters

2.156 'Carers and Comforters' are individuals who, other than as part of their occupation, 'knowingly and willingly' incur exposures to ionising radiation resulting from the support and comfort of another individual who is/was undergoing an exposure. They are often relatives or friends of those undergoing exposure. The role is now considered a medical exposure under IR(ME)R and, as such, is to be considered as part of the justification process. Note that the individual justification of exposures to carers and comforters is required in addition to the justification required for the associated medical exposure or non-medical imaging exposure. The person acting as the IR(ME)R Practitioner for the patient undergoing the examination/treatment may be different from the Practitioner for the carer/comforter. The Employer's procedures should be written to accommodate this, where required, and the individual acting in each role should be clearly identified (see appendix 14, *Issues to consider when writing IR(ME)R employers procedures*, procedure 'n' for more information).

2.157 Justification of the exposure of carers and comforters requires sufficient net benefit, taking into account:

- the likely direct health benefits to a patient; and
- the possible benefits to the carer or comforter. These benefits are likely to be psychological rather than physical (DHSC 2018a);
- the detriment that the exposure might cause.

2.158 The exposure of carers and comforters should be optimised and Regulation 6 (5)(d)(ii) of the IR(ME)R (IR(ME)R 2017) says that the Employer must establish dose constraints (not dose limits) with regard to the protection of carers and comforters. This is to be provided for through an Employer's procedure (see paragraphs 2.71–2.72) on the use of carers and comforters and guidance on minimising their exposure.

2.159 A dose constraint of 5 mSv has been recommended in the latest IR(ME)R guidance (DHSC 2018a). Normally, it should be possible to design procedures such that doses received are well below this level. However, it is accepted that there will be occasions when a higher value may be appropriate (such as where the carer or comforter is supporting the treatment of a vulnerable individual), and in those situations, dose constraints should be assessed and agreed upon on a case-by-case basis, making clear to the carer and comforter the risks involved (DHSC 2018a). The Employer's procedure should be flexible enough to accommodate the range of circumstances that may be encountered. Established dose constraints should be audited for suitability and compliance.

2.160 Carers and comforters must be informed of the risks involved in incurring exposures while supporting and comforting the patient (as well as the procedures to follow to restrict their exposure), and they must be willing to incur the exposure that they will receive. Although carers and comforters are not ‘patients’ under healthcare legislation, many of the principles and practices of the well-recognised system of ‘informed consent’ (National Health Service 2019, GMC 2020) may be useful in applying the requirements of IR(ME)R, that the carer or comforter is ‘knowingly and willingly’ receiving an exposure as part of another’s medical exposure.

2.161 Ideally, they would be adults on the grounds of reduced stochastic risk from exposure to radiation in older individuals and also because the ability to consent requires the capacity to understand risk, which may not be present in younger individuals. It would be appropriate to choose a dose constraint that is risk-based depending on the particulars of the carer and comforter chosen (see reference (IAEA 2009) for more on this, with a particular focus on the release of patients after administration of radionuclide therapy).

2.162 A record of those supporting a patient should be kept. Local arrangements, as set out in the Employer’s procedure, may include the use of a form or other documentation to record the information given to or received from the carer and comforter, e.g. pregnancy status, name, relationship to the individual exposed, etc.

2.163 The Employer can only do what is reasonable, both practically and within the bounds of patient confidentiality, to identify potential carers and comforters and ensure they are knowingly and willingly accepting the task (HSE 2003). Other measures must be used to minimise exposures to the general public or family members that have not been disclosed by the patient, for example, in procedures involving the administration of radioactive substances. The patient must be given information on restricting doses to others, and it is expected that they will use this information and inform others of the risks where applicable. If it is known that someone will go against the instructions provided to minimise dose through, for example, spending more time than necessary in close proximity of the patient, and if that would result in an unacceptable high risk to the carer, then the radiation facility may have little choice but to refuse treatment (as the medical exposure of the carer would not be justified) (HSE 2003).

2.164 Use of the 5 mSv dose limit in five consecutive years under IRR Schedule 3 (6) (IRR 2017) can be an alternative way of dealing with children who can’t give consent or those that are likely to be exposed but haven’t had the opportunity to consent and therefore can't be considered carers and comforters. However, an appropriate constraint should be used to optimise the exposure in any case.

2.165 In radiotherapy, it is strongly recommended that no person remains with the patient during external beam treatment (i.e. only in exceptional circumstances and not without a suitable and sufficient risk assessment). The use of carers and comforters may potentially arise during the use of therapeutic nuclear medicine and for some brachytherapy treatments where a patient cannot care for themselves and no alternative treatments are available for their condition.

2.166 Ideally, pregnant individuals should not be carers or comforters if it can be avoided; however, this may not always be practicable. The exposure must be justified, taking into consideration the individual circumstances, and in many optimised exposures, the dose may be negligible. The Employer's procedure should consider if/when pregnant carers or comforters are a concern.

2.167 Those exposed as a result of their occupations should not be considered carers and comforters. This includes those who do not work for an Employer, such as clerics, who can be at risk of significant exposure from certain ceremonies of significant importance to the patient (HSE 2003).

2.168 Although exposures to carers and comforters are now considered medical exposures, there is still useful information on this subject available, along with possible doses that can be expected in various exposure situations and measures to control them. In an HSE document, this is called 'Dose Constraints for Comforters and Carers (HSE 2003).

Repeat exposures and reject analysis

2.169 The need to repeat an exposure may arise for many reasons, such as patient movement or incorrect exposure factors that make an image undiagnostic. In some cases, additional images may be required due to cutoff of the desired anatomy. In these examples, there is no need for an IR(ME)R Practitioner to justify the repeat exposure as the original justification has not been fulfilled. A quality assurance programme should capture the frequency and cause of these repeats and assess trends. Additional training may be required for certain radiographers, or an equipment fault might be identified. These technical repeats add to the patient's radiation exposure and may be considered poor practice or, in some cases, may be considered externally (i.e. to the regulator) reportable incidents/exposures, depending on the circumstances. An MPE can help interpret the latest reporting criteria (CQC 2023). It is worth ensuring that the radiographic team understands local procedures so that they know when and how to log a repeat exposure or bring the occurrence to the attention of management.

2.170 Consideration should be given to how the initial, insufficient quality image(s) will be treated. If it is undiagnostic, the image should be annotated as rejected with an appropriate reason given for the purposes of the reject analysis audit.

This will avoid the reporting radiologist, for example, mistakenly using the image as the basis for clinical evaluation and diagnosis. Where the original image is not undiagnostic, it should be sent to PACS in case it contains useful information that the subsequent image does not, e.g. a poorly positioned chest x-ray shows a tumour in the arm that would not be seen in a well-positioned repeat image. This approach should be discussed with the clinical team and communicated to ensure consistent archiving and evaluation.

2.171 The dose from the 'rejected' exposures should be included in the patient dose record for the examination.

Suspected radiation incident

2.172 IR(ME)R Regulation 8 (IR(ME)R 2017) provides for a comprehensive system of analysis, recording, and reporting of accidental or unintended exposures.

2.173 Regulation 8(1) states that the Employer's procedures must provide that the Referrer, the Practitioner, and the individual exposed or their representative (if there is one) are informed of the occurrence of a clinically significant unintended or accidental exposure and of the outcome of the analysis of this exposure. The professional bodies have published the definition of 'clinically significant' in chapters 19 and 21 of their therapy (Radiotherapy Board 2020) and diagnostic imaging (BIR *et al* 2020) guidance, respectively. The regulation is in place so that the professionals involved are aware of the event and can ensure appropriate care for the patient in the future. It also supports being open and honest. The regulation allows for instances when it is not in the best interest of the exposed individual to be notified, although it is recommended that a representative of the patient be informed wherever possible (DHSC 2018a). See also paragraphs 2.183–2.188 on *Duty of candour*.

2.174 Regulation 8 (2) states that the Employer's QA programme must, in respect of radiotherapeutic practices, include a study of the risk of accidental or unintended exposures.

2.175 Regulation 8(3) states that the Employer must establish a system for recording analyses of events involving, or potentially involving, accidental or unintended exposures, proportionate to the radiological risk posed by the practice. The MPE should contribute to the analysis of events as well as other staff and, in particular, IR(ME)R duty holders, as appropriate. Many healthcare facilities have incident management systems that are used to track and record analysis of events. It is advisable that incidents are reviewed periodically with the intention of looking for trends, and, where appropriate, incidents are discussed at meetings, such as the MEC or general staff meetings, to share learning. Trends can be better identified when providers use standardised taxonomies at a national level. Such examples are available for radiotherapy (BIR *et al* 2008) and clinical imaging departments (RCR, IPEM, and SCoR 2019).

2.176 Regulation 8(4) states that where the Employer knows, or has reason to believe, that an accidental or unintended exposure has, or may have, occurred in which a person, while undergoing:

(a) any exposure, was or could have been exposed to levels of ionising radiation significantly greater than those generally considered to be proportionate in the circumstances;
(b) a radiotherapeutic exposure was, or could have been, exposed to levels of ionising radiation significantly lower than those generally considered to be proportionate in the circumstances;

the Employer must:

(i) undertake an immediate preliminary investigation of the incident (see paragraphs 2.180–2.182 on performing an investigation below);
(ii) unless that investigation shows beyond a reasonable doubt that no such exposure has occurred, immediately notify the relevant enforcing authority;
(iii) conduct or arrange for a detailed investigation of the circumstances of the exposure and an assessment of the dose received; and
(iv) notify the relevant enforcing authority, within the time period specified by the relevant enforcing authority, of the outcome of the investigation and any corrective measures adopted.

2.177 The English, Welsh, Scottish, and Northern Irish enforcement authorities published guidance on what is considered significantly different from what is generally considered proportionate in the circumstances and when to report to the regulator (CQC 2023).

2.178 The Employer's procedures should also make allowances for voluntary notifications in instances where the criteria for notification have not been met but where further learning can be achieved. See appendix 12 on incidents and shared learning for more on other reporting routes, such as to the MHRA and the radiotherapy National Reporting and Learning System.

2.179 Documentation relating to potential and actual incidents should be retained in line with relevant guidance, including that published by the enforcing authority (DHSC 2018a). For example, the current version of the NHS code of practice for record retention states that records should be kept for at least 10 years for non-serious incidents or 20 years for serious incidents (NHSX 2021).

Performing an investigation

2.180 Please refer to chapter 1, paragraphs 1.233–1.240, for general guidance on investigation. This section only focuses on additional considerations under IR

(ME)R. Detailed guidance on performing investigations can be found in (BIR *et al* 2008) and (NHS England 2015).

2.181 Any suspected accidental or unintended exposure (both greater than intended and, in the case of radiotherapeutic exposures, less than intended) or near miss should be investigated to a level appropriate to the risk. Make sure the right people are involved, i.e. those that have the first hand information. Those who hold roles under IR(ME)R (Referrer, Practitioner and Operator), equipment service engineers, other staff and patients may have helpful information. The MPE should contribute to the analysis of events. See paragraphs 2.172–2.179 for more on IR(ME)R incidents.

2.182 An investigation might include evidence about the following:

Procedural related

- content and suitability of the Employer's procedures, standard protocols or equipment quality assurance framework and adherence to these;
- training;
- staffing levels;
- pause and check;
- communication between those performing roles along of the patient pathway.

Equipment related

- the settings on the equipment along with any equipment generated log-files;
- recorded exposure parameters;
  - in diagnostic radiology: kV, mAs, DAP, DLP, etc;
  - in radiotherapy: treatment machine monitor units, beam time, beam modifiers, geometric settings, etc;
- other measures of exposure;
  - in diagnostic radiology: Detector Dose Indicator (DDI);
  - in nuclear medicine: count rate in a gamma camera image;
- information may be gained from any patient measurements undertaken as part of dose assessment programmes;
- QA records (including training records of those persons carrying out QA duties);
- training records of Operators;
- any fault reports and service records;
- any tests on equipment carried out for the purpose of the investigation;
- an account of what happened by the person operating the equipment.

Duty of candour

2.183 The duty of candour is not an IR(ME)R requirement. It is an additional consideration to those required under IR(ME)R in regard to accidental or unintended exposures (or near misses) that must be investigated and, as appropriate, be reported to the relevant regulator and if 'clinically significant', it must be shared with the Practitioner, Referrer and individual exposed (see paragraphs 2.172–2.179).

2.184 Every healthcare professional must be open and honest with patients, colleagues, Employers, the regulators, and other organisations as appropriate when something goes wrong with treatment or care that causes, or has the potential to cause, harm or distress (Nursing and Midwifery Council *et al* 2015). This is a professional duty of candour. There is also a statutory duty of candour on many health and social care providers in the UK, except in Northern Ireland (where there are plans to introduce a statutory duty). The regulations of relevance for:

- England are (*The Health and Social Care Act 2008 (Regulated Activities) Regulations 2014* 2014);
- Scotland are (*The Duty of Candour Procedure (Scotland) Regulations 2018* 2018);
- Wales are (*Health and Social Care (Quality and Engagement) (Wales)* 2020).

2.185 The duty of candour is designed to promote openness and transparency in the case of an incident that has the potential to (for example, based on the English regulations) cause moderate or severe harm, death due to treatment rather than natural causes, or psychological damage lasting more than 28 days.

2.186 Some of these moderate/severe harm incidents may also be considered 'clinically significant'. Guidance on what is considered 'clinically significant' under IR(ME)R is given by UK professional bodies in chapter 19 of their therapy guidance (Radiotherapy Board 2020) and chapter 21 of their diagnostic imaging guidance (BIR *et al* 2020). Threshold levels of significance are based on additional lifetime radiation-induced cancer risk, risk of radiation-induced childhood cancer for fetal exposures where pregnancy was not known, absorbed dose to the lens of the eye, heart or skin and psychological harm.

2.187 An example of psychological harm might be where a mother has been informed about an unintended fetal dose, irrespective of the dose received. As demonstrated by this example, the irony of openness and transparency is that, in certain circumstances, it may actually cause significant harm. This may also be the case where an individual is informed about an incident resulting in unnecessary exposure to ionising radiation and their perception of the risk is such that they have significant psychological harm despite the radiation risk being negligible. The negative perception of radiation risk for some individuals should not be underestimated by professionals. A record of whether or not a patient is informed about an incident should be included in the investigation and if the patient is not told, the reasons why should be included.

2.188 Detailed advice on providing openness and honesty can be found in various documents, such as (Nursing and Midwifery Council *et al* 2015) or (NPSA 2009).

Clinical evaluation

2.189 Regulation 12(9) (IR(ME)R 2017) states that the Employer must take steps to ensure that a clinical evaluation of the outcome of each exposure is recorded in accordance with the Employer's procedures (unless the exposure was of a carer or comforter), including, where appropriate, factors relevant to patient dose. The guidance (DHSC 2018a) recommends that the evaluation be accurate and timely so that it contributes appropriately to the patient's care. The evaluation might include the diagnostic findings, the therapeutic implications, whether the procedure was successful or abandoned, whether the delivered dose in a therapy was as intended or not, etc. If it is known prior to the exposure that no clinical evaluation will occur, then the exposure is not justified and should not take place.

2.190 The CQC wrote to the chief executives of all NHS Trusts in England in relation to compliance with this regulation in 2011. It was in response to a published audit (Kiu *et al* 2010), which showed that in a particular NHS radiology department, approximately 50% of radiographs that were provided to referring clinicians for their interpretation (rather than being reported within radiology) had no clinical evaluation report recorded in the patient's notes. Although the CQC acknowledges that this requirement of IR(ME)R is difficult to implement in practice, they expect Employers to take steps to ensure the regulation is complied with. This could be achieved through an audit, followed by appropriate action on the findings where compliance is less than 100%, followed by further audit some time later to assess the effectiveness of action taken and the new rates of compliance.

2.191 It may be useful for workforce planning where the audit identifies whether examinations were reported by radiologists, non-radiologist medical practitioners, or allied healthcare professionals such as radiographers.

2.192 Clinical evaluation is an IR(ME)R Operator task, and as such those who perform the role must be adequately trained and entitled. Organisations who outsource or delegate reporting to non-radiology clinical staff ('auto-reporting') must assure themselves that those individuals are appropriately trained and competent to perform the task (CQC 2018b):

- On outsourcing, organisations need to assure themselves that the radiologists (or other 'reporters') employed by the outsourcing companies are appropriately trained, that clinical audits of the quality of the reports are performed and that systems are in place to flag up urgent and unexpected findings.
- On auto-reporting, there is a potential risk of harm to patients associated with this, particularly for chest and abdomen x-rays. Organisations need to

assure themselves that non-radiology staff who are responsible for reporting images are aware of the support measures in place (such as a radiology opinion, if requested) and that they are competent to perform the task. Employers also need to make sure that audits are performed to make sure that reports are documented and accurate

2.193 Factors relevant to patient dose should be recorded such that an effective dose could be estimated at a later date, if required. This typically takes place alongside image evaluation through entries on a Radiology Information Systems (RIS) or as appropriate in the patient notes. With digital imaging modalities, factors are often available on the image sent to the PACS and are available in the DICOM header. Where images were not taken with standard factors as documented in the Employer's protocols, it is advisable to record this. See appendix 14, *Issues to consider when writing IR(ME)R employers procedures*, procedure 'e' for more.

2.194 For more information on the subject of clinical evaluation and a useful discussion on who is best placed to fulfil this role, see references (SCoR 2013, Royal College of Radiologists 2018, SCoR 2019e).

Clinical audit

2.195 IR(ME)R Regulation 7 requires that there be provision for clinical audit within the Employer's procedures. The IR(ME)R definition of clinical audit is 'a systematic examination or review of medical radiological procedures that seeks to improve the quality and outcome of patient care through structured review, whereby medical radiological practices, procedures and results are examined against agreed standards for good medical radiological procedures, with modification of practices, where indicated, and the application of new standards if necessary'. This quality improvement cycle should be consistent with national and professional guidance. It is expected that an annual clinical audit programme will be in place which will include radiation protection elements and demonstrate where results of audits have informed practice.

2.196 The whole patient pathway should be subject to clinical audit under the categories of structure, process, and outcome (European Society of Radiology 2011).

- *Structure*—includes lines of authority, professional roles and radiation protection responsibilities, premises, equipment and information systems.
- *Process*—justification and referral processes, protocols, optimisation procedures, patient dose assessment, image quality, emergency incident procedures and reliability of patient image/data transfer.
- *Outcome*—includes methods for follow-up of the outcome of examinations/procedures, over both short and longer term.

2.197 There are numerous ways to perform such an audit, an example of which can be found in appendix 18, *Example IR(ME)R audit: procedures and document control.*

2.198 Observational checks could be performed as part of clinical audits, such as spending time in a department watching the patient identification process, pregnancy checks, and recording of exposures over a given period of time/set number of procedures. There is some merit in observational checks; however it is sometimes more practical to perform retrospective reviews. For example:

- Auditing referrals (request cards or the electronic system), to assess:
    - the quality of the information in the initial referral against the requirements for patient identification and provision of clinical information to allow for the justification process.
    - if justification has been performed correctly by someone who is entitled to do so.
- Reviewing training records (see paragraphs 2.109–2.120).
- Reviewing entitlement documentation including documentation for non-medical Referrers.
- Establishing the number of unreported scans (i.e. no clinical evaluation—see paragraphs 2.189–2.194) or the length of reporting times.
- Establishing the quality of imaging. A useful method here is peer review and reject analysis.
- Patient dose audits (see paragraphs 2.146–2.154).

2.199 The EC (European Commission 2009), IAEA (International Atomic Energy Agency 2010), and ESR (European Society of Radiology 2022) have produced guidance on clinical audit in radiological practices.

2.200 The Royal College of Radiologists have some templates for IR(ME)R audit which can be searched for and accessed on their website, (RCR, AuditLive)

Special considerations

2.201 IR(ME)R regulation 11(3) requires special consideration in regard to the justification process of exposures of carers and comforters and asymptomatic individuals, and of the urgency of the exposure in cases involving breastfeeding individuals or where pregnancy cannot be excluded.

2.202 Regulation 12(8) requires the Practitioner and Operator to pay particular attention to the optimisation of exposures to children, those in a health screening programme, those receiving a high radiation dose, and exposures involving breastfeeding individuals or where pregnancy cannot be excluded.

2.203 The doses must be sufficient to be consistent with the intended purpose of the exposure, but as low as reasonably practicable. The need for extra care is due to the associated higher risk e.g. for paediatrics, and because not all exposures deliver a direct benefit to the individual exposed. The choice of exposure factors, the angulation of beams, and the choice of protective material might result in changes from standard practice for children and high dose procedures. Case reviews and peer reviewed articles are good sources of information to inform centres of best practice. Visiting centres of excellence for particular techniques also enables best practice to be shared more widely. Reviews of techniques, DRLs, and audits are common ways of demonstrating compliance with these requirements. Multidisciplinary input can be crucial.

Research

2.204 Please see appendix 19, *Guidance on medical research exposures*, for detailed advice on this topic.

References

ADR no date *About the ADR. Agreement concerning the International Carriage of Dangerous Goods by Road, UN Economic Commission for Europe website* (United Nations) https://unece.org/about-adr (accessed 12 April 2024)

ARSAC 2023 *Notes for Guidance on the Clinical Administration of Radiopharmaceuticals and Use of Sealed Radioactive Sources.* Administration of Radioactive Substances Advisory Committee https://gov.uk/government/publications/arsac-notes-for-guidance (accessed 25 February 2023)

BEIS 2019 *The Justification of Practices Involving Ionising Radiation Regulations 2004. Guidance on their Application and Administration* https://gov.uk/government/publications/the-justification-of-practices-involving-ionising-radiation-regulations-2004-guidance-on-their-application-and-administration (accessed 31 January 2023)

BIR 2022 *Guidance for Non-medical Referrers to Radiology, British Institute of Radiology Webpage* https://bir.org.uk/media-centre/position-statements-and-responses/guidance-for-non-medical-referrers-to-radiology/ (accessed 22 February 2023)

BIR, IPEM, NPSA, SCoR and RCR 2008 *Towards Safer Radiotherapy* (London: British Institute of Radiology, Institute of Physics and Engineering in Medicine National, Patient Safety Agency, Society and College of Radiographers, Royal College of Radiologists) https://www.rcr.ac.uk/our-services/all-our-publications/clinical-oncology-publications/towards-safer-radiotherapy/ (accessed 12 April 2024)

BIR, RCR, IPEM, SCoR and PHE 2020 *IR(ME)R Implications for Clinical Practice in Diagnostic Imaging, Interventional Radiology and Diagnostic Nuclear Medicine* (London) https://www.rcr.ac.uk/our-services/all-our-publications/clinical-radiology-publications/ir-me-r-implications-for-clinical-practice-in-diagnostic-imaging-interventional-radiology-and-diagnostic-nuclear-medicine/ (accessed 12 April 2024)

British Cardiovascular Society 2014 *Guidance on Revalidation for Cardiologists*

CIB 2019 *New Patient Information Posters on the Benefits and Risks of Imaging, RCR Website* (Clinical Imaging Board) https://rcr.ac.uk/posts/new-patient-information-posters-benefits-and-risks-imaging (accessed 5 December 2022)

COSHH 2002 *The Control of Substances Hazardous to Health Regulations 2002* https://legislation.gov.uk/uksi/2002/2677 (accessed 31 January 2023)

CQC 2014 *Care Quality Commission IR(ME)R Annual Report 2013* (Newcastle upon Tyne: Care Quality Commission) https://webarchive.nationalarchives.gov.uk/20161107115553/https://cqc.org.uk/content/key-findings-and-reports (accessed 31 January 2023)

CQC 2018a *Care Quality Commission IR(ME)R annual report 2017/18* (Newcastle upon Tyne: Care Quality Commission) https://cqc.org.uk/sites/default/files/20181115-IRMER-annual-report-2017-18-FINAL.pdf (accessed 31 January 2023)

CQC 2018b *Care Quality Commission Radiology Review* (Newcastle upon Tyne: Care Quality Commission) https://cqc.org.uk/sites/default/files/20180718-radiology-reporting-review-report-final-for-web.pdf (accessed 31 January 2023)

CQC 2023 *Ionising Radiation (Medical Exposure) Regulations (IR(ME)R). 19/4/23, CQC website* (London: Care Quality Commission) https://cqc.org.uk/guidance-providers/ionising-radiation/ionising-radiation-medical-exposure-regulations-irmer (accessed 10 May 2023)

Department of Health 2009 *Reference Guide to Consent for Examination or Treatment (2nd Edition)* https://gov.uk/government/publications/reference-guide-to-consent-for-examination-or-treatment-second-edition (accessed 31 January 2023)

Department of Health 2014 *Justification of Computed Tomography (CT) for Individual Health Assessment* https://assets.publishing.service.gov.uk/government/uploads/system/uploads/attachment_data/file/326572/IHA_-_June_Report.pdf (accessed 31 January 2023)

Department of Health 2016 *COMARE 16 DH Expert Working Party Response to the Committee on Medical Aspects of Radiation in the Environment's 16th Report: Patient Radiation Dose Issues Resulting from the Use of CT in the UK* https://gov.uk/government/publications/response-to-the-review-of-radiation-dose-issues-from-ct-scans (accessed 31 January 2023)

DHSC 2018a *Guidance to the Ionising Radiation (Medical Exposure) Regulations 2017, Gov.UK website* (Department of Health and Social Care) https://gov.uk/government/publications/ionising-radiation-medical-exposure-regulations-2017-guidance (accessed 31 January 2023)

DHSC 2018b *MPEs Recognition Scheme* (Department of Health and Social Care) https://gov.uk/government/publications/medical-physics-experts-recognition-scheme (accessed 31 January 2023)

European Commission 2009 *European Commission Guidelines on Clinical Audit for Medical Radiological Practices (Diagnostic Radiology, Nuclear Medicine and Radiotherapy)* (Luxembourg: European Commission) 159 https://op.europa.eu/en/publication-detail/-/publication/75688cc6-c9d3-4c43-9bfd-ce5cea0d8bcb (accessed 12 April 2024)

European Society of Radiology 2011 European Commission guidelines on clinical audit. Statement by the European Society of Radiology *Insights Imaging* **2** 97–8

European Society of Radiology 2022 *Esperanto ESR Guide to Clinical Audit in Radiology* 3rd edn (Vienna: European Society of Radiology) https://www.myesr.org/app/uploads/2023/08/Esperanto-ESR-Guide-to-Clinical-Audit-in-Radiology-3rd-Edition.pdf (accessed 12 April 2024)

Faculty of General Dental Practice 2018 General Dental Council (no date) *Recommended CPD topics, General Dental Council Website* https://gdc-uk.org/education-cpd/cpd/recommended-cpd-topics (accessed 9 February 2022)

GMC 2020 *Decision Making and Consent, Guidance on Professional Standards and Ethics for Doctors* (General Medical Council) https://gmc-uk.org/ethical-guidance/ethical-guidance-for-doctors/decision-making-and-consent (accessed 31 January 2023)

GMC 2008 *Consent: Patients and Doctors Making Decisions Together The Duties of a Doctor Registered with the General Medical Council* (General Medical Council) https://www.gmc-uk.

org/-/media/documents/GMC-guidance-for-doctors---Consent---English-2008---2020_pdf-48903482king-decisions-together-2008---2020_pdf-84769495.pdf?la=en (accessed 12 April 2024)

Health and Social Care (Quality and Engagement) (Wales) 2020 https://legislation.gov.uk/asc/2020/1 (accessed 31 January 2023)

HPA, RCR and CoR 2009 *RCE-9 Protection of Pregnant Patients during Diagnostic Medical Exposures to Ionising Radiation* (Health Protection Agency, Royal College of Radiologists, and College of Radiographers) https://www.rcr.ac.uk/our-services/all-our-publications/clinical-radiology-publications/protection-of-pregnant-patients-during-diagnostic-medical-exposures-to-ionising-radiation/ (accessed 14 February 2024)

HRA, HCRW, HSC and NRS 2020a *Clinical Radiation Expert (CRE) Review Procedure* Health Research Authority, Health and Social Care (Northern Ireland), Health and Care Research Wales, NHS Research Scotland https://myresearchproject.org.uk/Help/Help Documents/CRE_Review_Procedure_FINAL_v3_0_21_October_2020.pdf (accessed 31 January 2023)

HRA, HCRW, HSC and NRS 2020b Medical Physics Expert (MPE) Review Procedure. Health Research Authority, Health and Social Care (Northern Ireland), Health and Care Research Wales, NHS Research Scotland https://myresearchproject.org.uk/Help/Help Documents/Medical_Physics_Expert_Review_Procedure_FINAL_v3_0_21_October_2020.pdf (accessed 31 January 2023)

HSE 2003 Dose constraints for comforters and carers*Research Report 155* https://hse.gov.uk/research/rrpdf/rr155.pdf (accessed 31 January 2023)

HSE 2006 *Guidance Note PM77 (Third edition) Equipment Used in Connection with Medical Exposure* 3rd (Sudbury: Health and Safety Executive) https://hse.gov.uk/pubns/guidance/pm77.pdf (accessed 31 January 2023)

IAEA 2009 Release of patients after radionuclide therapy *Safety Reports Series No. 63. SRS No 63* (Vienna: International Atomic Energy Agency) https://iaea.org/publications/8179/release-of-patients-after-radionuclide-therapy (accessed 31 January 2023)

ICRP 2007 The 2007 Recommendations of the International Commission on Radiological Protection. ICRP Publication 103 *Ann. ICRP* **37** 1–337 (https://icrp.org/publication.asp?id=ICRP Publication 103) (accessed 31 January 2023)

International Atomic Energy Agency 2010 *Comprehensive Clinical Audits of Diagnostic Radiology Practices: A Tool for Quality Improvement* **4** (Vienna, Austria) https://iaea.org/publications/8187/comprehensive-clinical-audits-of-diagnostic-radiology-practices-a-tool-for-quality-improvement (accessed 31 January 2023)

IPEM 2005 *Report 91: Recommended Standards for the Routine Performance Testing of Diagnostic X-ray Imaging Systems* (York: Institute of Physics and Engineering in Medicine)

IPEM 2007 *Report 94: Acceptance Testing and Commissioning of Linear Accelerators* ed M Kirby, C Hall and S Ryde (York: Institute of Physics and Engineering in Medicine)

IPEM 2015 *Report 111: Quality Control of Gamma Cameras and Nuclear Medicine Computer Systems* (York: Institute of Physics and Engineering in Medicine)

IPEM 2018 Report 81: physics aspects of quality control in radiotherapy ed I Patel, S Weston, L A Palmer, W P M Mayles, P Whittard, R Clements, A Reilly and T J Jordan (York: Institute of Physics and Engineering in Medicine) 2nd edn

IR(ME)(A)R 2018 *The Ionising Radiation (Medical Exposure) (Amendment) Regulations* https://legislation.gov.uk/uksi/2018/121 (accessed 31 January 2023)

IR(ME)R(NI) 2018 *The Ionising Radiation (Medical Exposure) Regulations (Northern Ireland)* https://legislation.gov.uk/nisr/2018/17 (accessed 31 January 2023)

IR(ME)R 2017 *The Ionising Radiation (Medical Exposure) Regulations* www.legislation.gov.uk/uksi/2017/1322 (accessed 31 January 2023)

IRR(NI) 2017 *The Ionising Radiations Regulations (Northern Ireland)* http://legislation.gov.uk/nisr/2017/229 (accessed 31 January 2023)

IRR 1999 *The Ionising Radiations Regulations* https://legislation.gov.uk/uksi/1999/3232 (accessed 31 January 2023)

IRR 2017 *The Ionising Radiations Regulations.* Great Britain https://legislation.gov.uk/uksi/2017/1075 (accessed 31 January 2023)

JoPIIRR 2004 *The Justification of Practices Involving Ionising Radiation Regulations 2004* https://legislation.gov.uk/uksi/2004/1769 (accessed 31 January 2023)

JoPIIRR 2018 *The Justification of Practices Involving Ionising Radiation (Amendment) Regulations 2018* https://legislation.gov.uk/uksi/2018/430 (accessed 31 January 2023)

Kiu A, Bano F, Barnes R and Khan S H M 2010 IRMER regulations: compliance rate of radiograph reporting by non-radiology clinicians *Clin. Radiol.* **65** 984–8

Malone J *et al* 2012 Justification of diagnostic medical exposures: some practical issues. Report of an International Atomic Energy Agency Consultation *Br. J. Radiol.* **85** 523–38

MARS 1978 *The Medicines (Administration of Radioactive Substances) Regulations 1978.* https://legislation.gov.uk/uksi/1978/1006 (accessed 31 January 2023)

MHRA 2021 *Managing Medical Devices, Guidance for Health and Social Care Organisations, Health Policy.* Medicines and Healthcare Products Regulatory Agency https://gov.uk/government/publications/managing-medical-devices (accessed 8 February 2023)

National Health Service 2019 *Consent to Treatment, NHS Website* https://nhs.uk/conditions/consent-to-treatment/ (accessed 4 March 2023)

National Health Service Reform and Health Care Professions Act 2002 2002 https://legislation.gov.uk/ukpga/2002/17 (accessed 8 February 2023)

NHSBSP 2023 *Guidance for the Implementation of the IR(ME)R Regulations 2017, GOV.UK Website* (NHS England) https://gov.uk/government/publications/breast-screening-guidance-on-implementation-of-ionising-radiation-medical-exposure-regulations-2017/guidance-for-the-implementation-of-the-irmer-regulations-2017 (accessed 11 July 2023)

NHS England 2015 *Serious Incident Framework* https://england.nhs.uk/patient-safety/serious-incident-framework/ (accessed 8 February 2023)

NHSX 2021 *Records Management Code of Practice 2021 A guide to the management of health and care records.* August 202 https://nhsx.nhs.uk/information-governance/guidance/records-management-code/ (accessed 31 January 2023)

NPSA 2009 *Being Open: Saying Sorry When Things Go Wrong* (National Patient Safety Agency) https://qi.elft.nhs.uk/wp-content/uploads/2013/12/being-open-framework.pdf (accessed 8 February 2023)

Nursing and Midwifery Council, General Medical Council, Nursing & Midwifery Council and General Medical Council 2015 *Openness and Honesty When Things Go Wrong: The Professional Duty of Candour (GMC Guideline)* https://gmc-uk.org/ethical-guidance/ethical-guidance-for-doctors/candour---openness-and-honesty-when-things-go-wrong (accessed 8 February 2023)

PHE and FGDP 2020 *Guidance Notes for Dental Practitioners on the Safe Use of X-ray Equipment (2nd Edition)* (London: Public Health England, Faculty of General Dental Practice (UK)) https://cgdent.uk/wp-content/uploads/securepdfs/Guidance-notes-for-dental-practitioners-on-the-safe-use-of-x-ray-equipment-2020-online-version.pdf (accessed 12 April 2024)

Radiotherapy Board 2020 *Ionising Radiation (Medical Exposure) Regulations: Implications for Clinical Practice in Radiotherapy Guidance from the Radiotherapy Board* (London: Royal College of Radiologists) https://www.rcr.ac.uk/media/smmkkrsa/ionising-radiation-medical-exposure-regulations-implications-for-clinical-practice-in-radiotherapy.pdf (accessed 12 March 2024)

RCR (AuditLive) Audit and Quality Improvement—AuditLive https://www.rcr.ac.uk/career-development/audit-quality-improvement/auditlive-radiology/ (accessed 12 April 2024)

RCR 2014 R*adiotherapy Prescribing Framework for those not on the Specialist Register for Clinical Oncology*, Royal College of Radiologists website (London: Royal College of Radiologists) https://www.rcr.ac.uk/news-policy/latest-updates/radiotherapy-prescribing-framework-for-those-not-on-the-specialist-register-for-clinical-oncology/ (accessed 12 April 2024)

RCR 2017 *iRefer Guidelines: Making the Best Use of Clinical Radiology. Royal College of Radiologists Website* 8th edn (London: Royal College of Radiologists)) https://www.rcr.ac.uk/our-services/irefer/ (accessed 12 April 2024)

RCR, IPEM and SCoR 2019 *Learning from Ionising Radiation Dose Errors, Adverse Events and Near Misses in UK Clinical Imaging Departments* (London: Royal College of Radiologists, Institute of Physics and Engineering in Medicine Society and College of Radiographers) https://www.rcr.ac.uk/our-services/all-our-publications/clinical-radiology-publications/learning-from-ionising-radiation-dose-errors-adverse-events-and-near-misses-in-uk-clinical-imaging-departments-working-party-user-guidance/ (accessed 12 April 2024)

RCR 2022 *Implementing Ordercomms (electronic requesting) in radiology* (London, UK: Royal College of Radiologists) https://www.rcr.ac.uk/our-services/all-our-publications/clinical-radiology-publications/implementing-ordercomms-electronic-requesting-in-radiology/ (accessed 12 April 2024)

Royal College of Radiologists 2012 *Standards for Patient Consent Particular to Radiology* (London: Royal College of Radiologists) 2nd edn https://www.bsgar.org/static/uploads/BFCR(12)8_consent.pdf (accessed 12 April 2024)

Royal College of Radiologists 2018 *Standards for Interpretation and Reporting of Imaging Investigations* 2nd (London: Royal College of Radiologists) https://www.rcr.ac.uk/our-services/all-our-publications/clinical-radiology-publications/standards-for-interpretation-and-reporting-of-imaging-investigations-second-edition/ (accessed 12 April 2024)

RPA 2000 2014 The Renewal of RPA Certification Scheme, RPA 2000 Website http://rpa2000.org.uk/rpa-certification-scheme/ (accessed 8 February 2023)

RPA 2000 2023 List of Certificate Holders, RPA 2000 website http://rpa2000.org.uk/list-of-certificate-holders/ (accessed 31 August 2021)

SCIN 2019 *Risk Posters, Scottish Clinical Imaging Network Website* (Edinburgh: Scottish Clinical Imaging Network) http://scin.scot.nhs.uk/scin-work/risk-posters (accessed 5 December 2022)

SCoR 2013 *Preliminary Clinical Evaluation and Clinical Reporting by Radiographers: Policy and Practice Guidance* (Society and College of Radiographers) https://sor.org/learning-advice/professional-body-guidance-and-publications/documents-and-publications/policy-guidance-document-library/preliminary-clinical-evaluation-and-clinical-repor (accessed 8 February 2023)

SCoR 2017a *Have You Paused and Checked? IR(ME)R Referrers, Society and College of Radiographers Website* https://sor.org/learning-advice/professional-body-guidance-and-publications/documents-and-publications/posters/have-you-paused-and-checked-ir(me)r-posters (accessed 19 July 2021)

SCoR 2017b *Have You Paused and Checked? Radiotherapy, Society and College of Radiographers Website* (Society and College of Radiographers) https://www.sor.org/learning-advice/professional-body-guidance-and-publications/documents-and-publications/posters/have-you-paused-and-checked-radiotherapy-posters (accessed 8 February 2023)

SCoR 2018a *Obtaining Consent: A Clinical Guideline for the Diagnostic Imaging and Radiotherapy Workforce* (Society and College of Radiographers) https://sor.org/Learning-advice/Professional-body-guidance-and-publications/Documents-and-publications/Policy-Guidance-Document-Library/Obtaining-consent-a-clinical-guideline-for-the-dia (accessed 8 February 2023)

SCoR 2018b *The Diagnostic Radiographer as the Entitled IR(ME)R Practitioner* (Society and College of Radiographers) https://sor.org/Learning-advice/Professional-body-guidance-and-publications/Documents-and-publications/Policy-Guidance-Document-Library/The-Diagnostic-Radiographer-as-the-entitled-IR(ME) (accessed 8 February 2023)

SCoR 2019a *Communicating Radiation Benefit and Risk Information to Individuals Under the Ionising Radiation (Medical Exposure) Regulations (IR(ME)R)* https://sor.org/learning-advice/professional-body-guidance-and-publications/documents-and-publications/archive-documents/communicating-radiation-benefit-and-risk-infor-(1) (accessed 4 March 2023)

SCoR 2019b *Have You Paused and Checked? A Clinical Imaging Examination IR(ME)R Operator Checklist* (Society and College of Radiographers) https://www.sor.org/learning-advice/professional-body-guidance-and-publications/documents-and-publications/posters/have-you-paused-and-checked-ir(me)r-posters (accessed 12 April 2024)

SCoR 2019c *Have You Paused and Checked? IR(ME)R Operator Checklist for Administration of Radioisotopes for Molecular Imaging Procedures* (Society and College of Radiographers) https://www.sor.org/learning-advice/professional-body-guidance-and-publications/documents-and-publications/posters/have-you-paused-checked-checklist-for-administrati (accessed 12 April 2024)

SCoR 2019d Have you 'Paused & Checked'? IR(ME)R Operator checklist for Molecular Imaging Procedures: Image acquisition https://www.sor.org/learning-advice/professional-body-guidance-and-publications/documents-and-publications/posters/have-you-paused-checked-molecular-imaging-(mri) (accessed 12 April 2024)

SCoR 2019e *Ionising Radiation (Medical Exposure) Regulations; Briefing for Radiographers Who Undertake Commenting or Reporting* (Society and College of Radiographers) https://sor.org/learning-advice/professional-body-guidance-and-publications/documents-and-publications/policy-guidance-document-library/ionising-radiation-(medical-exposure)-regulations (accessed 8 February 2023)

SCoR 2019f *Student Radiographers & Trainee Assistant Practitioners as 'Operators' under IR(ME)R 2017 (2018 in Northern Ireland)* (Society and College of Radiographers) https://sor.org/learning-advice/professional-body-guidance-and-publications/documents-and-publications/policy-guidance-document-library/student-radiographers-trainee-assistant-practition (accessed 20 February 2023)

SCoR 2019g *The Impact of IR(ME)R 2017 IR(ME)R (NI) 2018 on Pregnancy Checking Procedures* (Society and College of Radiographers) https://www.sor.org/learning-advice/professional-body-guidance-and-publications/documents-and-publications/policy-guidance-document-library/inclusive-pregnancy-status-guidelines-for-ionising/the-impact-of-ir(me)r-2017-ir(me)r-(ni)-2018-on-pr (accessed 12 April 2024)

SCoR 2019h *The Radiographic Assistant Practitioner's Role in Quality Control of Radiological Equipment* (Society and College of Radiographers) https://www.sor.org/learning-advice/professional-body-guidance-and-publications/documents-and-publications/policy-guidance-document-library/the-radiographic-assistant-practitioner%E2%80%99s-role-in (accessed 12 April 2024)

SoR 2021 *Inclusive Pregnancy Status Guidelines for Ionising Radiation: Diagnostic and Therapeutic Exposures* (London: Society of Radiographers) https://sor.org/learning-advice/professional-body-guidance-and-publications/documents-and-publications/policy-guidance-document-library/inclusive-pregnancy-status-guidelines-for-ionising (accessed 8 February 2023)

The Duty of Candour Procedure (Scotland) Regulations 2018 (*Scotland*) https://legislation.gov.uk/ssi/2018/57 (accessed 8 February 2023)

The Hazardous Waste (England and Wales) Regulations 2005 (*England and Wales*) https://legislation.gov.uk/uksi/2005/894 (accessed 8 February 2023)

The Hazardous Waste Regulations (Northern Ireland) 2005 (*Northern Ireland*) https://legislation.gov.uk/nisr/2005/300 (accessed 8 February 2023)

The Health and Social Care Act 2008 (Regulated Activities) Regulations 2014 https://legislation.gov.uk/uksi/2014/2936/ (accessed 8 February 2023)

The Special Waste Amendment (Scotland) Regulations 2004 https://legislation.gov.uk/ssi/2004/112 (accessed 8 February 2023)

UKHSA 2022 *Guidance: National Diagnostic Reference Levels (NDRLs) from 13 October 2022, GOV.UK website* https://www.gov.uk/government/publications/diagnostic-radiology-national-diagnostic-reference-levels-ndrls/ndrl (accessed 7 November 2022)

WEEE 2013 *The Waste Electrical and Electronic Equipment Regulations 2013* https://legislation.gov.uk/uksi/2013/3113 (accessed 8 February 2023)

**John Saunderson, Mohamed Metwaly, William Mairs, Philip Mayles,
Lisa Rowley and Mark Worrall**

Chapter 3

Diagnostic radiology (excluding dental) and fluoroscopically guided interventions

Scope

3.1 This chapter contains guidance on radiation protection in diagnostic radiology and in fluoroscopically guided interventions. It does not include dentistry. It covers all applications of x-rays in radiography, fluoroscopy, mammography, CT, and bone densitometry. For convenience, the subject of the x-ray examination is referred to as 'the patient' in this guidance, where this term also includes an individual undergoing medical exposure for health screening, a volunteer participating in a medical research programme, asymptomatic individuals, and an individual undergoing non-medical imaging using medical radiological equipment (but does not include carers and comforters). The guidance also applies when the diagnostic x-ray equipment is used for training, in research, including the optimisation of examination techniques, for examinations of cadavers and pathological specimens, and for all other purposes, such as testing or measuring the radiation output in a clinical environment.

Principles of radiation protection

3.2 For medical exposures in diagnostic radiology and fluoroscopically guided interventions, applying the radiation protection principles of justification and optimisation (as required by the IR(ME)R (IR(ME)R 2017, IR(ME)R(NI) 2018) indicates, amongst other issues, that:

(a) a radiological examination must be carried out only after it has been justified and the authorisation recorded, and

doi:10.1088/978-0-7503-2332-1ch3

(b) each exposure must be optimised so that all patient radiation doses during x-ray examinations are kept as low as reasonably practicable (ALARP), consistent with the intended clinical objective.

Roles and responsibilities

Employer

3.3 Employers must develop referral guidelines for all their medical exposures. The RCR publication, *iRefer: Making the best use of clinical radiology* (RCR 2017), can help departments define their guidelines. This document does not cover all investigations that require referral guidelines, however, advice is available from other professional bodies (e.g. the Royal College of Physicians for bone mineral densitometry (NOGG 2021)). The Employer must inform Referrers of the criteria that apply to the investigations that are within their scope of referral, including information on radiation doses and associated risks, particularly for use in the case of paediatric patients, or persons who might be pregnant. In the case of pregnant patients, this information can be found in the HPA document, *Protection of Pregnant Patients during Diagnostic Medical Exposures to Ionising Radiation* (HPA, RCR and CoR 2009). The Employer must ensure the information is readily available to all Referrers.

Referrer

3.4 Referrers must be registered healthcare professionals in the UK; (in Northern Ireland, this also includes medical practitioners registered with the Medical Council of Ireland (DHSC 2017)). The Referrer and the scope within which they are entitled to refer must be detailed in an Employer's procedure. The Referrer must provide sufficient information about the patient to allow for their identification, their relevant medical history, and the reason for the request for the IR(ME)R Practitioner to justify if radiology is the appropriate diagnostic (or interventional) method for the patient. Referrals should be clear and unambiguous. If they are not, the Operator or Practitioner will have to seek additional information from the Referrer, which could delay the patient's examination.

3.5 Referrers should discuss requests not covered by the referral criteria directly with the IR(ME)R Practitioner who should decide if the examination is justified.

Practitioner

3.6 IR(ME)R Practitioners must be registered healthcare professionals in the UK and entitled by the Employer to a scope of practice. In the radiology department, radiologists are usually IR(ME)R Practitioners and it is common for adequately trained radiographers to be IR(ME)R Practitioners. Outside of radiology, there is a wide range of specialists that can be entitled as IR(ME)R Practitioners.

3.7 IR(ME)R Practitioners must take the data supplied by the Referrer into account when justifying examinations. Relevant clinical information may include previous diagnostic information, medical records, images from other investigations, and, for persons of childbearing potential, whether they are known to be or could possibly be pregnant. Employers must have adequate systems to provide Practitioners with previous images. If there is insufficient information in the referral, the IR(ME)R Practitioner should ask for more information from the Referrer or reject the request.

3.8 IR(ME)R Practitioners must further consider the benefit and risk of the exposure likely to be received by a carer and comforter, that is, an individual knowingly and willingly incurring an exposure to ionising radiation by helping in the support and comfort of individuals undergoing an exposure. Carers and comforters are subject to dose constraints that must be established in the Employer's procedures; the IR(ME)R Practitioner must consider, in addition to the likely direct health benefit to a patient, the possible benefits to the carer and comforter, as well as the detriment from the exposure to the patient and the carer and comforter alike.

3.9 All examinations must be justified by an IR(ME)R Practitioner. The justification should take into account recommendations by scientific societies and relevant bodies (i.e. NICE and screening programmes).

3.10 Before an examination can proceed, it is necessary for it to be authorised (the act of documenting that it has been justified). This is commonly performed by the Operator, e.g. by means of an electronic record or by signing an appropriate box on the request form, which then authorises the examination. Alternatively, when authorisation is made through the authorisation guidelines, entitled Operators can authorise specified diagnostic exposures by recording that the request complies with the agreed guidelines produced by the IR(ME)R Practitioner.

Operator

3.11 Any person who undertakes a medical exposure, or who performs practical aspects related to a medical exposure, must be entitled as an IR(ME)R Operator. Any member of staff who undertakes equipment-specific practical aspects which could affect a future medical exposure (e.g. equipment quality control (QC)) must also be entitled as an IR(ME)R Operator. Operators can only be entitled once they have been adequately trained. They may not perform any practical aspect of exposure without adequate training and entitlement as an Operator.

3.12 All IR(ME)R Operators must be adequately trained (see paragraphs 2.109–2.120, *Training*) and competent to operate their equipment. This training must include all the safety features and warning devices that may affect the patient's (and staff's) exposure, including emergency off-switches. It must also include how to use any particular features of a piece of equipment that enhance exposure optimisation, such as the selection of additional beam filtration or the use of pulsed fluoroscopy.

3.13 Operators are directly responsible for the practical aspects of any exposure that they perform. In addition to following the Employer's procedures and protocols, they must optimise the exposure by selecting the practical methods and, where possible, the equipment to ensure that the necessary diagnostic information is obtained for the lowest dose to the patient. They should clearly understand the extent of their own responsibilities in order to avoid accidental or unintended exposure of the patient. A trainee must be supervised by an Operator who is adequately trained in those practical aspects.

3.14 An Employer's procedures will also require Operators to record that other checks have been performed for individual patients, e.g. patient identification, authorisation of exposure as justified, and whether a person of childbearing potential may be pregnant.

Medical physics expert

3.15 Employers must appoint appropriate medical physics expert(s) (MPE(s)) to advise on optimisation of radiological practice. These formal appointments should be in writing and reviewed periodically to ensure the MPE(s) and the scope of their advice remain appropriate. The MPE(s) must be able to advise the Employer on all diagnostic and interventional modalities used. A list of MPEs with their scope of practice, should be available in or referenced in the Employer's procedures.

3.16 The role of the MPE is discussed in general in paragraphs 2.56–2.59 and in Appendix 15.

Radiation protection adviser

3.17 Employers must appoint an RPA to advise on compliance with the requirements of the IRR (IRR 2017, IRR(NI) 2017). The RPA for radiology services should have appropriate experience in the application of radiation protection principles in radiology. The formal appointment must be in writing and reviewed periodically to ensure the RPA and the scope of their advice remain appropriate. The role of the RPA is discussed in general in paragraphs 1.88–1.91. The RPA may be the same person as the MPE, though it should be recognised that their roles are separate and distinct.

Radiation protection supervisor

3.18 The role of the RPS, in general, is discussed in paragraphs 1.119–1.126. They must supervise the arrangements contained in the local rules. The RPSs can also provide a communication channel with the RPA (see Appendix 3, *The role of the radiation protection adviser*).

3.19 The number of RPSs appointed should be sufficient to ensure that work in all controlled areas is adequately supervised. It is not necessary for an RPS to always be present, but an RPS should work in a controlled area often enough to ensure that all staff members are aware of the local rules and that they are observed. It is for the Employer to decide how many RPSs are required, with due regard to the number of locations covered, the complexity of the work, presence in the department, and the advice of the RPA. Where systems of work, including local rules and contingency plans, require urgent action by an RPS in certain circumstances, the Employer should ensure that Operators have the means to contact an RPS during all working hours.

Facility design

3.20 New installations should be designed and constructed for optimal practice and workflow given the procedures to be undertaken. This design should consider the needs of all clinical users and the Employer's RPA and MPE must be consulted regarding design and radiation protection matters. NHS estates (NHS Estates 2001) has produced guidance on the design of a radiology department. More frequently, however, it is necessary to install new equipment and facilities within the footprint of an existing building, department, or room; or work involving ionising radiation may be undertaken in areas that were never previously intended for the use of ionising radiation (theatres, for example). Such modifications of existing areas can result in compromises having to be made; it is for the design team to ensure these do not negatively impact safety. The RPA must be consulted on the shielding, PPE, and work practices that must be implemented for the safe use of ionising radiation in any location. For new and existing facilities, the BIR has produced guidance on the appropriate shielding of diagnostic facilities (Sutton *et al* 2012).

3.21 In hospitals, all x-ray examinations should be carried out in a purpose-built facility whenever practicable. Examples of these include a diagnostic radiology department, an interventional suite, a dedicated facility for cardiac catheterisation, or a room for mammography. Sometimes, the condition of the patient may make it necessary for the examination to be carried out elsewhere, e.g. in a ward or an operating theatre. Appropriate control measures must be put in place to ensure that x-rays can be used safely in such locations.

3.22 Entrances to radiology departments and doors of x-ray rooms should be wide enough to allow beds and other ancillary equipment to pass through unless they are only going to be used by outpatients; this increases the number of patient examinations that can be undertaken in purpose-built facilities.

3.23 All x-ray rooms should include features to avoid unnecessary exposure to the room's occupants. In general, x-ray rooms should have a protective screen incorporating lead glass or lead acrylic viewing panels to protect the Operators while still enabling them to have a clear view of the patient. Where staff are not shielded by a fixed or mobile screen (such as for dual-energy X-ray absorptiometry (DXA) examinations, where this is not usually necessary), they should be able to stand at a large enough distance from the patient and equipment to minimise exposure.

3.24 Protected areas should normally be provided for staff at all control panels such that the radiation dose received there by a person not wearing PPE is not significant (an annual dose of less than 0.3 mSv should be attainable). Protective screens, including any lead glass or lead acrylic, should be marked with their lead equivalent thickness at an appropriate peak kilovoltage (kV_p).

3.25 Ideally, the Operator should be able to see all entrances to the x-ray room from the control panel. Where there is no direct line of sight, the use of mirrors can provide the visibility required. Alternatively, access should be controlled for any entrance that cannot be seen from the control panel. This could be achieved by locking the door from the inside during examinations (where clinical considerations allow) or restricting access to the door from the outside. Where this is not possible, there should be an appropriate contingency in place that covers someone entering the room during exposure. This must be considered in the risk assessment and included in the local rules.

3.26 An x-ray room should not be used for more than one radiological procedure at a time unless the room has been designed and built with this possibility in mind, e.g. a resuscitation area. In this case, there should be some means (e.g. a protective panel separating parts of the room) to ensure that there is no significant additional exposure, either of one patient from the radiography of another or of staff from examinations in which they themselves are not engaged.

3.27 If the structure of the x-ray room does not provide adequate shielding to persons in adjacent areas, the shielding should be improved. Until the necessary improvements are complete, an RPA should be asked to advise on the restriction of workload and beam direction necessary to ensure the protection of those in adjacent areas.

3.28 The room should not constitute a throughway from one place to another except in emergencies. Viewing areas should have separate access. Where there is no option (i.e. a poorly designed existing facility), this should be considered in the risk assessment.

3.29 A sign containing a radiation warning symbol and enough written information to indicate the type of radiation present and the risk is required at all entrances to a controlled area where the controlled area extends to the entrance. These signs should be placed at eye level to make them more likely to be seen. The radiation warning symbol and a recommended form of words for the signs are shown in Appendix 7, *Warning signs and notice*.

3.30 Where possible, these signs should be illuminated and wired to the equipment so as to indicate the presence of a controlled area when there is power connected to the equipment and the presence of x-rays during the exposure state. Some signs can be wired such that the exposure state is also indicated during the preparation for exposure stage, though this is not required by IEC 60601 (BSI 2014) and the preparation stage is not well defined for some modalities, so it should not be expected for all equipment. The controlled area warning should be illuminated yellow and the x-ray exposure warning red. The significance of the warning should be self-evident or be stated in a notice (e.g. 'Do not enter when the red light is on'; see Appendix 7, *Warning signs and notice*).

3.31 Illuminated signs and lights should give an immediate indication of exposure status by using filament bulbs or LEDs rather than fluorescent tubes. It is beneficial to use the bulb or LED with the longest projected lifespan to limit the potential for failure of the warning sign and to have spares on site for quick repair.

3.32 When a mobile fluoroscopy system is often used in a single location, e.g. a cardiac pacing room, endoscopy suite, theatre etc, it is recommended that temporary warning signs be placed at all entrances to the controlled area prior to an examination and removed at the end. This maintains the correct designation of the area at all times, allowing it to be used for non-radiation work when the mobile equipment is not in use.

Protection of staff and members of the public

3.33 Only persons whose presence is essential should remain in an x-ray room when radiation is being generated; where possible, they should stand well away from the radiation beam and, if one has been provided, behind a protective screen.

3.34 Any member of staff not behind protective screens should wear appropriate protective clothing, as identified by the risk assessment and as specified in the local rules. They should stand as far away from the patient as the effective discharge of their duties allows. These measures are particularly important in theatres and with cardiological and interventional procedures, where there are likely to be several people in the room.

3.35 Staff must not expose parts of their body, even if protected, to the unattenuated primary beam.

3.36 Staff training should cover the identification of both the source of scattered radiation and the direction in which the scatter is most likely to go. This provides staff with the knowledge they require to allow them to occupy the areas of least exposure in any given exposure situation. In general, the greatest intensity of scattered radiation will be from the side of the patient directly exposed to radiation (backscatter) rather than from the side nearest to the image receptor.

3.37 For examinations where the Operator's extremities or eyes may be close to the incident x-ray beam or in an area of greater intensity of scattered radiation, the risk assessment must determine whether there is a need for extremity and eye monitoring (see paragraphs 1.153–1.166, *Personal monitoring for external radiation*), and the advice of the RPA must be sought.

3.38 The doors of x-ray rooms should be closed during examinations and equipment testing. It may be appropriate to lock certain x-ray room doors (where clinical considerations allow), to prevent unauthorised entry where the door is remote from the Operator and outside their immediate field of view (see paragraph 3.25). This would also be necessary if the x-ray beam were ever directed towards the door.

3.39 The radiation beam size should be kept to a minimum in order to limit the patient dose and reduce scattered radiation. The beam should be directed away from adjacent occupied areas if they are not protected by adequate primary shielding. The unshielded primary beam should not be directed at doors, protected areas of the control panel, or mobile protective screens if at all possible. If this exposure geometry could potentially be used, it must be considered in the risk assessment.

3.40 If a carer or comforter is not available (see paragraphs 3.61–3.64), patient support may be provided by a member of staff. This scenario must be included in the radiation risk assessment, and consultation with the RPA is essential. The risk assessment must consider the possibility of the staff member being pregnant and estimate the frequency with which this practice is likely to be required.

3.41 Protection arrangements for hospital staff who support patients must be defined in the local rules. Particular attention should be given to the arrangement's high workload areas, such as paediatrics. The protection arrangements for restricting or monitoring the exposure of this staff must be based on a risk assessment for each type of procedure and involve consultation with the RPA.

3.42 Where it is necessary to leave a person who is untrained in radiation protection alone in an x-ray room, e.g. patients, those accompanying patients, contractors, etc, the equipment should be left in a 'safe' condition where the single-step activation of hand or foot exposure switches is disabled so it will not trigger x-rays.

Personal protective equipment

3.43 General advice on PPE, including the standards to which it must comply, its storage, and quality assurance (QA), can be found in paragraphs 1.72–1.78.

3.44 It is for the Employer's radiation risk assessment to determine what PPE is necessary for staff, however, it is likely that some kind of PPE will be necessary for all staff in the controlled area during interventional or fluoroscopy

Table 3.1. Suggested PPE requirements in radiology and fluoroscopically guided interventions. Reproduced from BIR (2016) with permission

Application	Annual whole body dose level (mSv)	Workload for individual employee (DAP) (Gy cm^2)	Recommended apron (mm Pb eq)	Recommended thyroid collar (mm Pb eq)
General Radiography[1]	<2 (under PPE)	<1200	0.25	None
General fluoroscopy[2]	<2 (under PPE)	<1200	0.25	None
Interventional radiology and cardiology[3]	<2 (under PPE) <20 (outside PPE)	<50 000	0.25	0.35
CT interventional work[4]			0.35	0.35/0.5
Operating theatre[5]	<2 (under PPE)	<1200	0.25	None

[1] Includes mobile radiography.
[2] If dose level of workload exceeded then thyroid collar recommended. If the average kV exceeds 100, then a 0.35 mm apron is required.
[3] If the dose level or workload is exceeded then additional shielding or a thicker apron is recommended. If the average kV exceeds 100, then a 0.35 mm apron is required.
[4] Assuming work is carried out at >100 kV
[5] If dose level or workload is exceeded, a thyroid collar is recommended.

procedures and for the Operator undertaking mobile radiography. PPE may also be required for carers and comforters (see paragraphs 3.61–3.64), or for members of staff providing patient assistance during any examination.

3.45 Gloves, aprons, thyroid shields, and glasses are designed to protect the wearer only from scattered radiation. They will not provide adequate protection from the primary beam.

3.46 Suggested PPE requirements in radiology and fluoroscopically guided interventions are given in the BIR publication *Personal Protective Equipment for Diagnostic X-Ray Use* (BIR 2016), which is reproduced with permission in table 3.1. Note that these suggestions do not preclude the application of ALARP to an individual employee but provide a suggested starting point for a risk assessment on PPE requirements.

3.47 A range of apron sizes must be available to suit all staff who may use them. Consideration should be given to providing half-body aprons (vests) in combination with a lead skirt to reduce the weight on the wearer's back for lengthy examinations. Only very rarely can the use of backless aprons be justified, as these leave large areas of the body unprotected from both direct scatter from the patient and tertiary scatter from walls, etc.

3.48 If required, gloves should be available with protection equivalent throughout both the front and back (including fingers and wrist) to not less than 0.25 mm lead for x-rays up to 150 kV (BIR 2016). Only very rarely will disposable surgical gloves with lesser amounts of lead shielding (e.g. 0.07 mm) be justified. At 70 kV_p, such thin gloves transmit at least seven times the dose of 0.25 mm lead gloves and can give a false sense of security, leading to poor practice and a higher skin dose.

3.49 Eye protectors must be used as indicated by the radiation risk assessment. They are most likely to be required in cardiology and interventional radiology, where the dose to the eye can be significant, and there are examinations where the ceiling-suspended eye shield cannot be used for clinical reasons. Prescription protective eyewear is available; the unique needs of the individual should be considered when purchasing protective eyewear.

Protection of the patient

3.50 While it is necessary to optimise the exposure of all patients, IR(ME)R requires that particular attention be paid to paediatric patients and patients who are (or may be) pregnant at the time of the examination.

3.51 For patients known or likely to be pregnant, where the examination has been justified on the basis of clinical urgency and involves irradiation of the abdomen, Operators must optimise the technique to minimise irradiation of the foetus. Radiography of areas remote from the foetus, e.g. the chest, skull, or hand, may be carried out safely at any time during pregnancy with good beam collimation and properly shielded equipment. The use of out-of-beam shielding for pregnant patients is not recommended (BIR 2020).

3.52 The beam should only be directed towards the body part under investigation. In particular, the gonads and breasts should be avoided, unless this is essential to the conduct of the examination. This should be borne in mind in the examination of the limbs, especially the hands, with the patient in the sitting position.

3.53 The direction in which the radiation passes through the patient should be optimised, since it can greatly influence the distribution of the absorbed dose to sensitive organs. For example, using posteroanterior (PA) rather than anteroposterior (AP) projections of the chest and left (rather than right) lateral lumbar spine views (left side of the patient closest to the detector) on adult patients can reduce the effective dose from these exams by 30%–40% (Nicholson *et al* 1999).

3.54 Strict limitation of field size (or of irradiated length in computed tomography (CT)) to the area necessary for the particular examination should be routinely practised. It is particularly important that the field size be restricted to the essential area for children (ICRP 2013b). The field size should always be smaller than the detector size to confirm collimation (except at the chest wall in mammography). When the equipment incorporates detector sensors to provide automatic collimation, Operators should confirm and optimise collimation using a light beam. All radiographic and fluoroscopic automatic sensing and collimation systems, particularly those with pre- and post-patient collimation, should be subject to routine QC and regular maintenance to ensure their correct operation. Active collimation should be used during helical CT when available.

3.55 Poor collimation in radiography, fluoroscopy, and CT leads to unnecessary irradiation of the patient as well as increased scattered radiation and reduced image quality. Automatic collimation of the image receptor and limitation of the beam by adjustable collimators such as light beam diaphragms (LBDs) will largely eliminate one of the major sources of unnecessary radiation. With fluoroscopy, laser positioning and virtual collimation devices should be used if available to enable radiation-free positioning and collimation.

3.56 Where a variety of equipment is available within a department, examinations of high-risk groups, such as paediatric patients and pregnant patients, should be undertaken on the equipment most capable of providing optimal performance for the examination. This may include preferentially using a direct digital x-ray unit over computed radiography or selecting a CT scanner with iterative reconstruction over one that uses filtered back projection.

3.57 CT scanners may have dose-reduction options available (Automatic Exposure Control, or AEC). CT AECs are now ubiquitous and modulate the exposure across the patient based on the initially Scanned Projection Radiograph (SPR). This has been demonstrated to reduce patient dose while maintaining image quality (McCollough *et al* 2009). The most sophisticated CT AEC option can modulate the exposure multiple times in a single rotation. Some can be instructed to reduce the dose in areas containing radiosensitive organs for further reduction of the patient's dose. Some CT scanners have an automated kV selection and will suggest a kV for the examination that will maintain the image contrast-to-noise ratio based on the initial SPR. This will often lead to a reduced patient dose (Bodelle *et al* 2015). Iterative reconstruction methods can produce diagnostic quality images with less exposure if set up to do so (Beister *et al* 2012). Consideration should be given to the routine use of these options where they are available and appropriate for the examination. An MPE must be consulted, as they will have knowledge of how the local system works and whether it will be beneficial with respect to the intended application.

3.58 The use of patient contact shielding in diagnostic radiology and fluoroscopically guided interventions is not generally recommended (BIR 2020). There will be very few specific situations where this does not apply; any use of patient contact shielding should be justified on a case-by-case basis, with the Operator using professional judgement to weigh the benefits and risks (BIR 2020).

3.59 The use of gonad shields is not recommended for any examination or any patient cohort (AAPM 2019, BIR 2020).

3.60 Human subjects must not be used for training in radiography or for optimising new examination techniques or imaging technology. Anthropomorphic phantoms and test objects should be used when necessary.

Protection of carers and comforters

3.61 Children, anaesthetised patients, uncooperative patients, and those patients with additional support needs sometimes need to be supported for radiography. The use of mechanical devices to ensure immobilisation should be avoided, as this challenges the patient's rights and freedom of action (SCoR 2018). Any proposed use of mechanical devices for immobilisation should be discussed within the multidisciplinary team (MDT), with the Referrer, and with the patient or their parent, guardian, or carer (SCoR 2018). If mechanical devices cannot be used or the support required is not of a physical nature, the patient's parent, guardian, or carer are the preferred persons to offer patient support and would act in the capacity of a carer and comforter. The exposure of a carer and comforter during an examination should be considered by the IR(ME)R Practitioner when justifying the examination, and a dose constraint must be set in the Employer's procedures. It is suggested that a dose constraint of 5 mSv can be considered appropriate for most circumstances (DHSC 2017).

3.62 It is preferable to avoid designating a pregnant individual as a carer and comforter; however, this may not always be practicable (BIR *et al* 2020). An MPE should assess the potential dose to the foetus as well as the carer and comforter to assist the IR(ME)R Practitioner with justification. It may be appropriate to have a reduced dose constraint for pregnant carers and comforters.

3.63 The carer and comforter must undertake this role knowingly and willingly. The benefits and risks must be explained to the individual undertaking the role prior to the examination by the Operator, or through some other means, such as an information sheet. The Employer must make suitable arrangements for providing this information in their procedures.

3.64 The carer and comforter must be given appropriate PPE and advice on where they should position themselves during the examination.

Diagnostic reference levels

3.65 The Employer must establish diagnostic reference levels (DRLs) for all standard diagnostic x-ray examinations and for interventional radiology procedures, where appropriate. While patient dose values for standard diagnostic x-ray examinations principally vary only with patient size, for interventional procedures, the wide variation in the complexity of cases often has a significant influence on the dose. Section 4 of ICRP Publication 135 (ICRP 2017) gives further useful guidance on the use of DRLs for interventional radiology. DRLs may be based on the representative patient doses measured within the QA programme (see paragraph 4.45), with due regard to European and national DRLs where these are available (UKHSA 2022). The dose quantity used for the DRL should be appropriate to the type of equipment and examination. For example, dose-area product (DAP) would be appropriate for radiographs, whereas DAP, air kerma at the patient entrance reference point, and exposure time could all be used for diagnostic fluoroscopy examinations (ICRP 2017). (*N.B. at diagnostic energies, air kerma is for all practical purposes equal to absorbed-dose-to-air, and so the terms kerma-area product (KAP) and dose-area product (DAP) are interchangeable in radiology*.) Exposure time is a poor metric for interventional procedures (ICRP 2017). Volume CT dose index (CTDIvol) and dose-length product (DLP) may be used in CT (ICRP 2017). An MPE must be asked to advise on the appropriate quantities and values for DRLs.

3.66 Local DRLs must be regularly reviewed (IR(ME)R 2017, IR(ME)R(NI) 2018) by the employer (e.g. at a medical exposures committee); where examinations using a particular x-ray unit or performed by an individual Operator regularly exceed the appropriate local DRL, this should be brought to the attention of the Operator, and IR(ME)R Practitioner, and the reason should be investigated, with the MPE if necessary. Consequential corrective action might involve improvements to equipment performance, additional Operator training, or a reassessment of the local DRL.

3.67 Where the revised local DRL exceeds the national DRL, an extensive review of examination practice and equipment capability should be undertaken. This should involve an appropriate multidisciplinary team (MDT), including IR(ME)R Practitioners, Operators and an MPE, all of whom should be familiar with the examination and the equipment on which it is performed. Attempts must be made to optimise the examination to ensure that patient doses are as low as reasonably practicable while achieving the clinical intent. Where the revised local DRL continues to exceed the national DRL, the reasons must be identified. Where the higher levels of dose are clinically justified (i.e. they involve a conscious decision relating to the level of image quality required to address the clinical question or to suit the experience of those reporting or using the images), this justification should be recorded. Where the reason is related to the performance of the x-ray equipment, a risk assessment should be undertaken in consultation with the MDT, considering radiation risk and clinical benefit from continual use of the system for the procedures in question. The MDT should then make an appropriate recommendation to the employer. This might include considering the following options:

(a) Continual clinical use of the equipment is justified (benefit outweighs risk), with a local DRL above the national DRL. Review regularly.

(b) Limit clinical use of the equipment to those procedures where benefit significantly outweighs risk or to those procedures where the national DRLs are not exceeded.
(c) Recommend to the employer the replacement of the equipment with equipment that will give doses below the national DRL and recommend that the risk be added to the Employer's risk register or equivalent. Continual use for all or a limited range of procedures may be justified until the equipment is replaced.
(d) Recommend suspension of the use of the equipment, and the risks to patients arising from this action should be added to the Employer's risk register or equivalent.

Patients of childbearing capacity

3.68 Patients of childbearing capacity having an examination of the lower abdomen or pelvis should be asked if there is any possibility of pregnancy. The age range for inquiries must be defined in the Employer's procedures. A range of 12–55 is recommended in the ARSAC *Notes for Guidance* (ARSAC 2023), and the IR (ME)R guidance (BIR *et al* 2020) confirms this is typical. However, the Employer may choose a wider range if they deem it appropriate. Guidance on pre-procedure pregnancy checking in Under 16s has been published by the Royal College of Paediatrics and Child Health (RCPCH 2012). The responsibility for this action should be defined within the Employer's 'pregnancy enquiry' procedure and should be that of an Operator. The action, response, and decision to proceed or otherwise must be recorded, as required by the Employer's procedures. If pregnancy is suspected, the Practitioner must consider this information specifically before authorising that the examination can proceed. This includes cases where justification has previously been made through a pre-arranged procedure.

3.69 The Employer's procedures must identify who has the responsibility for confirming pregnancy status prior to any examination involving a general anaesthetic or sedation. Emergency examinations involving patients who are unable to confirm their pregnancy status must also be addressed in the Employer's procedures.

3.70 Notices (with translations into the common local languages) should also be displayed in x-ray departments to alert patients to the importance of informing radiographers if they may be pregnant.

3.71 The procedures for identifying potentially pregnant patients should include a policy covering examinations that may give a dose to the foetus in excess of some tens of milligray (mGy). Such situations may indicate deferment of the examination to within a period of 10 days following the onset of menstruation, when conception is unlikely to have occurred, often known as the 'Ten Day Rule'. If deferment of the examination is not appropriate, e.g. due to the urgency of the procedure or the lack of a regular menstrual cycle, the

Practitioner should treat the patient as pregnant when justifying the exposure. The Employer's 'pregnancy enquiry' procedure should indicate the course of action to take. Further guidance on managing examinations for patients who are or may be pregnant is given in *Protection of Pregnant Patients During Diagnostic Medical Exposures to Ionising Radiation* (HPA, RCR and CoR 2009).

Common considerations for all modalities

3.72 In advance of a procedure taking place, where practicable, patients should be given adequate instructions as to what is expected of them for their part in the examination, in terms of positioning, breath holding, etc. This may be in the form of verbal advice or a written card with instructions in an appropriate language.

3.73 Attention should be paid to image viewing conditions. Monitors used for reporting must be well calibrated and tested as part of a QA programme to maintain optimal performance, and there should be sufficient control over the environment around the monitor to ensure optimal viewing conditions at any time of day and for the unique preferences of the individual.

3.74 A value that is representative of the patient's dose must be recorded after each medical exposure. The Employer's procedures must define the dose quantity that will be recorded (e.g. DAP, DLP etc.) and where it will be recorded (i.e. in a dose management system (DMS), radiology information system (RIS), or patient information system). In those cases where a standard protocol is followed with no variation in the amount of radiation used from patient to patient, no further information needs to be recorded other than an indication that the standard protocol was followed. This only applies where a record of the standard protocol exists.

3.75 Adequate commissioning tests must be performed on the picture archiving and communication system (PACS). In particular, this should include verification of the accuracy of any electronic measurement device (i.e. ruler, region of interest etc) used at this end point of the imaging chain.

Additional precautions

Radiography

3.76 Exposure factors and the selection of receptor (e.g. table or Bucky stand) should be checked by the Operator on each occasion before an examination is performed. Where AEC devices are used, the Operator should be aware of the typical ranges of post-exposure mAs values for the projections undertaken. Where anatomically programmed radiography (APR) is used, the Operator should check that the programmed factors are suitable for the individual patient. Where the equipment allows, the backup mAs values should be programmed for each APR so as to minimise the likelihood of excessive exposure in the event of an equipment malfunction or a patient or equipment positioning error. These should be set in consultation with an MPE when APRs are first created and reviewed if modified.

3.77 Protocols for each type of radiographic examination carried out in an x-ray room should be available in the room. These must be specific to the particular room. If more than one x-ray tube may be used to carry out a particular x-ray examination in a room, separate protocols must be defined for each piece of equipment. The protocol should include information on image receptor-focus distance, radiographic exposure factors, and the selection of equipment options, e.g. AEC, for each radiographic projection.

3.78 For fixed installations, the x-ray exposure should be initiated from the control panel, and the exposure switch should be so situated that the Operator cannot leave the protected area during the exposure. For mobile radiography, the exposure switch should allow the Operator to stand at least 2 m from the x-ray tube and patient.

3.79 The imaging system should be chosen to ensure that patient doses are kept ALARP and are appropriate for the production of the required diagnostic information. Doses to patients and staff should be reduced by using low attenuation materials such as carbon fibre in ancillary equipment, e.g. the table.

3.80 The focal spot-to-skin distance should never be less than 30 cm for any x-ray examination at any beam angle, and preferably not less than 45 cm when stationary equipment is used. For radiography of the chest, the distance should not be less than 60 cm (NRPB *et al* 1988).

3.81 Digital radiographic equipment, both computed radiography (photostimulable phosphor radiography) and direct digital radiography, allows images to be produced over a very wide range of doses per image. Regular QC tests on this equipment should include measurements of dose per image for all commonly used sets of programmed exposure factors, and Operators should be aware of these values.

3.82 Exposure factors should be chosen according to local protocols, and the equipment detector dose indicator (DDI) should be checked post-exposure against normal ranges for that examination to ensure an excessive exposure was not used. If clinical DDIs exceed the normal ranges regularly, the advice of an MPE should be sought.

Fluoroscopy

3.83 Fluoroscopy should be carried out only when indicated by the examination protocol, usually when radiography alone is not expected to provide the required information, e.g. in dynamic studies. Fluoroscopy should never be used as an aid to patient positioning for radiography.

3.84 Prior to the start of any examination using fluoroscopy, there may be merit in running through a pre-examination checklist to ensure doses are kept ALARP. This simple checklist could be used to ensure that all staff present are wearing appropriate PPE, are wearing all of the dose monitors they should be, that any in-room protective devices (ceiling-suspended eye shields, mobile shields, table side shields etc) are in use, and that staff are all appropriately positioned. Any important point from the local rules can be added to this checklist, which should be written with input from the RPA, MPE, and MDT. The checklist is likely to be of greatest benefit for complex fluoroscopy examinations involving lengthy exposures with many staff present.

3.85 Fluoroscopic equipment must be under the direct control of an adequately trained Operator. The Operator should communicate to others in the room in a clear and unambiguous manner when they are about to initiate an exposure and when it has terminated. This will enable other people to position themselves to ensure that their doses are ALARP.

3.86 Fluoroscopy should be conducted for short periods rather than continuously if this is clinically feasible. Total fluoroscopy dose can be reduced by effective use of last image hold and cineloop replay features, a low dose rate, and pulsed fluoroscopy options. Under no circumstance should the x-ray tube be energised when the person carrying out the examination is not looking at the display monitor, as this could involve the irradiation of the patient without attaining any diagnostic information.

3.87 Full use should be made of the DAP indication, air kerma/skin dose rate, and displayed exposure time to limit patient dose. Operators should monitor these values during each procedure in relation to the diagnostic reference level for that examination. Some indicator, be it an accumulated DAP, skin dose, and/or exposure time, should be recorded for each patient.

3.88 An evidence based review of fluoroscopic equipment testing found that the absorbed dose rate at the skin of a standard patient (18.5–20 cm water equivalent thickness) did not exceed 42 mGy min^{-1} for fixed image intensifier systems, 24 mGy min^{-1} for mobile image intensifier systems, and 35 mGy min^{-1} for fixed flat panel detector systems. These values were derived from data pertaining to any available field size for exposures using continuous fluoroscopy in a normal dose mode. An MPE should be consulted where local values exceed these for an equivalent test. Guidance on the skin entrance dose rates at which adequate image quality should be achievable for modern fluoroscopy systems is available.

3.89 As a guideline, the average skin entrance dose rate for a standard patient (18.5–20 cm of water) on the largest available field size using continuous fluoroscopy is shown in table 3.2, along with the average effect of applying a low and high dose setting for fixed image intensifiers and flat panel systems (Worrall *et al* 2019). Measurement and consideration of the skin entrance dose/frame in

Table 3.2. Guideline skin entrance dose rate limits for a standard patient (18.5–20 cm water), largest field size, continuous fluoroscopy.

	Guideline skin entrance dose rate limit ($mGy\ min^{-1}$)		
	Dose setting		
System type	Low	Normal	High
Flat panel detector systems	0.6 × normal	12.5	1.5 × normal
Image intensifier systems	0.6 × normal	10	2.0 × normal

digital subtraction angiography and/or digital cineradiography and the maximum obtainable skin entrance dose rates in fluoroscopy should also be included in the commissioning and optimisation process.

3.90 Automatic dose rate control systems feature multiple selectable kV versus mA curves; the optimum curve should be chosen for the examination, e.g. iodine or barium curves for examinations involving the use of contrast media. The Operator should be aware of the relative dose rates provided by the available automatic settings and should employ the lowest appropriate dose rate for the clinical objective. The values of the detector input and skin entrance dose rates should be measured during QC.

3.91 Care should be taken to minimise the area of the x-ray field consistent with the requirements of the examination. For image intensifier systems, collimation should be used rather than magnification, since reducing the field of view by using intensifier magnification increases the dose rate selected when using automatic dose rate control. It should be noted, however, that collimation down to very small areas may also produce unexpected increases in the automatic dose rate as the collimator may shield the sensor area of the automatic dose rate control system. Disabling the automatic dose rate control system by locking the fluoroscopic exposure factors is therefore recommended prior to using tight collimation for small areas, where possible.

3.92 Where thin body parts are being examined, e.g. in paediatrics, the literature is divided on whether the anti-scatter grid should be removed. Some papers (Lu *et al* 2005, Partridge *et al* 2006, King *et al* 2011) conclude that removal of the anti-scatter grid can result in a significant dose saving without significantly degrading the image. However, Strauss *et al* (2015) performed a study using anaesthetised piglets as a surrogate and concluded that the grid was necessary to achieve the highest levels of image quality and that this was related to advanced image processing. It is likely that the decision as to whether it is preferable to remove the grid is dependent upon the exact equipment in use and the nature of the examination. It is vital that an MPE be involved in this optimisation decision.

3.93 The ambient light level in the examination room should be maintained at a constant level, as the eye will take time to respond to any changes. Modern monitors have anti-reflective coatings and high brightness, so the historic practice of setting low ambient light levels is no longer necessary. Image display monitors should be positioned so that the person carrying out the examination can fully appreciate the detail in the image.

3.94 The focal spot-to-skin distance should never be less than 30 cm, and when stationary equipment is used, preferably not less than 45 cm (BSI 2009).

3.95 During fluoroscopy with the patient in the erect position, the radiologist (or other clinician), Operator and any other essential person(s) who are present should ensure that they are protected by the detector housing and, if provided, the protective drapes suspended from it.

3.96 During fluoroscopy, with the patient in the horizontal position, the use of an undercouch x-ray tube is recommended, with the detector as close to the patient as possible. If the examination being undertaken specifically requires the use of an overcouch or C-arm tube, then particular care is necessary both to prevent inadvertent insertion of the Operator's hands or head into the primary beam and to avoid exposure to scattered radiation, especially for the Operator's eyes.

3.97 Where lateral beams are used, staff should work on the beam exit side of the patient. Whenever possible, examinations should be conducted from the remote control panel. When it is necessary for persons to be near the table, e.g. for specialised examinations involving catheterisation, they should make every reasonable effort to shield themselves from scattered radiation by using lead aprons, eye and thyroid shields, or mobile or ceiling-suspended shields as appropriate. Doses to the lower limbs can be significant, so extra shielding to protect them may be necessary (e.g. table side lead drapes). Wherever possible, during examinations known to give high doses to patients or staff, staff should stand away from the patient during image acquisition 'runs'. The likely eye doses of such staff should be established, e.g., through initial scatter calculations and/or published data followed up with monitoring. If appropriate, the RPA may recommend that protective eyewear be worn.

3.98 During fluoroscopy, palpation with the hand should be reduced to a minimum. It should only be undertaken on the image receptor side of the patient and therefore should not be carried out at all with an overcouch tube. For overcouch tubes' settings, automatic compression devices should be available. Where a protective glove is worn during procedures in which the hands or forearms are close to the beam or in areas of high scatter, it should have a lead equivalent thickness of at least 0.25 mm for up to 150 kV (BIR 2016). The need to wear a fingertip dosemeter should be considered and discussed with the RPA (see paragraph 1.153–1.166, *Personal monitoring for external radiation*) (Martin *et al* 2018). The presence of the glove used for palpation can cause a significant increase in the absorbed dose rate when an automatic dose rate control system is used. Care should be taken to minimise this effect by keeping the glove outside the sensing area of the automatic dose rate control system (which is normally close to the centre of the field of view) and/or by locking the fluoroscopic factors before palpation begins, if possible.

Fluoroscopically guided interventions

3.99 The use of a pre-examination checklist, as discussed in paragraph 3.84, may be of particular benefit for fluoroscopically guided interventions. Its use should be discussed with the RPA, MPE, and MDT.

3.100 Complex interventional procedures involving long fluoroscopy times and the acquisition of numerous images can result in skin doses to the patient that exceed the threshold for radiation-induced skin injury (ICRP85 *Avoidance of Radiation Injuries from Interventional Procedures* (ICRP 2020)). For such examinations, standard operating procedures and clinical protocols should be established to minimise the likelihood of this outcome. Particular attention should be paid to optimising fluoroscopically guided interventions. The advice of an MPE should be sought regarding those examinations for which high skin doses are likely.

3.101 Dose assessment should be carried out for any examinations that risk radiation-induced skin injury. A system should be established for recording and evaluating the measured or estimated skin absorbed dose in the patient's record so that any necessary patient follow up can be arranged. The ICRP recommends that this system be triggered at a system displayed skin dose of 3 Gy, a dose-area product (DAP of 500 Gy cm^2) or an air kerma at the patient entrance reference point of 5 Gy (ICRP 2013a). An MPE must be involved in the implementation of this system, as it will require local knowledge of how the equipment's displayed skin dose or air kerma at the patient entrance reference point is calibrated. This also allows the cumulative dose from multiple examinations over a short period of time to be evaluated and the patient's follow up adjusted accordingly. The patient should be counselled regarding potential symptoms and risks prior to the procedure (ICRP 2013a).

3.102 For long and complex procedures, more than one projection should be used during the course of the procedure wherever possible to reduce the maximum localised radiation dose to the skin. Dose-reduction features, such as additional beam filtration (Nicholson *et al* 2000), pulsed fluoroscopy, cineloop, fluorograb, and virtual collimation, should be used where available.

3.103 The Employer must establish appropriate eye and extremity radiation monitoring arrangements for interventional radiology staff in consultation with the RPA and review the effectiveness of the measures for their radiation protection. Advice is given in *Guidance on the Personal Monitoring Requirements for Personnel Working in Healthcare* (Martin *et al* 2018).

3.104 Ceiling suspended eye shields are the most effective means of reducing the eye dose for the Operator. Different fluoroscopically guided interventions require different Operator positioning with respect to the patient; therefore, the suspended eye shield must cover a range of positions so as to be useful across all the fluoroscopically guided interventions undertaken locally. During installation, consideration should be given to the range of positions that the eye shield can reach. Where possible, the eye shield should not be installed on the same ceiling rails as any other item of equipment, such as a contrast pump, as this will restrict the range of positions the eye shield can reach. Where different fluoroscopically guided interventions have the Operator standing on different sides of the table, consideration should be given to installing two ceiling-suspended eye shields if one cannot reach both sides in order to maximise the number of examinations for which a ceiling-suspended eye shield can be used.

Precautions with mobile equipment

3.105 When mobile x-ray units are used, Operators should ensure that there are as few people in the vicinity as possible and that those that are in the vicinity of the patient are afforded adequate protection from the primary beam and from scattered radiation. Operators should wear a protective apron of at least 0.25 mm lead equivalent thickness (BIR 2016) and stand as far back as practicable when initiating the exposure.

3.106 The focal spot-to-skin distance should never be less than 30 cm (BSI 2009).

3.107 For radiography with mobile x-ray units on wards, in A&E departments, etc., the Employer, in consultation with the RPA, must devise a procedure to ensure adequate protection of the patient, the Operator and all those in the vicinity of the examination. This should normally be agreed upon with ward management and documented in the local rules. Particular care is necessary with regard to the direction (especially for lateral projections) and size of the radiation beam, as partition walls may not provide sufficient x-ray attenuation. Local shielding may be needed (e.g. a lead apron over the bed-head). The immediate vicinity of the patient may be a controlled area; control of access may be exercised by the Operator giving a verbal instruction and ensuring that staff and visitors leave the area. Adequate instruction for all staff groups likely to be in the vicinity must be given. Portable, free-standing radiation warning signs may also be used to avoid inadvertent access to the controlled area while the x-ray procedure is being performed. The RPA should advise on when, if ever, these are necessary.

3.108 Mobile equipment should not be capable of being used by unauthorised persons and should be left in a 'safe' condition when it is left unattended.

Precautions with computed tomography scanners

3.109 During 'warm up' and detector calibration procedures, persons should not be allowed to enter or remain in the scanner room. Local arrangements must be put in place to ensure that all staff in the vicinity are aware that warm up and detector calibration procedures are underway and that access to the room is prohibited until they are complete.

3.110 The Employer must consult the RPA about protection for staff who need to remain in the examination room during clinical procedures. This is particularly important for CT fluoroscopy and interventional techniques (see paragraphs 3.118–3.125).

3.111 Although CT scanners do not have an exposure switch that needs to be pressed continuously during exposure, an Operator should remain at the control panel of a CT scanner throughout the scanning sequence. From this position, the Operator should have a clear view of the patient and of all doors leading into the scanner room; where this is not possible, consideration should be given to locking the doors to prevent inadvertent entry (see paragraph 3.25).

3.112 In the event that someone does inadvertently enter the CT scanner room during an examination, the Operator should not terminate the exposure, as the dose to the patient from a partial or complete repeat of the examination is much higher than that received by the individual who entered the room. Such instances should be investigated afterwards in line with local procedures (see paragraphs 1.82–1.87 *Contingency plans*). The scenario must be addressed in the risk assessment.

3.113 It is advised that the optimisation of CT protocols be undertaken by a multidisciplinary team (COMARE 2014), which COMARE refers to as 'radiation protection champions'—an Imaging Optimisation Team that consists of a radiologist, a radiographer, and a medical physicist. The Department of Health recommends (Department of Health 2016) that these individuals be appropriately trained, command sufficient authority and seniority, and have details of their role, including available resources and management support, in writing from their Employer. An appropriate forum for the Imaging Optimisation Team would be as a sub-group of the MEC or the RSC.

3.114 Where available, full advantage should be taken of any capacity the CT scanner has for automated exposure factor adjustment (e.g. mA modulation and kV selection). Where the scanner has the option for iterative reconstruction, this allows for a significant patient dose reduction (Beister *et al* 2012) and should be used in preference to filtered back projection techniques, except where there is a time delay in reconstruction that could have clinical implications for the patient.

3.115 For cardiac imaging, prospective gating offers a significant dose savings to the patient and should be used in preference to retrospective gating where an appropriate heart rate can be achieved (McCollough *et al* 2009). In general, a cardiac CT service should operate using the CT scanner with the largest z-detector range size available to reduce the number of scans required for full cardiac imaging.

3.116 Care should be taken to minimise exposure to the eyes of the patient. The dose to the lens tissue can often be substantially reduced by the angulation of the gantry to exclude the eyes from the primary beam during head examinations (Nikupaavo *et al* 2015), but only where clinically feasible. Bismuth eye shields can be used, but these can result in image artefacts (McCollough *et al* 2009). An MPE should be consulted as to the best means of limiting the eye dose without interfering with the clinical aim of the examination.

3.117 If a dose alert system is available on the CT scanner, it must be set up with MPE involvement.

Computed tomography fluoroscopy, and interventional techniques

3.118 All Operators must receive adequate training before any work using CT fluoroscopy or interventional techniques begins. Training in the use of the CT scanner for non-fluoroscopy or interventional work alone cannot be considered adequate. As there may be Operators in the CT scanning room and at the CT scanning console who both have some control over the exposure, effective communication during the examination is essential.

3.119 The use of CT fluoroscopy or interventional techniques will require a dedicated and thorough risk assessment. The type of procedure, the examination technique, and the expected range of complexity must be accounted for. Patient doses have been shown to vary widely with procedure (Leng *et al* 2011, Yang *et al* 2018), which will result in a similar variation in scattered radiation for the Operator. An RPA must be consulted, and direct observation of procedures will be necessary for the production of an adequate risk assessment.

3.120 The use of remote injection facilities, appropriate spacers, and any options available on the scanner to modify the beam in appropriate locations during fluoroscopy will enable staff to stand away from the patient and restrict the dose to the hands.

3.121 There is a wide range of eye doses for the Operator reported in the literature. For example, Saidatul *et al* (Saidatul *et al* 2010) report dose rates of 20–40 μSv min^{-1} to the eye, with doses for individual procedures as high as 260 μGy. Measurements of scatter dose rates from phantoms have given dose rates of 2 μGy s^{-1} to the eyes with the Operator positioned 40 cm from the scan plane (Keat 2001). Olerud *et al* (2002) found the eye dose for an individual procedure to be as high as 400 μGy. Additional shielding in the scan plane and PPE for the Operator should be considered to reduce eye doses. Eye dose monitoring will be necessary in all cases initially and on a continuing basis (Martin *et al* 2018), if the RPA so recommends following the risk assessment and preliminary findings. Guidance on the personal monitoring requirements for personnel working in healthcare (Martin *et al* 2018) contains guidance on eye dose monitoring during flurosocopic and interventional CT, including suggested protection factors to be applied to the results where lead glasses are used.

3.122 There is a wide range of doses for the fingers reported in the literature. Nickoloff *et al* (2000) report dose rates of up to 1.5 mGy min^{-1}. Saidatul *et al* (2010) report doses as high as 354 μGy in a single procedure, whereas Olerud *et al* (2002) report a dose of 350 mGy in a single procedure. Very high finger doses can be reached in a short time when poor technique results in the Operator's hands being placed in the direct beam (Martin *et al* 2018). In these scenarios, extremity monitoring in fluoroscopic and interventional CT may be of limited value, as a ring dosemeter worn at the base of the finger or a finger stall worn at the tip could easily both miss the area of primary irradiation and report a dose much lower than that received (Martin *et al* 2018). It is therefore very important that the fingers do not enter the primary beam. The Operator must receive training specific to this point, and the use of needle holders should be considered. These allow the Operator to remain well outside of the primary beam while still being able to manipulate the needle. The risk assessment must include an assessment of the dose to the fingers as a result of accidental exposure to the primary beam (based on measurements made directly on the CT scanner).

3.123 The Operator should be aware of the potential for high patient skin doses in CT fluoroscopy when scanning in the same place continuously. As entrance surface dose rates are around 5–7 mGy s^{-1} (Olerud *et al* 2002), they should be mindful of the fluoroscopy time taken to reach a specified dose (e.g. 1 Gy).

3.124 A maximum limit on the mA selectable for CT fluoroscopy should be set locally, if not physically restricted in the equipment design. The RPA and MPE must both be consulted on this decision.

3.125 Under no circumstance should CT fluoroscopy be used when the person carrying out the examination is not looking at the monitor.

The use of equipment outside of specialist areas

3.126 It might be necessary to perform some work with x-rays in areas that were not specifically designed for it, for example, in a theatre. There are additional considerations to be made regarding this work, which can be listed as follows:

(a) Doses to individuals in adjacent areas: rather than radiology professionals, non-radiation workers are more likely to be present in these areas, and there may also be members of the general public. An account of doses should be taken for the individuals in these areas.

(b) Adequate instruction and training: areas outside of radiology will have staff groups that are not familiar with the risks of ionising radiation, the designation of areas, general work practices, warning signs, etc. Where any of these staff will be present in the controlled area, a programme of training will be necessary to provide training proportionate to their role. Often, these staff groups will be large with a high turnover of rotation; the training programme must account for this.

(c) Area designation and access arrangements: area designation must be unambiguous, especially as some or all of the area may retain its original non-radiation use. There must be no confusion as to when an area is designated as controlled and when the access arrangements are in force.

(d) Provision and storage of PPE: this must be provided for the additional staff groups present. The PPE must be available in a range of sizes and should offer protection that is adequate for the level of exposure each staff group is likely to receive (see paragraphs 3.43–3.49). Unless it is practicable for this PPE to be stored in radiology and moved to areas outside of radiology as and when required, provision must be made for the adequate storage of PPE where used.

(e) If an area is to be used for procedures involving x-rays for the long term, consideration should be given to providing as much in-room shielding as possible in the form of ceiling-suspended eye shields, table side lead drapes, mobile protective screens, etc where practicable. An RPA must be consulted.

(f) Where the radiation risk assessment for the area determines a maximum workload that can be safely performed in the area before design constraints are reached, a robust method for the ongoing assessment of workload is required.

(g) Careful consideration should be given to the selection of an appropriate RPS for the area (see paragraphs 1.119–1.126, *Radiation protection supervisor*).

The use of modular facilities

3.127 The use of modular CT and fluoroscopy facilities is becoming increasingly common as a means of temporarily increasing imaging capacity in the short to medium term. This can be useful for reducing patient waiting lists or providing a service during the replacement of existing equipment. There are two models on offer: the simple rental of a modular unit or a fully managed service provided by a third party supplier. In either case, it is essential that the Employer's RPA and MPE are both involved at the planning stage for the hiring of a modular facility.

3.128 All roles and responsibilities must be clearly defined between all parties, for example, the Employer under the IRR (IRR 2017, IRR(NI) 2017) and the IR (ME)R (IR(ME)R 2017, IR(ME)R(NI) 2018), all RPAs, MPEs and duty holders under the IR(ME)R. There should also be a clear means to communicate between Employers and the relevant RPAs and MPEs. Access to the unit by representatives of the manufacturer for any necessary planned preventative maintenance and repair work must be considered.

3.129 The supplier should be able to provide information on the extent of shielding within the modular facility and the results of past environmental monitoring to allow the RPA to advise on a safe location for the facility, the risk assessment, and the extent of the controlled area. Note that many modular facilities do not have shielding on the roof, which could limit the time they can be used adjacent to multistorey buildings (or preclude these locations altogether). The supplier should also provide a copy of the equipment's critical examination.

3.130 The supplier should be able to provide information on equipment performance, including the service schedule, acceptance data, commissioning data, and the results of subsequent QA testing. This will help the Employer's MPE advise on optimisation for the procedures undertaken in the modular facility. If the unit will be on site long enough to require QA, the supplier should provide a protocol for local user QA and the Employer's MPE must receive a copy. If the Employer's contracted physics service is to undertake any routine level B testing of the unit (IPEM 2005), they should have access to the commissioning data to give them something to compare the results to.

3.131 Where it is the Employer's staff that will use the unit, training on the use of the x-ray system must be provided for all operators, and documentation of the training must be kept.

References

AAPM 2019 *Position Statement on the Use of Patient Gonadal and Fetal Shielding 2019* (USA) http://aapm.org/org/policies/details.asp?id=468 (accessed 9 February 2023)

ARSAC 2023 *Notes for Guidance on the Clinical Administration of Radiopharmaceuticals and Use of Sealed Radioactive Sources* (Administration of Radioactive Substances Advisory Committee) https://gov.uk/government/publications/arsac-notes-for-guidance (accessed 25 February 2023)

Beister M, Kolditz D and Kalender W A 2012 Iterative reconstruction methods in x-ray CT *Physica Med.* **28** 94–108

BIR 2016 *Personal Protective Equipment for Diagnostic X-Ray Use* ed P Hiles, H Hughes, D Arthur and C Martin (London: British Institute of Radiology) https://birpublications.org/page/ppe (accessed 25 February 2022)

BIR 2020 *Guidance on Using Shielding on Patients for Diagnostic Radiology Applications* ed P Hiles *et al* (London: British Institute of Radiology) https://bir.org.uk/media/414334/final_-patient_shielding_guidance.pdf (accessed 9 February 2023)

BIR, RCR, IPEM, SCoR and PHE 2020 *IR(ME)R Implications for Clinical Practice in Diagnostic Imaging, Interventional Radiology and Diagnostic Nuclear Medicine* (London) https://www.rcr.ac.uk/our-services/all-our-publications/clinical-radiology-publications/ir-me-r-implications-for-clinical-practice-in-diagnostic-imaging-interventional-radiology-and-diagnostic-nuclear-medicine/ (accessed 14 February 2024)

Bodelle B, Beeres M, Scheithauer S, Wichmann J L, Nour-Eldin N-E A, Vogl T J and Schulz B 2015 Automated tube potential selection as a method of dose reduction for CT of the neck: first clinical results *Am. J. Roentgenol.* **204** 1049–54

BSI 2009 *BS EN 60601-2-54:2009 Particular Requirements for the Basic Safety and Essential Performance of X-ray Equipment for Radiography and Radioscopy* (London: British Standards Institute)

BSI 2014 *BS EN 60601 Medical Electrical Equipment and Systems* (London: British Standards Institute)

COMARE 2014 *Review of Radiation Dose Issues from the Use of CT in the UK* (Didcot: Public Health England) https://gov.uk/government/publications/review-of-radiation-dose-issues-from-the-use-of-ct-in-the-uk (accessed 9 February 2023)

Department of Health 2016 *COMARE 16 DH Expert Working Party Response to the Committee on Medical Aspects of Radiation in the Environment's 16th Report: Patient Radiation Dose Issues Resulting from the Use of CT in the UK* https://gov.uk/government/publications/response-to-the-review-of-radiation-dose-issues-from-ct-scans (accessed 31 January 2023)

DHSC 2017 Guidance to the Ionising Radiation (Medical Exposure) Regulations *Gov.UK Website* (Department of Health and Social Care) https://gov.uk/government/publications/ionising-radiation-medical-exposure-regulations-2017-guidance (accessed 31 January 2023)

HPA, RCR and CoR 2009 *RCE-9 Protection of Pregnant Patients during Diagnostic Medical Exposures to Ionising Radiation* (Health Protection Agency, Royal College of Radiologists, and College of Radiographers) https://www.rcr.ac.uk/our-services/all-our-publications/clinical-radiology-publications/protection-of-pregnant-patients-during-diagnostic-medical-exposures-to-ionising-radiation/ (accessed 14 February 2024)

ICRP 2013a Radiological protection in cardiology *Ann. ICRP* **42** Publication 120 https://www.icrp.org/publication.asp?id=ICRP Publication 120 (accessed 14 February 2024)

ICRP 2013b Radiological protection in paediatric diagnostic and interventional radiology *Ann. ICRP* **42** Publication 121

ICRP 2017 Diagnostic reference levels in medical imaging *Ann. ICRP* **46** Publication 135 https://www.icrp.org/publication.asp?id=ICRP Publication 135 (accessed 14 February 2024)

ICRP 2020 Avoidance of radiation injuries from interventional procedures *Ann. ICRP* **30** Publication 85 https://www.icrp.org/publication.asp?id=ICRP Publication 85 (accessed 14 February 2024)

IPEM 2005 *Report 91: Recommended Standards for the Routine Performance Testing of Diagnostic X-ray Imaging Systems* (York: Institute of Physics and Engineering in Medicine)

IR(ME)R 2017 *The Ionising Radiation (Medical Exposure) Regulations* www.legislation.gov.uk/uksi/2017/1322 (accessed 31 January 2023)

IR(ME)R(NI) 2018 *The Ionising Radiation (Medical Exposure) Regulations (Northern Ireland)* https://legislation.gov.uk/nisr/2018/17 (accessed 31 January 2023)

IRR 2017 *The Ionising Radiations Regulations* https://legislation.gov.uk/uksi/2017/1075 (accessed 31 January 2023)

IRR(NI) 2017 *The Ionising Radiations Regulations (Northern Ireland)* http://legislation.gov.uk/nisr/2017/229 (accessed 31 January 2023).

Keat N 2001 Real-time CT and CT fluoroscopy *Br. J. Radiol.* **74** 1088–90

King J M, Elbakri I A and Reed M 2011 Antiscatter grid use in pediatric digital tomosynthesis imaging *J. Appl. Clin. Med. Phys.* **12** 3641

Leng S, Christner J A, Carlson S K, Jacobsen M, Vrieze T J, Atwell T D and McCollough C H 2011 Radiation dose levels for interventional CT procedures *Am. J. Roentgenol.* **197** W97–103

Lu Z F, Nickoloff E L, Ruzal-Shapiro C B, So J C and Dutta A K 2005 New automated fluoroscopic systems for pediatric applications *J. Appl. Clin. Med. Phys.* **6** 88–105

Martin J C, Temperton D H D, Hughes A and Jupp T 2018 *Guidance on the Personal Monitoring Requirements for Personnel Working in Healthcare* (Bristol: IOP Publishing Ltd)

McCollough C H, Primak A N, Braun N, Kofler J, Yu L and Christner J 2009 Strategies for reducing radiation dose in CT *Radiol. Clin. North Am.* **47** 27–40

NHS Estates 2001 *Facilities for Diagnostic Imaging and Interventional Radiology. HBN 6* (Norwich: HMSO) http://gov.uk/government/publications/facilities-for-diagnostic-imaging-and-interventional-radiology (accessed 9 February 2023)

Nicholson R, Thornton A and Sukumar V 1999 Awareness by radiology staff of the difference in radiation risk from two opposing lateral lumbar spine examinations *Br. J. Radiol.* **72** 221

Nicholson R, Tuffee F and Uthappa M C 2000 Skin sparing in interventional radiology: the effect of copper filtration *Br. J. Radiol.* **73** 36–42

Nickoloff E L, Khandji A and Dutta A 2000 Radiation doses during CT fluoroscopy *Health Phys.* **79** 675–81

Nikupaavo U, Kaasalainen T, Reijonen V, Ahonen S-M and Kortesniemi M 2015 Lens dose in routine head CT: comparison of different optimization methods with anthropomorphic phantoms *Am. J. Roentgenol.* **204** 117–23

NOGG 2021 *Clinical Guideline for the Prevention and Treatment of Osteoporosis, National Osteoporosis Guideline Group Website* (UK: National Osteoporosis Guideline Group - UK) https://nogg.org.uk/full-guideline (accessed 9 February 2023)

NRPB, HSE, Department of Health and Social Security, Department of Health and Social Services (Northern Ireland), Scottish Home and Health Department and Welsh Office 1988 *Guidance Notes for the Protection of Persons against Ionising Radiations Arising from Medical and Dental Use* (Chilton: National Radiological Protection Board)

Olerud H M, Obberg S, Widmark A and Hauser M 2002 Physician and patient radiation dose in various CT guided biopsy protocols *Proc. of the 6th European ALARA Network Workshop on Occupational Exposure Optimisation in the Medical Field and Radiopharmaceutical Industry* **6** 26–35 www.eu-alara.net/images/stories/pdf/program6/Session C/HM_Olerudposter.pdf (accessed 9 February 2023)

Partridge J, McGahan G, Causton S, Bowers M, Mason M, Dalby M and Mitchell A 2006 Radiation dose reduction without compromise of image quality in cardiac angiography and intervention with the use of a flat panel detector without an antiscatter grid *Heart* **92** 507–10

RCPCH 2012 *Pre-procedure Pregnancy Checking in Under 16s: Guidance for Clinicians* (London: Royal College of Paediatrics and Child Health) https://www.rcpch.ac.uk/resources/pre-procedure-pregnancy-checking-under-16s-guidance-clinicians (accessed 9 February 2024)

RCR 2017 *iRefer Guidelines: Making the best use of clinical radiology (8th edition)* (London: Royal College of Radiologists Website) https://www.rcr.ac.uk/our-services/irefer/ (accessed 14 February 2024)

Saidatul A, Azlan C A, Megat Amin M S A, Abdullah B J and Ng K H 2010 A survey of radiation dose to patients and operators during radiofrequency ablation using computed tomography *Biomed. Imag. Interv. J.* **6** e2

SCoR 2018 *Consent: Guidance on Mental Capacity Decisions in Diagnostic Imaging and Radiotherapy* (London) https://sor.org/learning-advice/professional-body-guidance-and-publications/documents-and-publications/policy-guidance-document-library/guidance-on-mental-capacity-decisions-in-diagnosti (accessed 9 February 2023)

Strauss K J, Racadio J M, Abruzzo T A, Johnson N D, Patel M N, Kukreja K U, den Hartog M J H, Hoornaert B P A and Nachabe R 2015 Comparison of pediatric radiation dose and vessel visibility on angiographic systems using piglets as a surrogate: antiscatter grid removal vs. lower detector air kerma settings with a grid—a preclinical investigation *J. Appl. Clin. Med. Phys.* **16** 408–17

Sutton D G, Martin C J, Williams J R and Peet D J 2012 *Radiation Shielding for Diagnostic Radiology* 2nd edn (London: The British Institute of Radiology) https://birpublications.org/doi/10.1259/book.9780905749747 (accessed 31 January 2023)

UKHSA 2022 Guidance: National Diagnostic Reference Levels (NDRLs) from 13 October 2022, GOV.UK Website https://gov.uk/government/publications/diagnostic-radiology-national-diagnostic-reference-levels-ndrls/ndrl (accessed 7 November 2022)

Worrall M *et al* 2019 IPEM Topical Report: an evidence and risk assessment based analysis of the efficacy of quality assurance tests on fluoroscopy units—part I; dosimetry and safety *Phys. Med. Biol.* **64** 195011

Yang K, Ganguli S, DeLorenzo M C, Zheng H, Li X and Liu B 2018 Procedure-specific CT dose and utilization factors for CT-guided interventional procedures *Radiology* **289** 150–7

IOP Publishing

Medical and Dental Guidance Notes (Second Edition)

A good practice guide on all aspects of ionising radiation protection in the clinical environment: IPEM Report 113

John Saunderson, Mohamed Metwaly, William Mairs, Philip Mayles, Lisa Rowley and Mark Worrall

Chapter 4

Diagnostic radiology (excluding dental) and fluoroscopically-guided interventions x-ray equipment management

Scope

4.1 This chapter pertains to equipment designed for use in diagnostic and fluoroscopically guided interventions (with the exception of dedicated dental equipment, which is covered in chapter 6). It covers the specification, installation, maintenance, and testing of both mobile and stationary equipment. Dedicated dental equipment is covered in chapter 6.

General recommendations

4.2 Diagnostic x-ray equipment should be designed, constructed, and installed to comply with the relevant British and International Standards (e.g. BS EN 60601 *Medical Electrical Equipment. General Requirements for Safety* (BSI 2008b) together with the particular and associated collateral standards).

4.3 All medical devices placed on the market should meet the relevant essential requirements for safety and performance of the Medical Devices Directive (European Commission 1993), transposed into United Kingdom law by the Medical Devices Regulations (MDR 2002), and be conformity marked (e.g., CE, UKCA) as appropriate. Meeting the requirements of the Medical Devices Directive is a declaration that the device meets the essential requirements for device safety.

4.4 To ensure safety and the proper operation of the medical electrical equipment, the electrical installation should comply with the current version of BS 7671

doi:10.1088/978-0-7503-2332-1ch4

Requirement for Electrical Installations (BSI 2008a), including section 710—Medical Locations and other special location requirements (e.g., section 717—Mobile or Transportable Units). Guidance is also provided in the Health Technical Memoranda HTM 06–01 and STM 06–01 (HFS 2015, DH 2017). Any new or modified electrical installation should be certified according to BS 7671; a copy of the certificate should be available for inspection by those carrying out acceptance of the installation. Electrical safety testing of medical electrical equipment is outside of the scope of BS 7671 and requires the use of BS EN 62353 (BSI 2014). Where emergency switching-off mushroom buttons have been provided, their function should be tested to confirm power is removed from the associated medical electrical equipment and be in accordance with section 465 of BS 7671.

4.5 The mechanical safety of all fixed x-ray equipment should be ensured at installation. For mobile x-ray equipment, mechanical safety should be checked as part of the process of setting up the equipment after delivery.

4.6 Equipment should be maintained in accordance with the recommendations of the manufacturer (MHRA 2021).

Equipment specification

4.7 The Employer must involve a medical physics expert (MPE) in the technical aspects of equipment specification. The Employer must give the MPE adequate information relating to the intended use of the equipment so they can contribute to an appropriate technical specification. This may include information on the types of procedures that will be performed, the patient cohorts on which the equipment will be used, and the projected patient throughput.

Equipment design

X-ray source assembly

4.8 Every x-ray source assembly (comprising an x-ray tube, an x-ray tube housing, and a beam-limiting device) should be constructed so that, at every rating specified by the manufacturer for that x-ray source assembly, the air kerma from the leakage radiation at a distance from the focal spot of 1 m averaged over an area not exceeding 100 cm^2 does not exceed 1 mGy h^{-1} (BSI 2010). Reference should be made to the leakage technique factors of the x-ray tube that describe the maximum continuous ratings for the x-ray tube, e.g. 125 kV and 3.6 mA for a 450 W tube.

4.9 Every x-ray source assembly should be marked to identify the nominal focal spot position as in BS EN 60601-2-28 (BSI 2010). When an x-ray tube assembly is covered with additional protective or aesthetic covers, then the covers should also be marked with the nominal position of the focal spot. The exception to this is the gantry cover of a CT scanner.

Beam filtration

4.10 The permanent filtration of every x-ray tube assembly should be marked durably and clearly on the housing by the installer (BSI 2010). If the assembly is totally enclosed, e.g. within a CT gantry, a duplicate label must be placed on the exterior of the equipment.

4.11 Every added filter should be marked permanently and clearly in terms of its material composition (e.g. chemical symbol) and thickness (in mm). Materials other than aluminum, e.g. copper, molybdenum, rhodium, and rare-earth materials, may be used as alternative materials for x-ray beam filtration. The aluminium equivalence may also be marked for filters made of materials other than aluminium. However, the beam energy at which the Al equivalence is determined should also be stated. Where there is a means of adjusting the added filtration, either manually or motor-driven, and such facilities are not used in the examination protocols in the room, the means of altering the filtration from the standard setting should be disabled. The Al equivalence of any permanently installed dose area product (DAP) chamber should be marked.

4.12 The total beam filtration includes permanent filtration, any added filtration, and filtration afforded by attenuating material that always intercepts the beam, e.g. the mirror of a light beam diaphragm (LBD) or a DAP chamber. The IEC specifies the minimum permissible first HVL for x-ray tube voltages between 50 and 150 kV (BSI 2008b). These HVLs all correspond to a total filtration of 2.5 mm of aluminium (ICRP 1982). At 70 kV, the minimum permissible half-value layer (HVL) is 2.5 mm of aluminium (BSI 2008b). The total filtration (or its constituent parts plus, in the case of under-table tubes, the filtration of the tabletop) should be provided with the equipment. The total filtration should also be marked on the tube assembly.

4.13 Dedicated equipment should always be used for mammography. The total filtration should never be less than the equivalent of 0.5 mm aluminium (or 0.03 mm of molybdenum) (ICRP 1982). Most modern equipment incorporates systems that automatically change the factors (e.g. filter, anode material, and kV) during the patient exposure to optimise either the mean glandular tissue dose or the contrast rendition. The users must be aware of and understand these systems (if used clinically).

Beam size

4.14 Beam collimators should be designed to minimise extra-focal radiation. The maximum beam size should be permanently limited to that required in practice for each particular source assembly. The maximum cross-section should correspond to the largest dimension of the imaging device at the minimum focus to detector distance that would be used in clinical practice.

4.15 All radiographic x-ray equipment, including mobiles, should be provided with properly aligned adjustable LBD beam-limiting devices or, in special circumstances, e.g. skull units, cones to keep the radiation beam within the limits of the x-ray detector selected for each examination. Any cones should be marked with their coverage details, e.g. 15 cm diameter at 100 cm focus to detector distance (FDD).

4.16 Equipment for fluoroscopy should be provided with automated means to confine the radiation beam within the image receptor area. The collimation should automatically adjust for the FDD and the selected detector field of view. The use of virtual collimation is recommended. It should be possible for the Operator to adjust the field size during examinations down to the equivalent of 5 cm × 5 cm at 1 m from the focal spot.

4.17 For CT scanners, the radiation beam profile width in the 'z-axis' should be as close as possible to the total imaged thickness and be controlled by pre-patient collimation for all slice widths. The scanner operating instructions should state the radiation beam width and the imaged slice thickness for all widths. Where the scanner is to conform to BS EN 60601-2-44:2009 (BSI 2009a), the geometric efficiency should be displayed on the console if it is less than 70%.

Selection of materials for ancillary equipment

4.18 The attenuation of the x-ray beam between the patient and the image receptor should be minimised by the use of suitable materials for the construction of the tabletop (in the case of over-table tubes), the front of the detector, and the anti-scatter grid.

Dose-reduction features

4.19 Equipment may be sold with some dose-reduction packages included and others as options. An MPE must be consulted on the value of including dose-reduction packages in the purchase, with due regard given to the extent of the dose reduction and the use of the unit.

Exposure switches

4.20 Exposure switches on all x-ray diagnostic equipment should be arranged so that exposure continues only while continuous pressure is maintained on the switch and terminates if pressure is released (i.e. 'deadman') unless it is previously terminated by other means, e.g. at the end of the set exposure time (BSI 2008b). Exceptions are equipment that scans along the patient, e.g. DXA, some lithotripters, and CT scanners (other than in CT fluoroscopy mode).

4.21 Exposure switches relying on the remote control to initiate the exposure should contain all the safety features of conventional exposure switches with regard to exposure control and release, e.g. deadman. The exposure switch should be positioned at the control panel or at the positions intended to be occupied by the Operators.

4.22 For mobile equipment, the exposure switch should enable the Operator to be outside the radiation beam and at least 2 m away from both the tube housing and the patient.

4.23 Exposure switches should be designed to prevent the inadvertent production of x-rays. Detailed advice is given in BS EN 60601–1–3 (BSI 2008b). In particular, it should not be possible for fluids to enter the switch and cause a short circuit. Additionally, foot switches should be constructed so that exposure does not occur if they are accidentally overturned. Switches should be positioned so that they cannot be depressed accidentally during stowage. Ideally, a storage bracket should be provided and used. If resetting is automatic, it should be ensured that pressure on the exposure switch has to be released completely before the next exposure can be made. It is important that the remote control devices be designed so that other nearby devices do not unintentionally trigger exposures.

Equipment installation

4.24 New fixed radiological installations (or existing installations that have been relocated or substantially modified) require a critical examination to ensure that radiation safety features and warning devices function correctly (IRR 2017, IRR(NI) 2017). The critical examination is the responsibility of the installer and must be performed in consultation with a radiation protection adviser (RPA). The installer and purchaser should identify who will perform the examination. The content of the critical examination (IPEM 2012) and the results should be communicated to the purchaser, who should consult their RPA as to their adequacy.

4.25 For new mobile units, a critical examination is not required (IPEM 2012). The safety features and warning devices should still be checked prior to first use; these checks are the responsibility of the purchaser and not the provider. The RPA should be consulted on this testing.

4.26 Before radiological equipment enters clinical use, it must undergo acceptance testing to ensure that it performs to specification. The manufacturer should provide data outlining the results expected for a range of dosimetric and image quality tests, along with the tolerances used by the manufacturer and details on the method and equipment used to make the measurement (MHRA 2021).

4.27 Acceptance testing should be undertaken on behalf of the Employer by, or in consultation with, an MPE. Acceptance testing is best undertaken alongside a representative of the manufacturer; this will commonly be one of the installation engineers or an applications specialist. This arrangement allows for any discrepancy in measured results with acceptance data to be investigated straight away; the result could be real and require action, or it could be the result of a methodological difference in testing. Any changes that are deemed necessary can be undertaken by the manufacturer's representative and confirmed straight away by the MPE.

4.28 Where acceptance testing is being undertaken by an MPE or their representative and a representative of the manufacturer together, there must be cooperation between Employers regarding who is responsible for the controlled area. Those present could be employees or outside workers, and the Employer must ensure that arrangements regarding the provision of personal monitoring arrangements, PPE, and training are in place (see paragraphs 1.200–1.216, *Outside workers*). An RPA must be consulted regarding these arrangements.

4.29 Following acceptance, commissioning tests should be undertaken to provide baseline results for subsequent quality control (QC) tests. The timing of the commissioning tests requires some consideration. Where no further adjustments will be made to the equipment by the manufacturer or installer following acceptance, commissioning can be undertaken straight away. Where it is expected that adjustments will be made to clinical programs and/or image processing during the application training period, commissioning should be scheduled for after the application training period. An MPE must advise on the tests necessary as part of suitable commissioning and ongoing QC testing.

Specific considerations for mobile equipment

4.30 A key-operated switch, pin number, or password is only required for a mobile unit with a battery (BSI 2009b). With such a design, it is normal for powered movements and the generation of x-rays to only be possible with the system enabled, but for battery charging to be possible in any state. Where physical keys are used, they should be removed from the unit when not in use and stored in a secure location, and where pin numbers or passwords are used, the user should log off when leaving the unit unattended.

Equipment maintenance

4.31 Manufacturers, suppliers, and installers of x-ray equipment are required, under IRR (IRR 2017, IRR(NI) 2017) Regulation 32 and under the Medical Devices Regulations 2002 (MDR 2002), to provide the Employer with adequate information about the proper use and maintenance of that equipment.

4.32 The radiation safety features of equipment must be properly maintained. It cannot be considered safe from a radiation point of view unless it is in good order both mechanically and electrically. Maintenance and associated checks should be in accordance with the advice of the manufacturer, the supplier, and the RPA. Arrangements for the testing of radiation safety features, such as emergency stops, must be included in the radiation risk assessment.

4.33 All equipment should be appropriately maintained to the specification and at the frequency recommended by the manufacturer by engineers who are suitably trained to do so.

4.34 A record of maintenance should be kept for each item of x-ray equipment. This record should include information on any defects found by users (fault log), interim remedial action taken (e.g. changes to AEC sensitivity settings), subsequent repairs made, and the results of testing before equipment is reintroduced to clinical use. When a maintenance log is provided, it is the legal duty of the Employer to keep it up-to-date (HSE 2014).

4.35 Representatives of the manufacturer or contracted service provider should be asked to complete the Association of x-ray Equipment Manufacturers (AXREM, now called the Association of Healthcare Technology Providers for Imaging, Radiotherapy, and Care) (AXREM) or local equivalent controlled area and equipment handover form (AXREM 2018) prior to commencing work on any item of equipment. This transfers responsibility for the controlled area over to them for the duration, for which their own radiation risk assessment and local rules apply. Upon completion, the controlled area should be handed back to the equipment owner via the same form.

4.36 A person who carries out modifications or maintenance that could affect the radiation dose to patients or staff or affect image quality should immediately inform a designated trained and entitled responsible person before the equipment is returned to clinical use. This information should also be recorded in the controlled area handover form, which also provides a solution for communicating changes when the work is performed out of hours. This can be used instead of or in addition to a notice drawing attention to the modification or maintenance attached to the equipment. In all cases, a clear written report of the changes and current operational status should be given, including the name of the person concerned and the date. This should include details of any changes affecting dose or image quality that may require further measurements. Note that these changes can involve hardware or software.

4.37 After work affecting equipment performance, the designated responsible person should seek advice from the MPE regarding necessary actions. This advice could be in the form of a protocol written by or in conjunction with the MPE. The MPE will decide whether the changes are sufficient to require a survey of the performance and safety of the equipment. Note that where the changes could have affected the equipment's safety features, it will be necessary to consult an RPA, as a critical examination may be required. Again, this advice could be in the form of a protocol written by or in conjunction with the RPA. Any testing required by the in-house quality assurance (QA) programme should be performed and the results recorded before the equipment is returned to clinical use.

4.38 Planned maintenance work should ideally be scheduled to ensure that there is sufficient time for any necessary testing before the equipment is returned to clinical use.

4.39 Where changes are made to the equipment, it may be necessary to alter operating procedures, equipment exposure protocols, or local rules. Such decisions must be made in consultation with the MPE and/or RPA, as appropriate. All Operators should be informed of these changes. It might be necessary to reassess the representative patient doses; this decision must be taken in consultation with the MPE. If deemed necessary, the patient dose audit process should begin shortly after the changes have been implemented.

4.40 Maintenance records should be kept for 11 years after the decommissioning of the equipment (NHSX 2021).

Equipment testing

4.41 Employers are required to establish a QA programme for all x-ray equipment and any associated components required to create an image (e.g. detectors) under IR(ME)R Regulation 15 (IR(ME)R 2017, IR(ME)R(NI) 2018). An appropriate QA programme should consist of tests that are frequently performed by local staff and are designed to detect gross changes in performance. These should be supplemented by less frequent but more complex testing that requires specialist equipment; these would normally be undertaken by a physics department or qualified person. Guidance on the content of such tests is given in IPEM Report 91 (IPEM 2005) and the IPEM Report 32 series (IPEM 2016), but the MPE must be consulted.

4.42 The Employer must have a procedure that ensures QA programmes for equipment are followed. A standard operating procedure (SOP) should govern the QA performed by local staff members. The SOP should include details on how to undertake the test, the expected test result under normal circumstances, action levels in the event that the result of the test deviates from the expected, and instructions on what to do in the event that the action levels are exceeded. The SOP should also detail how to record the result of the test in a QA record.

4.43 It is for each department to decide how best to manage the local equipment QA programme. Different staff groups can be involved, provided appropriate education and training are provided and they are properly entitled as Operators. An appropriately trained and entitled individual should have overall responsibility for the QA programme; this provides a clearly defined line of accountability for reporting the results of the QA programme.

4.44 The local QA programme is important for identifying problems with equipment before they become serious or affect too many patients. It is crucial to conduct tests at the MPE's recommended frequency. The best practice is to conduct an ongoing audit of the local QA programme. This should include ensuring the timeliness of the testing, the recording of results, and the follow-up of results that do not meet the recommended levels are appropriate. Arrangements should be in place locally to decide how often and by whom this audit will be undertaken. Where the audit identifies deficiencies, these must be addressed, and improvements should be made in line with the SOP describing the process.

4.45 The QA programme should include some indication of the doses administered to persons undergoing medical exposure. The MPE can advise on the best means of achieving this, as it is dependent upon the systems that are available locally. Patient dose management systems offer a convenient means of assessing patient doses. Patient dose data can also be downloaded from a radiology information system and analysed to derive indicative patient doses or collected prospectively at the time of the patient's examination using the methods outlined in IPEM Report 88 (IPEM 2004).

4.46 Representative measurements of the doses administered to persons undergoing medical exposure, made during equipment testing using standard phantoms or tissue equivalent material, can still give useful and quick comparative information on the performance of a unit. Guidance on making representative measurements is provided in IPEM reports (IPEM 2005, 2016), NHSBSP Report 0604 (NHSBSP 2009), PHE NHS Breast Screening Programme Equipment Report 1407 (PHE 2015), and the National Protocol for Patient Dose Measurements in Diagnostic Radiology (IPSM, RPB, and CoR 1992).

4.47 Whenever practicable, mobile equipment should be maintained and tested in adequately shielded x-ray rooms. If this is not practicable, a temporary controlled area should be designated, with suitable measures to restrict access to the area. An RPA must be consulted on the designation of the area and the restriction of access.

4.48 All staff involved in maintaining and testing x-ray equipment must take care to avoid exposing themselves to the radiation beam or scattering from test objects. In particular, they must avoid exposing their hands to the beam while performing beam alignment procedures or QA tests on fluoroscopy equipment.

4.49 For any equipment, the Operator should check that the exposure warning light and, where provided, any audible warning signals operate at each exposure and cease at the end of the intended exposure. If the exposure does not appear to have terminated as intended, the unit must be immediately disconnected from the mains electricity supply, labelled as defective and out of service, and the MPE and RPA should be consulted.

4.50 If there is reason to think that the exposure control is defective, the exposure warning does not operate, or that there may be some other fault (e.g. signs of damage, excessive x-ray tube temperature), the equipment should be disconnected from the supply, labelled as defective and out of service, and not used again until it has been checked and repaired by a service engineer.

References

AXREM 2018 *Radiation Controlled Area and Equipment Handover Form* (UK: AXREM) www.axrem.org.uk/resource/ensuring-compliance-with-radiation-safety-regulations/ (acessed 31 January 2023)

BSI 2008a *BS 7671:2008 Requirements for Electrical Installations* (London: British Standards Institute)

BSI 2008b *BS EN 60601-1-3: 2008 General Requirements for Basic Safety and Essential Performance. Collateral Standard. Radiation Protection in Diagnostic X-ray Equipment* (London: British Standards Institute)

BSI 2009a *BS EN 60601-2-44: 2009 Medical Electrical Equipment. Particular Requirements for the Basic Safety and Essential Performance of X-ray Equipment for Computed Tomography* (London: British Standards Institute)

BSI 2009b *BS EN 60601-2-54:2009 Particular Requirements for the Basic Safety and Essential Performance of X-ray Equipment for Radiography and Radioscopy* (London. UK: British Standards Institute)

BSI 2010 *BS EN 60601-2-28:2010 Medical Electrical Equipment. Particular Requirements for the Basic Safety and Essential Performance of X-ray tube Assemblies for Medical Diagnosis* (London: British Standards Institute)

BSI 2014 *BS EN 62353:2014 Medical Electrical Equipment. Recurrent Test and Test After Repair of Medical Electrical Equipment* (London: British Standards Institute)

DH 2017 *(HTM 06-01) Electrical Services Supply and Distribution* (London: Department of Health) https://england.nhs.uk/publication/electrical-services-supply-and-distribution-htm-06-01 (acessed 22 March 2023)

European Commission 1993 Council Directive 93/42/EEC of 14 June 1993 concerning medical devices *Off. J. L 169* https://eur-lex.europa.eu/legal-content/EN/TXT/?uri=CELEX%3A31993L0042 (acessed 31 January 2023)

HFS 2015 *Electrical Services Supply and Distribution (SHTM 06-01)* (Scotland: Health Facilities Scotland) https://nss.nhs.scot/publications/electrical-services-supply-and-distribution-shtm-06-01 (acessed 22 March 2023)

HSE 2014 *L22: Safe Use of Work Equipment. Approved Code of Practice and Guidance for the Provision and Use of Work Equipment Regulations 1998* (Norwich: Health and Safety Executive) https://hse.gov.uk/pubns/books/l22.htm (acessed 22 March 2023)

ICRP 1982 *Protection Against Ionizing Radiation from External Sources Used in Medicine. ICRP Publication 33. Ann. ICRP 9 (1)* (Vienna, Austria: International Commission on Radiation Protection) www.icrp.org/publication.asp?id=ICRP Publication 33 (Accessed 14 February 2024)

IPEM 2004 *Report 88: Guidance on the Establishment and use of Diagnostic Reference Levels for Medical X-ray Examinations* (York: Institute of Physics and Engineering in Medicine)

IPEM 2005 *Report 91: Recommended Standards for the Routine Performance Testing of Diagnostic X-ray Imaging Systems* (York: Institute of Physics and Engineering in Medicine)

IPEM 2012 *Report 107: The Critical Examination of X-ray Generating Equipment in Diagnostic Radiology* (York: Institute of Physics and Engineering in Medicine)

IPEM 2016 *Report 32: Measurement of the Performance Characteristics of Diagnostic X-Ray Systems. Part I to Part VII* (York: Institute of Physics and Engineering in Medicine)

IPSM, RPB and CoR 1992 *National Protocol for Patient Dose Measurements in Diagnostic Radiology* (Chilton: National Radiological Protection Board)

IR(ME)R(NI) 2018 *The Ionising Radiation (Medical Exposure) Regulations (Northern Ireland).* https://legislation.gov.uk/nisr/2018/17 (acessed 31 January 2023)

IR(ME)R 2017 *The Ionising Radiation (Medical Exposure) Regulations.* www.legislation.gov.uk/uksi/2017/1322 (acessed 31 January 2023)

IRR(NI) 2017 *The Ionising Radiations Regulations (Northern Ireland).* http://legislation.gov.uk/nisr/2017/229 (acessed 31 January 2023)

IRR 2017 *The Ionising Radiations Regulations* https://legislation.gov.uk/uksi/2017/1075 (acessed 31 January 2023)

MDR 2002 *The Medical Devices Regulations 2002* http://legislation.gov.uk/uksi/2002/618/pdfs/uksi_20020618_en.pdf (acessed 31 January 2023)

MHRA 2021 *Managing Medical Devices, Guidance for Health and Social Care Organisations, Health Policy* (Medicines and Healthcare Products Regulatory Agency) https://gov.uk/government/publications/managing-medical-devices (acessed 8 February 2023)

NHSBSP 2009 *Equipment Report 0604. Commissioning and Routine Testing of Full Field Digital Mammography Systems, gov.uk website* (NHS Breast Screening Programme) https://gov.uk/government/publications/breast-screening-digital-mammography-commissioning (acessed 22 March 2023)

NHSX 2021 *Records Management Code of Practice* 2021 *A Guide to the Management of Health and Care Records* AUGUST 202. NHSX https://nhsx.nhs.uk/information-governance/guidance/records-management-code/ (acessed 31 January 2023)

PHE 2015 *NHS Breast Screening Programme Equipment Report 1407. Routine Quality Control Tests for Breast Tomosynthesis (Physicists), Gov.uk Website* (London: Public Health England) https://gov.uk/government/publications/breast-screening-routine-quality-control-tests-for-breast-tomosynthesis (acessed 22 March 2023)

IOP Publishing

Medical and Dental Guidance Notes (Second Edition)

A good practice guide on all aspects of ionising radiation protection in the clinical environment: IPEM Report 113

John Saunderson, Mohamed Metwaly, William Mairs, Philip Mayles, Lisa Rowley and Mark Worrall

Chapter 5

Dental radiography

Scope

5.1 This chapter applies to the use of equipment specifically designed for radiography of the teeth or jaws, including radiography using an intra-oral image receptor (or, with the same equipment, an extra-oral receptor), panoramic radiography with an extra-oral x-ray tube, cephalometric radiography, and dental cone beam computed tomography (CBCT). While primarily concerned with the use of the equipment for the examination of patients, the guidance is also relevant during equipment testing, measurement of the radiation produced, staff training, research into examination techniques, the examination of volunteers in approved research projects, and other uses at the place where the equipment is normally used. Reference should be made to chapter 3 if general-purpose x-ray equipment is used. The principles of radiation protection detailed in section 3.2 apply equally to dental radiology.

Registration

5.2 Dental radiography is a registerable practice under IRR (IRR 2017, IRR(NI) 2017) (working with a radiation generator). A single Employer requires only one registration for all radiation generators across all its sites. Dental practices that are part of a larger Employer (for example, a community dental practice within an NHS Employer) will have been included in the main Employer's registration. A private dentist will require its own registration, but if it has more than one site, these can all be included under a single registration.

5.3 The Employer must notify the HSE/HSENI of any material changes to its registration. In general, a material change is one that would change one or more

doi:10.1088/978-0-7503-2332-1ch5

of the answers given to the questions during the registration process (HSENI no date, HSE no date). These relate to the address of the Employer, the number of UK-based employees, the number of classified employees, the use of portable ionising radiation sources, and the number of fixed sites on which the Employer carries out work with ionising radiation. The Employer should consult their radiation protection adviser (RPA) if they are in any doubt as to whether they should inform the HSE/HSENI of a material change.

5.4 If the Employer stops working with ionising radiation and no longer requires registration, they should inform the HSE/HSENI (HSENI no date, HSE no date).

5.5 If the Employer changes its name, this requires notification of cessation of work for the old Employer and a new application in the name of the new Employer (HSENI no date, HSE no date).

Roles and responsibilities

5.6 The general advice given in chapters 1 and 2 applies equally to users of dental x-ray equipment.

Radiation protection adviser

5.7 Employers will need to consult an RPA on matters listed in appendix 3 from time to time. There is no exemption from this requirement for dental x-ray equipment users, and when advice is required, an RPA must be formally appointed (see paragraphs 1.88 to 1.91). The appointment of an RPA does not have to be a permanent arrangement. However, a formal ongoing appointment is recommended since it ensures that advice is always immediately available and provides continuity of advice. The formal appointment should be reviewed periodically to ensure the RPA and the scope of their advice remain appropriate.

Radiation protection supervisor

5.8 The role of the RPS, in general, is discussed in paragraphs 1.119 to 1.126. In the dental radiology setting, they must supervise the arrangements contained within the local rules. The RPSs can also provide a communication channel with the RPA (see appendix 3).

5.9 The number of RPSs appointed should be sufficient to ensure that work in all controlled areas is adequately supervised. It is not necessary for an RPS to always be present, but an RPS should work in a controlled area often enough to ensure that all staff members are aware of the local rules and that they are observed. It is for the Employer to decide how many RPSs are required, with due regard to the number of locations covered, the complexity of the work, absences from the department, and the advice of the RPA.

Medical physics expert

5.10 Employers must appoint appropriate MPE(s) to advise on optimisation of radiological practice. These formal appointments should be in writing and reviewed periodically to ensure the MPE(s) and the scope of their advice remains appropriate. The MPE(s) must be able to advise the Employer on all diagnostic dental examinations undertaken. A list of MPE(s) together with their scope of practice should be available in or referenced in the Employer's procedures. An MPE may be the same person as the RPA, though it should be recognised that their roles are separate and distinct.

5.11 The role of the MPE is discussed in general in paragraphs 2.56 to 2.59 and in appendix 15.

Referrer

5.12 For dental radiography, the Referrer will be a registered healthcare professional who is entitled in accordance with the Employer's procedures. This could include a dental or medical Practitioner, a dental hygienist, and a dental therapist. In settings other than hospital radiology departments, it will be common for the Referrer and IR(ME)R Practitioner to be the same person, in which case the formal exchange of clinical information is unnecessary. Referral guidelines must always be provided, even when the IR(ME)R Practitioner is also the Referrer. When establishing such guidelines, the Employer may wish to make use of evidence-based selection criteria guidelines such as the booklet published by the Faculty of General Dental Practitioners—Royal College of Surgeons (FGDP-RCS), *Selection Criteria for Dental Radiography* (Faculty of General Dental Practice 2018), or similar guidelines such as *Guidelines for the Use of Radiographs in Clinical Orthodontics* (BOS 2015) or *Cone Beam CT for Dental and Maxillofacial Radiology* (SEDENTEXCT Project 2012) and relevant NICE guidelines.

5.13 At present, dental CBCT is not sufficiently covered in undergraduate dental programmes to allow the newly qualified dentist to be considered adequately trained to refer for dental CBCT examinations. Dental CBCT is similarly largely absent from the professional qualifications of other dental care professionals currently. Further training is required for anyone wishing to refer patients for dental CBCT examinations to ensure their referrals are appropriate and in accordance with current referral guidelines for dental CBCT (PHE and FGDP 2020).

IR(ME)R practitioner

5.14 The IR(ME)R Practitioner for dental radiology will be a registered healthcare professional, usually a dental Practitioner, oral surgeon, orthodontist, or radiologist, but may be another trained and qualified healthcare professional, such as a dental hygienist or dental therapist, trained and entitled for this role for specified examinations within the Employer's procedures. The primary role of the IR(ME)R Practitioner is to undertake the justification of individual exposures. All IR(ME)R Practitioners must be adequately trained to undertake this function. Adequate training for a dentist to act as an IR(ME)R Practitioner for intra-oral, panoramic, and cephalometric radiography comprises:

(a) for UK graduates, an undergraduate degree conforming to the requirements for the undergraduate dental curriculum in dental radiology and imaging (GDC 2002) and the core curriculum in dental radiography and radiology for undergraduate dental students (NRPB and RCR 1994).
(b) for non-UK graduates, the Employer should establish whether the IR (ME)R Practitioner's undergraduate degree matches the above requirements. Where the IR(ME)R Practitioner is also the Employer they should seek the advice of the dental practice adviser and/or the postgraduate dental dean. The GDC's assessment of the staff member's application onto the GDC register will also indicate whether their degree is considered substantially equivalent in content to a UK degree. The requirements for adequate training of non-dentists to become IR(ME)R Practitioners are outlined in Schedule 3 of IR(ME)R (IR(ME)R 2017, IR(ME)R(NI) 2018). It is for the Employer to complete and retain individual records of training for each IR(ME)R Practitioner. These records should record the nature of the training received, the date of the training, and the subjects detailed in Schedule 3 of IR(ME)R that were addressed by the training. When the training record demonstrates that all of the relevant subjects detailed in Schedule 3 of IR(ME)R have been adequately met, the Employer can entitle the individual as an IR(ME)R Practitioner. The Employer can consult the MPE on whether any training on offer adequately meets those subjects in Schedule 3 of IR(ME)R concerned with radiation protection. As with all IR(ME)R Practitioners, the Employer must ensure that these entitled IR(ME)R Practitioners undertake continuing education and training. Their training records must be updated accordingly.

For dental CBCT, adequate training for Practitioners is presented in the Guidance Notes for Dental Practitioners on the Safe Use of X-ray Equipment—2nd edn (PHE and FGDP 2020). Theoretical training for Practitioners is outlined in a position paper prepared by the European Academy of DentoMaxilloFacial Radiology (Brown et al 2014) and in Cone Beam CT for Dental and Maxillofacial Radiology (SEDENTEXCT Project 2012). Adequate training must include a practical component that is proportionate to the IR(ME)R Practitioner's involvement with the system (i.e. patient doses and the effect of physical and programmable changes on patient dose, the use of the software for implant planning, etc). It is for the Employer to establish appropriate training and retain training records. An MPE must contribute to the radiation protection aspects of this training.

5.15 Before an examination can proceed, it is necessary for it to be authorised (the act of documenting that it has been justified). This is commonly performed by the Operator, e.g. by means of an electronic record. Alternatively, when authorisation is made through authorisation guidelines, entitled Operators can authorise specified diagnostic exposures by recording that the request complies with the agreed guidelines produced by the IR(ME)R Practitioner.

Operators

5.16 In dental radiology, it is common for the Referrer and IR(ME)R Practitioner to be the same person. They may also act as Operators, particularly for dental equipment located in departments outside of radiology. However, many dental nurses or other dental care professionals (DCPs) will also perform some of the functions of an Operator (with exceptions; nurses cannot undertake clinical evaluation, and dental hygienists and dental therapists should only do so where it is relevant to their scope of practice). All Operators must be adequately trained to perform these functions.

For intra-oral, panoramic, and cephalometric radiography, adequate training needs to address two groups of Operators:

(a) *Operators whose duties include selecting exposure parameters and/or positioning the detector, the patient, and the tube head:*

- Dental Practitioners should fulfil the requirements as for IR(ME)R Practitioners (see paragraph 5.14).
- Dental nurses should possess a Certificate in Dental Radiography, conforming to the syllabus prescribed by the National Examining Board for Dental Nurses (NEBDN) (NEBDN 2019).
- Dental hygienists and dental therapists should have received an equivalent level of training to that of dental nurses.

(b) *Other Operators*

- Dental nurses (and any other DCPs) whose duties are limited to patient identification, image processing, and quality assurance should preferably possess a Certificate in Dental Nursing (or the equivalent) (NEBDN 2020). Failing this, they must have received adequate and documented training specific to the tasks that they undertake; this training may be provided 'in-house'.
- Dental nurses who do not hold the Certificate in Dental Radiography (and any other DCPs without equivalent training) and who initiate the exposure as part of a patient examination that has been physically set up by an adequately trained Operator, may only do so in the continued presence and under the direct supervision of that Operator. They must have received the appropriate documented instructions for this task.
- For dental CBCT, adequate training for Operators is outlined in Cone Beam CT for Dental and Maxillofacial Radiology (SEDENTEXCT Project 2012) and the *Guidance Notes for Dental Practitioners on the Safe Use of X-ray Equipment*—2nd edn (PHE and FGDP 2020).

The trained and entitled Operator remains responsible for the exposure throughout.

Facility design and area designation

5.17 An RPA and MPE must be consulted at the planning stage when any new dental x-ray equipment is being considered. They must be provided with all relevant information, including the equipment to be installed, the current or intended design of the facility, including room dimensions and wall construction materials (where known), the examinations that are to be undertaken with the equipment, the estimated number of examinations in a day and week, and the use of all adjacent areas. Advice on facility design for all dental installations can be found in the BIR publication *Radiation Shielding for Diagnostic Radiology* (Sutton *et al* 2012).

5.18 Persons in areas outside an x-ray room should be adequately protected, meaning that the controlled area should not extend beyond the x-ray room. It is only acceptable for adjacent areas outside the room to be designated as controlled if access can be effectively restricted during radiography (e.g. a locked storeroom).

5.19 For existing facilities, the current shielding of the walls, floor, and ceiling of the room may be enough to ensure the controlled area does not extend beyond the room. Where the RPA advises that additional shielding is necessary, additional protective materials such as lead ply or materials containing barium sulphate can be added. It may be the case that an additional layer of plasterboard would provide sufficient shielding. The RPA should specify the material and thickness necessary, or a range of materials, so the most cost-effective can be selected.

5.20 For new facilities, the RPA can advise on the most cost-effective material to use to ensure the required level of protection.

5.21 The RPA must be consulted on the designation of the controlled area. The decision to designate a controlled area either around the unit or encompassing the whole room is dependent on several factors, including the room layout, operator position, the type of x-ray unit in use, its operating kilovoltage, and the workload.

5.22 In considering the occupational dose to any individual, the relative contribution from all dental imaging units and all other potential sources of occupational exposure must be taken into account.

Fixed intra-oral x-ray units

5.23 Intra-oral radiography should be carried out in a room (the x-ray room) from which all persons whose presence is unnecessary are excluded during radiography. This room, which should ideally be a dental surgery or a separate examination room, should not be used for other work or as a passageway while radiography is in progress.

5.24 Intra-oral equipment should be installed such that, when used, the useful beam is directed away from any door. It should also be directed away from any window if the space immediately beyond the window is frequently occupied.

5.25 The Operator's position during radiography should ideally be outside the controlled area in a position that allows them to see the patient and restricts access to the room. The exposure switch should be installed with this in mind.

5.26 Isolation of the power to the x-ray unit is an appropriate contingency plan in the event of a failure of the unit to terminate the exposure at the end of the pre-set time. For fixed installations, the mains isolator should ideally be positioned at, or within easy reach of, the Operator's position.

5.27 If more than one x-ray unit is sited in any room (e.g. in open-plan accommodation), arrangements must be made in consultation with an RPA to ensure that patients, staff, and all other persons are adequately protected. In particular, it should not be possible for an Operator to inadvertently energise the wrong x-ray unit or to accidentally irradiate people working independently in another part of the room. This is best achieved by engineering controls, such as the provision of a clearly labelled selector switch that will only permit power to be supplied to one x-ray unit at any time.

5.28 If it is possible from a single Operator position to initiate the production of x-rays from more than one x-ray tube, separate warning lights should be provided that are visible from the Operator's position and that is clearly labelled to indicate which x-ray unit has the mains power switched on.

Hand-held intra-oral x-ray units

5.29 Hand-held units can be used to share between surgeries in a single practice to reduce the overall amount of equipment required, or they can be used in multiple remote locations. In the former case, the Employer must consult an RPA as to the extent of the controlled area in each surgery. In the latter, the RPA should advise on how the Operator should assess the suitability of each location for intra-oral radiography on a case-by-case basis (PHE and FGDP 2020). This is likely to involve additional training for the Operator.

5.30 Hand-held units require specific working methods to be devised to restrict the dose to the Operator. An RPA must be consulted, and the working methods must be detailed in the local rules. These should include the use of backscatter shields. More information on the aspects that need to be considered is given by PHE (PHE 2016, PHE and FGDP 2020).

5.31 Hand-held units should be supplied with rectangular collimators. Where they are supplied with both rectangular and circular collimators, the rectangular collimators should be used.

5.32 For hand-held units, the appropriate contingency plan in the event of a failure of the unit to terminate the exposure at the end of the pre-set time must be devised at the risk assessment stage before any use of the unit. The contingency plan will vary with the design of the equipment, but in all cases, the first action should be to direct the x-ray beam away from all persons present and towards a shielded wall before switching off the power either by means of removing the battery or using the controls.

5.33 The Employer must consider the security of the hand-held unit when it is not in use or in transit. Appropriate arrangements should be in place to ensure it is not accessible to anyone other than trained employees.

5.34 Arrangements must be made to ensure the safety of the hand-held unit when in use and in transit in order to prevent damage that might compromise the unit's safety.

Panoramic and cephalometric

5.35 Panoramic and cephalometric examinations should be carried out in a dedicated x-ray room from which all persons whose presence is unnecessary are excluded while x-rays are being produced. The x-ray room should be adequately shielded; the Employer must consult an RPA on adequate shielding.

5.36 An optimal design should have no unshielded windows in the controlled area if the space immediately beyond the window is frequently occupied.

5.37 The Operator's position during radiography should ideally be outside the room, but if the room is sufficiently large, it may be appropriate for the Operator to remain within the room; an RPA can advise. The exposure switch should be installed with the Operator position in mind. The patient should be visible to the Operator throughout the examination; depending on the Operator's position, this can be achieved through a direct line of sight through a suitably shielded window, fixed mirrors, or CCTV.

5.38 Isolation of the power to the x-ray unit is an appropriate contingency plan in the event of a failure of the unit to terminate the exposure at the end of the pre-set time. The mains isolator switch should ideally be positioned outside the controlled area and close to the Operator's position for this purpose.

Cone beam computed tomography

5.39 Dental CBCT examinations should be carried out in a dedicated x-ray room from which all persons whose presence is unnecessary are excluded while x-rays are being produced. The x-ray room should be adequately shielded, including the door, if necessary. An RPA must be consulted on adequate shielding.

5.40 Since dental CBCT systems exhibit higher levels of scattered radiation compared with any other form of dental x-ray equipment, it will normally be necessary to designate the whole room in which the dental CBCT unit is situated as a controlled area during radiography.

5.41 All shielding assessments will require estimates as to the likely workload for the system. This should be discussed with the local users, but it should be no lower than 10 exposures per week for a dental practice and no lower than 50 per week for an installation in a hospital.

5.42 The Operator position could be adjoining the room with appropriate transparent screen protection offering full visibility of the patient at all times, or within the room behind a shielded barrier; an RPA can advise. Where it isn't possible to have clear visibility of the patient during an exposure in an adjoining room, consideration should be given to other methods (i.e. mirrors or CCTV) to ensure the Operator can see the patient throughout the exposure. The exposure switch should be installed with the Operator position in mind.

5.43 Isolation of the power to the x-ray unit is an appropriate contingency plan in the event of a failure of the unit to terminate the exposure at the end of the pre-set time. The mains isolator switch (and emergency stop where provided with the dental CBCT unit) should be positioned such that they are outside the controlled area or within reach of the Operator without leaving their protected exposure position. In particular, it should be possible to isolate the equipment from the main power supply without having to enter the controlled area.

Room warning signs and signals

5.44 For fixed intra-oral units, the need for warning signs or lights is dependent upon the controlled area designation. An RPA must advise.

5.45 For hand-held intra-oral units, it is not practicable to use a warning light, as this would have to be manually operated. If there are doors leading into the controlled area, these should be locked. If the Operator cannot adequately restrict access to the controlled area, a portable warning sign could be used, or a trained colleague should act as a sentry to restrict access.

5.46 Panoramic equipment and dental CBCT systems generally support the use of a two-stage automatic illuminated warning signal that indicates when the controlled area exists (i.e. when there is power to the equipment) and when radiation is being emitted. The Employer should consult an RPA as to whether or not this is required, as this is dependent upon the characteristics of the installation.

Protection of staff and members of the public

Classification of staff and personal dosimetry

5.47 Sensible application of radiation protection measures should ensure that all staff involved in dental radiography receive an annual effective dose of considerably less than 6 mSv, with annual equivalent doses to the eyes and extremities likewise considerably less than 6 mSv. Consequently, it will seldom, if ever, be necessary for staff to be designated as classified persons solely because of work involving dental radiography. However, any non-classified staff who enter a controlled area must be subject to suitable written arrangements designed to ensure that, if adhered to, exposures will not exceed levels above which classification would be required.

5.48 The risk assessment must determine whether there is any need for personal monitoring for non-classified staff; an RPA can advise. Personal monitoring need not be continuous; monitoring could be undertaken for a trial period, after which the RPA can review the results and advise whether any further monitoring is necessary.

5.49 The IPEM report *Guidance on the Personal Monitoring Requirements for Personnel Working in Healthcare* (Martin *et al* 2018) recommends an initial period of monitoring using a collar dosemeter if the number of intra-oral radiographs taken per week exceeds 100 scans, if the number of panoramic examinations per week exceeds 50 scans, or if the combined workload exceeds the pro-rata total. It is recommended that dental staff operating CBCT equipment wear a collar dosemeter for an initial period to establish the effective dose and dose to the eyes.

5.50 The dosemeter wear period may be up to three months; the advice of the supplier should be followed. Results should be recorded and periodically discussed with an RPA to ensure that doses are kept ALARP and below the Employer's dose investigation level.

Patient support

5.51 If there is ever the intention that a member of staff could provide support for a patient during an examination, this scenario must be included in the risk assessment and a suitable procedure included in the local rules. Consultation with the RPA is essential. The risk assessment must consider the possibility of the staff member being pregnant and estimate the frequency with which this practice is likely to be required. Those providing support should have adequate instruction and be informed of the level of risk involved. Employers must make suitable arrangements to provide this information.

5.52 Personal protection equipment (PPE) is not required for anyone providing support for a patient in dental radiology.

Carers and comforters

5.53 The patient's parent, guardian, or carer are the preferred persons to offer patient support and would act in the capacity of a carer and comforter. The exposure of a carer and comforter during an examination must be considered by the IR(ME)R Practitioner when justifying the examination, and a dose constraint must be set in the Employer's procedures.

5.54 It is preferable to avoid designating a pregnant individual as a carer and comforter; however, this may not always be practicable (BIR *et al* 2020). An MPE should assess the potential dose to the foetus as well as the carer and comforter to assist the IR(ME)R Practitioner with justification. It may be appropriate to have a reduced dose constraint for pregnant carers and comforters.

5.55 The carer and comforter must undertake this role knowingly and willingly. The benefits and risks must be explained to the individual undertaking the role prior to the examination by the Operator, or through some other means, such as an information sheet. The Employer must make suitable arrangements for providing this information in their procedures.

5.56 PPE is not required for anyone acting in the capacity of carer and comforter in dental radiology.

Personal protective equipment

5.57 No fixed installation should require the Operator to wear a lead apron or any other PPE during radiography. For radiography using a hand-held intra-oral unit, however, there may be some views that require the Operator to wear PPE (PHE 2016). This must be considered in the risk assessment. The Employer must consult an RPA.

5.58 If, based on RPA advice, it is necessary for an Employer to provide protective aprons, the Employer must provide the means for these to be stored appropriately and they should be routinely inspected as discussed in paragraphs 1.72 to 1.78, *Personal protective equipment*.

Protection of the patient

5.59 The IR(ME)R (IR(ME)R 2017, and IR(ME)R(NI) 2018) require that the patient be given information relating to the benefits and risks prior to their examination, wherever practicable. Posters relating to this information have been developed for use in England, Wales, and Northern Ireland by the Clinical Imaging Board (CIB 2019) and in Scotland by the Scottish Clinical Imaging Network (SCIN 2019). These posters should be displayed somewhere patients will see them, such as in a waiting area or next to the x-ray unit, and brought to the attention of any patients that will undergo an x-ray examination.

5.60 An assessment of patient dose must be included in a quality assurance (QA) programme for dental equipment, as discussed in paragraph 6.40. Local measurements must be compared to the relevant National Diagnostic Reference Levels (DRLs).

5.61 National DRLs for dental radiography are reviewed and updated regularly following national dose audits undertaken by the UK HSA. The Department of Health and Social Care (UKHSA 2022) offers online access to the most recent values. These should be used as a starting point on which to base local DRLs. It will be adequate for representative patient doses to be assessed as a part of routine testing, provided that the QA programme is able to confirm an acceptable ongoing quality of radiographs.

5.62 The use of patient contact shielding in dental radiography is not generally recommended (BIR 2020, PHE and FGDP 2020).

5.63 Where inclusion of the thyroid in any dental view is unavoidable, the use of a thyroid shield must be discussed with the MPE. This should only be relevant for very few intra-oral views (such as the anterior oblique occlusal) and dental CBCT using a large field of views that cannot be collimated to remove the thyroid (BIR 2020, PHE and FGDP 2020).

5.64 Where equipment provides a choice of beam or field sizes, the smallest reasonably practicable size should be used, consistent with the radiographic procedure. This will reduce patient dose and could improve image resolution.

General procedures

5.65 The Operator should check that either the exposure warning light or, where provided, any audible warning signal operates at each exposure and ceases at the end of the intended exposure. If the exposure does not appear to have terminated as intended, the unit must be immediately disconnected from the mains electricity supply. A well-designed hand-held intra-oral unit should have a means of switching off the power or removing the battery (see paragraph 5.32). Where there is a failure of the beam to terminate, the MPE must be consulted, and a preliminary investigation should be undertaken to ascertain whether the incident warrants further investigation and possible reporting. If the failure to terminate could have affected staff or a member of the public, the RPA must be consulted.

5.66 If there is reason to think that the exposure control is defective, the exposure warning does not operate, or there may be some other fault (e.g. signs of damage, excessive x-ray tube temperature), the equipment should be disconnected from the power supply and not used again until it has been checked and repaired by a service engineer. It should be labelled to indicate that it is out of service until repaired and accepted back into service in accordance with the RPA's and MPE's recommendations.

5.67 The exposure settings should be chosen and checked by the Operator on each occasion before an examination is made. This is especially important where different exposure settings are routinely used (e.g. to make allowance for the use of a long or short cone, variable kV, mA, and time settings, different image receptor speeds, or exposure programmes).

Additional precautions, by modality

Intra-oral radiography

5.68 Whenever practicable, techniques using image receptor holders incorporating beam-aiming devices should be adopted for bitewing and periapical radiography. Rectangular collimation should be used, for which a beam-aiming device is essential for accurate alignment with the intra-oral image receptor. Attention is drawn to the probable need for additional Operator training in the use of image receptor holders when moving from circular to rectangular collimation.

5.69 The equipment should have an open-ended rectangular beam collimator conforming to the recommendations of paragraph 6.16. When a choice of beam collimator is provided, the one most suited to the technique to be employed should be fitted, ideally just covering the image receptor. The open end of the collimator should be placed as close as possible to the patient's head to minimise the size of the incident x-ray beam. If it is beneficial to use a longer focus-to-surface distance (FSD), then a longer collimator should be employed.

5.70 The image receptor should only be held by the patient when it cannot otherwise be kept in position. It should not normally be held by anyone else. Exceptionally, it may be held by someone other than the patient using a pair of forceps or other appropriate holders to avoid direct irradiation of their fingers, e.g. when a child or a patient with additional needs cannot hold the receptor themselves.

5.71 Where extra-oral or vertex occlusal views are taken, it is recommended that a left and/or right marker be used on the image receptor to confirm which side of the patient has been examined.

Intra-oral radiography using fixed units

5.72 The tube housing should never be held by hand during exposure by the Operator or the patient. The Operator should stand outside the controlled area and, preferably, at least 1.5 m away from the x-ray tube and patient and away from the direction of the primary beam. The Operator should make use of the full length of any cable to the exposure switch.

Intra-oral radiography using hand-held units

5.73 Hand-held intra-oral units should be mounted on a tripod with a remote or wired exposure switch where practicable. The Operator should then stand at least 1.5 m away from the x-ray tube and patient, making full use of the length of any cable to the exposure switch, and away from the direction of the primary beam. Where not practicable, including cases where there is no compatible tripod, remote, or wired exposure switch supplied with the unit, it should always be used with a backscatter shield to reduce the dose to the Operator. The beam should be aimed horizontally in order for the backscatter shield to effectively protect the operator, which may require tilting the patient's head for some views.

Panoramic radiography

5.74 Where panoramic equipment features a number of different rotational modes (e.g. TMJ mode), a check should be made before every exposure to ensure that the correct mode has been selected.

5.75 Patients should be positioned using a light beam or lasers to ensure the correct area of interest is captured. Immobilisation devices such as headrests, chin cups, and head straps should be used where appropriate to ensure patient movement is at a minimum throughout the image acquisition.

5.76 If the rotational movement of the tube head or the scanning movement of the image receptor fails to start or stops before the full exposure is completed, the exposure switch should be released immediately to avoid any highly localised exposure of the patient. The reason for any such failure must be investigated, and any faults must be rectified by an engineer before the equipment is used again for clinical purposes.

Cephalometry

5.77 To minimise magnification effects, the focus-to-detector distance (FDD) should be greater than 1 m and ideally within the range of 1.5–1.8 m.

5.78 A cephalostat should position the patient in relation to the x-ray field. The Operator should use the positioning markers incorporated in the cephalostat to ensure correct alignment of the image receptor with the selected collimator unless the equipment is of a type that performs this alignment automatically.

5.79 Scanning digital systems offer significant patient dose savings over static systems (Holroyd 2011).

5.80 In the case of units where a narrow x-ray beam scans across the patient's head, if the scanning movement fails to start or stops before the examination is completed, the exposure switch should be released immediately to avoid any high localised exposure of the patient. The reason for any such failure must be investigated, and any faults must be rectified by an engineer before the equipment is used again for clinical purposes.

Dental cone beam computed tomography

5.81 A dental CBCT image may include anatomical information in regions outside of the area of interest. These must also be evaluated, which may require the services of a dental and maxillofacial radiologist or a clinical radiologist if they extend beyond the dentoalveolar region. Employers must ensure that such arrangements are in place before any exposures are undertaken (PHE and FGDP 2020).

5.82 Patients should be positioned using a light beam or lasers to ensure the correct area of interest is captured. If available, a scout view is a low-dose means of ensuring the correct anatomy is in the field of view.

5.83 Immobilisation devices such as headrests, chin cups, and head straps should be used where appropriate to ensure patient movement is at a minimum throughout the image acquisition.

5.84 It is likely the dental CBCT system will offer a range of acquisition volume sizes; it is important that the smallest volume required to answer the clinical question is used.

Image receptors, image viewing, and film processing optimization

Digital imaging systems

5.85 Digital images should be reported on a monitor of adequate quality that is included in a QA programme, and therefore subjected to checks of condition and calibration that are in line with current guidance (AAPM 2019). The viewing environment should be suitable (e.g. ambient lighting, etc (RCR 2019)). The MPE must be consulted on an appropriate QA programme, see paragraphs 6.36 to 6.45, *Equipment maintenance and testing*.

5.86 In selecting digital equipment, it is necessary to ensure that the chosen system offers the field sizes that are clinically required. Field sizes should be available in a range that is comparable with dental film and compatible with the x-ray systems with which it is to be used.

5.87 The sensitivity of the image receptor system must be compatible with the x-ray unit(s) for which it is to be used. The x-ray unit used should allow for the selection of low exposure times or output to optimise the exposure by making the best use of the sensitivity of the digital imaging system.

5.88 Exposure settings should be reduced to the minimum compatible with the diagnostic quality of the image.

5.89 Digital radiographs can be retaken easily; therefore, it is essential to ensure that all retakes are properly justified and the rejected images are recorded and included in QA statistics.

Intra-oral film

5.90 Intra-oral films should be of ISO speed group F. The use of 'instant process films' should be limited to specific, essential situations (e.g. during surgery or endodontics). It should be noted that, in situations where 'rapid images' are routinely required, conventional films with rapid processing chemistry will generally give better results than instant processing films.

Extra-oral, panoramic, and cephalometric film

5.91 The fastest available film and intensifying screen combination consistent with satisfactory diagnostic results should be used. The speed of the system should be at least 400. The light sensitivity of the film must be correctly matched with the intensifying screens. The condition and effectiveness of the screens should be confirmed at regular intervals as part of the QA procedures. Provided that screens are handled carefully during routine use, cleaning should only be required infrequently. When cleaning is necessary, it should be done in accordance with the manufacturer's instructions.

Film processing

5.92 Strict attention should be paid to correct and consistent film processing so as to produce good-quality radiographs and avoid the necessity for examinations to be repeated. Where automatic processing is used, the processor should be properly cleaned and maintained in accordance with the manufacturer's instructions for use. In the case of manual processing, the temperature of the developer should be checked prior to film processing and the development time adjusted in accordance with the film manufacturer's instructions for use. The developer should be changed at regular intervals in accordance with the manufacturer's instructions for use.

Film viewing

5.93 In order to extract full diagnostic information from the films, it is essential to have dedicated viewing facilities. A specially designed light-box should be installed in an area where the ambient lighting can be adjusted to appropriate levels. Suitable film masking should be used to optimise viewing conditions by cutting out stray light. For viewing dense areas of a radiograph, the incorporation of a high-intensity light source in the light-box is recommended. The provision of magnification by a factor of two would be beneficial.

References

AAPM 2019 *Display Quality Assurance; The Report of AAPM Task Group 270* Alexandria, VA, American Association of Physicists in Medicine ed N B Brevin, M S Silosky, A I Walz-Flannigan and A Badano

BIR *Guidance on Using Shielding on Patients for Diagnostic Radiology Applications* 2020 (London: British Institute of Radiology) https://bir.org.uk/media/414334/final_patient_shielding_guidance.pdf (accessed 9 February 2023)

BIR, RCR, IPEM, SCoR and PHE 2020 *IR(ME)R Implications for Clinical Practice in Diagnostic Imaging, Interventional Radiology and Diagnostic Nuclear Medicine* (London: British Institute of Radiology) https://www.rcr.ac.uk/our-services/all-our-publications/clinical-radiology-publications/protection-of-pregnant-patients-during-diagnostic-medical-exposures-to-ionising-radiation/ (accessed 14 February 2024)

BOS 2015 *Guidelines for the Use of Radiographs in Clinical Orthodontics* 4th edn (London: British Orthodontic Society) www.bos.org.uk/BOS-Homepage/News-Publications/Orthodontic-Radiographs-Guidelines (accessed 11 February 2023)

Brown J, Jacobs R, Jäghagen E L, Lindh C, Baksi G, Schulze D and Schulze R 2014 Basic training requirements for the use of dental CBCT by dentists: a position paper prepared by the European Academy of DentoMaxilloFacial Radiology *Dentomaxillofac. Radiol.* **43** 20130291

CIB 2019 *New Patient Information Posters on the Benefits and Risks of Imaging, RCR Website* (Clinical Imaging Board) https://rcr.ac.uk/posts/new-patient-information-posters-benefits-and-risks-imaging (accessed 5 December 2022)

Faculty of General Dental Practice 2018 *Selection Criteria for Dental Radiography, College of General Dentistry Website* https://cgdent.uk/selection-criteria-for-dental-radiography/ (accessed 4 March 2023)

GDC 2002 *The First Five Years. The Undergraduate Dental Curriculum* 2nd edn (London: General Dental Council)

Holroyd J R 2011 National reference doses for dental cephalometric radiography *Br. J. Radiol.* **84** 1121–4

HSE no date *HSE no date Notify, registration or consent for work with ionising radiation, HSE website.* UK: Health and Safety Executive https://www.hse.gov.uk/radiation/ionising/notifiy-register-consent.htm (accessed 14 February 2024)

HSENI no date *Overview - Notification, registration or consent for work with ionising radiation, HSENI website.* Belfast: Health and Safety Executive for Northern Ireland https://www.hseni.gov.uk/overview-notification-registration-or-consent-work-ionising-radiation (accessed 14 February 2024).

IR(ME)R(NI) 2018 *The Ionising Radiation (Medical Exposure) Regulations (Northern Ireland)* https://legislation.gov.uk/nisr/2018/17 (accessed 31 January 2023)

IR(ME)R 2017 *The Ionising Radiation (Medical Exposure) Regulations* www.legislation.gov.uk/uksi/2017/1322 (accessed 31 January 2023)

IRR(NI) 2017 *The Ionising Radiations Regulations (Northern Ireland)* http://legislation.gov.uk/nisr/2017/229 (accessed 31 January 2023)

IRR 2017 *The Ionising Radiations Regulations.* Great Britain. https://legislation.gov.uk/uksi/2017/1075 (accessed 31 January 2023).

Martin J C, Temperton D H D, Hughes A and Jupp T 2018 *Guidance on the Personal Monitoring Requirements for Personnel Working in Healthcare* (Bristol: IOP Publishing Ltd)

NEBDN 2019 *Training in Dental Radiography Nursing: Intended Learning Outcomes* (Preston: National Examining Board for Dental Nurses) https://nebdn.org/app/uploads/2019/05/Dental-Radiography-Syllabus-1.pdf (accessed 11 February 2023)

NEBDN 2020 *The NEBDN National Diploma in Dental Nursing: Curriculum/Syllabus* 2nd edn (Preston: National Examining Board for Dental Nurses) https://nebdn.org/app/uploads/2020/03/NEBDN-National-Diploma-Syllabus-2020-Website-Version-V2.pdf (accessed 11 February 2023)

NRPB and RCR 1994 *Guidelines on Radiology Standards for Primary Dental Care, Documents of the NRPB* (Didcot: National Radiological Protection Board)

PHE 2016 *PHE-CRCE-023: Guidance on the Safe Use of Hand-Held Dental X-ray Equipment* ed A D Gulson and J R Holroyd. (Didcot: Public Health England) https://gov.uk/government/publications/hand-held-dental-x-ray-equipment-guidance-on-safe-use (accessed 11 February 2023)

PHE and FGDP 2020 *Guidance Notes for Dental Practitioners on the Safe Use of X-ray Equipment (2nd Edition)* 2nd edn (London: Public Health England, Faculty of General Dental Practice (UK)) https://www.rqia.org.uk/RQIA/files/44/449bdd1c-ccb0-4322-b0df-616a0de88fe4.pdf (accessed 31 January 2023)

RCR 2019 *Picture Archiving and Communication Systems (PACS) and Guidelines on Diagnostic Display Devices* 3rd edn (London: Royal College of Radiologists) https://www.rqia.org.uk/RQIA/files/44/449bdd1c-ccb0-4322-b0df-616a0de88fe4.pdf (accessed 11 February 2023)

SCIN 2019 *Risk Posters, Scottish Clinical Imaging Network Website* (Edinburgh: Scottish Clinical Imaging Network) http://scin.scot.nhs.uk/scin-work/risk-posters (accessed 5 December 2022)

SEDENTEXCT Project 2012 *The SEDENTEXCT Project. Radiation Protection No. 172: Cone Beam CT for Dental and Maxillofacial Radiology; Evidence-Based Guidelines* (Luxembourg: European Commission) https://sedentexct.eu/content/guidelines-cbct-dental-and-maxillofacial-radiology (accessed 11 February 2023)

Sutton D G, Martin C J, Williams J R and Peet D J 2012 *Radiation Shielding for Diagnostic Radiology.* 2nd edn (London: The British Institute of Radiology) https://birpublications.org/doi/10.1259/book.9780905749747 (accessed 31 January 2023)

UKHSA 2022 *National Diagnostic Reference Levels (NDRLs) from 19 August 2019*. 24/11/2022, *GOV.UK website* https://gov.uk/government/publications/diagnostic-radiology-national-diagnostic-reference-levels-ndrls/ndrl (accessed 5 December 2022)

John Saunderson, Mohamed Metwaly, William Mairs, Philip Mayles, Lisa Rowley and Mark Worrall

Chapter 6

X-ray equipment for dental radiography

Scope

6.1 This chapter applies to equipment specifically designed for radiography of the teeth or the teeth and jaws. This includes equipment for intra-oral and extra-oral radiography, panoramic tomography, cephalometric radiography, and dental cone beam computed tomography (CBCT). The chapter includes information on the design, specification, radiation protection features, installation, maintenance, and testing of dental x-ray equipment.

General recommendations

6.2 Dental x-ray equipment should be designed, constructed, and installed to comply with recognised British, European, or international standards of construction that will enable the recommendations in this chapter to be met. Medical devices placed on the market in the European Community should meet the relevant essential requirements for safety and performance of the Medical Devices Directive (European Commission 1993), be transposed into United Kingdom law by the Medical Devices Regulations (MDR 2002), and be CE-marked or UKCA as appropriate. Meeting the requirements of the Medical Devices Directive is a declaration that the device meets the essential requirements for device safety.

6.3 The electrical installation for new or re-installed fixed (permanently installed) x-ray equipment should comply with the current version of BS 7671 Wiring Regulations (BSI 2008a) with particular regard to section 710—Medical Locations. Testing and certification of the electrical installation should be performed according to BS 7671, which excludes medical electrical equipment. Electrical safety testing of medical electrical equipment should be performed

doi:10.1088/978-0-7503-2332-1ch6

according to the current version of BS EN 62353 (BSI 2014). Prior to installation, the supplier should provide a pre-installation specification to the purchaser (to include mechanical and electrical safety and installation requirements) to facilitate the safe installation, use, and maintenance of the equipment. The purchaser should identify who does the necessary pre-installation work and confirm that it is done to specification as part of the acceptance process. The electrical safety certificate should be available for inspection by the first user of the equipment, usually concerned with the critical examination or acceptance testing of the equipment, or someone on their behalf.

6.4 Equipment should be maintained in accordance with the recommendations of the manufacturer (MHRA 2021).

Equipment specification

6.5 The Employer must involve an MPE in the technical aspects of equipment specification. The Employer must give the MPE adequate information relating to the intended use of the equipment so they can contribute to an appropriate technical specification. This may include information on the types of procedures that will be performed, the patient cohorts on which the equipment will be used, and the projected patient throughput.

Equipment design

X-ray source assembly

6.6 For fixed intra-oral units, the x-ray source assembly (comprising an x-ray tube, an x-ray tube housing, and a beam-limiting device) should be constructed so that, at every rating specified by the manufacturer for that x-ray source assembly, the air kerma from the leakage radiation at a distance from the focal spot of 1 m averaged over an area not exceeding 100 cm^2 does not exceed 0.25 mGy in 1 h (BSI 2013).

6.7 For hand-held dental x-ray equipment, the x-ray source assembly should be constructed so that the annual effective dose to the operator is unlikely to exceed 0.25 mSv and the annual equivalent dose to the operator's hands is less than 10 mSv during normal use, under all reasonably foreseeable circumstances (PHE 2016).

6.8 For all other dental equipment (panoramic, cephalometric, and dental CBCT units), at every rating specified by the manufacturer for that x-ray source assembly, the air kerma from the leakage radiation at a distance from the focal spot of 1 m averaged over an area not exceeding 100 cm^2 does not exceed 1 mGy in 1 h (BSI 2010).

6.9 The x-ray source assembly should be marked to identify the nominal focal spot position (BSI 2010).

Beam filtration

6.10 The total filtration of the beam (made up of the inherent filtration and any added filtration) should be equivalent to not less than the following (BSI 2010):
(a) 1.5 mm aluminium for x-ray tube voltages up to and including 70 kV, or
(b) 2.5 mm aluminium, of which 1.5 mm should be permanent, for x-ray tube voltages above 70 kV.

6.11 The value of the permanent filtration should be marked clearly on the tube housing, preferably in millimetre aluminium (mm Al) equivalent (BSI 2010). Every added filter should also be clearly marked with its filtration capacity in aluminium equivalent. Where materials other than aluminium have been used as filters, the x-ray tube should be clearly marked with the chemical symbol and thickness in mm of the filter, or marked with the equivalent filtration in mm Al.

X-ray tube operating parameters

6.12 Equipment for dental radiography should incorporate adequate provision for adjustment of exposure factors (a suitable range of kV, mA, and exposure time) to allow for the required range of views, patient size, and detector technology.

6.13 For intra-oral radiography, the nominal tube potential should be within the 60–70 kV range.

6.14 For panoramic and cephalometric radiography with manual control, a range of tube potential settings should be available, preferably from 60 to 90 kV. There should be provision for the selection of a range of tube currents so that full advantage can be taken of the sensitivity of the detector technology being used.

6.15 For dental CBCT radiography, a range of tube potential settings should be available, preferably from 60 to 120 kV. There should be provision for the selection of a range of tube currents, exposure times, and fields of view to allow optimisation of the exposure with regard to patient size, required image resolution, and the area of clinical interest.

Beam size and distance control

Intra-oral radiography

6.16 Rectangular collimation should be provided on new equipment and should be retrofitted to existing equipment at the earliest opportunity, where possible. Rectangular collimation should be combined with beam-aiming devices and image receptor holders, since this not only reduces patient dose but will also improve the diagnostic quality of radiographs and reduce the proportion of rejected images. Rectangular collimators should be designed so that the beam size at the tip of the collimator does not exceed 40 mm by 50 mm (i.e. does not overlap the dimensions of the standard ISO detector size 2 by more than 5 mm at any edge) and preferably does not exceed 35 mm by 45 mm (i.e. no more than a 2.5 mm overlap at any edge).

6.17 Where circular x-ray beams continue to be used, this must be supported by a written justification prepared in consultation with the MPE. The beam diameter should not exceed 60 mm at the end of the beam collimator/director, with a maximum tolerable error of +3 mm. It is stressed that beam diameters less than this and rectangular collimation will significantly reduce patient dose.

6.18 Beam collimators/directors should be open-ended and should provide a minimum focus-to-skin distance (FSD) of 200 mm.

Panoramic radiography

6.19 Equipment must be provided with patient positioning aids, which may be in the form of positioning lasers, light beams, camera imaging, and guidance lines. Some systems may offer AI assistance for improved accuracy.

6.20 Equipment must be provided with the capability to select suitable exposure protocols for different-sized patients, particularly paediatric patients.

6.21 Field size limitations can significantly reduce patient exposure. New equipment should be provided with the capability to select specific regions of the jaw opening for radiography (e.g. sectional views of wisdom teeth) in cases where diagnostic information is only required for those regions. The equipment should collimate the beam automatically.

6.22 The beam height at the received slit or secondary collimator should be restricted (automatically or manually) to no greater than that required to expose the area of diagnostic interest and certainly no greater than the detector in use (normally 125 mm or 150 mm). All primary slits (more than one may be selectable) should be accurately aligned with the receiving slit.

Cephalometry

6.23 Equipment must be capable of ensuring the precise alignment of the x-ray beam, detector, and patient. A means should be provided to allow the beam to be accurately collimated to include only the diagnostically relevant area (BSI 2008b).

6.24 Where a wedge filter is provided to facilitate the imaging of soft tissues, it should be provided at the x-ray tube head in preference to the detector.

6.25 Equipment must be provided with the capability to select suitable exposure protocols for different-sized patients, particularly paediatric patients.

6.26 Limitations on field size can substantially reduce patient exposure. In instances where diagnostic information is only required for select regions of the field aperture, new equipment should be supplied with the capacity to choose specific parts of the jaw openings for radiography. The equipment should collimate the beam automatically based on the area of clinical interest selected by the operator, although manual selection is acceptable for existing equipment.

Dental CBCT

6.27 Equipment must be provided with patient positioning aids, which may be in the form of positioning lasers, light beams, camera imaging, and guidance lines. Multiple scout views, taken from different angles, are beneficial for patient positioning. Some systems may offer AI assistance for improved accuracy.

6.28 Equipment must be capable of a range of field of view (FoV) sizes to fit the full range of diagnostic tasks the system could be used for. These FoV should be automatically collimated, not manually (systems that allow the manual fine adjustment of collimation based on a scout image are acceptable).

Radiation protection features

6.29 For all dental x-ray equipment, the release of the exposure switch should result in the immediate termination of the exposure and, in the case of panoramic, cephalometric, and dental CBCT units, stop the rotation or scanning motions (BSI 2008b, 2009).

6.30 In the case of hand-held intra-oral units, where the exposure switch is necessarily close to the tube and patient, a well-designed unit will incorporate a shield near the tube output to protect the operator from patient scatter and sufficient levels of tube shielding to protect the operator from tube leakage. It should also provide a means of easily switching off the power or removing the battery in the event of a timer failure (PHE 2016).

6.31 When purchasing new panoramic, cephalometric, or dental CBCT equipment, it is recommended that equipment be chosen that is designed to abort the exposure automatically on sensing a failure or interruption of rotational or scanning movement, thereby avoiding an unnecessary and high localised skin dose to the patient. When an exposure is interrupted, the unit should be unable to restart the exposure from the interrupted position.

Equipment installation

6.32 New fixed radiological installations (or existing installations that have been relocated or substantially modified) require a critical examination to ensure that safety features and warning devices function correctly (IRR 2017, IRR (NI) 2017). The critical examination is the responsibility of the installer and must be performed in consultation with the radiation protection adviser (RPA). The content of the critical examination (IPEM 2012) and the results should be communicated to the purchaser, who should consult their own RPA as to their adequacy.

6.33 For new mobile and hand-held units, a critical examination is not required (IPEM 2012). The safety features and warning devices should still be checked prior to first use—these checks are the responsibility of the purchaser and not the provider. The RPA should be consulted on this testing.

6.34 Before radiological equipment enters clinical use, it must undergo acceptance testing to ensure that it performs to specification. The employer must consult an MPE regarding acceptance testing (IR(ME)R 2017, IR(ME)R(NI) 2018).

6.35 Following acceptance, commissioning tests should aim to provide baseline results for subsequent quality assurance (QA) tests. The employer must consult an MPE regarding the QA programme (IR(ME)R 2017, IR(ME)R(NI) 2018).

Equipment maintenance and testing

6.36 Suppliers, erectors, and installers of dental x-ray equipment and the associated automatic processors, computed radiography (photostimulable phosphor plate, often called 'PSP' or 'phosphor plate') readers, and direct digital radiography (often called 'sensor plate') systems are required, under IRR (IRR 2017, IRR(NI) 2017) Regulation 32, and under the Medical Devices Regulations 2002 (MDR 2002), to provide the Employer with adequate information about the proper use, maintenance, and testing of that equipment.

6.37 The radiation safety features of equipment must be properly maintained. It cannot be considered safe from a radiation point of view unless it is in good order both mechanically and electrically. Maintenance and associated checks should be in accordance with the advice of the manufacturer, the supplier, and the RPA. Arrangements for the testing of radiation safety features must be included in the radiation risk assessment.

6.38 Employers are required to establish a QA programme for all dental equipment under IR(ME)R (IR(ME)R 2017, IR(ME)R(NI) 2018) Regulation 15. An appropriate QA programme includes frequently performed, simple tests designed to detect gross changes and trends in performance, usually undertaken by local users. This is supplemented by independent and more complex testing, usually undertaken by a contracted physics department or qualified expert. An MPE must be consulted on the content of the QA programme, which should include everything in the imaging chain, including the x-ray equipment, film processors, phosphor plate readers, sensor plate detectors, and viewing monitors. Guidance on the content of such tests is given in PHE-CRCE-023 *Guidance on the Safe Use of Hand-held Dental X-ray Equipment* (PHE 2016), IPEM Report 91 *Recommended Standards for the Routine Performance Testing of Diagnostic X-Ray Systems* (IPEM 2005), *Guidance Notes for Dental Practitioners on the Safe Use of X-ray Equipment—2nd Edition* (PHE and FGDP 2020), and chapter 6 and appendix 4 of *Radiation Protection No.* 172 *Cone Beam CT for dental and maxillofacial radiology (evidence-based guidelines)* (SEDENTEXCT Project 2012).

6.39 Any display monitor that is used for the reporting of patients' images must be routinely checked to ensure it remains suitable for reporting. There are systems readily available for appropriate QA that involve scoring or making measurements on displayed test screens (AAPM 2019, Brevin *et al* 2020). This display QA should also consider whether the viewing conditions for the monitor are optimal.

6.40 The QA programme must include measurements designed to assess representative patient doses and enable their comparison to local and national diagnostic reference levels (DRLs). The most up-to-date National DRLs can be found on the Department of Health and Social Care website (UKHSA 2022).

6.41 The QA programme should include rejection analysis. The rejection rate for all digital imaging should be less than 5%. For film-based imaging, it should be less than 10% (PHE and FGDP 2020). Where the reject rate is found to be in excess of these values, an urgent review, examining both practice and equipment, is required. This review must involve an MPE.

6.42 A standard operating procedure (SOP) ought to be in place for the QA that local staff performs. The SOP should include details on how to undertake the test, the expected test result under normal circumstances, action levels in the event that the result of the test deviates from the expected, and instructions on what to do in the event that the action levels are exceeded. The SOP should also detail how to record the result of the test in a QA record.

6.43 An MPE must be consulted on the frequency of testing for all equipment. The frequency of the complex testing undertaken by a contracted physics department or qualified expert should be at intervals not exceeding three years (PHE and FGDP 2020).

6.44 Electrical and mechanical faults could give rise to inadvertent radiation exposure, e.g. a faulty cable to a hand switch or failure/malfunction of the rotational or scanning movement mechanisms on panoramic, cephalometric, or dental CBCT equipment. It is expected that these kinds of faults will be identified and rectified by the employer's overall programme of managing work equipment safely. For further advice, see '*L22: Safe use of work equipment. Approved code of practice and guidance for the Provision and Use of Work Equipment Regulations* 1998' (PUWER) (HSE 2014).

6.45 A record of maintenance, including any defects found and their repair, should be kept for each item of x-ray equipment and relevant auxiliary equipment. Following maintenance, the service engineer should provide a written report prior to handing the equipment back for clinical use. This should detail any changes that may affect radiation dose (to patients or staff) or image quality. The RPA and/or MPE should be consulted as necessary. When a maintenance log is provided, there is a legal duty to keep it up to date (HSE 2014).

References

AAPM 2019 *Display Quality Assurance; the report of AAPM Task Group 270* ed N B Brevin *et al* (Alexandria, VA: American Association of Physicists in Medicine) https://www.aapm.org/pubs/reports/detail.asp?docid=183 (accessed 14 February 2024)

Brevin N B *et al* 2020 Practical application of AAPM Report 270 in display quality assurance: a report of Task Group 270 *Med. Phys.* **47** e920–8 (accessed 23 March 2023)

BSI 2008a *BS 7671:2008 Requirements for Electrical Installations* (London: British Standards Institute)

BSI 2008b *BS EN 60601-1-3: 2008 General Requirements for Basic Safety and Essential Performance. Collateral Standard. Radiation Protection in Diagnostic X-ray Equipment* (London: British Standards Institute)

BSI 2009 *BS EN 60601-2-54:2009 Particular Requirements for the Basic Safety and Essential Performance of X-ray Equipment for Radiography and Radioscopy* (London: British Standards Institute)

BSI 2010 *BS EN 60601-2-28:2010 Medical Electrical Equipment. Particular Requirements for the Basic Safety and Essential Performance of X-ray Tube Assemblies for Medical Diagnosis* (London: British Standards Institute)

BSI 2013 *BS EN 60601-2-65: 2013. Particular Requirements for the Basic Safety and Essential Performance of Dental Intra-Oral X-ray Equipment* (London: British Standards Institute)

BSI 2014 *BS EN 62353:2014 Medical Electrical Equipment. Recurrent Test and Test After Repair of Medical Electrical Equipment* (London: British Standards Institute)

European Commission 1993 Council Directive 93/42/EEC of 14 June 1993 concerning medical devices *Off. J. L169* https://eur-lex.europa.eu/legal-content/EN/TXT/?uri=CELEX%3A31993L0042 (accessed 31 January 2023)

HSE 2014 *L22: Safe Use of Work Equipment. Approved Code of Practice and Guidance for the Provision and Use of Work Equipment Regulations 1998* (Norwich: Health and Safety Executive) https://hse.gov.uk/pubns/books/l22.htm (accessed 22 March 2023)

IPEM 2005 *Report 91: Recommended Standards for the Routine Performance Testing of Diagnostic X-ray Imaging Systems* (York: Institute of Physics and Engineering in Medicine)

IPEM 2012 *Report 107: The Critical Examination of X-ray Generating Equipment in Diagnostic Radiology* (York: Institute of Physics and Engineering in Medicine)

IR(ME)R(NI) 2018 *The Ionising Radiation (Medical Exposure) Regulations (Northern Ireland)* https://legislation.gov.uk/nisr/2018/17 (accessed 31 January 2023)

IR(ME)R 2017 *The Ionising Radiation (Medical Exposure) Regulations* www.legislation.gov.uk/uksi/2017/1322 (accessed 31 January 2023)

IRR(NI) 2017 *The Ionising Radiations Regulations (Northern Ireland)* http://legislation.gov.uk/nisr/2017/229 (accessed 31 January 2023)

IRR 2017 *The Ionising Radiations Regulations.* Great Britain https://legislation.gov.uk/uksi/2017/1075 (accessed 31 January 2023)

MDR 2002 *The Medical Devices Regulations 2002* http://legislation.gov.uk/uksi/2002/618/pdfs/uksi_20020618_en.pdf (accessed 31 January 2023)

MHRA 2021 *Managing Medical Devices, Guidance for Health and Social Care Organisations, Health Policy* (Medicines and Healthcare Products Regulatory Agency) https://gov.uk/government/publications/managing-medical-devices (accessed 8 February 2023)

PHE 2016 *PHE-CRCE-023: Guidance on the Safe Use of Hand-Held Dental X-ray Equipment* ed A D Gulson and J R Holroyd (Didcot: Public Health England) https://gov.uk/government/publications/hand-held-dental-x-ray-equipment-guidance-on-safe-use (accessed 11 February 2023)

PHE and FGDP 2020 *Guidance Notes for Dental Practitioners on the Safe Use of X-ray Equipment (2nd Edition)* 2nd edn (London: Public Health England, Faculty of General Dental Practice (UK)) https://www.rqia.org.uk/RQIA/files/44/449bdd1c-ccb0-4322-b0df-616a0de88fe4.pdf (accessed 31 January 2023)

SEDENTEXCT Project 2012 *The SEDENTEXCT Project. Radiation Protection No. 172: Cone Beam CT for Dental and Maxillofacial Radiology; Evidence-Based Guidelines* (Luxembourg: European Commission) https://sedentexct.eu/content/guidelines-cbct-dental-and-maxillofacial-radiology (accessed 11 February 2023)

UKHSA 2022 *Guidance: National Diagnostic Reference Levels (NDRLs) from* 13 October 2022, *GOV.UK Website* https://gov.uk/government/publications/diagnostic-radiology-national-diagnostic-reference-levels-ndrls/ndrl (accessed 7 November 2022)

IOP Publishing

Medical and Dental Guidance Notes (Second Edition)

A good practice guide on all aspects of ionising radiation protection in the clinical environment: IPEM Report 113

John Saunderson, Mohamed Metwaly, William Mairs, Philip Mayles, Lisa Rowley and Mark Worrall

Chapter 7

Radiotherapy

Scope

7.1 This chapter describes the duties and responsibilities of staff in relation to radiotherapy, the written procedures required for clinical radiotherapy, and the design of radiotherapy treatment rooms in relation to the following techniques:

(a) external beam radiotherapy using a collimated beam of ionising radiation (x-rays, gamma rays, beta rays, electrons, or protons), including intra-operative radiotherapy;

(b) brachytherapy using remotely controlled after-loading equipment, which transfers sealed sources from a storage container into applicators pre-positioned at a treatment site and withdraws the sources after treatment. (See chapter 9 for other sealed source brachytherapy, such as permanent seed implants.)

7.2 For the purposes of these guidance notes, remotely controlled after-loading equipment is divided into two classes, high dose rate (HDR) and low dose rate (LDR), giving an instantaneous absorbed dose rate, with the source in the patient, greater or less than 10 mGy h^{-1} at 1 m, respectively. Pulsed Dose Rate (PDR) after-loading equipment is classified as HDR in terms of the potential hazards involved; however, risk assessments should take into account the special control systems involved. It is noted that LDR equipment is being phased out, so only passing reference is made to it.

7.3 Stand-alone C-arm imaging devices, including those in brachytherapy suites, have mostly been replaced by computed tomography (CT) scanners, where the considerations are no different from those of diagnostic scanners, although the number of scans per hour may be less. The radiotherapy imaging room should be designed with adequate protection as defined in paragraphs 1.104–1.115,

doi:10.1088/978-0-7503-2332-1ch7

Controlled and supervised areas, and appendix 6, *Designation of controlled and supervised areas*, following the relevant guidance given in chapters 3 and 4. Particular attention should be given to the differences in equipment parameters and operational procedures between a radiotherapy simulator and a diagnostic fluoroscopic unit, and members of staff operating the equipment should be trained appropriately for the specific uses of the equipment.

7.4 The installation, management, and safe operation of radiotherapy equipment for these techniques are described in chapter 8, *Radiotherapy and brachytherapy equipment*. Particular arrangements for equipment on loan are detailed in paragraphs 2.95–2.98, *Equipment on loan*. A bibliography detailing British Standards relating to radiotherapy equipment is given in appendix B of IPEM Report 81, Physics Aspects of Quality Control in Radiotherapy, 2nd edition (IPEM 2018), together with other national and international documents. In addition to these, the reader is also referred to the Radiotherapy Board's guidance on the implications of IR(ME)R for clinical practice in radiotherapy (Radiotherapy Board 2020) for more detail on clinical aspects of compliance with IR(ME)R.

Roles and responsibilities of duty holders in radiotherapy

7.5 The Employer's duties are discussed in chapters 1 and 2. The Employer is ultimately responsible for the safe use of ionising radiation and for ensuring that the actions discussed below are carried out accordingly.

7.6 The Referrer, as discussed in chapters 1 and 2, must be a registered health care professional (IR(ME)R 2017, IR(ME)R(NI) 2018), and in the radiotherapy process, it will usually be a medical practitioner. Even if the proposal to treat with radiotherapy arises from a multidisciplinary team (MDT) there should still be an individual identified as the IR(ME)R Referrer (such as the clinical/radiation oncologist present at the MDT). The Referrer for concomitant exposures (see paragraphs 2.18 and 2.39–2.44) will depend on local procedures. The Referrer must provide sufficient information about the patient for the IR(ME)R Practitioner to decide if radiotherapy is an appropriate treatment for the patient. The Referrer should be identified in the patient's record. A list of entitled Referrers should be available, but note that this can include references to generic groups of Referrers where the individual identities of Referrers are unlikely to be known. Referral guidelines must be provided that also indicate the scope of practice to which particular Referrers are entitled.

7.7 The IR(ME)R Practitioner in the radiotherapy process will normally be a clinical oncologist, and for brachytherapy will also need an IR(ME)R licence (see paragraphs 9.8–9.10). (Often in this context, the Referrer and the Practitioner are the same person). It should be possible for the staff treating the patient to be clear as to who the IR(ME)R Practitioner is: a list of entitled Practitioners should be maintained and should be available to all clinical staff associated with patient treatment. For paper-based systems, a list of signatures is helpful. The IR(ME)R Practitioner should:

(a) assess the information provided by the Referrer and, if necessary, ask for more information from the Referrer;
(b) decide if radiotherapy is an appropriate treatment for the patient, balancing the risks and benefits;
(c) justify in writing (or in the electronic record) and authorise the medical exposures necessary to deliver the course of radiotherapy treatment, if indeed radiotherapy is the appropriate treatment.

7.8 The IR(ME)R Practitioner will also prescribe the course of radiotherapy treatment (in most cases according to an agreed local protocol). The justification and prescription should be recorded in the patient's record. Justification of a course of radiotherapy treatment may include justification of all the medical exposures involved, including localisation/verification and treatment-related imaging, and CT (or simulator) planning. Another possibility is for the planning exposures to be justified separately. The number of verification images that are authorised should be indicated in the relevant protocols or otherwise made specific. The responsibilities for justification and authorisation must be made clear in the appropriate local protocols. Treatment planning exposures are unlikely to be justified except as preparation for treatment. These exposures should therefore not proceed until the results of any tests that will affect the decision to treat are available, and a system should be in place to ensure this.

7.9 Where patient exposure involves the administration of radioactive substances, the Employer must be licensed by the Licensing Authority for each site (i.e. hospital), and the licence may include research using radionuclides. The Practitioner must also be licensed for the particular use of the radionuclide. This practitioner licence is not tied to a particular location or Employer. Instead, it will be the responsibility of the employer to entitle the Practitioner to operate in the employer's facility. The Licensing Authority (see appendix 22, *Authorities and organisations of interest*) is guided by the Administration of Radioactive Substances Advisory Committee. (See appendix 19, *Guidance on medical research exposures*, for more guidance on research exposures.)

7.10 IR(ME)R Operators in the radiotherapy process will be those appropriately trained personnel who initiate the exposure in either treatment or imaging, carry out or check treatment planning, calibrate, service, or repair radiotherapy equipment, or carry out any other tasks that could affect the safety or effectiveness of the radiotherapy treatment. This includes, for example, identifying the patient. Any individual responsible for any step within the radiotherapy process becomes an IR(ME)R Operator and the Operator responsible for each step should be clearly identified. Staff in training for Operator roles should be under the direct supervision of an Operator. Any treatment accessories (such as immobilisation shells, boluses, or lead cutouts) manufactured within the radiotherapy department should be checked by an appropriately entitled Operator before being used in the treatment of a patient. Operators in a radiotherapy department may be medical staff, radiographers, nurses, physicists, clinical technologists, or other appropriately trained staff. A list of entitled Operators should be available, together with details and dates of their training. Operators may authorise exposures in accordance with guidelines issued by the Practitioner. Particular care must be taken in determining training needs where radiotherapy equipment is used because of the potentially severe consequences of errors. Training should be given and recorded for the specific equipment to be used, and where staff have used similar equipment at another centre, training may still be required and is recommended, as equipment may be configured, set up, and used differently in different departments.

7.11 Medical Physics Experts (MPEs) in radiotherapy must be appropriately certificated (see appendix 15, *Role of the MPE*) and entitled by the employer, and a list of MPEs should be available together with their entitled scope of practice. MPEs in radiotherapy must be closely involved 'in every radiotherapeutic practice other than standardised therapeutic nuclear medicine practices', i.e. in all procedures related to the radiotherapy techniques described in this chapter and chapter 9. This will involve providing advice on and taking oversight of all aspects of dosimetry, including absolute dosimetry of equipment and physical measurements for evaluation of the dose delivered. They will also give advice on the optimisation and safety of treatment and treatment planning, on the quality assurance programme (including quality control procedures), and on any other matters relating to the safety of the radiotherapy equipment or treatment. An MPE should also be consulted before complex radiotherapy equipment or new treatment techniques are introduced. They will contribute in particular to the matters listed in section 3 of Regulation 14 of IR(ME)R, including:

- oversight of the preparation of technical specifications and acceptance testing of new or modified radiotherapy equipment, its commissioning and definitive calibration (see IPEM Report 81 (IPEM 2018)), including selection of appropriate test equipment;
- Employer licence applications under IR(ME)R for brachytherapy procedures;
- collaboration with the RPA (see paragraph 7.12) for aspects of safety of patients, public, and staff in radiotherapy, including, for example, when the installer undertakes the critical examination of new equipment or major upgrades of existing equipment;
- training of Practitioners, Operators and other staff in radiation protection;
- analysis of events involving, or potentially involving, accidental or unintended medical exposures or potentially clinically significant incidents;
- advice to the employer on compliance with IR(ME)R.

7.12 A Radiation Protection Adviser (RPA) for the radiotherapy department should have appropriate experience in the application of radiation protection principles in radiotherapy (see appendix 3, *Role of the RPA*). The RPA should be consulted on all matters specified in Schedule 4 of IRR (2017)/IRR(NI) (2017) relevant to the radiotherapy department, including the local rules for the use of ionising radiation required for all controlled areas within the radiotherapy department. In addition to the specific matters set out in Schedule 4, Employers are required to consult RPAs where advice is necessary for compliance with the Regulations. This should include: (a) the radiation risk assessment required by Regulation 8; (b) the designation of controlled and supervised areas as required by Regulation 17, except where there is good reason to consider that such areas are not required, for example, based on advice from the supplier of the radiation source or written guidance from an authoritative body; (c) the handling of the various investigations required by the Regulations; (d) the drawing up of contingency plans required by Regulation 13; and (e) the dose assessment and recording required by Regulation 22. Due to the higher dose rates and hence higher risks involved in radiotherapy, it is important that an RPA is available to consult and provide the advice required from an RPA in a suitably timely manner, and this should be considered when deciding the number and availability of RPAs appointed.

7.13 The local rules must be appropriate for the radiotherapy equipment and practices used in the areas and should contain contingency plans relating to the equipment or operations, as appropriate, in the event of a radiation incident or emergency (see also paragraph 7.57). Such incidents may include failure to terminate the exposure normally, unusual machine indications, and perhaps a member of staff being left in the room when the machine is turned on.

7.14 An adequate number of Radiation Protection Supervisors (RPSs) should be appointed by the employer and available to supervise the work with ionising radiation and to ensure that the local rules are adhered to. A ratio of one RPS to every 20 radiation workers should be regarded as the minimum. The RPSs should be involved in preparing the local rules for designated areas and should establish and maintain a communication channel with the RPA. The name(s) of the RPS responsible for each area must be recorded in the local rules.

Radiotherapy procedures

7.15 Written procedures, with appropriate document control, are required for the radiotherapy techniques specified in paragraph 7.1. For clinical radiotherapy techniques, there may be several written protocols to allow for differences in procedure between different treatment sites, different complexities of treatment techniques, etc. Clinical procedures should include:

(a) Procedures for patient identification confirmation should involve positive questioning (e.g. ‘what is your name’ rather than ‘is your name …’. Three identifiers should be used. The procedure should specify how the patient should be identified. The procedure should also provide methods of identification where the patient is not able to respond.

(b) Identification of individuals entitled to act as Referrers, IR(ME)R Practitioners, and Operators and their scope of practice. This may be by profession, grade, or individual name. A list of those entitled should be available. Where entitlement is identified by the group, any training requirements should be defined, and it must be possible for the Employer to confirm that an individual is a member of the group. For Practitioners and Operators, the list must include the date on which training was completed and the nature of the training (IR(ME)R 2017; IR(ME)R(NI) 2018 Regulation.17(4)).
(c) Procedures to assess and record treatment plans and treatment exposures, together with estimates of the associated concomitant doses.
(d) Procedures for implementing quality assurance (QA) programmes for written procedures and protocols, and for equipment.
(e) Procedures for providing appropriate information and written instructions to patients who are receiving radionuclide therapy.
(f) Procedures to establish pregnancy status prior to treatment.
(g) Procedures for carers and comforters.
(h) Procedures for medical or biomedical research exposures (see appendix 19, *Guidance on medical research exposures*, for more information on research exposures) including procedures for providing patients with adequate information on the risks and benefits of their treatment.
(i) Procedures to minimise risks of accidental or unintended exposures, and procedures to inform the relevant parties of any such incidents if appropriate.
(j) If applicable, procedures relating to carers and comforters.

See also chapter 2.

7.16 Written procedures are also required for some of the equipment-related procedures described in chapter 8, such as equipment acceptance, calibration, and equipment QA.

7.17 If radioactive sources are being used for therapy, additional procedures will be required:
(a) procedures for ordering and controlling sources, including waste management;
(b) security of sources;
(c) record keeping.

7.18 Where IR(ME)R Practitioners authorise a course of radiotherapy according to a generic local clinical protocol for a particular tumour type or treatment site, that protocol shall provide details of the dose, fractionation scheme, specific treatment techniques to be implemented, and treatment aids that might be used, as well as any associated imaging procedures, including the extent of such imaging, so that what the Practitioner has authorised is clear. The IR(ME)R Practitioner will also provide and sign (either in a manuscript or digitally) a radiotherapy prescription for each patient. This signature will normally be the required evidence of authorisation for the exposures. The clinical protocol should specify any circumstances that might arise where an Operator must refer to the IR(ME)R Practitioner, such as when additional imaging is needed or when to consult an MPE, before proceeding with the treatment.

7.19 Where a local clinical protocol does not exist or is not appropriate for a specific patient, the IR(ME)R Practitioner must ensure that a clear description of the intended treatment is accessible to all Operators in the radiotherapy process. The documentation should also specify any circumstances that might arise where an Operator must refer to the IR(ME)R Practitioner, or consult an MPE, before proceeding with the treatment. It is good practice for a second clinical oncologist to review such off-protocol treatments.

7.20 Particular care should be taken to ensure that all IR(ME)R Practitioners, Operators and MPEs are fully informed and trained when new techniques or protocols are introduced or when there are any changes in these protocols.

Radiotherapy treatment rooms

7.21 This section applies to equipment for external beam radiotherapy and HDR (or PDR) remotely controlled after-loading.

7.22 Radiotherapy equipment should be installed in a treatment room designed for this purpose within a radiotherapy department. Information on the design of treatment rooms is available in IPEM75 *The Design of Radiotherapy Treatment Room Facilities, 2nd edition* (IPEM 2017). Where radioactive substances are involved, consultation with the relevant environment agency (EA, NIEA, NRW, or SEPA) and the Counter Terrorism Security Adviser (CTSA) should take place at an early design stage of the facility.

7.23 Normally, two or more items of therapy equipment should not be installed in the same treatment room. If, however, a second item of therapy equipment is installed in a treatment room (e.g. an HDR remotely controlled after-loader installed in a kilovoltage treatment room), then there must be suitable engineering controls to ensure that only one item of equipment can operate at a time. If such appropriate design precautions are implemented and a suitable risk assessment shows that they are acceptable, it may be reasonable to house two systems in the same room, but the room must be designed to the specification of the equipment associated with the greatest potential hazard.

7.24 The treatment room should be designed with adequate shielding as described in paragraphs 1.104–1.115, *Controlled and supervised areas*, and appendix 6, *Designation of controlled and supervised areas*. The instantaneous dose rate (IDR) for collimated beam equipment should be measured outside each primary barrier, with the radiation beam pointing directly at that barrier. The thickness of shielding should allow for the fact that the area of a wall, floor, or ceiling that produces scattered radiation is much greater when the useful beam is uncollimated (as is the case for after-loading equipment).

In estimating adequate protection at the design stage, the following possible future developments should be considered:

(a) increases in dose rates;
(b) increases in the number of beams per hour and in the dose per beam;
(c) increased use of high-dose-rate unflattened beams and the consequent effect on time-averaged dose rate (TADR);
(d) increase in the use of higher energy photon beams;
(e) increases in the number of total body and hemi-body treatment techniques involving beam projection against a specific wall;
(f) changes in the use and occupancy of adjacent areas;
(g) The dose to secondary barriers for linacs is primarily leakage radiation, and so it depends more on beam-on time (or MU) than on isocenter dose.

Further advice on the design of radiotherapy treatment rooms is available (IPEM 2017). A review of shielding adequacy should take place before any planned modifications of treatment techniques that would affect any of the criteria used to determine the parameters of the shielding are installed.

7.25 It should be noted that the instantaneous dose rate (IDR) averaged over 1 min is measured in order to establish whether special precautions are required in that area in conformance with section 297 of the Approved Code of Practice (ACOP) for IRR (HSE 2018). On its own, IDR is not an appropriate determinant of shielding adequacy, as the dose a person might receive in the area is equally dependent on the beam-on time per day. This is particularly true for flattening-filter-free accelerators, which have very high IDRs for a correspondingly shorter time. Account should also be taken of the increasing use of IMRT, which implies an increase in the beam-on time (and MU) and consequently the required thickness of the secondary barrier without affecting the primary barrier. During machine commissioning, it may be necessary to impose additional restrictions to account for the continuous running of the beam while gathering beam data. Controlled areas should always be designated where the dose rate averaged over the working day (TADR) exceeds 7.5 μSv h^{-1}, or where the IDR exceeds 100 μSv h^{-1} (L121 paragraph 298 (HSE 2018)), which should be easily avoidable for all radiotherapy facilities if appropriate guidelines have been followed for shielding design to keep doses as low as reasonably practicable (ALARP) (IPEM 2017). If new treatment techniques are planned that might affect the beam-on time, then accumulated doses in adjoining areas should be reassessed before the new techniques are implemented.

7.26 Adjacent areas where there may be a radiation risk while equipment is in use, such as roofs, basements, and upper parts of external walls, may not need to be considered controlled or supervised areas if they are declared as prohibited areas and if effective arrangements have been made that physically prevent access whenever the equipment is, or may be, in operation (L121 paragraph 310 (HSE 2018)). Adjacent plant or equipment rooms that are designated as controlled or supervised areas may also be declared prohibited areas, with access for servicing following a system of work detailed in the local rules. All designated areas should be specifically described in the local rules. All systems of work for access, and arrangements for prohibiting access, should be regularly reviewed in consultation with the RPA in conjunction with the appropriate RPS and the local manager. However, consideration must be given to dose rates at accessible points beyond the prohibited area (such as on the floor above a plant room). Where such areas are controlled by employees of other employers, such as building management companies for PFI buildings having control of roof spaces, there should be formal cooperation between the radiotherapy employer and the other employer to ensure safe working for all.

7.27 All treatment rooms will be considered controlled areas. The control panel should be located outside the treatment room. Ideally, the control panel should be positioned so that the treatment Operator can view the entrance to the treatment room. Where this is not possible, a closed-circuit television system should be used to view the entrance to the treatment room. For the special case where the control panel for kilovoltage equipment used for superficial therapy, operating at no more than 100 kV, is inside the treatment room, there should be a protective panel installed to provide adequate protection (see paragraphs paragraph 1.104–1.115, *Controlled and supervised areas*, and appendix 6, *Designation of controlled and supervised areas* and 7.24, but also 7.55) for the treatment Operator. The local rules should detail procedures to ensure that all staff are behind this protective panel before radiation is initiated. In the special case of a PDR machine, it may be appropriate to have the programming console within the treatment room, with the start and stop controls outside the room.

7.28 A similar special case exists when radiation devices are used for intra-operative radiotherapy. An RPA must be consulted to enable safe working practices to be established that are consistent with the clinical needs of the operation.

7.29 If a maze entrance is designed, the protection provided by the door to the treatment room may be reduced, or the door may be replaced by another type of barrier. Light screen barriers are effective but should ensure that a child cannot crawl beneath them. It may also be appropriate to fix a notice advising caution in any maintenance or contingency procedure, because of the weight of the door.

7.30 Electrically operated doors in treatment rooms should have an emergency mechanical or battery-operated mechanism for opening them in the case of electrical power failure. This mechanism should be tested regularly and appropriate staff training provided.

7.31 Observation windows should not be installed in external beam megavoltage radiotherapy treatment rooms. Observation windows in kilovoltage treatment rooms, or HDR after-loading equipment treatment rooms, should provide the same degree of protection as that required of the walls or doors in which they are located.

7.32 Effective interlocks should be provided to restore equipment to a safe state when doors are opened or other access barriers are interrupted or passed. These interlocks should not be reset simply by closing the door or restoring the barrier. All interlock switches must fail to be safe. Where redundant interlocks of the same type are employed, they should operate in opposite modes so that a single failure in the 'on' state will not result in a loss of redundancy.

7.33 A door interlock reset switch should be provided. This should be positioned in a position from which a person leaving the treatment room and pressing the switch has a clear view of all areas of the room and can ensure that no one other than the patient remains in the room. This switch should have a delayed action sufficient to permit the person to exit the room. During this time delay, a second action has to be completed, either by closing the room door or operating a switch outside the room. The local rules should make it clear that the last person to leave the treatment room should check visually that no one but the patient is in the room; only then should that person operate both these switches in series as described. Where there are enclosed spaces within the room, additional measures (e.g. a key interlock) should be in place to ensure the beam cannot be activated if someone is inadvertently hidden from view. Such measures may include a key interlock and/or a warning sound activated when the first switch is pressed. Any such warning sound must not be so alarming as to cause the patient to move from their correct treatment position.

7.34 The access interlock, when reset, should return the equipment from the safe state to a preparatory state and not to a radiation state. These states are defined in the appropriate part of BS EN 60601 Medical Electrical Equipment and Systems—see Report 81(IPEM 2018). Advice on the design of interlocks for equipment safety is given in PD5304:2019 (BSI 2019).

7.35 Means must be provided, normally by a high-quality closed-circuit colour television system, for the treatment Operator to observe the patient from the control panel during treatment. Where parts of the radiotherapy equipment move during treatment, it will be necessary to have more than one television camera to ensure that the patient is visible at all times during treatment, irrespective of the gantry angle. It must also be possible for the treatment Operator to hear and speak to the patient during treatment. This may be done by means of a switched intercom system. It is advisable to have one monitor that has a wider view of the treatment room to detect any unusual occurrences. These systems for communication with the patient are essential for safe radiotherapy, and they must be properly maintained.

7.36 Warning notices, indicating that the room is a controlled area and the nature of the radiation source (Appendix 7, *Warning signs and notices*), should be fixed to treatment room doors or clearly displayed at the entrance to the controlled area. The local rules should be readily available to staff working in the controlled area.

7.37 Illuminated warning signs should be installed at the entrance to the treatment room. These signs should be controlled by the radiation equipment (or, in the case of some after-loading equipment, by an independent radiation detector as discussed in paragraph 7.50) and should normally indicate two states of the radiation equipment: (i) when the equipment is in a preparatory or ready state, and (ii) when radiation is being emitted (the beam-on state). The preparatory or ready state should be indicated by the radiation trefoil and, where appropriate, a 'Controlled Area' or 'Radiation Hazard' legend. The controlled area circuit can use fluorescent, tungsten, or LED bulbs. Alternatively, a three-state system can be used, with 'Controlled Area' being illuminated when the machine is powered on, a second sign labelled 'Do Not Enter' or 'Ready' when the interlock system is completed prior to treatment, and finally 'Radiation On'. The radiation beam-on state should be indicated as 'Radiation On' or 'X-rays On', in red lettering on a black background, using tungsten bulbs or LEDs rather than fluorescent bulbs to ensure immediate indication. One possible arrangement for these signs is shown in appendix 7, *Warning signs and notices*. The legend on an illuminated sign should not be visible when the lamp is off, and ambient lighting adjacent to the sign should be sufficiently dim that the beam-on light is clearly visible.

7.38 Similar warning signs should be installed within the treatment room (and any adjacent designated area), operating in conjunction with those at the entrance to the treatment room. These should be augmented with a continuous audible indication of the radiation state. The audible signal may be a sound that is produced by the equipment when in the radiation beam-on state, or it may be an independent audible signal modulated at a frequency of 500 Hz–800 Hz or another appropriate sound. Modern equipment may be less audible than in the past, so in the former situation, consideration must be given to whether the change in sound is sufficiently clear. When an alarm device is used, the warning sound must be sufficient to indicate that radiation is being emitted, but not be so alarming as to cause the patient to move from their correct treatment position. The operation of these illuminated signs and audible signals should be checked before clinical treatments commence every day. If these signs or audible signals fail to operate, the treatment Operator should stop and report the situation to the Operator responsible for the technical maintenance of the equipment. The equipment should not be used until the signs and audible signals are operational again.

7.39 Emergency-stop switches should be provided at the control panel, at the entrance to the treatment room, within the treatment room, and in any adjacent controlled area. Operation of these switches should cause the equipment to stop emitting radiation and also stop any dynamic motions of the equipment that might put the patient at risk. The switches within the treatment room should be located so that any person accidentally in the room when radiation commences can easily reach an emergency stop switch without passing through the radiation beam and would not be tempted to cross the primary beam in order to reach one. Emergency-stop switches should be of the lock-on type so that the equipment remains in a safe state until the switch is released by an appropriately entitled Operator who will ensure that it is safe to return the equipment to the preparatory state. The procedure to be followed after an emergency-stop switch has been used should be clearly identified in the local rules. Each emergency stop switch should be regularly tested as part of the equipment quality control (QC) procedures. It may be helpful to number each emergency-stop switch and label it accordingly to assist in the recording of the testing. Any required maintenance of emergency-stop switches should be considered within the radiation risk assessment.

Additional requirements for treatment rooms containing electron-beam equipment

7.40 Normally, a treatment room designed to provide adequate protection for x-ray-generating equipment and satisfying the recommendations in the preceding paragraphs will provide adequate protection for electron-beam-generating equipment. The shielding of the treatment room barriers in the directions in which the useful electron beam can be directed should, however, take into account the production of bremsstrahlung in the shielding barrier.

7.41 Treatment rooms designed for electron beam therapy should be well-ventilated, with adequate airflow to remove ozone formed by the irradiation of oxygen in the air. This is particularly important for whole-body electron treatments using long treatment distances and high electron dose rates. Recommendations for ventilation requirements to remove ozone vary from three changes per hour (NCRP 2005) to 15 changes per hour (Elekta) with the IAEA recommending eight changes per hour. Manufacturers' recommendations should be followed, as these include requirements for heat extraction (IPEM 2017), but if total skin electron irradiation is envisaged, more ventilation may be required than if photon-only use is expected. The exposure limits for ozone are given in the HSE publication EH40/2005 *Workplace Exposure Limits 4th Edition* 2020 (HSE 2020). Advice on the health hazards of ozone is given in the HSE publication EH38 *Ozone: Health Hazards and Control Measures 3rd Edition* 2014 (HSE 2014). The legal workplace exposure limit is 0.2 ppm in the air averaged over a 15-minute period, but it is recommended that it be below 0.1 ppm.

Additional requirements for treatment rooms operating with x-ray energies at 10 MeV or higher

7.42 Neutrons may be produced adventitiously by therapeutic beams with an x-ray energy spectral component of 10 MeV and above and by bremsstrahlung from electron beams of similar energy. It should be noted that the nominal energy of a therapeutic beam may not indicate the maximum energy in the beam.

Neutrons may, consequently, form a significant fraction of the stray radiation from megavoltage equipment. This should be taken into account in the design of the treatment room to ensure that adequate protection as defined in paragraphs 1.104–1.115, *Controlled and supervised areas*, appendix 6, *Designation of controlled and supervised areas*, and 7.24 is achieved. Neutron-absorbing materials can produce gamma rays, and these may require additional shielding.

7.43 Induced radioactivity will occur in high-energy accelerators. Materials used for collimation and shielding should be chosen to avoid long-lived activity. Following appropriate measurements and a risk assessment, it may be necessary to delay entry into the room or parts of the room or to restrict service access to the equipment to avoid a hazard from short-lived activity, especially when work is to be done on the treatment head. These restrictions should be included in the local rules (see also paragraph 8.41). The possibility of activated components should be considered when disposing of therapeutic equipment and in compliance with the Radioactive Substances Regulations (RSA 1993, EPR 2016, EA(S)R 2018) as well as the Ionising Radiations Regulations (IRR 2017, IRR(NI) 2017). The matters discussed in paragraphs 7.42 and 7.43 should be specifically considered and referred to in the radiation risk assessment (IRR 2017, IRR(NI) 2017 Regulation 8).

Additional requirements for proton treatment rooms

7.44 High-energy proton beams are a relatively new addition to the radiotherapy armamentarium in the UK. As such, responsible local RPAs and MPEs should consult with those with more experience in this modality to establish the necessary safety precautions.

7.45 Proton beams produce secondary neutrons, and neutron monitoring of personnel is therefore necessary (see also paragraph 8.75).

7.46 Activation by protons and neutrons is a greater potential hazard than that described in paragraph 7.43. Objects within the beam path in the treatment room will become activated, including shells, range shifters, QC devices, and phantoms. Components of the accelerator, energy selection system, and beamline also become highly activated. The risk assessment must include arrangements for the assessment of dose rates, access restrictions, and safe decay storage of these items. A radioactive waste adviser (RWA) should be consulted regarding the storage and disposal of activated materials, and an Environment Agencies (EA, NIEA, NRW or SEPA) permit for the storage and disposal of such radioactive materials will be required.

7.47 Personal protective equipment (gloves and aprons) is required when handling activated water phantoms due to the contamination risk. A contamination monitor should also be available.

7.48 Commissioning, QA and service work should be planned to minimise the risk from activated items. Procedures that involve access to such components (e.g. changing a modulator wheel) must be risk assessed with this in mind, and dose estimates must be obtained and verified with monitoring.

Additional requirements for treatment rooms containing after-loading or gamma-ray equipment

7.49 HDR after-loading equipment, even that designed as mobile equipment, should always be used in a properly shielded and designated treatment area.

7.50 Treatment rooms containing after-loading (e.g. Ir-192) or teletherapy (e.g. Co-60) sources should have a radiation monitor permanently installed in the room. The radiation monitor should have a 'supply on' indicator and should be independently connected to the electricity supply rather than directly to the radiotherapy equipment. Ideally, it should also have a battery supply that automatically provides electrical power in the event of a supply failure. An audible indication of high dose rates is advisable. Installed radiation monitors should be tested, preferably daily, but at least weekly, as part of the equipment QA procedures.

7.51 The action to be taken if a radiation emergency occurs, such as the source transfer mechanism sticking or failing to operate (including shutter closure for gamma-ray equipment), should be specified in the local rules as a local equipment contingency plan (including the issues referred to in paragraphs 7.52 and 7.53). The local rules should also provide a list of people who must be informed and their contact details. A summary of the procedure and contact information should be clearly displayed on the control panel and at the entrance to the treatment room. All treatment Operators should be conversant with this procedure. The local manager, RPA, RPS, and an MPE with knowledge of the equipment must always be consulted about the plan. The emergency plans should be rehearsed and recorded as part of staff training, whenever the operating team changes, and at least once a year for all staff, with a record kept of the rehearsals (see paragraph 7.58). Source changes provide a useful opportunity to do this.

7.52 Any equipment or tool required for use in a radiation emergency must be kept outside the treatment room, close to the control panel, and in a clearly marked location so that anyone entering the room to rectify the emergency does not have to pause to obtain the tool. However, heavy equipment, such as a lead storage pot, is best placed where it will be most easily reached. Such pots may be supplied by the manufacturer of remote after loading equipment, but if this is not the case, an appropriate pot should be kept for emergency use. The equipment or tool should only be used by authorised staff trained in its use, such as the RPS and MPE, and identified in the contingency plan. Every day, before clinical use of the machine commences, a check must be made to ensure that this equipment or tool is in the correct location and that the Operator has been trained in its use.

7.53 A portable radiation monitor should be available and should be used in the event of a failure of the source return mechanisms during the treatment of a patient to identify the probable position of the source or the hazard level so that the patient can be safely removed as quickly as possible. The portable monitor should be battery-operated and tested at least annually. The monitor should only be used by appropriately trained staff, such as the RPS, MPE, or others identified in the contingency plan. Every day, before clinical use of the machine commences, a check must be made to ensure that this monitor is in the correct location and operational.

7.54 Warning notices indicating a controlled area and the nature of the radiation source (Appendix 7, *Warning signs and notices*) should be fixed to treatment room doors and consideration should be given to the radiation associated with the source when it is in the 'safe' position.

Additional requirements for portable x-ray units operating at or below 100 kV

7.55 'Portable' x-ray therapy units normally operating at around 50 kV are available, which may be used for contact therapy or intra-operative radiotherapy (IORT). While dose rates with such devices in contact with the patient may be acceptable without fixed barriers, consideration must be given to the risk of exposure if patient contact is broken. Exposures made for QA purposes may be made at the surface of a phantom, leading to higher scatter doses than in the clinical set-up, so this must also be considered. A detailed risk assessment must be carried out if they are to be used with staff inside the room. In general, hand-held use is not recommended. The requirements of IRR and IR(ME)R are no less relevant if such equipment is used outside the radiotherapy department.

Radiotherapy accidents and incidents

7.56 A risk assessment to identify all hazards with the potential to cause a radiation accident within the radiotherapy department must be undertaken before any new activity involving radiation equipment or sources. This should be undertaken by the employer and the management of the radiotherapy department in cooperation with the RPA and the appropriate MPEs. The risk assessment should be reviewed in cooperation with the RPA at regular intervals or in the event of changes to the equipment, working practices or controls (see Paragraph 78 of L121(HSE 2018)). Detailed guidance on risk assessment is available in Chapter 1, *General measures for radiation, protection* and 19, *Contingency planning and emergency procedures*, and on the HSE website, www.hse.gov.uk/simple-health-safety/risk/index.htm (HSE no date). The risk assessment needs to consider the items listed in paragraphs 70 and 71 of Work with Ionising Radiation (L121 (HSE 2018)) and the requirements of IR(ME)R regulation 8(2), where they are relevant.

7.57 Where the radiation risk assessment shows that a particular type of radiation accident is reasonably foreseeable or foreseeable with serious consequences, such as a source sticking or becoming detached, failure to terminate a beam, or a fire outbreak, the Employer and the management of the radiotherapy department, in cooperation with the RPA, RPS, and the appropriate MPEs, should prepare a contingency plan for the radiotherapy department. The contingency plan should be designed to limit the consequences of any accident that does occur. The contingency plan may comprise different sections relating to different types of equipment or operations, as mentioned in paragraph 7.51, and should detail the immediate action to be taken to limit the consequences of the accident. Appropriate details of the contingency plan should be incorporated into the local rules. Rehearsals of the arrangements in the plan should be undertaken at suitable intervals, depending on the probability and severity of the accident. The contingency plan should be reviewed in cooperation with the RPA at regular intervals, at least annually. The indiscriminate use of emergency stop buttons may cause unnecessary stress to the equipment, and it is not necessary to actually press the button when rehearsing a contingency plan. Radiation risk assessments should identify whether staff who may be involved in such radiation accidents and the implementation of contingency plans need to be classified as persons under IRR regulation 21. General HSE guidance on what constitutes 'reasonably foreseeable' can be found in section 6 of Reducing Risks Protecting People HSE's Decision Making Process (HSE 2001) and specific guidance for radiation sources in Paragraphs 238–240 of L121 (HSE 2018).

7.58 Rehearsals of contingency plans should be recorded and the records kept for a minimum of two years. The actual implementation of a contingency plan should trigger an investigation by the employer with advice from the RPA as described in paragraph 247 of L121 (HSE 2018), and the records of the investigation should be kept for a minimum of two years—see paragraph 1.85 for details.

7.59 Where any accident or other occurrence takes place that is likely to result in a person receiving an effective dose of ionising radiation greater than 6 mSv or an equivalent dose greater than 15 mSv for the lens of an eye or greater than 150 mSv for the skin or the extremities, an investigation must be carried out as described in IRR Regulations 24 and 26 and a report submitted to the appropriate authority under IRR (Appendix 22, *Authorities and organisations of interest*). The unexpected dispersal or loss of a sealed source must also be reported (Regulation 31).

7.60 In the case of loss, theft, or attempted theft of a source, the relevant environment agency (EA, NIEA, NRW, or SEPA) and the police should be contacted. In the case of a fire or flood, the relevant environmental agency should be contacted.

7.61 Significant Accidental or unintended exposures to the patient, whether as a result of equipment malfunction or procedural failure, should be reported to the appropriate authority under IR(ME)R/IR(ME)R(NI) (Appendix 22, *Authorities and organisations of interest*). This includes significant over-doses, under-doses and geometric misses. Guidance from the four UK IR(ME)R regulators on what constitutes such an exposure is available from the CQC website (CQC 2023), and additional advice has been published by the Radiotherapy Board (Radiotherapy Board 2020) and details of reporting codes are available (CQC, RQIA, HIS and HIW 2020).

7.62 Safety or performance issues with equipment should also be reported to the appropriate authority, as listed in appendix 22, *Authorities and organisations of interest*. See also appendix 12, *Incidents—shared learning*.

7.63 An Environmental Permit (EPR permit) is required for the use and disposal of after loading sources. Additional requirements, such as security arrangements and arrangements for disposal, including financial provision, are necessary for High Activity Sealed Sources (HASS).

References

BSI 2019 *PD 5304:2019 Guidance on Safe Use of Machinery* (London: British Standards Institute)

CQC, RQIA, HIS and HIW 2020 *Significant Accidental and Unintended Exposures Under IR(ME)R: Guidance for Employers and Duty-Holders. Version 2, Care Quality Commission* (The Regulation and Quality Improvement Authority, Healthcare Improvement Scotland, Healthcare Improvement Wales) https://cqc.org.uk/guidance-providers/ionising-radiation/saue-criteria-making-notification (accessed 4 February 2024)

CQC 2023 *Ionising Radiation (Medical Exposure) Regulations (IR(ME)R). 19/4/23, CQC website. 19/4/23* (London: Care Quality Commission) https://cqc.org.uk/guidance-providers/ionising-radiation/ionising-radiation-medical-exposure-regulations-irmer (accessed 4 February 2024)

EA(S)R 2018 The Environmental Authorisations (Scotland) Regulations 2018 http://legislation.gov.uk/ssi/2018/219 (accessed 4 February 2024)

EPR 2016 *Environmental Permitting (England and Wales) Regulations. England and Wales* https://legislation.gov.uk/uksi/2016/1154 (accessed 4 February 2024)

HSE 2001 *Reducing Risks Protecting People HSE's Decision Making Process* (Norwich: Health and Safety Executive) https://www.hse.gov.uk/enforce/expert/r2p2.htm (accessed 4 February 2024)

HSE 2014 *EH38 Ozone: Health Hazards and Control Measures* 3rd edn (Norwich: Health and Safety Executive) http://hse.gov.uk/pubns/eh38.htm (accessed 4 February 2024)

HSE 2018 L121 Work with ionising radiation Ionising Radiations Regulations 2017 *Approved Code of Practice and guidance* 2nd edn (Norwich: Health and Safety Executive) https://hse.gov.uk/pubns/books/l121.htm (accessed 4 February 2024)

HSE 2020 *EH40/2005 Workplace Exposure Limits* 4th edn (Norwich: Health and Safety Executive) http://hse.gov.uk/pubns/books/eh40.htm (accessed 4 February 2024)

HSE no date *Managing Risks and Risk Assessment at Work, HSE Website* www.hse.gov.uk/simple-health-safety/risk (accessed 4 February 2024)

IPEM 2017 *Report 75: Design and Shielding of Radiotherapy Treatment Facilities* ed P Horton and D Easton (Bristol: Institute of Physics) 2nd edn

IPEM 2018 *Report 81: Physics Aspects of Quality Control in Radiotherapy* ed I Patel, S Weston, L A Palmer, W P M Mayles, P Whittard, R Clements, A Reilly and T J Jordan (York: Institute of Physics and Engineering in Medicine) 2nd edn

IR(ME)R(NI) 2018 *The Ionising Radiation (Medical Exposure) Regulations (Northern Ireland)* https://legislation.gov.uk/nisr/2018/17 (accessed 4 February 2024)

IR(ME)R 2017 *The Ionising Radiation (Medical Exposure) Regulations* www.legislation.gov.uk/uksi/2017/1322 (accessed 4 February 2024)

IRR(NI) 2017 *The Ionising Radiations Regulations (Northern Ireland)* http://legislation.gov.uk/nisr/2017/229 (accessed 4 February 2024)

IRR 2017 *The Ionising Radiations Regulations.* Great Britain https://legislation.gov.uk/uksi/2017/1075 (accessed 4 February 2024)

NCRP 2005 *Report No. 151: Structural Shielding Design and Evaluation for Megavoltage X- and Gamma-Ray Radiotherapy Facilities* (Bethesda, USA: The National Council on Radiation Protection and Measurements)

Radiotherapy Board 2020 *Ionising Radiation (Medical Exposure) Regulations: Implications for Clinical Practice in Radiotherapy Guidance from the Radiotherapy Board* (London: Royal College of Radiologists) https://www.rcr.ac.uk/our-services/all-our-publications/clinical-oncology-publications/ionising-radiation-medical-exposure-regulations-implications-for-clinical-practice-in-radiotherapy/ (accessed 4 February 2024)

RSA 1993 *Radioactive Substances Act 1993* (http://legislation.gov.uk/ukpga/1993/12) (accessed 4 February 2024)

IOP Publishing

Medical and Dental Guidance Notes (Second Edition)
A good practice guide on all aspects of ionising radiation protection in the clinical environment:
IPEM Report 113
**John Saunderson, Mohamed Metwaly, William Mairs, Philip Mayles,
Lisa Rowley and Mark Worrall**

Chapter 8

Radiotherapy and brachytherapy equipment

Scope

8.1 This chapter is concerned with the following radiotherapy equipment:
- (a) accelerators for megavoltage x-ray and electron radiotherapy and associated record and verification systems;
- (b) kilovoltage x-ray equipment for superficial and orthovoltage radiotherapy;
- (c) gamma-ray therapy equipment (e.g. Co-60 radiosurgery);
- (d) remotely controlled after-loading equipment;
- (e) radiotherapy x-ray imaging devices, including simulators, computed tomography (CT) scanners, and on-treatment verification imaging;
- (f) intraoperative radiotherapy;
- (g) treatment planning equipment and computers.

8.2 For the purposes of these guidance notes, remotely controlled after-loading equipment is divided into two classes, high dose rate (HDR) and low dose rate (LDR), giving an instantaneous absorbed dose rate greater or less than 10 mGy h^{-1} at 1 m, respectively. Pulsed Dose Rate (PDR) after-loading equipment is classified as HDR (see paragraph 7.2). As discussed in paragraph 7.2, only passing reference is made to LDR equipment.

8.3 The guidance in this chapter also applies when equipment intended primarily for therapy is being tested or calibrated at its place of use or used for *in vitro* irradiation.

8.4 Guidance on other types of radiotherapy equipment (such as those involving neutron beams) has not been included in this chapter. For advice on all these types of radiotherapy equipment or techniques, appropriate advice should be sought from the radiation protection adviser (RPA), medical physics expert (MPE), and, where appropriate, the radioactive waste adviser (RWA). Advice may also be

doi:10.1088/978-0-7503-2332-1ch8

available from the HSE, the relevant Health Department and the appropriate professional bodies.

Equipment inventory

8.5 An inventory of all radiotherapy equipment that delivers ionising radiation to a patient or that directly controls or influences the extent of such exposure must be maintained by the Employer. For each piece of radiotherapy equipment, IR (ME)R requires that the following information be recorded:

(a) name of manufacturer;
(b) model number;
(c) serial numbers or other unique identifiers;
(d) year of manufacture;
(e) year of installation

The type of equipment (following the classification of paragraph 8.1) and date of acceptance should also be recorded. The inventory should include: CT scanners, accelerators, kV x-ray machines, after-loading units, planning systems, and record and verify systems. It is not necessary to include dosimetry equipment, with the exception of dose calibrators used to determine the therapeutic radioactivity to be delivered to the patient.

8.6 For equipment containing a radioactive source, details of the source, source activity, and date of installation should be recorded, along with a copy of the radioactive substances regulations' permit or authorisation for the site (RSA 1993, EPR 2016, EA(S)R 2018). There are additional requirements for high activity sealed sources (HASS) (see 8.19). A copy of this information should be available to staff who need to know it. This information will usually be available on the supplier's calibration certificate. For equipment controlled by computer software or firmware, details of the software modules, versions, and dates of installation should be recorded.

8.7 The inventory information must be updated whenever a major upgrade or source change is undertaken. When equipment is decommissioned or removed from the department, this should also be recorded in the inventory. The inventory for the radiotherapy department must be available for inspection by the appropriate authority (see addresses in appendix 22, *Authorities and organisations of interest*).

General radiation equipment safety recommendations

8.8 Radiotherapy equipment should comply with all sections of BS EN 60601 2015 that are relevant to the safe operation of the equipment. Appendix B of IPEM Report 81 Physics Aspects of Quality Control in Radiotherapy 2nd edition (IPEM 2018) lists the relevant sections of BS EN 60601 (note that BS 5724 has been retitled BS EN 60601). See also paragraphs 8.61 to 8.70.

8.9 For all types of radiotherapy treatment equipment, with the exception of equipment described in paragraph 7.55, *Additional Requirements for*

Portable x-ray units operating at or below 100 kV, it should only be possible to commence an exposure from the control panel. The control panel should give a clear and unambiguous indication of the treatment mode selected. The treatment Operator should be required to confirm that the correct mode has been selected. The operational state of the equipment should also be clearly and unambiguously indicated on the control panel. Operational states are defined in the appropriate section of BS EN 60601–2–1 (IPEM 2018, appendix B). The control panel should also indicate the choice and correct location of optional treatment accessories such as electron applicators, etc. Where there is a choice of available accessories, a select and confirm system should be provided. The control panel should provide means of initiating the radiation exposure, interrupting the radiation exposure, and terminating the radiation exposure.

8.10 It should be possible to inhibit or disable the facility to initiate radiation exposure from the control panel by means of either a key switch with a removable key or a password entered via a keyboard at the control panel. A clear indication should be available at the control panel if the treatment has been terminated by any event other than the operation of the primary dosimetry or timer system.

8.11 The control panel should indicate the dose delivered during the emission of radiation. In order for the magnitude of any exposure to be obvious to the treatment Operator, all dose indicators should count up from zero. All equipment should be provided with two or more detectors for monitoring and displaying dose and dose rate, as well as a back-up timer. This requirement is not appropriate for after-loading equipment. Details of the requirements for these systems for different types of radiotherapy equipment are given in the relevant sections of BS EN 60601 (see IPEM 2018 appendix B).

8.12 Wherever practicable, interlocks and safety tripping mechanisms should be designed so that their operation or non-operation is evident. The manufacturer should offer facilities so that testing of the operation of interlocks and safety tripping mechanisms can take place.

8.13 Suppliers and manufacturers of equipment should provide appropriate operating manuals for the equipment. They should also provide technical manuals and, where appropriate, circuit diagrams, giving details of calibration and QC procedures. These manuals, or other documents provided, must contain adequate information about safe use, testing, and maintenance of the equipment. Manuals and circuit diagrams should be updated when appropriate as part of the quality assurance (QA) programme.

Radiation risk assessment and authorisation to begin practice

8.14 A new activity involving the therapeutic use of radiation may not begin until a radiation risk assessment has been made to identify hazards, assess the risks and determine controls, and has been recorded in writing or electronically.

This assessment must be kept up-to-date if there is any significant change in the equipment or its operational use. Risk assessments should also be reviewed at suitable intervals, including a review of all assumptions on workload, techniques, etc used in the assessment. Further guidance on risk assessments is available in Work with Ionising Radiation L121 (under Regulation 8) (HSE 2018) and in Regulation 3 of the Management of Health and Safety at Work Regulations 1999 (MHSWR 1999) (see paragraphs 1.44 to 1.54, Risk Assessment). Risk assessments should include an estimate of doses and dose rates likely to be received in connection with operating the equipment, including in an accident scenario, and should be reviewed annually.

Notification, registration, and consent

8.15 Details of the notification, registration, and consent processes are available from the HSE and HSE(NI) websites (HSENI no date, HSE no date) and in chapter 1.

8.16 Notification of work with ionising radiation must be made to the HSE in Great Britain or the HSE(NI) in Northern Ireland. In practice, all use of radiotherapy equipment will require either registration or consent from HSE or HSE(NI).

8.17 Registration is required for the operation of a radiation generator or a radioactive source (not including a High Activity Sealed Source (HASS) for which consent is required). This category includes kilovoltage x-rays for both imaging and treatment. Only one registration is required per employer. The requirements for registration are:

- completion of a radiation risk assessment;
- existence and rehearsal of contingency plans;
- estimation of employee doses and action to minimise these;
- appointment of an RPA;
- provision of adequate training including update training;
- designation of controlled and/or supervised areas and creation of local rules and appointment of Radiation Protection Supervisors (RPSs) as required.

8.18 Consents from HSE or HSE(NI) are required for the administration of radioactive substances (brachytherapy and radionuclide therapy), for the operation of an accelerator, and for the use of a HASS source. This will apply to most radiotherapy departments. Consent is associated with an Employer not a site. The requirements for consent in addition to those in paragraph 8.17 are:

- a programme to monitor and audit compliance with the ionising radiation regulations;
- nomination of a manager to be responsible for radiological protection
- control of radiation sources no longer in use;
- appropriate engineering controls to maintain staff safety;
- training of employees not engaged in the practice;
- knowledge of expected doses to staff and public as a result of the practice;
- an equipment quality assurance programme.

8.19 HDR after-loading equipment generally contains a source whose activity is defined as an HASS. In addition to HSE or HSE(NI) consent, HASS sources also have to be permitted by the relevant environment agency, which will impose certain requirements. These will include:

(a) the requirement to report whenever a source is changed;
(b) five yearly reporting;
(c) appropriate security as defined by National Counter Terrorism Security Office (NaCTSO) guidelines;
(d) an appropriate financial arrangement to ensure that the source can be disposed of when it is no longer in use;
(e) appropriate security arrangements for staff.

Installation and commissioning of radiotherapy equipment

8.20 Before new fixed equipment is used clinically, Regulation 32(2) (IRR 2017, IRR(NI) 2017) requires a representative of the installer to carry out a critical examination of safety features in consultation with an RPA in respect of its nature, extent, and results. The RPA could be appointed by the installer or by the employer engaged in work with ionising radiation. Specific attention should be paid to leakage from the radiation head, Operator-controlled beam and motion termination, machine-controlled beam and motion termination, visible and audible indications of the operational state of the radiation equipment, and the operation of safety-critical interlocks. A report of the examination should be made to the employer following a review by an RPA on the outcome and should be kept for reference during the working life of the equipment. The critical examination should be repeated after any major upgrade to the equipment.

8.21 An MPE should be responsible for the acceptance testing of new or modified radiotherapy equipment (see paragraphs 2.88 to 2.94, *Acceptance*, commissioning and routine testing, and 1.246 to 1.260 *Critical examination*). Radiotherapy equipment should comply with the appropriate sections of BS EN 60601 (see appendix B of IPEM Report 81(IPEM 2018)) relevant to the safety of equipment. If the manufacturer or supplier has an acceptance testing protocol, this should also be followed. Further guidance on the acceptance testing and commissioning of radiotherapy equipment is given in IPEM Report 94, Acceptance Testing and Commissioning of Linear Accelerators (IPEM 2007), and IPEM Report 81, Physics Aspects of Quality Control in Radiotherapy, 2nd edition (IPEM 2018). The radiotherapy network should be maintained in accordance with IPEM Report 93, Guidance for the Commissioning and Quality Assurance of a Networked Radiotherapy Department (IPEM 2006).

8.22 All available combinations of treatment mode and treatment accessory should be tested as part of the acceptance procedure. If any combination is not tested or accepted, then the MPE should ensure that all Operators are informed that these combinations cannot be used clinically until they have been accepted. Equipment such as treatment planning systems and dosimetry equipment, which can affect the safety of radiotherapy treatment, should also undergo acceptance testing.

8.23 Before the equipment is brought into clinical use, adequate testing of the equipment must be carried out to establish baselines for the QA programme; all appropriate data should be recorded. All safety-related features of the equipment should be examined and, where appropriate, demonstrated or tested, including those associated with the treatment room.

8.24 An MPE should be responsible for the commissioning of the necessary data for planning and checking radiotherapy treatments. Appropriate testing should be undertaken to ensure the integrity of equipment data transferred to and from subsidiary equipment, such as treatment planning systems. Systems or software set up to check radiotherapy treatments should, as far as possible, be independently commissioned.

8.25 Where equipment can be customised by modifying accessories or the control system software, an MPE should be responsible for ensuring that details of all customisation are recorded. All the appropriate Operators should be informed by the MPE of the implications of any customisation both for treatment techniques and for patient safety. Any such changes must be authorised by the manufacturer in order for the equipment to retain its regulatory compliance.

Calibration

8.26 A definitive calibration should be performed before the equipment is first used clinically, as described in appendix A of IPEM Report 81, 2nd edition, Physics Aspects of Quality Control in Radiotherapy (IPEM 2018). When source changes are undertaken in after-loading equipment, appropriate measurements should be taken to confirm source strength and properties. These measurements should be checked by an MPE, who will then be responsible for issuing appropriate source data for clinical use.

8.27 The appropriate Code of Practice for radiotherapy dosimetry should be used for all measurements. The Codes of Practice used should be those recommended by the Institute of Physics and Engineering in Medicine (IPEM 2018), or if no IPEM code yet exists, then a suitable international code. In general, a department should adopt new Codes of Practice within three years of their recommendation. Details of the current Codes of Practice are listed in the bibliography of IPEM Report 81 (IPEM 2018). A new IPEM Code of Practice for megavoltage beams was published in 2020 (IPEM 2020).

8.28 An MPE should ensure that the calibration of all radiotherapy equipment is checked and recorded at regular intervals. Recommendations on minimum frequencies of dosimetry calibration are given in IPEM Report 81, 2nd edition (IPEM 2018). The date and results of calibration measurements and recalibrations should be entered in a calibration record as part of the QA system. These records should be signed or countersigned by an appropriate physics Operator and should be available for inspection by the appropriate authority as given in appendix 22, *Authorities and organisations of interest.*

8.29 Any significant changes in output or absorbed dose rate should be reported to, and the causes investigated by, an MPE, who should ensure that any necessary action is taken before the equipment is used clinically again (IPEM 2018). This is particularly important for any equipment that does not have an integrated dosemeter, such as some equipment for superficial therapy.

8.30 Whenever service or maintenance procedures or repairs have been undertaken that might have involved adjustments or changes to the equipment that could alter the output calibration for any reason, the equipment should be checked according to the principles in appendix A of IPEM Report 81 (IPEM 2018) before it is used clinically.

8.31 Dosemeters and dosimetry equipment used for calibrations and output checks must be maintained in good condition and appropriate tests made to ensure that their sensitivity remains within acceptable limits. Their sensitivity (traceable to a national primary standard) should be calibrated according to appendix A of IPEM Report 81 (IPEM 2018) and checked at least annually over the range of radiation qualities normally used within the department. Further guidance on the QA of dosimetry equipment is given in IPEM Report 81, 2nd edition (IPEM 2018).

Quality assurance

8.32 QA testing of all radiotherapy equipment should be included in the radiotherapy department's QA programme. An MPE must be closely involved in, and will usually be delegated responsibility for, ensuring that appropriate QA tests are undertaken at suitable frequencies. Guidance on QC programmes and the frequency of testing is given in IPEM Report 81 (IPEM 2018).

8.33 The MPE should ensure that appropriate corrective action is taken if the results of any test indicate that the safety of the patient or the accuracy of the treatment may be compromised. Records of QA testing should be signed or countersigned by an appropriately trained Operator and be available for inspection by the appropriate authority, as given in appendix 22, *Authorities and organisations of interest.*

8.34 The QA testing programme should include programmable electronic control systems that control the operation of radiotherapy equipment. Guidance on programmable electronic systems in safety-related applications is available in BS EN 61508 2010 (BSI 2010). Particular attention should also be given to the QA of systems such as treatment record and verify systems, computer media, and networks that manipulate, store, or transfer radiotherapy data used for specifying patient treatments (IPEM 2006).

8.35 The QA testing programme should also include any non-radiation equipment, such as treatment planning systems and dose measuring equipment, that can affect the safety of radiotherapy treatment. This may include ancillary equipment such as contrast injectors and gating systems. Guidance on the QA of treatment planning equipment is given in IPEM Report 81, 2nd edition (IPEM 2018).

8.36 Radiotherapy simulators, CT scanners, and on-treatment imaging should be included in the QA testing programme. Special attention should be paid to the accuracy of the mechanical parameters as well as the fluoroscopic and radiographic parameters.

Maintenance and servicing of radiotherapy equipment

8.37 Treatment and pre-treatment equipment Operators should report all faults that could compromise the safety or accuracy of patient treatments to the appropriate staff. These should be investigated and, if necessary, corrected before the equipment is used again clinically. Local procedures should identify the action to be taken and the conditions under which clinical use of the equipment should stop.

8.38 A record of defects and maintenance should be kept for each item of equipment. This record should be examined regularly to identify degradations in equipment performance or systematic faults. Further information on maintenance logs is available in the Guidance for Provision and Use of Work Equipment Regulations (PUWER 1998, PUWER(NI) 1999).

8.39 A clear indication should be available at the control panel when equipment is being serviced or repaired. This may be done by means of a notice clearly indicating that the equipment is being serviced. There should be a procedure specifying clear handover arrangements between treatment and service staff. Safety precautions for maintenance and repair procedures should also be described in the local rules (e.g. paragraph 8.41).

8.40 When gamma-beam teletherapy equipment is being maintained or repaired, the source or shutter should be locked in the 'off' position. As the source is still emitting in this position, the room should be entered only by staff for the minimum time associated with the maintenance or repair of the equipment.

8.41 When maintenance or repair work has been done in the vicinity of the electron or proton path, the x-ray target, the flattening filter or scattering foils, or the magnet system of any electron accelerator operating at or above 10 MeV or any proton accelerator, monitoring for possible induced radioactivity should be carried out. (Neutron yields at 10 MeV are about 30 times less than at 15 MeV.) Such monitoring should take account of any components present in the equipment that are made from uranium (see paragraph 8.59). If significant activity is detected, protective measures such as remote handling and the use of personal protective equipment (PPE) or clothing should be taken, or further work should be postponed until the short-lived activity has decayed. Equipment and components for which these precautions are required should be identified in the local rules, which should also detail the appropriate safety procedures. Anyone carrying out maintenance or repair work should be in full compliance with IRR before the work starts; there must be a specific radiation risk assessment in place for the work.

8.42 Any maintenance or service operation that might alter the radiation output, quality of the radiation, or shielding of the radiation source should be notified to the staff responsible for the technical operation of the equipment, and there should be a written procedure for this. The equipment should not be used clinically until the appropriate parameters have been checked. A local QA procedure should be available to ensure that these checks are undertaken and that the machine is not used clinically until the appropriate calibration testing has been undertaken and recorded in the calibration record. An MPE should have oversight of this procedure.

8.43 When maintenance or repair work is carried out by external service engineers, great care should be taken to ensure that the equipment is safe for use when this work is completed. A local QA procedure for the handover of equipment should be used. It is of particular importance that the appropriate procedures discussed in paragraph 8.42 are undertaken. The handover process must contain a clear statement of the work done that is sufficient to enable the MPE to decide if the equipment requires QA or recalibration. Handover forms should be kept for future reference. An appropriate form has been agreed upon between HSE and the AXREM, and this (or a similar form) should be used. There should be clear information on any handover form showing who is responsible for the controlled area while any work is being carried out and that all relevant aspects of IRR 2017/IRR(NI) 2017 are in place.

8.44 Before treatments are commenced following any maintenance or experimental work, a test exposure should be made to ensure that the equipment is functioning correctly and that all safety interlocks are operational.

Loading or exchanging sources in therapy equipment within a hospital

8.45 Great care is essential when loading or exchanging sources in therapy equipment, in view of the high activity of these sources and because it may be necessary to override safety interlocks. A system of work for this procedure should be set out in writing, taking into account the following paragraphs. A suitably experienced medical physicist, such as a radiotherapy MPE, should supervise radiation safety during such a procedure.

8.46 The procedure to be adopted should be drawn up by the employer responsible for the source change in consultation with their RPA. Any technical instructions provided by the equipment manufacturer should be strictly followed. Since the work in the controlled area will not be covered by the hospital's clinical work local rules, it is appropriate for the manufacturer to take responsibility for the controlled area, and it is then their responsibility to ensure its safety. However, it is appropriate for the procedures to be agreed upon with the hospital, as, if things go wrong, the hospital staff may need to be involved in the solution. If the hospital is responsible for radiation safety, an RPS should be nominated to ensure compliance with the written procedure.

8.47 Before the work is commenced, a risk assessment should be carried out, covering any reasonably foreseeable complications and including estimates of the expected associated doses, including in accident situations. Based on this, a written contingency plan should be drawn up in consultation with the RPA and, if appropriate, the RPS supervising the procedure. All equipment required to implement any contingency plan should be available prior to the work starting. All parties involved should understand the plan and undertake appropriate rehearsals.

8.48 Loading and unloading should be carried out by at least two people (who may need to be designated as classified radiation workers) who are properly trained and experienced. They should be the only people authorised to enter or remain in the room during the procedure, apart from the RPS, as discussed in paragraph 8.52.

8.49 If the work is carried out by an external contractor, either with his or her own engineers or in collaboration with hospital staff, responsibility should be clearly defined in writing. These operations are not frequently carried out in some departments, and they require good co-ordination between individuals who do not usually work together (i.e. contractors and hospital staff).

8.50 The transfer container should be positioned so that all persons remain adequately shielded from the source throughout the transfer operation and personal doses are as low as reasonably practicable.

8.51 Each person involved in source changing should wear a direct-reading electronic personal dosemeter covering the anticipated dose range and a range one order of magnitude higher, in addition to the normal passive personal dosemeter and any audible alarm.

8.52 The RPS supervising the operation should be in addition to those concerned with the actual loading or unloading. The operation should be monitored and timed, and, in case of difficulty, instructions should be given for a pre-arranged contingency plan to be followed.

8.53 As part of the critical examination required when the operation has been completed, a check should be made to ensure that all interlocks function and that all other safety features are fully operational. The results of the critical examination need to be recorded and consulted upon with an RPA.

8.54 Appropriate arrangements must be in place for the secure storage of the source (s) when not contained in the therapy machine. Advice should be sought from the local Counter Terrorism Security Adviser (CTSA)

8.55 When the old source is being returned to the manufacturer, it is essential that the transport staff be personally identified to avoid the high activity source falling into the wrong hands. Transport Regulations must be followed and the sources appropriately consigned by a trained member of staff (see chapter 18, *Keeping, accounting for & moving radioactive substances (including transport)*). Advice from a Dangerous Goods Safety Adviser (DGSA) should be sought (CDG 2009, CDG(NI) 2010, ADR 2022).

8.56 Sealed radioactive sources for gamma-beam therapy or remotely controlled after-loading equipment should have been tested by the manufacturer before delivery and should be accompanied by a leakage test certificate. If this is not so, the source should be tested for leakage before it is loaded into the equipment or the equipment is brought into clinical use. Sources should be retested at least once every two years, but for those that are in frequent use for remotely controlled after-loading, leakage tests should be made at least once a year. These sources may be subject to mechanical wear, and there would be a particular hazard to a patient or staff if leakage occurred. A test should be made immediately if any damage to the source is suspected. It is usually unnecessary and likely to be more hazardous to make direct leakage tests on installed gamma sources. Instead, the equipment should be tested for leakage at the sites indicated in table 8.1. Where equipment contains components made from uranium (see paragraph 8.59), there is a need to carry out leakage tests for these components using an instrument able to detect alpha-emitting material. BS EN ISO 2919–2014 (BSI 2014b) contains guidance on sources. It recommends that each source have a recommended working life designated by the manufacturer. If this life is exceeded, then a qualified body (e.g. the manufacturer) should inspect it and carry out a technical assessment, including leakage testing. Leakage test methods are described in BS ISO 9978 (BSI 2020).

8.57 If, in a leakage test, the activity measured on the swab is less than 200 Bq, the source(s) may be considered leak free (BSI 2020). Immediate steps should be taken to prevent the spread of contamination if the measured activity is greater than 200 Bq, followed later by the removal of the leaking source and decontamination of the equipment. All leakage tests should be recorded as part of the QC testing programme.

Decommissioning of radiotherapy equipment

8.58 The target and collimators in accelerators operated at or above 10 MeV may become activated during use. It is necessary to determine whether, when these items become waste, they require a permit, fall within an environmental regulations permit exemption, are out-of-scope, or, in Scotland, require a

Table 8.1. Leakage testing.

Equipment	Place to be wipe-tested
Gamma-beam therapy equipment	The surface of the radiation head including the treatment aperture
Remote controlled after-loading equipment	The internal surfaces of the transit tubes after ensuring that all the sources have recently passed through the transit tubes

permit, registration, notification, or fall within general binding rules. Guidance on these issues for England, Wales, and Northern Ireland can be found in the following guidance document: Scope of and Exemptions from the Radioactive Substances Legislation in England, Wales, and Northern Ireland (BEIS, DEFRA, Welsh Government and DAERA 2018) and for Scotland in the Authorisation Guide for Radioactive Substances Activities (SEPA 2020).

8.59 Steps should be taken to ascertain whether accelerators or gamma-beam teletherapy equipment contain depleted uranium shielding or collimators, although this is now unlikely. If depleted uranium is present, the items concerned should be permitted or registered under the relevant Environment Permitting Regulations (RSA 1993, EPR 2016, EA(S)R 2018 as appropriate) and disposed of according to the guidance in chapter 19, *Disposal of radioactive waste*.

8.60 The decommissioning of proton beam equipment is a more significant issue than for other sources because of the induced activity likely to be generated in the shielding structure in addition to the components of the accelerator. Of particular concern will be any steel (or other high atomic number material) in the shielding structure of the walls. The Environment Agency permit will require a decommissioning plan to exist as part of the permit conditions, and this should be reviewed every five years. The plan will need to consider the potentially increasing level of radioactivity in the components of the accelerator. Some information on decommissioning can be found in the IAEA publication TRS414 (IAEA 2003), which refers to an EU publication (European Commission 1999) that provides information relating to the decommissioning of a 40 MeV cyclotron.

The safe operation of radiotherapy equipment

8.61 The treatment Operators and all other persons except the patient should be outside the controlled area when the equipment is delivering the therapy. If, for compelling clinical reasons, it is necessary for a 'carer or comforter' as defined by IR(ME)R (see chapter 2) to be in the controlled area during kilovoltage treatment (e.g. 50 kV treatments), the MPE should be consulted before the treatment course commences. For other people, such as employees, the RPA should be consulted. A written system of work identifying the individual person, other than the patient, who is to be in the controlled area and any associated restrictions or precautions that must be observed (based on appropriate dose constraints), should be available at the treatment room for this specific case. For a carer or comforter, the resulting exposures must be justified by a Practitioner demonstrating a benefit to the patient proportionate to the potential detriment to the carer. A person other than the patient should never be in the treatment room during megavoltage treatments.

8.62 The treatment Operator should make sure that only the patient is in the controlled area and that the entrance is unoccupied before setting the door or entrance interlock. Particular care is needed if there are areas within the treatment room or maze in which a person could be hidden from view, and consideration should be given to a more appropriate position (see paragraph 7.33).

8.63 To avoid errors in patient treatment, a strict procedure for the operation of the equipment is essential. This may be done by means of a work instruction in the QA system. The work instructions should clearly define the extent of the responsibility of the different Operators. The work instructions should include the checking of the identity of the patient and the operating conditions and set-up parameters by the treatment Operator before each treatment. Before treatments are commenced following any maintenance or experimental work, a test exposure should be made to ensure that the equipment is functioning correctly and that all safety interlocks are operational.

8.64 If there is a potential hazard to the treatment Operator from activated treatment accessories, e.g. removable wedges in accelerators operated at high photon energies or modulator wheels in proton facilities, the hazard and any necessary precautions should be fully considered in the radiation risk assessment and identified in the local rules.

8.65 The treatment Operator(s) should report immediately to the person specified in the local rules or governing procedure any of the following circumstances:

(a) if the treatment is terminated or interrupted by any machine control interlock (i.e. excluding human actions such as an interrupt button or the door interlock) other than the primary dose-integrating system (or primary timer in the case of timer-controlled therapy equipment);
(b) if any interlock or trip switch is observed not to function correctly;
(c) if any emergency stop switch is activated;
(d) if any other parameter falls outside locally defined limits.

8.66 The equipment should not be used again until the circumstances have been investigated and a safe mode of operation has been confirmed or re-established.

8.67 Where a back-up timer is provided, it should be pre-set to a time 10% greater than the estimated duration of treatment or, on the advice of the MPE, to an appropriate value that takes into account variations in source transit times or x-ray tube ramp up.

8.68 All modern radiotherapy equipment with a nominal beam energy greater than 1 MeV (installed since the implementation of the revised IR(ME)R in 2018) must have a device or other feature to verify key treatment parameters. Usually, the Oncology Information System will be able to fulfill this.

8.69 Whenever radiotherapy equipment is left unattended, it should be left in a safe state. The local rules should describe arrangements for access to treatment rooms by cleaners and hospital building maintenance staff.

8.70 All radiotherapy doses and treatment parameters should be recorded. Details of the length of time for which treatment records must be kept are given in appendix 10, *Record keeping*.

Additional considerations for whole-body electron treatments

8.71 Where an accelerator is used for whole-body high-dose-rate electron therapy, it should incorporate appropriate design features and interlocks to ensure safe operation. The control panel should give a clear and unambiguous indication when a high dose rate electron mode is selected. The treatment Operator should be required to confirm that this mode has been selected.

8.72 Before whole-body electron treatments involving extended treatment distances and high electron dose rates, a dummy run should be performed to check that the dose rate is within acceptable limits.

Particular considerations for proton accelerators

8.73 Protons can be generated in a cyclotron, synchrotron, or synchrocyclotron accelerator. A cyclotron produces protons at a specific energy, and these are then degraded when a lower energy is required, whereas a synchrotron can be tuned to produce a beam at a lower energy directly, while a synchrocyclotron is a frequency-modulated (FM) cyclotron designed to overcome the energy limitation in classical cyclotrons. The accelerator will normally be housed in a vault, with access restricted to a limited number of people. During machine running, the dose rate of neutrons will be significant, and this causes activation of the components of the accelerator and the walls of the room. To prevent the escape of thermal neutrons, access to the vault will be through a maze-structured access system. A second significant source of neutrons specific to high energy cyclotrons is the Energy Selection System (ESS).

8.74 Access restrictions are always necessary for the accelerator enclosure in order to allow activated components to decay to a level that gives an acceptable dose rate in the areas to be accessed. It may be prudent to set a higher acceptable dose rate limit for emergency work, provided this is allowed alongside a suitable time restriction, in order to restrict personal dose to as low as reasonably practicable (ALARP). Wearing electronic personal dosemeters with alarms for both dose and dose rate is advisable when entering these areas. It may be considered advisable to monitor the dose rate in the accelerator enclosure remotely and to provide an audible warning of high dose rates. The dose rate will vary depending on how recently the machine has been running—much of the induced activity has a short half-life. An audible warning system, including last-man-out buttons, should be provided (see paragraph 7.33). As the accelerator enclosure is not a patient area, the audible warning can be considerably louder than in a treatment room, especially since entering and exiting will not be a routine occurrence that takes place many times each day.

8.75 Monitoring neutron dose rates is much more complex than monitoring photon dose rates because the measurement of neutron sieverts needs to account for the variable quality factor at different neutron energies. There will always be an associated photon dose rate, so a photon detector will also have some sensitivity to neutrons. The choice of instrumentation will depend on the radiation field being measured, and expert advice should be sought.

8.76 Access to the accelerator for maintenance must be carefully controlled. Where it is necessary to disassemble the accelerator, a detailed safety plan should be developed in consultation with an RPA. This should include monitoring activity levels and allowing appropriate time for the decay of short-lived radionuclides.

8.77 In designing shielding structures, advice may be sought from the manufacturer as well as from IPEM Report 75 (IPEM 2017b), which also references PTCOG Report 1 (PTCOG 2010) and NCRP Report 144 (NCRP 2003). It is pointed out that the accuracy of shielding calculations for proton accelerators has much greater uncertainty than shielding calculations for photon accelerators. This is because neutron levels are very dependent on the energy and intensity of the beam, in addition to the type of accelerator. Consequently, it is important to keep the actual workload in comparison with the design workload under review.

8.78 The Environment Agency (EA) permits will require reports to be made of gaseous discharges to the atmosphere. These are likely to be a lot lower than those in cyclotrons used for the production of PET radionuclides and will need to be based on calculations rather than measurements. ^{15}O with a half- life of 2 min will be produced in air and in cooling water, but longer-lived components such as ^{41}Ar may be more important.

The safe operation of kilovoltage equipment (fixed location)

8.79 Where a dose-rate monitor is not fitted, the output under standard conditions should be checked at least once each working day. This is particularly important for sets operating below 50 kV, as small spectral changes can lead to significant output changes. Any variations from the standards specified in appendix A of IPEM Report 81, 2nd edition, Physics Aspects of Quality Control in Radiotherapy (IPEM 2018) should be reported to the appropriate Operator identified in the quality system.

8.80 The treatment Operator, and any other person who needs to be in the controlled area during kilovoltage therapy up to 100 kV should wear a protective apron and, if the hands are likely at any time to be close to the radiation beam, protective gloves. They should also make use of protective panels. See also paragraphs 7.55, *Additional Requirements for Portable x-ray units operating at or below 100 kV*, and 8.61. The suitability of personal protective equipment should be considered in the radiation risk assessment.

8.81 The x-ray tube or associated support or stand should never be held by any Operator while high voltage is applied.

8.82 For kilovoltage equipment, an interlocked system should be employed to control the maximum and minimum kilovoltage that may be used with a particular filter thickness.

8.83 Great care should be taken in the identification and positioning of all filters, especially wedge filters, and any other beam modifying devices if the equipment has no electrical facility to check and indicate the selection of these devices.

The safe operation of gamma-ray teletherapy equipment

8.84 There must be a clear indication to anyone about to enter the treatment room if the equipment is not in a safe condition. In addition to normal passive personal dosemeters, treatment Operators should wear audible, electronic personal alarms to detect high dose rates if the source has failed to retract to its shielded position. Such alarms give further indications of the source positions as well as room and equipment warning signals. The electronic personal monitors should be switched on throughout the working period and should be capable of giving a recognisable signal up to the maximum possible dose rate for the source being monitored. The audible indication bleep-rate is preferred to be proportional to dose-rate. Electronic personal monitors should also be worn for non-routine activities such as maintenance (L121, Paragraph 139 of HSE 2018).

8.85 The actions and steps required to deal with the source not being in a safe condition when the exposure has been terminated should be described in detail within the contingency plans (local rules) and rehearsed at suitable intervals.

8.86 All Operators should know how to use the emergency manual means to return the machine to the 'off' position with the least practicable exposure to themselves and the patient. A daily check should be made to ensure that any tool required for this operation is in the proper accessible position near the entrance to the room.

8.87 If it is not possible, even by emergency manual means, to return the source to the 'off ' position, it will be necessary to enter the treatment room in order to remove the patient. The patient should be removed as quickly as possible, with the greatest care taken to avoid any exposure to the radiation beam. If it is possible to close the collimators from the control panel, this should be done before entering the room; otherwise, it may be advisable to rotate the head away from the route of entry and then close the collimators immediately after entering the room. This action should be rehearsed from time to time.

8.88 All Operators should be conversant with the emergency procedure displayed at the entrance to the treatment room (see chapter 7). Further information on emergency procedures is given in chapter 20, *Contingency planning and emergency procedures for radioactive substances activity*.

8.89 As a therapy source is still emitting radiation when in the 'off' position, the room should be entered only by staff for the minimum time associated with the maintenance or repair of the equipment and other essential activities.

8.90 The shutter or source should be locked in the 'off' position when the equipment is not in use to protect persons who may enter the treatment room while the equipment is unattended.

The safe operation of remotely controlled after-loading equipment

8.91 To minimise the possibility of sources sticking in transfer tubes and applicators, excessive bends should be avoided, and the manufacturer's recommendation on the minimum radius of curvature should be followed. Transfer tubes and applicators should be examined for kinks before each treatment.

8.92 Operators should know how to use the emergency manual means provided on HDR equipment to return sources to the storage container with the least practicable exposure to themselves and the patient, and this procedure should be practised regularly.

8.93 In the event of failure of all the systems provided on the equipment for the return of sources, it may be necessary to enter the treatment room and withdraw the loaded applicators from a patient manually, using long-handled forceps or other instruments if necessary. Details of, or reference to, the action to be taken if such a radiation emergency occurs should be specified in the local rules. The local rules should also provide a list of people who must be informed and their contact details. A summary of the procedure and contact information should be clearly displayed at the entrance to the treatment room. All treatment Operators should be conversant with this procedure. The RPA, RPS, and an MPE with knowledge of the after-loading equipment must always be consulted after a failure or near miss to identify and report the equipment fault, estimate any additional exposure to patients or staff, and re-evaluate the emergency procedure.

8.94 Any tool and protected receptacle required for use in a radiation emergency must be kept at the entrance to, or just outside, the treatment room, close to the entrance, and in a clearly marked location. A check must be made that this tool is in the correct location every day before clinical use of the machine commences.

8.95 In addition to normal passive personal dosemeters, treatment Operators should wear audible personal alarms to detect high dose rates if the source is not retracted. Such alarms give some indication of the source positions as well as the room and equipment warning signals. The electronic personal monitors should be switched on throughout the working period and should be capable of giving a recognisable signal up to the maximum possible dose rate for the source being monitored. The audible indication should preferably be dose-rate dependent. Electronic personal monitors should also be worn for non-routine activities such as maintenance (L121 Paragraph 139 HSE 2018).

8.96 A portable monitor should be available and should be used to identify the probable position of the source in the event of a failure of the source return mechanisms. The portable monitor should be battery operated. In the case of LDR equipment, the portable monitor should be used after each treatment if it is not otherwise possible to check that all sources have been returned to the source container.

8.97 For LDR treatment, medical, nursing, and auxiliary ward staff should not remain unnecessarily in the vicinity of patients during treatment (they must not be present during HDR treatment). Where it is clinically acceptable, after-loading treatment should be interrupted and the sources withdrawn during nursing procedures. Patients should have visitors only when the sources are withdrawn.

8.98 As the sources are still emitting radiation even when retracted to the storage container, the room should not be used for other purposes unless a comprehensive risk assessment has been carried out that shows that dose rates are safe even in possible fault conditions.

The safe operation of radiotherapy x-ray imaging devices

8.99 Diagnostic reference levels (DRLs) and dose constraints for the patient will not usually apply to radiotherapy simulators, CT scanners, or on-treatment verification imaging when used for radiotherapy. However, the principle of ALARP must still be applied. National Dose Reference Levels for Radiotherapy Planning CT Scans have been issued by Public Health England (UKHSA 2022) and can be found on the UK Government website (UKHSA 2022). The guidance given in paragraphs 2.39 to 2.44 relating to concomitant exposures (i.e. all exposures within a course of radiotherapy other than the treatment exposures) should be noted, and the general guidance given in chapter 4 will apply).

8.100 Interlocks, warning signs, and emergency stops should apply equally to x-ray imaging when used alongside therapy devices (paragraphs 7.32–7.39), although shielding for kV imaging will be adequately covered by existing measures for MV equipment.

8.101 Operators should ensure that the beam is collimated to protect the patient so far as is compatible with the imaging process. Care should be taken to ensure that the image intensifier or detector intercepts the primary beam at all times.

8.102 Operators should minimise fluoroscopic exposure times to protect the patient so far as is compatible with the simulation process. It is not necessary for radiotherapy simulators to be fitted with dose area product (DAP) meters as the imaging field is necessarily larger than the treatment field, with organs adjacent to the treatment field being identified and exposed during simulation to avoid subsequent exposure during treatment. An estimate of time and exposure rate is sufficiently accurate when concomitant exposures are compared with subsequent therapy exposures (see paragraph 2.43). Equipment installed after 6th February 2018 must have a device or other feature capable of informing the Practitioner, at the end of an exposure, of relevant parameters for assessing the dose, and radiotherapy CT scanners must have the capacity to transfer this information to the record of a person's exposure. Linac imaging systems may also be used in fluoroscopic mode, and similar considerations apply.

The safe operation of mobile or intraoperative radiotherapy (IORT) devices

8.103 Mobile and/or Intraoperative Radiation Therapy (IORT) systems will be used outside the radiotherapy department in environments where staff members are less familiar with high dose radiation; therefore, careful consideration should be given to appropriate safe practice and a radiation risk assessment made prior to work starting. If an external supplier provides the service, then responsibilities must be clearly split out for each employer, and the co-operation between them should be fully documented. An MPE must be closely involved and therefore should be contactable during treatment, either on- or off-site. (The duties of the MPE are covered in paragraph 7.11.)

8.104 Many IORT systems can be used in operating theatres without permanent modification because of the low energy used (around 50 kV) or integral beam-stoppers for electron linacs. However, the RPA must be consulted, and prospective dose surveys should be performed in each operating theatre before clinical use. Environmental surveys should be repeated at regular intervals to confirm the sufficiency of protection measures for clinical patients. It may be prudent to designate the whole theatre as a controlled area, including any side rooms if they are difficult to restrict access to separately. Regard should be given to the local practices of the working theatres, including access routes for the movement of IORT equipment and storage.

8.105 Doors should be clearly signed and, where possible, locked to prevent unauthorised access. Appropriate training should be conducted for local theatre staff, and local rules must be clearly available. During irradiation, all non-essential staff should leave the theatre. Typically, just the two Operators and an anaesthetist would remain to monitor the patient, usually behind a mobile screen or in a side room. Lead aprons could also be used, but they are cumbersome to wear and do not provide whole-body protection. Personal electronic monitors (electronic personal dosimeters, EPDs) may be distributed to provide a live dose measurement during the procedure. Following the placement of appropriate shielding sheets and mobile screens, the instantaneous dose rate adjacent to the Operators should be confirmed either via EPD or a survey meter to confirm the efficacy of the positioning of the control measures. Another possibility would be to use a theatre that is already used for higher dose radiation procedures such as fluoroscopy. Specific risk assessments are still required for IORT, but the existing shielding is likely to be sufficient. In this case, Operators may stand outside the main room and observe through a window.

8.106 Electron linacs (4 MeV-12 MeV) may require restrictions on workload to limit exposure in adjacent areas, and commissioning or QA is better performed in a more shielded room such as a linac bunker; however, neutron production should be minimal even above 9 MeV.

8.107 Mobile superficial units are sometimes described as needing no additional protection measures because of the low kilovoltage energies used; however, previous guidance should be considered with reference to the location and use of the unit, e.g. in a dermatology or ophthamology clinic. The security of mobile units must also be considered. Risk assessments must include testing and calibration conditions, where scattered dose rates may be higher, as well as treatment conditions.

8.108 IPEM Report 75, second edition (IPEM 2017) can be consulted for further information.

Electrical safety

8.109 Although not strictly a radiation issue, the electrical installation for new or re-installed fixed (permanently installed) x-ray equipment should comply with the current version of BS 7671(BSI 2018) Wiring Regulations with particular regard to section 710—Medical Locations. Testing and certification of the electrical installation should be performed according to BS 7671, which does not cover the medical electrical equipment itself. Electrical safety testing of medical electrical equipment should be performed according to the current version of BS EN 62353 (BSI 2014a).

References

ADR 2022 *ADR 2023—Agreement concerning the International Carriage of Dangerous Goods by Road, UN Economic Commission for Europe* (United Nations) https://unece.org/transport/standards/transport/dangerous-goods/adr-2023-agreement-concerning-international-carriage (accessed 4 February 2024)

BEIS, DEFRA, Welsh Government and DAERA 2018 *Scope of and Exemptions from the Radioactive Substances Legislation in England, Wales and Northern Ireland Guidance Document* (London) https://gov.uk/government/publications/guidance-on-the-scope-of-and-exemptions-from-the-radioactive-substances-legislation-in-the-uk (accessed 4 February 2024)

BSI 2010 *BS EN 61508 Functional Safety of Electrical/Electronic/Programmable Electronic Safety-related Systems* (London: British Standards Institute)

BSI 2014a *BS EN 62353:2014 Medical Electrical Equipment. Recurrent Test and Test After Repair of Medical Electrical Equipment* (London: British Standards Institute)

BSI 2014b *BS EN ISO 2919:2014 Radiological Protection: Sealed Radioactive Sources. General Requirements and Classifications* (London: British Standards Institute)

BSI 2018 *BS 7671:2018+A1:2020 Requirements for Electrical Installations. IET Wiring Regulations* (London: British Standards Institute)

BSI 2020 *BS ISO 9978:2020 Radiation Protection: Sealed Sources. Leakage Test Methods* (London: British Standards Institute)

CDG(NI) 2010 *The Carriage of Dangerous Goods and Use of Transportable Pressure Equipment Regulations (Northern Ireland) 2010* https://legislation.gov.uk/nisr/2010/160 (accessed 4 February 2024)

CDG 2009 *The Carriage of Dangerous Goods and Use of Transportable Pressure Equipment Regulations 2009* https://legislation.gov.uk/uksi/2009/1348 (accessed 4 February 2024)

EA(S)R 2018 *The Environmental Authorisations (Scotland) Regulations 2018* http://legislation.gov.uk/ssi/2018/219 (accessed 4 February 2024)

EPR 2016 *Environmental Permitting (England and Wales) Regulations* (England and Wales) https://legislation.gov.uk/uksi/2016/1154 (accessed 4 February 2024)

European Commission 1999 *Nuclear Safety and the Environment, Evaluation of the Radiological and Economic Consequences of Decommissioning Particle Accelerators* (Luxembourg: European Commission) https://op.europa.eu/en/publication-detail/-/publication/7321365e-d716-445c-9ab2-bcd35ec73e38) (accessed 4 February 2024)

HSE no date *Notify, registration or consent for work with ionising radiation* https://www.hse.gov.uk/radiation/ionising/notifiy-register-consent.htm (accessed 4 February 2024)

HSE 2018 L121 Work with ionising radiation Ionising Radiations Regulations 2017— *Approved Code of Practice and Guidance* 2nd edn (Norwich: Health and Safety Executive) https://hse.gov.uk/pubns/books/l121.htm (accessed 4 February 2024)

HSENI no date *Ionising Radiation, HSENI Website*. Health and Safety Executive for Northern Ireland https://www.hseni.gov.uk/topic/ionising-radiation (accessed 4 Feb 2024)

IAEA 2003 *Technical Report Series no. 414. Decommissioning of Small Medical, Industrial and Research Facilities, Technical Report Series* (Vienna, Austria: International Atomic Energy Agency) https://iaea.org/publications/6573/decommissioning-of-small-medical-industrial-and-research-facilities (accessed 4 February 2024)

IPEM 2006 *Report 93: Guidance for the Commissioning and Quality Assurance of a Networked Radiotherapy Department* ed M Kirby, D Carpenter, G Lawrence, S Poynter and P Studdart (York: Institute of Physics and Engineering in Medicine)

IPEM 2007 *Report 94: Acceptance Testing and Commissioning of Linear Accelerators* ed M Kirby, C Hall and S Ryde (York: Institute of Physics and Engineering in Medicine)

IPEM 2017 *Report 75: Design and Shielding of Radiotherapy Treatment Facilities* 2nd edn P Horton and D Easton (Bristol: Institute of Physics)

IPEM 2018 *Report 81: Physics Aspects of Quality Control in Radiotherapy. 2nd edn, IPEM Report 81* ed I Patel, S Weston, L A Palmer, W P M Mayles, P Whittard, R Clements, A Reilly and T J Jordan (York: Institute of Physics and Engineering in Medicine)

IPEM 2020 *IPEM code of practice for high-energy photon therapy dosimetry based on the NPL absorbed dose calibration service* ed D J Eaton, G Bass, P Booker, J Byrne, S Duane, J Frame, M Grattan, R A S Thomas, N Thorp and A Nisbet (York: Institute of Physics and Engineering in Medicine)

IRR(NI) 2017 *The Ionising Radiations Regulations (Northern Ireland).* http://legislation.gov.uk/nisr/2017/229 (accessed 4 February 2024)

IRR 2017 *The Ionising Radiations Regulations* (Great Britain) https://legislation.gov.uk/uksi/2017/1075 (accessed 4 February 2024)

MHSWR 1999 *The Management of Health and Safety at Work Regulations* www.legislation.gov.uk/uksi/1999/3242/regulation/3 (accessed 4 February 2024)

NCRP 2003 *Report No. 144: Radiation Protection for Particle Accelerator Facilities* (Bethesda, USA: National Council on Radiation Protection & Measurements)

PTCOG 2010 *PTOG Report 1 Sub-Committee Task Group on Shielding Design and Radiation Safety of Charged Particle Therapy Facilities* (Villigen) https://www.ptcog.site/index.php/ptcog-publications (accessed 4 February 2024)

PUWER(NI) 1999 Provision and Use of Work Equipment Regulations (Northern Ireland) 1999. http://legislation.gov.uk/nisr/1999/305 (accessed 4 February 2024)

PUWER 1998 *Provision and Use of Work Equipment Regulations* https://legislation.gov.uk/uksi/1998/2306 (accessed 4 February 2024)

RSA 1993 *Radioactive Substances Act 1993* http://legislation.gov.uk/ukpga/1993/12 (accessed 4 February 2024)

SEPA 2020 *Authorisation Guide for Radioactive Substances Activities. 1.2* (Stirling: Scottish Environmental Protection Agency) https://sepa.org.uk/media/371985/rs-authorisation-guide.pdf (accessed 4 February 2024)

UKHSA 2022 National Diagnostic Reference Levels (NDRLs) from 19 August 2019. 24/11/2022 https://gov.uk/government/publications/diagnostic-radiology-national-diagnostic-reference-levels-ndrls/ndrl (accessed 4 February 2024)

IOP Publishing

Medical and Dental Guidance Notes (Second Edition)
A good practice guide on all aspects of ionising radiation protection in the clinical environment: IPEM Report 113
John Saunderson, Mohamed Metwaly, William Mairs, Philip Mayles, Lisa Rowley and Mark Worrall

Chapter 9

Brachytherapy sources

Scope

9.1 This chapter presents the necessary precautions to be taken in the use of small sealed sources, or other solid sources, for intracavitary or interstitial radiotherapy other than by remotely controlled after-loading. This chapter includes the preparation, sterilisation, testing, and use of these sources. Chapter 18, *Keeping, accounting for and moving radioactive substances (including transport)*, should be consulted for guidance on storage arrangements for, and chapter 19, *Disposal of radioactive waste*, on disposal of, these sources. In the UK, the only sealed sources routinely used are iodine-125 (I-125) seeds and eye plaques using strontium-90 (Sr-90) or ruthenium-106 (Ru-106). Other sources that may be introduced from time to time can be considered based on these principles in consultation with an RPA. Unsealed source therapy is covered in chapter 13, Therapeutic uses of unsealed radioactive substances.

9.2 A sealed source is a radioactive substance whose structure is such as to prevent, under normal conditions of use, any dispersion of radioactive substances into the environment. 'Solid source' in these guidance notes means any non-dispersible solid source (e.g. a beta ray plaque) other than a sealed source. Note that non-dispersible sources (such as sealed sources whose radioactive cores are non-dispersible under emergency conditions) are exempt from the requirements of the Radiation (Emergency Preparedness and Public Information) Regulations (REPPIR 2019, REPPIR(NI) 2019).

9.3 All sealed sources used for brachytherapy should conform to BS EN ISO 2919–2014 Sealed Radioactive Sources (BSI 2014). If any source that does not conform to the standard is to be used, the RPA should be consulted before it is used, and the clinical justification for its use should be recorded in writing.

doi:10.1088/978-0-7503-2332-1ch9

9.4 Procedures for the use of all sources should take into account the supplier's recommendations on working life and the environment of use. Copies of these recommendations should be referenced in the local rules and made available to all staff involved in their use.

9.5 Brachytherapy sealed sources will often be classified as radioactive High Activity Sealed Sources (HASS(NI) 2005, EPR 2016, EA(S)R 2018), with special requirements for security and governance.

Roles and responsibilities of duty holders in brachytherapy

9.6 The Employer's duties are discussed in chapter 2, *Radiation protection of persons undergoing exposures*. The Employer is ultimately responsible for the safe use of ionising radiation and for ensuring that the actions discussed below are carried out accordingly. The Employer must be licensed to undertake brachytherapy exposures, as detailed below.

9.7 The Referrer for a brachytherapy procedure must be a registered healthcare professional and will always be a medical practitioner. The Referrer must provide sufficient information about the patient for the clinical oncologist acting as the IR(ME)R Practitioner to decide if radiotherapy is an appropriate treatment for the patient. The Referrer should be identified in the patient's record. A list of authorised Referrers should be available. Referral guidelines must also be provided that indicate the scope of practice to which particular Referrers are entitled.

9.8 The IR(ME)R Practitioner in the brachytherapy process must be a person licensed under IR(ME)R (IR(ME)R 2017, IR(ME)R(NI) 2018) (i.e. the holder of an ARSAC licence) and will normally be a radiation oncologist, but other medical professionals such as ophthalmologists may also use brachytherapy. A list of Practitioners should be recorded in the local protocols for brachytherapy treatment. A Practitioner must:

(a) assess the information provided by the Referrer and, if necessary, ask for more information from the Referrer;
(b) decide if brachytherapy is an appropriate treatment for the patient;
(c) justify and authorise in writing (or using a traceable electronic method) the medical exposures necessary to deliver the course of brachytherapy treatment, if indeed radiotherapy is the appropriate treatment. The record of authorisation of the treatment is evidence of the justification (but see paragraph 9.10).

9.9 The IR(ME)R Practitioner will also prescribe the brachytherapy treatment (in most cases according to an agreed local protocol). The authorisation and prescription should be noted in the patient's records. Justification of a course of brachytherapy treatment will generally imply justification of all the medical exposures involved, including any localisation and verification imaging, but the extent of this must be clear.

9.10 Radioactive substances may only be administered to patients if the administration has been justified by a person licensed under IR(ME)R (IR(ME)R 2017, IR(ME)R(NI) 2018) and where the employer holds a licence for the site (e.g. hospital) where the administration will take place. The administration must be authorised by a licensed Practitioner or, if this is not possible, by an appropriately trained Operator following a protocol issued by the licensed Practitioner (IR(ME)R 2017, IR(ME)R(NI) 2018, ARSAC 2023). The IR (ME)R Practitioner (or in some circumstances the Operator) should explain to the patient, and if appropriate, their relatives, prior to treatment the possible need to remove permanent implants surgically in the unlikely event of their death, or to accept burial rather than cremation, and should obtain their written consent. If the patient lacks mental capacity, the IR(ME)R Practitioner may decide it is more appropriate for a relative to give written consent. The precautions necessary in the event of death should be recorded in the patient's notes and communicated to the patient's GP (see paragraph 17.4).

9.11 IR(ME)R Operators in the brachytherapy process will include those members of staff who initiate the treatment exposure, take localisation, or verification imaging, carry out or check treatment planning, calibrate, service, or repair brachytherapy equipment, or carry out any other task that could affect the safety of the brachytherapy treatment. Anyone responsible for any such step within the brachytherapy process becomes an IR(ME)R Operator. Staff who are not Operators should be under the direct supervision of an Operator. Any treatment aids manufactured within the radiotherapy department should be checked by an Operator before being used in the treatment of a patient. Operators in a brachytherapy department may be medical staff, radiographers, nurses, physicists, clinical technologists, or other appropriately trained staff. A list of Operators and a record of their training must be available for inspection by the regulator.

9.12 Medical Physics Experts (MPEs) in radiotherapy must be closely involved in all procedures related to the brachytherapy techniques described in this chapter. This will involve providing advice on all aspects of source and patient dosimetry, on the optimisation and safety of treatment and treatment planning, on the QA programme (including quality control procedures) and on any other matters relating to the safety of the radiotherapy equipment or treatment. An MPE should also be consulted before new treatment techniques are introduced. A list of MPEs should be available, along with their scope of practice. MPEs will also oversee the acceptance testing of new or modified brachytherapy equipment, its commissioning, and definitive calibration (see appendix A of IPEM Report 81 (IPEM 2018)). The MPE will be involved with the RPA in the critical examination of new or modified equipment. (See also paragraph 7.11 for more detail on the role of the MPE.)

9.13 The Radiation Protection Adviser (RPA) for the radiotherapy department should have appropriate experience in the application of radiation protection principles in brachytherapy. The RPA should be consulted on all matters specified in Schedule 4 of the IRR (IRR 2017, IRR(NI) 2017) relevant to brachytherapy within the radiotherapy department. One or more RPSs should be appointed by the employer to ensure that all staff are aware of the local rules and that the local rules are observed. The roles of the RPA and RPS are discussed in paragraphs 7.12 to 7.14.

Preparation and cleaning room

9.14 A dedicated room should be provided for the preparation, cleaning, and storage of sources and applicators. This room should only be used for this work and should only be occupied during such work. Arrangements for restricted access should be specified in the local rules. The room should be designated a controlled area because of the radiation hazard. Dose rates can often be reduced significantly by the use of appropriate local shielding, and this should be considered in the design of the room. The design should also take into account the possibility of contamination from the handling of sealed sources and be designed for easy cleaning and decontamination. Workbenches, floors, and walls should have smooth impervious surfaces, and flooring material should be coved at all walls to avoid the possibility of contamination leaking under coverings. Sinks should have suitable traps so that sources cannot be lost. It is advisable, when designing new rooms, to identify the location of drainage pipework in case lost sources need to be traced. The condition and integrity of these safety features should be checked preferably monthly, e.g. by the RPS, and confirmed annually by the Employer who could delegate this to the RPA.

9.15 The entrance of the room should be marked with an appropriate warning sign to indicate both the presence of radioactive sources and the existence of a controlled area (see appendix 7, *Warning signs and notices*).

9.16 Sources should be stored in a shielded safe, preferably in the preparation room, to minimise radiation doses arising from the movement of sources. Source storage needs to be considered as part of the radiation risk assessment. The security level must adhere to that specified in the site's radioactive substances permit or registration. Information on the storage of sources is given in chapter 18 *Keeping, accounting for, and moving radioactive substances (including transport)*.

Source calibration and testing

9.17 Sealed radioactive sources will usually have been tested by the manufacturer or supplier and be accompanied by a leakage test certificate. Test certificates must be filed for future reference. If there is no certificate, the source should be tested for both leakage and surface contamination before it is used clinically for the first time. Sources should be retested at least once per year and whenever damage or leakage is suspected. This is necessary because brachytherapy sources are subject to wear through frequent use, cleaning, and sterilisation procedures. Because of the particular hazard to a patient, if leakage occurs, pathways for intake should be comparatively short. Leakage test methods are described in ISO 9978 (BSI 2020). BS EN ISO 2919:2014 (BSI 2014) contains guidance on sources. It recommends that each source have a recommended working life designated by the manufacturer. If this life is exceeded, then a qualified body (e.g. the manufacturer) should inspect it and carry out a technical assessment, including leakage testing.

9.18 The activity of all sources should be measured and compared with the calibration certificate supplied by the supplier before being administered to a patient. Sources should be measured individually, but those that have been produced collectively in a common irradiation container, with each source having all dimensions less than 5 mm (e.g. I-125 seeds), may be measured collectively. The activity of sources containing long-lived radionuclides need not be checked on each occasion of use. All sources and calibration certificates should be kept and controlled as described in chapter 18, *Keeping, accounting for, and moving radioactive substances (including transport)*.

9.19 A record must be kept of the issue, distribution, and return of all sources and also of the administration of permanent implants. Further advice on record keeping is given in chapter 18, *Keeping, accounting for, and moving radioactive substances (including transport)*, and appendix 10, *Record keeping*.

9.20 If, in a leakage test, the activity measured on the swab (or in the liquid used for an immersion test) is less than 200 Bq, the source may be considered leak-free (BSI 2020). Activity greater than 200 Bq can also be due to surface contamination arising from a different source; therefore, the source should be decontaminated and retested before a leak is confirmed. In carrying out leakage tests of beta sources, care is essential to avoid damaging the 'window', if there is one, through which beta radiation is emitted.

9.21 Whenever it is believed that a source is, or might be, leaking, it should be sealed in an airtight container and kept separate from other sources. The RPS, RPA, and local manager should be informed, and an inspection by a competent authority should be arranged as soon as possible. Arrangements should then be made for repair or disposal by an authorised route. The RPA and RPS, together with the local manager, should investigate the possibility of staff and other persons having been contaminated as a result of the leak, including the possibility that some contamination may have occurred prior to detection.

General operating procedures and local rules

9.22 Special tools or surgical instruments should always be used when sources are being prepared for or administered to patients. These should be constructed so as to provide the maximum handling distance or shielding compatible with effective manipulation. All Operators should be trained not to pick up sources or load source applicators by hand under any circumstances, and this should be emphasised in the local rules and any work instructions. Methods of work and work instructions should be reviewed regularly by the RPS and Operators, and at least annually by the RPA. Potential staff doses and the controls necessary to minimise them should be fully considered in the radiation risk assessment.

9.23 The local rules should emphasise that work on equipment and applicators should be carried out, as far as possible, before the insertion of the source. If work has to be carried out on equipment or appliances with a source inserted, then precautions to minimise the exposure of any staff must be taken. The local rules should also provide detailed guidance on arrangements for monitoring during source preparation, with particular reference to hand monitoring. Consideration should be given to finger and eye dose monitoring for staff who are regularly handling radioactive sources. This does not need to be routine once it has been shown that doses are acceptable, but should be reviewed from time to time as an audit.

9.24 The local rules should explicitly prohibit the placing of objects in the mouth, eating, drinking, using nasal sprays, taking snuff (or similar), or the application of cosmetics within rooms in which radioactive sources are being manipulated. The local rules should also only permit the use of disposable handkerchiefs. These precautions are necessary because the tools for handling sources may become contaminated, creating a risk of contamination entering the body.

9.25 When needles, capsules, or other applicators of the same appearance but of different activities are used at the same time, they should be easily distinguishable, e.g. by different coloured threads, beads, or markings. Details of any colour code used should be clearly displayed wherever the sources are handled. Copies of any colour code should also be in the local rules. These should be checked regularly by the RPS.

9.26 The number and position of removable sources in or on the patient should be regularly checked by an Operator, using a procedure that minimises doses to staff. A radiation monitor should be used to confirm that no sources remain in the patient or in the treatment area or ward at the conclusion of treatment. The operation of the monitor should be tested by placing it close to the patient before the sources are removed. Dressings, wipes, and excreta from patients receiving treatment should not be disposed of until monitoring has shown that they are not contaminated by radioactive material and that all sources have been accounted for. To reduce the possibility of sources being mislaid, they should be inspected, cleaned, and returned to storage (or identified as sent for disposal) without delay. Advice on inspection is given in this chapter and chapter 18, *Keeping, accounting for, and moving radioactive substances (including transport)*, and on source disposal in chapter 19, *Accumulation and disposal of radioactive waste*. These procedures should be the responsibility of a specific Operator or the RPS.

9.27 A shielded container should be placed near the bed of a patient being treated with removable sources. If a source becomes accidentally displaced, it should be immediately transferred to the container using forceps. The incident should be reported immediately to the IR(ME)R Practitioner (ARSAC licence holder), to the clinician in charge of the patient, to the RPS, and to the local manager. This incident procedure should be described in detail in the local rules. Details of the incident should be recorded.

9.28 Steps should be taken to ensure that sources that might be mislaid or lost by a patient do not leave the ward or treatment area, e.g. for the hospital laundry or refuse incinerator. All containers, such as rubbish bins, soiled dressing bins, and laundry baskets, coming from a ward, side ward, theatre or other area where such sources are employed, should be tested for radioactivity with a monitoring instrument. An additional check can be provided using a permanent alarm installed in a doorway or corridor through which outgoing bins, baskets, and trolleys have to pass. Where permanent alarms are installed, the person responsible for ensuring they are regularly checked and recording that they are functioning correctly should be the RPS. If a source of activity greater than that shown in column 6 of Schedule 7 of IRR2017/IRR(NI)2017 (e.g. 10 MBq of I-125, 1 MBq of Ru-106 or 0.1 MBq of Ir-192) is indeed lost, there will be a requirement to notify HSE/HSE(NI) and the Environment Agency.

9.29 Sources used for brachytherapy should be cleaned and inspected before being returned to the store so as to minimise subsequent sterilisation and disinfection difficulties.

Permanent implant of sealed sources

9.30 Radioactive sources (e.g. I-125) can be permanently implanted in the patient. Guidance on radiation protection for such patients can be found in ICRP Report 98 (ICRP 2005) and IPEM Report 106 (IPEM 2012), which have detailed risk assessments for various situations. See also paragraph 9.55 and chapter 17, *Precautions after the death of a patient to whom radioactive substances have been administered.*

9.31 According to Que (2001), the dose rate at 1 m from the abdomen of a patient with an I-125 prostate implant is about 1.4 μSv h^{-1} and about 16 μSv h^{-1} in contact with the abdomen. IPEM Report 106 points out that the dose rate on the posterior surface can be as high as 170 μSv h^{-1}. If close clinical care is required in the hospital immediately after the implant, lead aprons can reduce the dose to negligible levels. If the patient is an in-patient, they should be asked to sieve their urine for the first three days after the implant.

9.32 Patients being discharged from the hospital must be given a guidance card, as discussed in chapters 16 and 17. For I-125 seed implants, the ICRP estimates that carers and comforters are unlikely to receive a dose greater than 1 mSv during the course of a year (ICRP 2005). However, it is not advisable for children to sit on the patient's lap for more than a few minutes, and the patient should avoid close contact with pregnant women for extended periods or have sex during the first two months. Patients should be advised to wear a condom for the first five ejaculations. Seeds can be flushed away in the lavatory pan, and it is not necessary to sieve urine at home. Advice should be sought from the radioactive waste adviser (RWA) if the patient is being returned to premises other than a domestic dwelling.

9.33 Cremation and embalming of patients who have recently been implanted with radioactive seeds create a potential hazard for the associated staff. An appropriate risk assessment must be carried out and advice given to the staff concerned. The employers may need to gain registration from HSE/HSE(NI) and consult an RPA. Embalming need not be forbidden; provided an incision is not made in the abdomen, the hazard is likely to be small. ICRP 98 recommends that cremation not take place within one year of the implant, while IPEM Report 98 recommends 10 half-lives (i.e. 20 months). If necessary, the prostate could be removed and the whole organ returned to the implanting hospital (see paragraph 9.34). A glass pot will provide significant attenuation. Retention of tissue requires consent in order to conform to the Human Tissue Act (Human Tissue Act 2004, Human Tissue (Scotland) Act 2006).

9.34 If other surgical procedures are required, the advice of the RPA should be sought. Surgery in the abdomen is likely to present a particular hazard unless it can be done quickly. Transurethral resection should ideally not be carried out within six months of the implant.

Sterilisation, disinfection and cleaning of small sources

9.35 When sterilising or disinfecting small sources, the manufacturer's or supplier's instructions should be consulted. Appropriate precautions should be taken to avoid:

(a) unnecessary radiation exposure of nursing and other staff;
(b) damage to sources;
(c) loss of sources.

9.36 If autoclaves, hot air ovens, and other equipment are used for sterilising or disinfecting sources, they should be adequately shielded and designed to prevent damage to or the loss of a source from the equipment during use.

9.37 The following special precautions need to be taken when sterilising or disinfecting sources:

(d) sterilisers should be fitted with a cutout that will prevent the temperature of the source rising above 180 °C;
(e) applicators containing sources should not be sterilised or disinfected if found to be damaged;
(f) disinfecting solutions that do not attack identification marks should be used.

9.38 When sources need to be cleaned before being returned to storage, particular care should be taken with thin-walled sources.

9.39 Before sources are cleaned, the manufacturer's or supplier's instructions should be consulted. Abrasive substances (such as metal cleaners and polishes) should never be used, and sources should never be allowed to come into contact with mercury or mercury salts, iodine, solutions of hypochlorites, or corrosive substances. Immersion in a solution such as normal saline will aid in the removal of blood and tissues from appliances. The following cleaning methods are particularly suitable for sealed sources:
(a) soaking for at least an hour in a suitable disinfectant, or in a dilute solution of hydrogen peroxide, to remove dried blood, or in an organic solvent (such as xylene) to remove moulding material;
(b) thorough rinsing in warm or boiling water;
(c) ultrasonic cleaning using a low power generator.

9.40 Following sterilisation, disinfection, or cleaning, an Operator should check that identification marks or colour coding have not been damaged.

Loss or breakage of a source

9.41 Procedures for action to be taken in the case of the suspected loss or breakage of a radioactive source should be detailed in the local rules and contingency plan. Notices outlining the procedure should be displayed in each room, theatre or ward where such sources are handled or used (see the guidance in chapter 20, *Contingency planning and emergency procedures for radioactive substances activity*). Contingency plans should be rehearsed at appropriate intervals, normally annually. A record of these rehearsals should be kept for a minimum of two years. If a contingency plan is activated, this should trigger an investigation by the employer with advice from the RPA as described in paragraph 247 of L121 (HSE 2018), and the records of the investigation should be kept for a minimum of two years. In the case of loss, theft, or attempted theft of a source, the Environment Agency and the police must be informed.

9.42 Special care should be taken with sources of long half-life radionuclides with high or medium toxicity. The RPA should be consulted about the appropriate action to take in the event of any suspected leak or contamination.

Safe use of beta sources

9.43 Suitable shields or baffles should be provided to ensure adequate protection when manipulating beta radiation sources. A transparent plastic plate, e.g. poly(methyl methacrylate) (PMMA), of adequate thickness should be mounted or worn between the source and the face of the Operator, in order to prevent the head of the Operator from being placed too near the source and to protect the eyes and face from beta radiation. Particular care should be taken when using high-activity beta sources (3 GBq or greater), such as for the treatment of pterygia. A plastic shield should be in position on the handling rod to protect the Operator's hands, as the beta surface dose rates from such sources can be very high.

9.44 In the case of large activity beta sources, bremsstrahlung, characteristic x-rays and annihilation radiation may present a hazard that should be evaluated and the necessary precautions taken; e.g. sources should be kept well away from material of high atomic number. Some beta sources also emit gamma radiation (see paragraphs 9.46 and 9.47). The RPA should be consulted about suitable precautions.

9.45 Most sources intended for the utilisation of beta radiation outside the container have a thin 'window'. When the sources are not in use, this window should be covered using a low atomic number or plastic shield of sufficient thickness to stop all beta radiation, minimise bremsstrahlung radiation, and protect the window. When cleaning the sources, the precautions referred to in paragraphs 9.35–9.40 should be observed, and care should be taken to avoid damage to the window.

Safe use of gamma sources

9.46 Benches used for the preparation, assembly, and cleaning of gamma radiation source capsules and appliances should be provided with adequate protection for the Operator and for other persons either associated with the work or in adjacent areas. The RPA should be consulted on the protective measures required.

9.47 Mobile protective barriers, mounted on wheels and provided, where necessary, with sterile drapes, should be used in operating theatres and other treatment rooms. These barriers should be so designed to give protection in all directions where people are usually stationed during brachytherapy procedures. Where possible, the optimum positions for the shields should be clearly marked on the floor of the room. In some situations, however, the precise position of shields will depend on the specific source or procedure employed. Gamma sources should remain behind protective shielding as long as possible and be removed only when required for application to the patient. In all cases, expeditious handling and the use of suitable instruments (see paragraphs 8.93 and 9.27) will reduce the hazard. Mobile barriers should be clearly identified for use with specific sources and marked with their shielding properties in millimetres of lead equivalent. Mobile protective barriers should be regularly examined to ensure the integrity of the protection and the mechanical stability of the barrier's transport mechanism. The RPS will usually be the appropriate person to do this, as the way in which barriers are to be used must be specified in the local rules.

Protection of persons in proximity to patients undergoing brachytherapy

9.48 Most areas in which patients are treated will need to be designated as controlled areas, except where only low-activity beta sources are used. This should be determined in the radiation risk assessment.

9.49 Wherever possible, treatment should be carried out in rooms with only one bed or at most two beds. People in adjoining rooms (whether on the same level or on floors above and below) should be protected using adequate shielding (see paragraphs 1.104–1.115, *Controlled areas and supervised areas*).

9.50 Where treatment with low-activity sources is carried out in a general ward, the beds of patients under treatment should be positioned according to the advice given by the RPA to minimise radiation doses to other patients, particularly those not undergoing brachytherapy treatment. In general, this would require that any patient not being treated be at least 2.5 m from the centre of any bed occupied by a patient under treatment. Written arrangements should be available in the local rules for nursing access and for restricting access to these patients by other patients or visitors. For beta plaques, the nursing staff should confirm the presence of the source during each shift (ideally at shift handovers) using an appropriate radiation monitor.

9.51 Mobile protective shielding should be used around the beds of patients being treated with gamma or neutron sources, except possibly for I-125 seeds. It is advisable that their optimum shielding position around a treatment bed be clearly marked on the floor. Circumstances, where protective shielding cannot be used, should be included in a written system of work, detailed in the local rules, together with the alternative precautions to be undertaken. Guidance on the use of mobile shielding is given in paragraph 9.47.

9.52 Beds in which there are patients undergoing treatment with radioactive sources should carry a notice that includes a radiation warning sign (see appendix 7, *Warning signs and notices*). The nursing staff should be given written details of the number and nature of sources, their total activity, the time and date of application and intended removal, and relevant nursing instructions. The RPS should ensure that these details are kept up-to-date.

9.53 The maximum dose rate at a distance of 1 m from each patient undergoing treatment should be determined and recorded as part of the risk assessment conducted in consultation with the RPA.

9.54 The local rules should include:

(a) safe working procedures for all staff involved in the treatment of the patient;
(b) written arrangements providing entry to the controlled area for staff (with particular attention to domestic and housekeeping staff);
(c) written arrangements for visitors.

9.55 Patients with sources in or upon their bodies should not normally leave the ward or treatment room without the approval of the medical staff responsible for their treatment and the RPS. The nursing staff should keep records in the patient's notes of when patients leave the ward and if or when they return. Guidance on when patients may leave the hospital after administration of radioactive substances is given in chapter 16, *Patients leaving the hospital after administration of radioactive substances.* Guidance on the actions necessary in the event of the death of a patient is given in chapter 17, *Precautions after the death of a patient to whom radioactive substances have been administered.*

9.56 Nursing staff and any other persons, including visitors, should not remain unnecessarily in the vicinity of patients undergoing treatment with gamma or neutron sources. Where possible, medical tests and nursing procedures should be postponed until after the sources have been removed. Guidance on the role of the patient's comforters and carers is given in chapter 1, *General measures for radiation protection.*

9.57 It may be considered desirable to reallocate the duties of staff who have informed their management that they are pregnant. An appropriate risk assessment should be carried out, and the RPA should be consulted for advice.

References

ARSAC 2023 *Notes for Guidance on the Clinical Administration of Radiopharmaceuticals and Use of Sealed Radioactive Sources* (Administration of Radioactive Substances Advisory Committee) https://gov.uk/government/publications/arsac-notes-for-guidance (accessed 25 February 2023)

BSI 2014 *BS EN ISO 2919:2014 Radiological Protection: Sealed Radioactive Sources. General Requirements and Classifications* (London: British Standards Institute)

BSI 2020 *BS ISO 9978:2020 Radiation Protection: Sealed Sources. Leakage Test Methods* (London: British Standards Institute)

EA(S)R 2018 *The Environmental Authorisations (Scotland) Regulations 2018* http://legislation.gov.uk/ssi/2018/219 (accessed 31 January 2023)

EPR 2016 *Environmental Permitting (England and Wales) Regulations* https://legislation.gov.uk/uksi/2016/1154 (accessed 31 January 2023)

HASS(NI) 2005 *The High-activity Sealed Radioactive Sources and Orphan Sources Regulations 2005* http://legislation.gov.uk/en/uksi/2005/2686 (accessed 11 February 2023)

HSE 2018 L121 Work with ionising radiation Ionising Radiations Regulations 2017 *Approved Code of Practice and Guidance* 2nd edn (Norwich: Health and Safety Executive) https://hse.gov.uk/pubns/books/l121.htm (accessed 31 January 2023)

Human Tissue Act 2004 https://legislation.gov.uk/ukpga/2004/30 (accessed 11 February 2023)

Human Tissue (Scotland) Act 2006 https://legislation.gov.uk/asp/2006/4 (accessed 11 February 2023)

ICRP 2005 Radiation safety aspects of brachytherapy for prostate cancer using permanently implanted sources *Ann. ICRP* **35** Publication 98 https://journals.sagepub.com/doi/pdf/10.1177/ANIB_35_3 (accessed 4 February 2024)

IPEM 2012 *Report 106: UK Guidance on Radiation Protection Issues following Permanent Iodine-125 Seed Prostate Brachytherapy* ed P Bownes. (York: Institute of Physics and Engineering in Medicine)

IPEM 2018 *Report 81: Physics Aspects of Quality Control in Radiotherapy. IPEM Report 81* ed I Patel, S Weston, L A Palmer, W P M Mayles, P Whittard, R Clements, A Reilly and T J Jordan (York: Institute of Physics and Engineering in Medicine) 2nd edn

IR(ME)R 2017 *The Ionising Radiation (Medical Exposure) Regulations* www.legislation.gov.uk/uksi/2017/1322 (accessed 31 January 2023)

IR(ME)R(NI) 2018 *The Ionising Radiation (Medical Exposure) Regulations (Northern Ireland)* https://legislation.gov.uk/nisr/2018/17 (accessed 31 January 2023)

IRR 2017 *The Ionising Radiations Regulations*. Great Britain https://legislation.gov.uk/uksi/2017/1075 (accessed 31 January 2023)

IRR(NI) 2017 *The Ionising Radiations Regulations (Northern Ireland)* http://legislation.gov.uk/nisr/2017/229 (accessed 31 January 2023)

Que W 2001 Radiation safety issues regarding the cremation of the body of an I-125 prostate implant patient *J. Appl. Clin. Med. Phys.* **2** 174–7

REPPIR 2019 *The Radiation (Emergency Preparedness and Public Information) Regulations 2019* http://legislation.gov.uk/uksi/2019/703 (accessed 31 January 2023)

REPPIR(NI) 2019 *The Radiation (Emergency Preparedness and Public Information) Regulations (Northern Ireland)* http://legislation.gov.uk/nisr/2019/185 (accessed 31 January 2023)

IOP Publishing

Medical and Dental Guidance Notes (Second Edition)
A good practice guide on all aspects of ionising radiation protection in the clinical environment:
IPEM Report 113
**John Saunderson, Mohamed Metwaly, William Mairs, Philip Mayles,
Lisa Rowley and Mark Worrall**

Chapter 10

Diagnostic uses of open radioactive substances

Scope

10.1 This chapter applies to:

(a) the use of open (dispersible) radioactive substances that are administered to human subjects for diagnosis, health screening or research into diagnostic techniques;

(b) the use of open radioactive substances for *in vitro* studies made for the purpose of clinical diagnosis or research; and

(c) the use of radioactive substances for testing and calibrating equipment used under (a) and (b).

Additional information on the use of positron emitting radionuclides is detailed in chapter 11 *Positron Emission Tomography*.

Working with open radioactive substances

10.2 Under the Ionising Radiations Regulations (IRR 2017, IRR(NI) 2017) Part 2 (7), any practice where:

(a) radioactive products are administered to people or animals for medical, treatment or research purposes;

(b) radiopharmaceuticals are manufactured;

(c) cyclotrons are used;

(d) significant amounts of radioactive substances (gaseous or liquid) are discharged into the environment (i.e. greater than quantity for notification of occurrences in column 5 of part 1 of schedule 7 of IRR (IRR 2017, IRR(NI) 2017), or 100 GBq for Ga-67 (HSE 2018a, HSE no date);

requires consent from the Health and Safety Executive (HSE or HSENI), with information on matters to consider in Schedule 2 of the IRR (IRR 2017, IRR(NI) 2017). Consent must be obtained before the work is undertaken, and the HSE or HSENI must be informed of any material changes to the work or if the consented practice has stopped and will not be performed again.

10.3 Under Regulation 5 of the Ionising Radiation (Medical Exposure) Regulations 2017 (IR(ME)R 2017, IR(ME)R(NI) 2018) (IR(ME)R), both the Employer and the Practitioner must hold a valid licence where radioactive materials are administered to people for diagnosis, treatment or research.

10.4 The guidance in this chapter applies to work involving quantities of radionuclides that require IRR notification or registration. Quantities which are below the concentration levels in IRR Schedule 7 Part 1 column 2 or below the activity levels in IRR Schedule 7 Part 1 column 3 need not be notified and can be disregarded for the purposes of radiation protection under IRR (HSE 2018c) A table of the quantities requiring notification or registration for some commonly used radionuclides in nuclear medicine is given in table 10.1. These apply to all work carried out by an Employer and are not site specific, unlike permits and licences issued under radioactive substances and medical exposures regulations (RSA 1993, EPR 2016, EA(S)R 2018) (IR(ME)R 2017, IR (ME)R(NI) 2018). Note that IRR regulation 2(4) requires that where more than one radionuclide is involved, the sum of the quantity and/or concentration ratios must be determined in accordance with IRR Schedule 7 Part 3, and if this exceeds one, then the quantity or concentration limit is exceeded (IRR 2017, IRR(NI) 2017).

10.5 Often, hospital laboratories such as haematology or pathology only perform tests involving exempt levels of activity. Chapter 14 *Ionising Radiation in General Laboratories* has been written for these users. Specific advice on working with tritium is also given in chapter 14.

10.6 Any sites that are likely to discharge (in a single event) radioactivity into the environment in excess of the levels indicated in column 5 of table 10.1 will need to obtain prior consent from the HSE or HSENI under regulation 7(1)(i) (IRR 2017, IRR(NI) 2017, HSE 2018a, HSE no date). This is unlikely to be an issue for the most commonly used radionuclides in nuclear medicine. However, those substances that are not listed in Schedule 7 of the IRR must use the default values listed in the bottom row of table 10.1, which are very restrictive. Hence, centres planning to work with unusual or novel radionuclides may need to obtain consent from the HSE beforehand.

10.7 All sites holding radioactive sources or accumulating/disposing of/managing radioactive waste must hold a relevant permit/authorisation/registration from the appropriate environment agencies (The Environment Agency (England), Natural Resource Wales (Wales), the Northern Ireland Environment Agency (NIEA) and the Scottish Environment Protection Agency (SEPA)) under the Radioactive Substances Regulations (RSR) (The *Environmental Permitting (England and Wales) Regulations 2016* as amended (EPR 2016), the *Radioactive Substances Act* 1993 (Northern Ireland) (RSA 1993), the

Table 10.1. Quantity for notification of work (MBq) and registration (MBq/g) of some commonly used radionuclides under schedule 7 of IRR (IRR 2017, IRR(NI) 2017)].

1	2	3	4	5	6
Regulation:	r.5(1), Schedule 1 para 1(a), r.6(2)(f)	r.5(1), Schedule 1 para 1(b)	r.6(2)(e)	31(1)	31(3)
Radionuclide	Concentration for notification (<1000 kg) or Registration (>1000 kg) (Bq/g)	Quantity for notification (MBq)	Concentration for registration (kBq/g)	Quantity for notification of accidental releases (MBq)	Quantity for notification of source loss (MBq)
^{3}H	100	1000	1000	1 000 000	10 000
^{14}C[a]	1	10	10	100 000	100
^{15}O	0.01	1000	0.1	10 000	–
^{18}F	10	1	0.01	10 000 000	10
^{32}P	1000	0.1	1	10 000	1
^{35}S[b]	100	100	100	100 000	1000
^{51}Cr	100	10	1	1 000 000	100
^{57}Co	1	1	0.1	100 000	10
^{58}Co	1	1	0.01	10 000	10
^{59}Fe	1	1	0.01	10 000	10
^{68}Ga	0.01	0.1	0.01	10 000 000	1
^{68}Ge	0.01	0.1	0.01	10 000	1
^{75}Se	1	1	0.1	100 000	10
^{81}Kr	0.01	10	10	100 000	–
^{89}Sr	1000	1	1	10 000	10
^{90}Y	1000	0.1	1	100 000	1
^{93}Zr	10	10	1	1000	100
^{99}Mo	10	1	0.1	100 000	10
^{99m}Tc	100	10	0.1	10 000 000	100
^{111}In	10	1	0.1	100 000	10
^{123}I	100	10	0.1	1 000 000	100
^{125}I	100	1	1	10 000	10
^{131}I	10	1	0.1	10 000	10
^{153}Sm	100	1	10	100 000	10
^{177}Lu	100	10	1	100 000	100
^{201}Tl	100	1	0.1	1 000 000	10
^{223}Ra	1	0.1	0.1	10	1
Radionuclides not in full IRR Schedule 7 list	0.01	0.001	0.0001	0.1	0.01

[a] Not carbon monoxide or dioxide.
[b] Not oganic sulphur.

Environmental Authorisations (Scotland) Regulations 2018 (EA(S)R 2018)), or operate under a RSR exemption (EPR 2016, BEIS *et al* 2018) or general binding rules (see schedule 9) of EASR (EA(S)R 2018)). Further details may be found in chapter 18 *Keeping, Accounting for, and Moving Radioactive Substances (including Transport)* and chapter 19 *Disposal of Radioactive Waste*.

Hazards from open radioactive substances and principles for their control

10.8 A radiation hazard may arise from open radioactive substances, either through external irradiation of the body or through the entry of radioactive substances into the body. The main precautions required in dealing with external irradiation are similar for both open and sealed sources and depend on the physical characteristics of the radiation emitted, the total activity, the physical half-life of the radionuclide, and the intended use. Open radioactive substances may produce a further external radiation hazard as a result of contamination.

10.9 When open radioactive substances enter the body, the internal radiation dose will depend on factors such as the physical and chemical form of the material, the activity, the physical half-life, the mode of entry, and the biokinetics. Biokinetics of commonly used nuclear medicine radiopharmaceuticals are described in the International Committee for Radiological Protection (ICRP) Report 128 (ICRP 2015); however, caution is needed when applying this data to occupational dosimetry as the route of entry is likely to be different.

10.10 The design of all rooms, equipment and procedures/systems of work involving open radioactive substances should be aimed at;

(a) minimising occupational irradiation;
(b) optimising medical exposures;
(c) minimising the irradiation of members of the public (such as visitors and patients who are not being examined, or workers in adjacent areas); and
(d) minimising releases to the environment, including radioactive contamination, and controlling the spread should it occur.

10.11 The risk to any person from any work with open radioactive substances should be assessed prior to the start of the work and kept under review. General information on risk assessments and setting dose constraints is given in chapter 1 *General Measures for Radiation Protection* and appendix 5 *Radiation risk assessment pro-forma*.

10.12 Work with open radioactive substances should be governed by clearly understood and documented rules or procedures so that efficient practices can be organised and the control of hazards becomes an established routine. These will include local rules, procedures, protocols, and other types of instructions. A clear indication of the roles and responsibilities of staff must also be set out in procedures wherever open radioactive substances are used.

10.13 Staff involved in the use of open radioactive substances must receive training to ensure they can undertake their duties safely. Requirements for information, instruction, and training under regulation 15 of IRR (IRR 2017, IRR (NI) 2017) are discussed in L121 paragraphs 262 to 271 (HSE 2018c), Regulation 6(3) of IR(ME)R (IR(ME)R 2017, IR(ME)R(NI) 2018)) and are also detailed in the conditions of permits under RSR (RSA 1993, EPR 2016, EA(S)R 2018). Staff undertaking procedures involving the administration of radioactive substances to patients should have received appropriate practical training (BNMS 2010) (see paragraphs 2.109 to 2.120 *Training*). Adequate supervision must be available for staff training to undertake procedures or for less experienced staff members.

Pregnant or breastfeeding staff

10.14 The employee is responsible under IRR to inform the Employer in writing of their pregnancy. Employers should assess the likely dose to the foetus of a pregnant employee from each work activity (HSE 2015). The risk assessment should examine previous staff dose measurements and the likelihood of incidents leading to external or internal irradiation of the foetus. Doses should always be as low as reasonably practicable (ALARP) but if the foetus might receive more than 1 mSv (from internal and external sources combined) over the declared term of pregnancy (see paragraphs 1.24, 1.53, 1.64, 1.97, 1.134 and 1.157), a change of work activity should be discussed and agreed upon with the pregnant employee. For external irradiation from ^{99m}Tc or ^{131}I, a dose of 1 mSv to the foetus can be assumed if the dose measured at the surface of the maternal abdomen is 1.3 mSv (Mountford 1997), however, it is recommended that this be limited to 1 mSv (Temperton 2009). For external irradiation from positron emitters, the dose to the foetus may be similar to the dose at the surface of the maternal abdomen, and most staff working in PET will require a change in working practice, typically stopping roles relating to injecting patients, escorting patients to the toilet, and the unpacking of radiopharmaceuticals (Temperton 2009). Pregnant staff should not take part in any work activity involving significant risk of body contamination, including internal uptake (National Radiological Protection Board 2001), as determined by risk assessment. This would generally include the bulk dispensing of radiopharmaceuticals, clean-up of spills, and the care of very ill patients (Harding and Mountford 1993).

10.15 If a member of staff is breastfeeding, they should not take part in procedures or work in areas where there is a significant risk of bodily contamination, e.g. cleaning up a large spill of radioactivity, particularly of a longer-lived radionuclide. Airborne contamination from the use of ^{99m}Tc nebulisers from lung ventilation studies is unlikely to result in a significant dose to the foetus of a pregnant worker (Temperton 2009); however, an assessment should be undertaken of the potential radiation dose to the breastfeeding infant resulting from a chance inhalation by the mother (as a member of staff) of aerosols and other radioactive gases arising from their work. Appropriate action should then be taken to restrict this dose as necessary. Breastfeeding status must be determined on return to work following maternity leave, and duties must be altered accordingly, as laid out in the risk assessment.

Roles and responsibilities of duty holders in nuclear medicine

10.16 Paragraphs 10.17 to 10.31 should be read in conjunction with chapter 2 *Radiation protection of persons undergoing medical exposures*, which gives a general overview of IR(ME)R. Further details may be found in the guidance document *IR(ME)R Implications for clinical practice in diagnostic imaging, interventional radiology and diagnostic nuclear medicine* (BIR *et al* 2020). Duty holders under the IRR (IRR 2017, IRR(NI) 2017) are described in chapter 1 *General measures for radiation protection.*

10.17 Employers must develop referral guidelines for all their standard nuclear medicine investigations by consulting with and taking advice from the professionals involved. The British Nuclear Medicine Society (BNMS, no date), European Association of Nuclear Medicine procedure guidelines (EANM 2017), and the Royal College of Radiologists (RCR) guidelines *iRefer: Making the best use of clinical radiology* (RCR 2017) are available online and may help departments define their criteria. For cardiology, the British Nuclear Cardiology Society (BNCS no date) and the European Society of Cardiology (ESC no date) have guidelines available on their websites. The Society of Nuclear Medicine and Molecular Imaging (SNMMI) guidelines are also available online (SNMMI no date); however, it should be noted that some of the standard administered activities used in the USA are much higher than the ARSAC Diagnostic Reference Levels (DRLs) (ARSAC 2024).

Referrers

10.18 Referrers should be informed of the criteria that apply to the investigations they may request, including information on radiation doses and associated risks, particularly in the case of child patients or individuals of childbearing potential. The information should be given to Referrers in writing or electronic form. Further detail is available in *IR(ME)R: Implications for clinical practice in diagnostic imaging, interventional radiology and diagnostic nuclear medicine* (BIR *et al* 2020). Feedback should also be provided for inappropriate referrals.

10.19 The Referrer to a nuclear medicine facility will usually be a medical practitioner. The Referrer must provide sufficient information about the patient for the IR(ME)R Practitioner to decide if nuclear medicine is the appropriate diagnostic procedure for the patient. Referrers and referral groups (e.g. GPs) must be identified in the Employer's procedures for the studies to which they may refer. They must have sufficient instructions or training to ensure they understand their role as a Referrer, as well as the referral criteria, procedures for making the request and cancelling the request if no longer appropriate, the need for clinical evaluation, and the limitations of their entitlement. Non-medical Referrers should be identified by name.

10.20 Referrals should be clear and unambiguous, and the Practitioner should consult the Referrer if insufficient clinical information has been provided to enable them to justify the medical exposure involved in the examination. Relevant clinical information may include previous diagnostic information, medical records, images from other investigations, the higher radiosensitivity of an individual (due to genetic predisposition or disease), or the fact that a patient is known to be pregnant or breastfeeding. Employers should have adequate systems to provide Practitioners with previous images. Referrers should discuss requests not covered by the normal referral criteria directly with the Practitioner.

IR(ME)R practitioners

10.21 The Practitioner in nuclear medicine must have an appropriate IR(ME)R licence for each procedure to be undertaken, with the Employer's licence covering the site where the radiopharmaceutical is administered. The Practitioner should:

(a) assess the information provided by the Referrer and, if necessary, ask for more information from the Referrer;
(b) decide if nuclear medicine is an appropriate modality for the patient; and
(c) justify in writing or by electronic signature (see paragraph 10.23) the medical exposures necessary to fulfil the clinical request.

10.22 All examinations must be justified by a Practitioner and authorised by either the Practitioner or an appropriately entitled Operator following authorisation guidelines issued by the Practitioner.

10.23 Before an examination can proceed, it is necessary to identify that it has been justified (see paragraphs 2.27 to 2.34 and 2.38, *The practitioner and justification*). The Practitioner may, for example, record the authorisation by signing the request form or the prescription for the radioactive medicinal product to be used. Alternatively, in circumstances in which justification is made following authorisation guidelines, the appropriately entitled Operator can authorise it by recording that the request complies with the authorisation guidelines. Employer's procedures should also require Operators to record that other checks have been performed for individual patients, e.g. patient identification and individuals of childbearing potential.

IR(ME)R operators

10.24 All staff involved in the practical aspects of nuclear medicine procedures (Operators) must be adequately trained (see paragraphs 2.109 to 2.120 *Training*) and undertake continuing education in accordance with the Employer's procedure (IR(ME)R Schedule 2 (IR(ME)R 2017, IR(ME)R (NI) 2018) for the task(s) they undertake. Tasks to be performed by Operators must be specified in an Employer's procedure. If the Practitioner performs any of these practical aspects, they also take on the role of an Operator, and must be entitled appropriately. Operators who may generate the report of an investigation must be specified in an Employer's procedure.

10.25 Operators are directly responsible for the practical aspects of any part of a nuclear medicine procedure that they perform. In addition to following the Employer's procedures and protocols, they should ensure that the radiopharmaceutical administered to the patient is the one intended and that the activity administered is within the range specified in the protocol. Allowance may have to be made for dead space in the syringe or the adherence of the material to the delivery system. If the Practitioner wishes to exceed the DRL (see paragraph 10.26) in an individual case, the reasons for this should be fully documented. Operators should select the practical methods and, where possible, the equipment to ensure that as much diagnostic information as may be necessary is obtained once the radiopharmaceutical has been administered to the patient. Operators should clearly understand the extent of their own responsibilities. A trainee must be supervised by staff who are adequately trained. In this situation, the supervising Operator retains full responsibility as the IR(ME)R Operator for each task.

10.26 The national DRL for each procedure is tabulated in the ARSAC Notes For Guidance (ARSAC 2024). CT DRLs for hybrid nuclear medicine procedures are also published (Iball *et al* 2017, UKHSA 2022). Departments should set their own DRLs, taking local circumstances into account, such as software and equipment. These local DRLs may be lower than those quoted in the guidance; however, if these are increased beyond the national DRL, it may be necessary to obtain an amendment to the Employer licence. Further details are specified in the ARSAC Notes For Guidance (ARSAC 2024).

10.27 All the dispensed activities should be checked by two people prior to administration to the patient to ensure that the locally set DRL is not exceeded. The activities administered should be reviewed periodically, with the involvement of the medical physics expert (MPE) and Practitioner, as part of the quality assurance (QA) programme, to ensure that the optimum balance between maximising information (benefit to the patient) and minimising dose (risk to the patient) is maintained.

10.28 The level of MPE involvement depends on the service provided (IR(ME)R 2017, IR(ME)R(NI) 2018). The MPE helps to ensure the overall scientific and technical quality of the investigations by providing advice on a range of issues, such as the form and frequency of QA checks, patient dosimetry (including dosimetry assessments in research protocols), analysis, display and presentation of results, including images. MPEs would normally also be responsible for, or at least heavily involved in, the specification of new equipment, the acceptance testing or commissioning of new or modified equipment (including equipment on loan; see paragraphs 2.88 to 2.94, *Acceptance, commissioning and routine testing*, and 2.95 to 2.98, *Equipment on loan*) and optimisation of parameters and techniques for all studies undertaken. The role of the MPE is separate from that of the radiation protection advisor (RPA) and the radioactive waste advisor (RWA), although each informs the other. The role of the MPE is further detailed in appendix 15.

10.29 The RPA for the nuclear medicine facility should have appropriate experience in the application of radiation protection principles to radioactive sources. The RPA should be consulted on all matters specified in Schedule 4 of the IRR (IRR 2017, IRR(NI) 2017) relevant to the facility (see appendix 3, *The role of the radiation protection adviser*). These matters will include the correct designation of areas as controlled or supervised, the content of local rules, and other matters as detailed in the appendix.

10.30 The RWA for the nuclear medicine facility must be consulted on all matters related to the permit/authorisation held by the facility under the relevant RSR (SEPA *et al* 2020) (see appendix 21 *The role of the radioactive waste adviser*).

10.31 The Employer must ensure that all staff are aware of and understand the local rules. The task may often be delegated to the radiation protection supervisor (RPS) although the responsibility cannot. The RPS provision to supervise compliance with the local rules must be sufficient to cover peripatetic work away from the base location, sickness, and holidays. The RPSs can also provide a communication channel with the RPA (see chapter 1, *General measures for radiation protection*).

Design of nuclear medicine departments

10.32 The requirements for medical establishments using open radioactive substances vary with the type of medical procedures being undertaken. The RPA should be consulted about the design of laboratories and other work areas, and detailed plans should be drawn up in collaboration with the clinician responsible for the service and the MPE. Changes that may become necessary from time to time should also be discussed with the RPA and the RWA, where appropriate. Advice on the design of equipment and facilities is given in the IAEA publications *Nuclear Medicine Physics* (IAEA 2014), ICRP52 *Protection of the Patient in Nuclear Medicine* (ICRP 1987), and ICRP57 *Radiological Protection of the Worker in Medicine and Dentistry* (ICRP 1989). HBN6 *Facilities for Diagnostic Imaging and Interventional Radiology*

(NHS Estates 2001) and the IAEA reports *Applying radiation safety standards in Nuclear Medicine* (IAEA 2005) and *Radiation Protection and Safety in Medical Uses of Ionizing Radiation* (IAEA 2018) also give guidance, with further health building notes available from the Department of Health and Social Care (NHS Englandno date).

10.33 Where open radioactive substances are administered to patients, a risk assessment specifying the nature and extent of the clinical procedures to be undertaken should be made. It should include an assessment of the need for the following areas:
(a) radiopharmacy;
(b) blood cell labelling facility;
(c) radioactive waste storage and disposal;
(d) radionuclide administration to patients;
(e) separate patient waiting, changing and uptake areas and toilets;
(f) clinical measurements or imaging;
(g) sample measurements;
(h) area for decontamination of articles or people;
(i) considerations for receiving and dispatch areas for delivery or transport of radiopharmaceuticals;

as well as office and reporting areas and staff facilities such as cloakrooms, showers, and toilets.

10.34 Where radioactive substances are used solely for *in vitro* diagnostic tests, chapter 14 *Ionising radiation in general laboratories* may be more helpful. The design of the radiopharmacy is discussed in chapter 12 *Preparation of radiopharmaceuticals* and PET/CT departments in chapter 11 *PET–CT*.

10.35 Special consideration should be paid to ensuring that activities in one area do not interfere with work in adjoining areas. For example, radioactive samples or waste should not be handled or stored where they might interfere with imaging or counting procedures.

10.36 The appropriate designation of areas where open radioactive substances are stored or handled should be determined by risk assessment (see also paragraphs 1.44 to 1.54, *Risk assessment*, paragraphs 1.104 to 1.115, *Controlled areas and supervised areas*, and appendix 6, *Designation of controlled and supervised areas*). The risk assessment must be reviewed if there is a reason to suspect that it is no longer valid or if there has been a significant change in the matters to which it relates (L121 paragraphs 78 and 79 (HSE 2018c)), and should be reviewed periodically. The specific circumstances of the radioactive substances storage area that need to be taken into account are:
(a) whether special procedures must be followed to restrict significant exposure;
(b) the external radiation hazard;
(c) the risk and level of contamination hazard;
(d) the control of access;

(e) whether staff untrained in radiation protection need to enter;
(f) the length of time over which persons need to remain in the area; and
(g) whether the only radioactive substances present are within the body of a person, in which case restrictions based on instantaneous dose rate (IDR) may be relaxed (see L121 paragraphs 297 and 312 (HSE 2018c)).

10.37 All surfaces of the room where radionuclides are used or stored, including benches, tables, and seats, should be smooth and non-absorbent so that they can be cleaned and decontaminated easily. This is often difficult to implement fully, particularly in clinical areas. In such cases, articles that become contaminated should be removed from use. Further guidance is given in paragraphs 10.90 to 10.95, *Procedures in wards and clinics*. Areas where radioactive material is used should be clearly demarcated from those where it is not.

10.38 Floors and other surfaces should offer a good compromise between infection control, patient safety, and maintaining compliance with the relevant RSR. Materials not affected by dampness, chemicals, or heat should be chosen. Bench surfaces should be coved against walls and lipped at the edges. This includes non-slip flooring and surfaces that are smooth, continuous, and non-absorbent, such as Corian®, that can be easily cleaned and decontaminated (DH 2013). Floor coverings should be coved against walls and removable if necessary. Walls should be finished with a good, hard-gloss paint. The user can assure themselves that surfaces may be easily decontaminated by testing samples with radiopharmaceuticals to be used in the areas prior to installation. The choice of surface materials should take account of the effects of solvents and cleaning materials likely to be used. For further information on decontamination strategies, see paragraphs 10.101–10.112, *Decontamination of areas, surfaces, and equipment.*

10.39 The floor and benches, including worktops inside enclosures, should be strong enough to support the weight of any necessary shielding materials and radionuclide generators. A square metre of lead brick wall 50 mm thick weighs 570 kg. The Manual Handling Operations Regulations (*The Manual Handling Operations Regulations (Northern Ireland)* 1992, HSE 2016) require lifting tasks to be assessed and lifting equipment to be supplied for heavy loads, e.g. radionuclide generators. Remote handling equipment may be required under the Lifting Operations and Lifting Equipment Regulations (*The Lifting Operations and Lifting Equipment Regulations (Northern Ireland)* 1999, HSE 2018b)).

10.40 Work, other than clinical ventilation studies (see paragraph 10.86), that may cause airborne contamination should be carried out under conditions that prevent the inhalation of radioactive substances. A risk assessment will determine whether local exhaust ventilation should be provided. The work should be done in a contained workstation if significant contamination may occur (see paragraphs 12.23 to 12.27, *Enclosures and extract systems*).

10.41 One or more wash-hand basins fitted with foot, knee, or elbow operated taps and with hot and cold water supplies should be provided close to and preferably at the exit from each area where radioactive substances are handled. Disposable towels (and a nearby waste bin) should be provided. In radiopharmacies, these hand-washing and drying facilities will have to be provided in a way that does not contravene aseptic regulations (see paragraph 12.17). There should also be a means of monitoring people and items upon leaving the area, and recording the results if contamination is detected.

10.42 In designing an area where radioactive substances are to be handled, the following list of special features and requirements should be considered initially and independently of the likely and final designation of the area:

(a) The need to prevent entry by unauthorised persons.
(b) Adequate storage space so that essential equipment used in the laboratory can be kept there, thus minimising the risk of spreading contamination to other areas.
(c) One or more contained workstations designed for ease of decontamination.
(d) An entry lobby, or an area near the entrance, where protective clothing (see paragraphs 10.60–10.65, *Provision of protective equipment and clothing*) can be put on, taken off, and kept when not in use.
(e) A secure and shielded store for radioactive substances (see paragraph 18.8 *Design of stores*), ensuring that all security arrangements are satisfied, including ‘out-of-hours’ delivery, if applicable.
(f) A shielded temporary store for solid radioactive waste; it is convenient to use disposable plastic liners of distinctive colours and suitably marked to aid disposal.
(g) A limited number of places, designated and clearly marked, for the disposal of radioactive liquid waste. Where direct disposal of low activity waste to drains is authorised, the drain should be connected as directly as possible to the main sewer; drainage system materials should take account of the possible build-up of contamination on surfaces; U-bends flush more completely than traps, but if traps are used, they should be small, clearly labelled and accessible for monitoring.
(h) Appropriate shielding is required to protect the worker from unnecessary exposure to external radiation and to protect persons in adjacent rooms or corridors from external radiation that penetrates the floor, walls, or ceiling (see paragraphs 1.105).
(i) A wash-up area for contaminated articles such as glassware: the drainage should comply with (g) above; subject to the approval of the RPA/RWA, the wash-up area may be combined with (g) above but not used for hand-washing.
(j) Pipework through which radioactive materials flow should be clearly marked to ensure that monitoring precedes any maintenance.

10.43 In hospitals, excreta is not normally collected for specific disposal as radioactive waste. However, the collection of samples is sometimes needed for diagnostic tests. The store used to contain them should be locked and conform to the guidance in chapter 18, *Keeping, accounting for, and moving radioactive substances*.

10.44 Occasionally, some patient specimens that are sent for analysis (e.g. to a pathology laboratory) cannot be disregarded for the purposes of radiation protection. A risk assessment should be undertaken for each type of situation and take into account:

(a) the external dose rate and ease of shielding;
(b) the risk of contamination from the procedures to be performed;
(c) the time for which the sample must be worked; and
(d) the storage and disposal conditions.

10.45 Even when hazards to staff are minimal, the radioactivity in samples may still interfere with the analysis. Appropriate labelling of the sample will be required and should be used to warn other staff of this possibility.

10.46 The initial risk assessment should include a recommendation as to whether or not the receiving laboratory needs to consult an RPA/RWA and any other legislative requirements under RSR (RSA 1993, EPR 2016, EA(S)R 2018) or carriage of dangerous goods regulations (CDG 2009, CDG(NI) 2010) (see paragraph 14.24 concerning cooperation between employers). If the laboratory receiving radioactive samples is on a site not covered by an existing HSE notification, then the Employer at that site should be advised by their RPA if such notification is required (see table 10.1). The proper disposal of any radioactive waste generated should also be addressed (see chapter 19 *Accumulation and disposal of radioactive waste*).

10.47 Radioactive samples may also occur as a result of surgery (see paragraph 10.96, *Procedures in operating theatres*) and should be considered in the radiation risk assessment. This has been extensively investigated for sentinel node procedures (Waddington *et al* 2000, Klausen *et al* 2005).

10.48 Where an area for decontamination of persons is provided, a large wash-hand basin with a detachable spray head may be useful for carrying out procedures discussed in paragraphs 10.113–10.119, *Decontamination of persons*. The design of decontamination areas in casualty departments for use in major incidents is discussed in paragraphs 20.46 to 20.52.

10.49 In addition to the warning signs required for controlled areas, all work areas where open radioactive substances are present or stored, should be marked with a sign or the radiation trefoil, indicating ionising radiation (see appendix 7, *Warning signs and notices*).

Enclosures and extract systems

10.50 For details on the requirements for enclosures and extract systems, including those used for aseptic work and iodination procedures, see chapter 12 *Preparation of radiopharmaceuticals*.

Equipment: installation, maintenance and quality assurance

10.51 All commercial medical devices must carry the CE mark (prior to 30 June 2023) or UKCA marking to indicate that they conform to the Medical Devices Regulations (MHRA 2022). The enforcement authority for this directive in the United Kingdom is the Medicines and Healthcare Products Regulatory Agency (MHRA). The most common items of equipment in nuclear medicine facilities are the gamma camera and the radionuclide calibrator. The National Physical Laboratory (NPL) provides services to ensure that the performance of radionuclide calibrators is traceable to national and international standards.

10.52 All equipment that emits radiation will require a critical examination in accordance with IRR (IRR 2017, IRR(NI) 2017). This includes such items as the radioactive line sources or CT equipment attached to gamma cameras and PET scanners. This examination ensures that safety features and warning devices function correctly and that the radiation protection arrangements for the installation are satisfactory (see paragraphs 1.245 to 1.259 *Critical examination* and 13.4). The installation of radionuclide calibrators and gamma cameras implies work involving radioactive sources and radioactive patients in that area; the radiation protection arrangements of the designated areas where such equipment has been installed should therefore be checked.

10.53 Before any piece of nuclear medicine equipment enters clinical use, it should undergo acceptance testing to ensure that it operates safely and performs to specification. Manufacturers of gamma cameras specify performance in relation to standards laid down by the National Electrical Manufacturers' Association (NEMA). Guidance for QA and test frequency is detailed in IPEM Report 111 (IPEM 2015) with further guidance available from the EANM (Busemann Sokole *et al* 2010a, 2010b).

10.54 Further commissioning tests should aim to provide baseline results for subsequent QA tests and will provide information that can be used to define optimum acquisition parameters for all clinical procedures.

10.55 All departments should undertake QA to ensure the continual production of optimum quality images and diagnostic information using the minimum amount of activity administered to the patient. QA programmes should include checks and test measurements on all parts of the imaging and computing systems at appropriate time intervals. There must be a QA programme in place to check the performance of the radionuclide calibrator, with guidance available from the NPL (NPL 2006). A QA programme must also be in place to review the activities administered to patients; see paragraph 10.27. The MPE should be consulted when planning these programmes and they should be involved in regular reviews of their effectiveness. QA of radiopharmaceuticals is discussed in chapter 12 *Preparation of radiopharmaceuticals*.

10.56 All equipment should be maintained to the specifications and at the frequency recommended by the manufacturer.

10.57 A record of maintenance, including QA, should be kept for each item of equipment. This should include information on any defects found by users (fault log), interim remedial action taken, subsequent repairs made, and the results of testing before equipment is reintroduced to clinical use.

10.58 All persons carrying out modifications or maintenance that could affect image quality or the assay of radioactivity should sign a handover form, taking responsibility for that equipment (HSE 2006, AXREM 2018), and also any designated areas if they are working under their own risk assessments and local rules (see chapter 1, *General measures for radiation protection*). They should immediately inform the person responsible for the use of the equipment before it is returned to clinical use. Where this is not possible, e.g. when the work is performed out-of-hours, this should be clearly marked on the handover form, and this should be signed back, after appropriate checks, before clinical use. In all cases, a clear written report of the changes and current operational status should be given, including the name of the person concerned and the date, either in the maintenance log or in the form of a handover questionnaire. This should include details of any changes affecting image quality or the assay of radioactivity that may require further measurements, or any changes that could affect the dose received by the patient.

10.59 After work affecting equipment performance (see paragraph 10.58), staff must perform any in-house testing required by their QA programme before the equipment is returned to clinical use. In some cases, the programme may indicate that advice must be obtained from the RPA or MPE before proceeding. If alterations in procedures, clinical protocols, or local rules are required, these changes must be communicated to all users, and the process for this must be documented in the QA procedures.

Provision of protective equipment and clothing

10.60 Laboratories and other work areas used for the manipulation of open radioactive substances should be provided with equipment specifically for this purpose. This equipment may include:

(a) apparatus for maximising the distance of the worker from the source, e.g. tongs, forceps;
(b) containers for radioactive substances that incorporate the necessary shielding as close to the source as possible;

(c) double-walled containers (the outer wall being unbreakable) for liquid samples;
(d) a drip tray;
(e) disposable tip automatic pipettes; alternatively, hypodermic syringes to replace pipettes;
(f) syringe shields;
(g) lead walls or 'castles' for secondary shielding;
(h) lead barriers with lead glass windows;
(i) barriers incorporating Perspex (PMMA) for work with beta emitters;
(j) radiation and contamination monitoring equipment;
(k) equipment for the assay of stock solutions and radioactive substances prepared for administration to patients;
(l) carrying containers, wheeled if necessary, for moving radioactive substances from place to place;
(m) equipment and materials to deal with spills; and
(n) needle re-sheathing blocks.

10.61 Protective clothing should be used in work areas where there is a risk of contamination, both to protect the body or clothing of the wearer and to help prevent the transfer of contamination to other areas, as determined by the radiation risk assessment. The clothing should be monitored and removed before leaving designated areas. When moving between supervised areas such as the camera room and the injection area, it may not be necessary to change the protective clothing unless a spill is suspected. In any case, protective clothing should be removed prior to leaving the designated areas, e.g. when visiting the staff room. The clothing may include:
(a) laboratory coats, protective gowns or disposable aprons;
(b) gloves which are waterproof and, where low-energy beta emitters are handled, thick enough to protect against external beta radiation;
(c) overshoes; and
(d) caps and masks for aseptic work.

10.62 Respirators are not normally needed for medical work, though they may be required in some emergencies. The respirators must be adequately maintained, and records of maintenance must be kept (see appendix 10 *Record keeping*). People who may need to wear them should be suitably trained in their use.

10.63 Protective aprons, arm-sleeves and gloves should be worn when there is any direct contact with the patient or contaminated clothing, bed linen, or other articles. The method of removing gloves should be based on surgical technique in order to avoid transferring activity to the hands. If a risk assessment has shown that reusable gloves are acceptable, the same technique of removal will avoid contaminating the inner surfaces of the gloves. There are different considerations where the reusable gloves are attached to an isolator, see paragraph 12.29.

10.64 Persons working with open radioactive substances or those nursing high-activity patients should wash their hands thoroughly with mild soap and water before leaving work areas. Particular attention should be paid to cleaning fingernails. Hands should be monitored for residual contamination after washing.

10.65 Protection against external contamination should be provided for patients who receive oral administrations of radioactive materials or inhale radio-active aerosols. A plastic apron is normally sufficient to protect clothing.

General procedures in nuclear medicine departments

10.66 The advice of the RPA, RWA, and MPE should be sought, where appropriate, before new procedures are introduced or major changes are made to existing procedures. Extremity monitoring may be of particular use in planning and reviewing operational procedures. New or changed procedures should be rehearsed, where possible, without using radioactive substances.

10.67 For staff working in diagnostic nuclear medicine with gamma emitting radionuclides, whole-body dosimeters worn at the waist are recommended (Martin *et al* 2018). Extremity monitoring is recommended for staff administering radiopharmaceuticals regularly and those working in radiopharmacy. For doses approaching 100 mSv, finger stalls are recommended on the index finger tip when this does not interfere with manipulations; otherwise, ring dosimeters may be used with a multiplying factor applied. A factor of 6 should be used for dosimeters at the base of the finger and a factor of 2 for those worn on the second phalanx (Martin *et al* 2018); however, classified staff must wear the dosimetry as directed by the Approved Dosimetry Service and record the results as reported. It is unlikely that the dose to the lens of the eye will exceed any dose limits, and estimation of the eye dose can be made from the $H_p(3)$ or, if unavailable, the $H_p(10)$ measurement from a whole-body dosemeter on the chest or collar (Martin *et al* 2018).

10.68 Working procedures should be designed to prevent spillage and, in the event of spillage, to minimise the spread of contamination from the work area. This is necessary not only in the interests of the safety of people but also to prevent interference with the assay of samples containing radioactivity. Dispensing of radioactive substances should be done at a contained workstation, especially if particulates, aerosols, vapours, or gases are involved. All manipulations should be carried out over a drip tray in order to minimise the spread of contamination due to breakages or spills. No object should be introduced into a contained workstation if it interferes unacceptably with the pattern of airflow. Contingency plans for reasonably foreseeable incidents (IRR regulation 13 (IRR 2017, IRR(NI) 2017)) should be drawn up and referred to or included in the local rules (see also paragraphs 12.53 and 13.19). Such plans should include scheduling rehearsals and updating contingency procedures.

10.69 No food or drink (except that used for medical purposes), cosmetics, or smoking materials should be brought into an area where open radionuclides are used, nor should they be stored in a refrigerator used for open radioactive substances. In addition, crockery and cutlery (except those used for medical purposes) should not be brought into the laboratory for washing or storage.

10.70 Handkerchiefs should never be used in these areas; an adequate supply of paper tissues should be provided (unless in an aseptic area, see paragraph 12.29).

10.71 Any cut or break in the skin should be covered before a person enters an area where open radioactive substances are handled. Dressings should incorporate waterproof, adhesive strapping. If a cut or break in the skin is sustained while in the area, refer to paragraph 10.117.

10.72 Equipment provided specifically for the safe handling of open radioactive substances should always be used. Such equipment should not be removed from the work area. Never use your mouth to operate a pipette. Remote-handling equipment should be provided and used wherever it can be effective in reducing radiation doses.

10.73 The Health and Safety (Sharp Instruments in Healthcare) Regulations 2013 (*The Health and Safety (Sharp Instruments in Healthcare) Regulations (Northern Ireland) 2013*, HSE 2013) require assessment to ensure the risk from sharp injuries is suitably controlled, and individual Employers will have different policies in place to mitigate this risk, such as the use of retractable needles and re-sheathing blocks. Where it is not possible to dispose of sharps immediately after use, for example when measuring doses in the radiopharmacy or weighing syringes for GFR calculations, or where there is a high risk of contamination, re-sheathing blocks may be used for one-handed re-sheathing, as assessed in the local risk assessment (HSE 2013).

10.74 The work area should be kept tidy and free of articles not required for the work. It should be cleaned often enough to ensure minimal contamination. Cleaning methods should be chosen in order to avoid raising dust or spreading contamination. Articles used for cleaning controlled or supervised areas should be restricted to these areas. They should be monitored periodically, and if found to be contaminated, reference should be made to paragraphs 10.101–10.112, *Decontamination of areas, surfaces and equipment*. Cleaning and contamination control may be simplified by using disposable items and (except where this conflicts with aseptic requirements; see chapter 12 *Preparation of radiopharmaceuticals*) by covering benches and the interior of drip trays with disposable material such as plastic-backed absorbent paper.

10.75 Shielding should always be considered for any radioactive source. The radiation risk assessment should identify what shielding is required and what type and form it should take. Figure 10.1 may be helpful in this process. Appropriate shielding may be obtained using a variety of materials, such as lead, lead glass, lead composite, tungsten, or aluminium. Shields

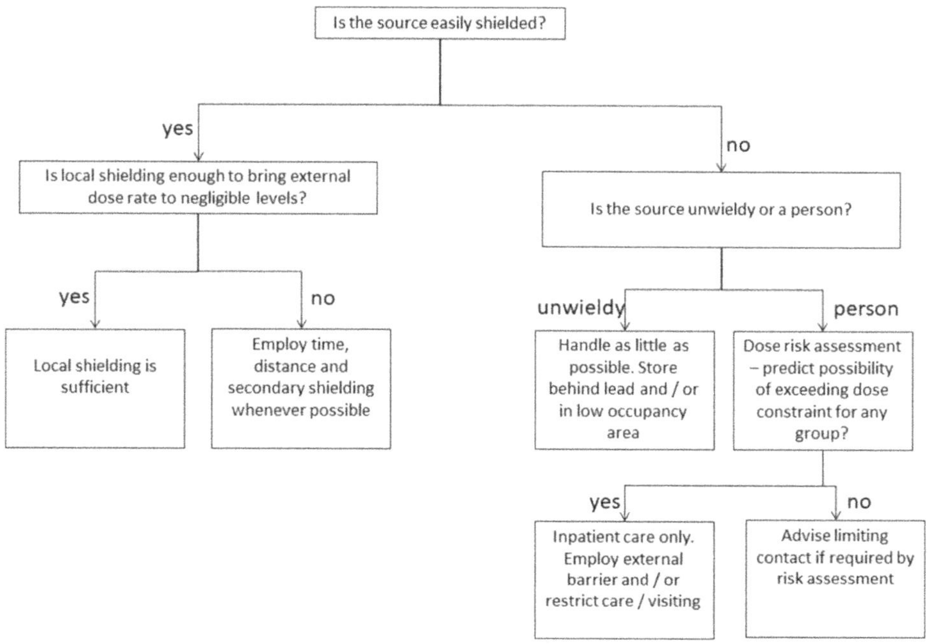

Figure 10.1. Shielding decision chart.

incorporating acrylic materials such as Perspex may be more suitable for beta emitters. Lead should be painted with a suitable gloss finish or otherwise encased to provide a safe, cleanable surface.

- Vials of most radiopharmaceuticals are easily shielded by being placed in lead pots, but the pots themselves may still need to be kept behind lead walls when the vials are not being used.
- It is generally impossible to avoid some direct handling of large flood phantoms (sealed or open) used for testing gamma cameras or to avoid using them while unshielded. Protection of staff is optimised by handling as little as possible and not lingering near the source when it is unshielded.
- Recommended values for dose constraints are given in paragraphs 1.60 to 1.64 Dose constraints, 2.156 to 2.168 *Carers and comforters.*
- Additional shielding may be required for sources kept near low level counting areas.

10.76 Perspex is usually the preferred shielding material for beta emitters, as this material minimizes the production of Bremsstrahlung radiation; however, Perspex syringe shields can be bulky. If Perspex cannot be used for this reason, a standard tungsten syringe may be used instead, which offers a similar level of protection (Hooper 2011).

10.77 Syringes used for handling radioactive liquids should be shielded wherever practicable from beta and gamma emitters. An assessment of potential doses should be carried out for alpha emitting radionuclides with additional gammas to determine if syringe shields are required. The use of a syringe shield (tungsten or lead glass) is recommended for the dispensing of ^{223}Ra (UK Radiopharmacy Group, IPEM NM SIG, and BNMS 2017). The distance between the fingers and the radioactive substance should be as large as can be safely achieved. Consideration should be given to the need for extremity monitoring, particularly in the case of new techniques, inexperienced staff, and increased workloads.

10.78 The activities of each radionuclide handled or stored in a particular work area should not exceed the values for which the area has been designed. These activity limits should be specified in the local rules. The RPA should be consulted in particular cases where it is desired to use larger activities than those specified. In all cases, the activities used must be within the Permit issued under RSR (RSA 1993, EPR 2016, EA(S)R 2018) (see chapter 18 *Keeping, accounting for, and moving radioactive substances*).

10.79 Radioactive substances should be clearly labelled, indicating the radionuclide, chemical form, activity at a given date and time, and batch number. The expiration date and time should be added if the substance is for human use. Records of stocks, administrations, and disposals should be kept (see chapter 18 *Keeping, accounting for, and moving radioactive substances* and appendix 10 *Record keeping*).

10.80 Systems must be in place to detect when maintenance work is to be performed on equipment that might be contaminated, particularly enclosures for controlling airborne activity, ventilation trunking, sinks, and waste pipes. In hospitals, this might take the form of information on potential areas being provided to the estates department and an instruction to inform the RPS whenever such work is planned. The Employer should carry out a risk assessment that takes account of the dose rate or activity likely to be encountered in the work and the nature of the necessary precautions to keep exposures ALARP. This will enable a decision to be made on a suitable system of work and the level of supervision needed. Direct supervision of the work may be required if, for example, the dose rate or activity is likely to be high, or complicated instructions are needed to keep exposures ALARP. Otherwise, it will be sufficient to issue a permit to work detailing the precautions to be taken by the maintenance staff. The maintenance staff should sign that they have read and understood the precautions before work can start.

Administration of radiopharmaceuticals

10.81 The activity administered to a patient, whether for diagnosis or research, must be determined by a person holding an appropriate Practitioner licence or under a protocol authorised by such a person, in the context of local DRLs and protocols. Any advice necessary to restrict the exposure of other persons (see paragraphs 2.155, *Information and instructions for nuclear medicine patients*, and 10.11) should be discussed with the patient in advance (see paragraph 13.12 for fuller details). Operators who administer the activity must be suitably trained, competent, and entitled.

10.82 Records should be kept of all administrations of radioactive substances, as indicated in paragraph 10.79. These records must be kept in such a way as to:

(a) facilitate accounting of stock under RSR (RSA 1993, EPR 2016, EA(S)R 2018) and IRR (IRR 2017, IRR(NI) 2017) (see chapter 18, *Keeping, accounting for and moving radioactive substances*);

(b) allow all components of any individual administration to be traced retrospectively (as required by the Human Medicines Regulations (*The Human Medicines Regulations* 2012));

(c) allow review of DRLs, clinical audit, and estimation of the effective dose to an individual patient as required under IR(ME)R (IR(ME)R 2017, IR (ME)R(NI) 2018).

(d) ensure that information on the administration is made available to persons who need to be advised, as part of the procedure adopted to restrict their exposure; and

(e) Identify the IR(ME)R Operators that may affect patient dose.

10.83 Equipment used for checking the activity prior to administration should be checked daily using a test sealed source and should be calibrated at regular intervals with sources traceable to national standards (see paragraphs 10.51–10.59, *Equipment: installation, maintenance and quality assurance*). In situations where calibration of activity is difficult, users should make every effort to ensure that the system they adopt is as robust as possible and developed with the advice of the MPE. Users may have to rely on calculations by volume, making the necessary decay corrections to the radioactive concentration printed on the vial by the manufacturer.

10.84 If it is suspected that a patient has received a significant, accidental, or unintended exposure (SAUE), an investigation of the circumstances must be undertaken immediately (see paragraphs 2.172 to 2.179, *Suspected radiation incidents*, and 2.180 to 2.182, *Performing an investigation)*, and the MPE should be involved. If the investigation confirms the suspicion, a detailed report must be made, and action should be taken to prevent a recurrence. The Practitioner must also be informed. If the incident is notifiable, the appropriate bodies must be contacted. If the patient's procedure was compromised or the exposure is classified as significant accidental or unintended exposure (CQC 2023), the patient should be informed under the duty of candour (The Health and Social Care Act 2008 (Regulated Activities) Regulations 2014 (2014). The MPE will normally perform this task, but it can be delegated if specified in the Employer's procedure. If the exposure was significantly greater than intended, it must be reported under IR(ME)R (IR(ME)R 2017, IR(ME)R(NI) 2018) to the relevant enforcing authority. In those situations where equipment malfunction was central to the overexposure, the incident should also be reported to the appropriate Health Department: the Manufacturer's On-line Reporting Environment at the MHRA, the Medicines and Healthcare Products Regulatory Agency (MHRA) (for England and Wales), Health Facilities Scotland (for Scotland), and the Northern Ireland Adverse Incident Centre (for Northern Ireland) where there are risks to patients concerning medical devices (see appendix 22 *Authorities and organisations of interest*).

10.85 If radioiodine in certain forms is going to be administered to a patient, a thyroid blocking agent may be used to lower the amount of radiation that goes to the patient's thyroid, as suggested in the ARSAC Notes for Guidance (ARSAC 2024) and the relevant clinical guidelines. This is particularly important in paediatric nuclear medicine, where the radiation risk to the thyroid is significantly higher than in the standard population (Wall *et al* 2011).

10.86 When radioactive gases or aerosols are administered, a build-up of airborne activity within the room should be avoided by conducting exhaled breath directly out of the building if this is permitted by the waste disposal authorisation under RSR (RSA 1993, EPR 2016, EA(S)R 2018), or by trapping it in a shielded leak-free container or filter. These precautions may not be necessary with very short half-life gases (e.g. ^{81m}Kr). Adequate ventilation should be provided for all rooms where radioactive gases and aerosols are stored, released, or used. The need for, or frequency of, air monitoring should be decided with advice from the RPA/RWA and described in the local rules. A gas is out of scope of RSR (RSA 1993, EPR 2016, EA(S)R 2018) if the substance has a half-life under 100 s.

10.87 Persons to whom open radioactive substances are administered (and/or their carers, as appropriate) should receive instructions and information on precautions to be taken and, if necessary, on any hazards involved (see chapter 16 *Patients leaving hospital after the administration of radioactive substances* for more detail). The instructions and information may be given verbally and/or in written form (see paragraphs 2.155, *Information and instructions for nuclear medicine patients* and paragraphs 2.156 to 2.168, *Carers and comforters*). The responsible persons for taking the necessary actions must be identified in the Employer's procedures.

10.88 In most cases, the radiation dose received by family and friends due to diagnostic procedures should be minimal. Chapter 16, *Patients leaving hospital after the administration of radioactive substances*, gives more information on the assessment of risk for these groups and any precautions that need to be taken when the patient goes home.

Extravasation

10.89 When administering radiopharmaceuticals intravenously, there is the risk of extravasation, resulting in a high localised dose and tissue damage.

Van der Pol *et al* (2017) conducted a review of published literature and found no reported adverse effects for ^{99m}Tc, ^{123}I, ^{68}Ga or ^{18}F extravasation, although follow up for these patients was rare. Symptoms, including radiation ulcers and radio-necrosis, were reported following ^{201}Tl, ^{131}I, and other therapeutic agents. See chapter 13 *Therapeutic uses of unsealed radioactive substances.*

Procedures in wards and clinics

10.90 If diagnostic radionuclides, including PET tracers, are administered to inpatients, the need for advice for staff on wards and in other departments should be considered as part of the risk assessment. Information, advice, and training must be provided as necessary for ward staff. It may also be necessary to record the administration and any advice regarding radiation protection in the patient's notes so that ward staff are fully informed.

10.91 Ward staff will usually be sufficiently protected from contamination from body fluids if standard hygiene precautions are followed, e.g. use of gloves and plastic aprons. The staff on the ward should be issued with storage and handling instructions to be followed if bedding or clothing becomes contaminated. In most situations involving diagnostic levels of activity, a risk assessment will show that contaminated items can be sent directly to the hospital laundry. Where higher activities or longer-lived radionuclides have recently been administered to a patient, the items may need secure storage in a plastic bag for an appropriate period to reduce the contamination to a level at which the items can be sent to the laundry. Consideration should also be given to disposal as radioactive waste (see paragraphs 10.104 to 10.107).

10.92 The external radiation dose received from the nursing of these patients will not normally be such that any special precautions will be required unless the patient needs intensive nursing. A risk assessment should be made to identify the situations in which sharing of nursing duties may be required in order to address this problem. The use of side rooms may also help reduce the dose given to other patients on a ward. The need for such advice is suggested, particularly for nurses on general wards, where staff are unlikely to have much experience working with radiation and where an RPS is unlikely to have been appointed. The situation is therefore different from that of nurses who are employed in nuclear medicine departments. However, both situations should have been covered in the radiation risk assessment.

10.93 Written advice should be returned to the ward with the patient's notes. It may be helpful to confirm this advice via telephone to the ward to ensure the nursing staff are aware of any restrictions or special procedures in place.

10.94 The dose received by pregnant or breastfeeding staff caring for or escorting diagnostic nuclear medicine patients is likely to be very small; however, risk assessment will determine if alternative duties are required, taking into account factors such as workload.

10.95 If radionuclides are to be administered outside the nuclear medicine department, such as on wards or in clinics, the local rules for the department must be extended to cover these areas. The ward or clinic area should, where possible, meet the requirements of paragraphs 10.37, 10.38, 10.39, 10.41, and 10.42. If this is not the case, special precautions should be taken. Consideration should be given to covering surfaces that may become contaminated, as detailed in paragraph 10.107(a). Administrations should not be undertaken in areas where the likelihood of maladministration is increased (e.g. very busy wards). Checks should also be made in advance so that a spill can be controlled and access to it prevented. Following the administration, the area should be monitored for contamination if there has been a suspected spill of radioactive material and decontaminated if necessary. A decontamination kit should be readily available, and there must be a means of summoning help from the RPS, nuclear medicine, or nursing staff. Surfaces must facilitate easy cleaning, and there must be a hand-washing basin nearby to allow personal decontamination. Procedures to deal with any contamination of the ward or clinic area should be contained in the local rules.

Procedures in operating theatres

10.96 When an operation takes place on a patient who is undergoing an investigation (e.g. sentinel node localisation) or has recently undergone a diagnostic nuclear medicine test (e.g. bone scan), the external radiation risk will not normally be such that any special precautions are required. The risk of contamination will depend on the distribution of the radiopharmaceutical within the body. Standard procedures in operating rooms should protect the staff from any contamination. If a significant risk is identified from the prior assessment, staff should be monitored. If, exceptionally, no radiation risk assessment exists, the RPA and RWA should be consulted for advice. Paragraphs 10.44–10.46 provide information on procedures to be adopted if contaminated samples or tissues are to be transferred to a laboratory for examination. Normal cleaning and sterilisation procedures should be adequate to remove any contamination from the equipment. If any potentially contaminated waste is generated, it should be collected in a clearly labelled plastic bag and removed by trained staff for disposal by an authorised route (see chapter 19 *Accumulation and disposal of radioactive waste*).

Monitoring of work areas and persons

10.97 Laboratories and other areas in which work with open radioactive substances is undertaken should be monitored on a systematic basis, both for external radiation and for surface contamination. For each controlled or supervised area, there should be a suitable monitoring schedule; for other areas, occasional monitoring is acceptable. Monitors used must be suitable for the task and regularly tested. A battery check and background level should be performed each time prior to using the monitor. Action levels must be determined. The person who undertakes the monitoring must be adequately trained. Records of the monitoring (made as a requirement of RSR) must be kept for the life of the relevant permit (see appendix 10 *Record keeping*), and should include the date, type of monitor, person undertaking the task, areas, results, and what action was taken if action levels were exceeded.

10.98 Contamination should be kept As Low As Reasonably Practicable (ALARP). Normal monitoring and decontamination procedures should aim to keep the level of contamination below levels that are detectable with the most sensitive method available. Procedures must be developed to deal with the situation if contamination is detected (see paragraphs 10.101 to 10.112, *Decontamination of areas, surfaces and equipment*).

10.99 Contamination monitoring will be required for:

(a) all working surfaces, waste storage areas (including the interior of enclosures), tools, equipment, handles (doors and cupboards), the floor and any items removed from the area; also during maintenance of contained workstations, ventilation systems and drains; direct monitoring should be used in preference to wipe testing unless high dose rates in the area would limit unacceptably the level of contamination that would be detected by direct monitoring; unless other information is available, it should be assumed that wipe testing will only remove 10 percent of the total contamination present on the swabbed area (NPL 2014);

(b) protective and personal clothing, including shoes, particularly when persons leave an area that is controlled due to the risk of contamination; a suitable instrument should be placed near the exit to enable such monitoring;

(c) clothing and bedding of patients where contamination is suspected; and

(d) the hands of staff before leaving an area where they have been handling open radioactive substances.

10.100 An instrument for monitoring the hands should be available where the hands are washed; this monitoring should extend to other skin areas, e.g. the face, if there is any reason to suspect that these areas may have become contaminated; low levels of skin contamination can be difficult to detect against even fairly modest background levels, depending on the design of the monitor; care should be taken to monitor in an area where the background is suitably low. Records of personnel monitoring should also be kept, and all personnel contamination incidents should be recorded.

Decontamination of areas, surfaces and equipment

10.101 If buildings, fittings, or articles become contaminated, prompt action should be taken to prevent the dispersal of the activity, particularly its transfer to the clothing or body of any person. This may be difficult to achieve in clinical areas. However, monitoring devices should be available in the vicinity and used if contamination is suspected and before leaving the wider area or eating or drinking. Contaminated equipment and areas of walls and floors should be isolated, and small objects should be transferred to containers using tongs and gloves. An assessment of the amount of activity involved and the associated radiation hazard must be made.

10.102 Once as much contamination as practicable has been removed, the remaining radioactivity and the hazards of it remaining there should be assessed. Acceptable levels of contamination are not specified in the regulations but should be determined as per the radiation risk assessment. Values of surface contamination that could result in exceeding the occupational dose limits under certain circumstances have been published (NRPB 1979, Delacroix *et al* 2002, IAEA 2014), and these may be helpful in deriving suitable action levels for different areas and surfaces, taking the case to scale for appropriate dose constraint. Measurements should be averaged over the most appropriate area, not exceeding 1000 cm^2 for floors, walls, and ceilings and 300 cm^2 for all other surfaces (except body surfaces, where 1 cm^2 is more likely to be appropriate) (HSE 2018c). The RPA/RWA may provide advice.

10.103 Care should be taken to ensure that the proposed action is in full compliance with the requirements of RSR (RSA 1993, EPR 2016, EA(S)R 2018) for the accumulation or disposal of radioactive waste (see chapter 19 *Accumulation and disposal of radioactive waste*). If the possibility of contravention of the RSR (RSA 1993, EPR 2016, EA(S)R 2018) does arise, the RWA should be contacted immediately for advice. Guidance is available from SEPA on decommissioning non-nuclear facilities (SEPA 2022).

10.104 When contamination is due to short-lived radionuclides and the risk of biological contamination is insignificant, it may be feasible to close off the area or store the article in an impermeable bag for a sufficient time to allow the radioactivity to decay away naturally. In any other circumstance, positive action to decontaminate will be necessary.

10.105 When storage is not practicable but suitable laundering facilities are available, any washable articles contaminated with short- or long-lived radionuclides may be given a series of washes in a proprietary cleansing agent, alternating with rinses (as provided by an automatic washing machine). Hot water may bind the contamination to the material. Machines may need to be checked for and rinsed clear of residual contamination afterwards, and outflow pipes should be marked as in paragraph 10.42(j).

10.106 If contaminated clothing or bedding is to be sent to public laundries, the items should be decay stored first until the activity is out of scope for RSR and exempt from the radiation requirements of the Carriage of Dangerous Goods Act 2009 (as amended) (CDG 2009, CDG(NI) 2010).

10.107 Proprietary cleaning agents are normally adequate for the removal of contamination. Agents that will not damage the material or interfere with its essential properties (e.g. non-slip or anti-static) should be chosen. Water and any other agent used for decontamination should be treated as radioactive waste (see chapter 19 *Accumulation and disposal of radioactive waste*). If cleaning fails to decontaminate adequately, the following actions should be taken:

(a) for short-lived radionuclides, the surface should be covered with impermeable coated paper: tar paper is not robust in wet conditions; strong polythene sheet may substantially increase the risk of slipping when used as a floor covering. Packexe® or similar brands are easy to apply, waterproof, and non-slip. The temporary covering should be clearly marked as radioactive and a date given for removal; and

(b) for longer-lived radionuclides, surfaces may have to be removed and replaced.

10.108 Glassware and porcelain can usually be cleaned with a suitable proprietary agent (e.g. Decon™, Milton™) or bleach. They are best cleaned immediately after use. The solutions used for cleaning should never be returned to the stock bottle.

10.109 Dilute nitric acid can be used to clean some plastics, since it will usually be effective without damaging the material. Care should be taken to avoid the use of ketonic solvents and certain chlorinated hydrocarbons, which will dissolve some plastics. No organic solvent has any effect on polythene.

10.110 Washing with a heavy-duty detergent, followed if necessary by specific chelating agents, may clean contaminated metal tools, trays, sinks, and equipment.

10.111 If articles cannot be decontaminated satisfactorily, or stored, they must be treated as radioactive waste. Articles should be monitored before reuse to ensure that decontamination has been successful.

10.112 Notification of occurrences to the HSE is detailed in L121 paragraphs 600–605 (HSE 2018c) and in chapter 20 *Contingency planning and emergency procedures for radioactive substances activity*.

Decontamination of persons

10.113 Hands should be washed if contamination is suspected, after completing work with open radioactive substances, and before leaving an area that is controlled due to the risk of contamination. Mild liquid soap should be provided unless aseptic considerations require an alternative cleaner. Swabbing may be more effective if the contamination is confined to a relatively small area. If detectable contamination remains on the hands after simple washing, using a surfactant or chelating agent specific to the chemical form of the contaminant may be more successful. The hands should not be rubbed or brushed vigorously, nor should harsh chemical treatments be attempted, as this might injure the skin and allow the radioactivity to enter the body. Non-abrasive single-use nailbrushes should only be used if contamination persists after simple washing.

10.114 When contamination of parts of the body, other than the hands, is suspected or when the procedures described above for decontamination of the hands are ineffective, the line manager and the RPS must be notified and consideration given to any further action. Some pointers are given below. The RPA may need to be contacted for advice

10.115 Special care needs to be taken in the decontamination of the face to restrict the entry of radioactive material into the eyes, nose, ears, or mouth. Decontamination of skin is best achieved as detailed for the hands in paragraph 10.113. Hair should be decontaminated similarly before being washed with shampoo. In all cases, care should be taken not to spread the contamination further.

10.116 Contaminated clothing should be removed as soon as practicable, taking care not to spread the contamination. If it cannot be left securely for the radioactivity to decay, the clothing should be dealt with as detailed in paragraph 10.106. If the contamination is long-lived, the clothing may have to be treated as radioactive waste (see chapter 19 *Accumulation and disposal of radioactive waste*).

10.117 If the skin is broken or a wound is sustained in conditions where there is a risk of radioactive contamination, the injury should be irrigated with water as soon as possible, taking care not to wash contamination into the wound or spread contamination to other areas of the body. As soon as the first aid measures have been taken, the person should seek further treatment, including decontamination if necessary. In such cases, it is essential that details of the incident be recorded and the head of department informed. The RPA should be contacted for advice. It is usual to keep the RPS informed. Further guidance on the procedures to be adopted in the case of serious incidents (including those involving members of the public and under the NAIR scheme) is given in chapter 20 *Contingency planning and emergency procedures for radioactive substances activity.*

10.118 All staff working with open sources should be trained in the locally agreed procedures for dealing with accidents, spills, or contaminated persons, with refresher training provided at appropriate intervals. A copy of the procedures should be readily available in the work area for reference. The procedures, a summary of them, or an indication of where they are kept, must be included in the local rules.

10.119 An assessment of skin dose in the event of personnel contamination should be performed if there is potential for a high dose, e.g. in the event of a large spill, non-removable contamination, or contamination that is not detected immediately. A crude approximation of dose can be performed using skin dose rate estimates from Delacroix *et al* (2002) or VARSKIN software (RAMP 2022) and imaging to assess activity and area. This should be averaged over 1 cm^2 for comparison with IRR dose limits (HSE 2018c).

References

ARSAC 2024 *Notes for guidance on the clinical administration of radiopharmaceuticals and use of sealed radioactive sources* March 2024. Administration of Radioactive Substances Advisory Committee https://www.gov.uk/government/publications/arsac-notes-for-guidance (accessed 18 April 2024)

AXREM 2018 *Radiation Controlled Area and Equipment Handover Form* (UK: AXREM) www.axrem.org.uk/resource/ensuring-compliance-with-radiation-safety-regulations/ (accessed 31 January 2023)

BEIS, DEFRA, Welsh Government and DAERA 2018 *Scope of and Exemptions from the Radioactive Substances Legislation in England, Wales and Northern Ireland Guidance Document* https://gov.uk/government/publications/guidance-on-the-scope-of-and-exemptions-from-the-radioactive-substances-legislation-in-the-uk (accessed 11 February 2023)

BIR, RCR, IPEM, SCoR and PHE 2020 *IR(ME)R Implications for Clinical Practice in Diagnostic Imaging Interventional Radiology and Diagnostic Nuclear Medicine* https://www.rcr.ac.uk/our-services/all-our-publications/clinical-radiology-publications/ir-me-r-implications-for-clinical-practice-in-diagnostic-imaging-interventional-radiology-and-diagnostic-nuclear-medicine/ (accessed 16 April 2024)

BNCS no date *British Nuclear Cardiology Society, BNCS Website* https://bncs.org.uk/ (accessed 16 July 2022)

BNMS 2010 *Certificate for the administration of Intravenous Radiopharmaceuticals by Healthcare Professionals Working in Nuclear Medicine Report January* https://bnms.org.uk/page/Administration-of-Intravenous-Radiopharmaceuticals (accessed 25 February 2023)

BNMS no date *British Nuclear Medicine Society Guidelines, BNMS Website* (Nottingham: British Nuclear Medicine Society) https://bnms.org.uk/page/Guidelines (accessed 22 May 2022)

Busemann Sokole E, Płachcínska A and Britten A 2010a Acceptance testing for nuclear medicine instrumentation *Eur. J. Nucl. Med. Mol. Imaging* **37** 672–81

Busemann Sokole E, Płachcínska A, Britten A, Lyra Georgosopoulou M, Tindale W and Klett R 2010b Routine quality control recommendations for nuclear medicine instrumentation *Eur. J. Nucl. Med. Mol. Imaging* **37** 662–71

CDG 2009 *The Carriage of Dangerous Goods and Use of Transportable Pressure Equipment Regulations 2009* https://legislation.gov.uk/uksi/2009/1348 (accessed 31 January 2023)

CDG(NI) 2010 *The Carriage of Dangerous Goods and Use of Transportable Pressure Equipment Regulations (Northern Ireland) 2010* https://legislation.gov.uk/nisr/2010/160 (accessed 31 January 2023)

CQC 2023 *Ionising Radiation (Medical Exposure) Regulations (IR(ME)R). 19/4/23, CQC website. 19/4/23* (London: Care Quality Commission) https://cqc.org.uk/guidance-providers/ionising-radiation/ionising-radiation-medical-exposure-regulations-irmer (accessed 10 May 2023)

Delacroix D, Guerre J P, Leblanc P and Hickman C 2002 Radionuclide and radiation protection data handbook 2nd edition (2002) *Radiat. Prot. Dosim.* **98** 1–168

DH 2013 *Health Building Note 02-01 Cancer Treatment Facilities* (London) https://england.nhs.uk/publication/cancer-treatment-facilities-planning-and-design-hbn-02-01/ (accessed 25 February 2023)

EA(S)R 2018 *The Environmental Authorisations (Scotland) Regulations 2018* http://legislation.gov.uk/ssi/2018/219 (accessed 31 January 2023)

EANM 2017 *European Association of Nuclear Medicine Guidelines, European Association of Nuclear Medicine Website* https://eanm.org/publications/guidelines/ (accessed 19 April 2022)

EPR 2016 *Environmental Permitting (England and Wales) Regulations* https://legislation.gov.uk/uksi/2016/1154 (accessed 31 January 2023)

ESC no date *European Society of Cardiology Guidelines and Scientific Documents, European Society of Cardiology Website* https://escardio.org/Guidelines (accessed 26 May 2022)

Harding L and Mountford P J 1993 Pregnant employees in a nuclear medicine department *Nucl. Med. Commun.* **14** 345–6

Hooper S 2011 *Examination of the Effectiveness of Different Syringe Shields for Beta Emitting Radionuclides and the Implications for Staff Finger and Whole Body Doses* (Cardiff University) http://orca.cf.ac.uk/23322/ (accessed 25 February 2023)

HSE no date *Work with ionising radiation that requires consent, HSE website* (Norwich: Health and Safety Executive) https://www.hse.gov.uk/radiation/ionising/consent.htm (accessed 14 February 2024)

HSE 2006 *Guidance Note PM77 (Third edition) Equipment used in connection with medical exposure.* 3rd edn (Sudbury: Health and Safety Executive) https://hse.gov.uk/pubns/guidance/pm77.pdf (accessed 31 January 2023)

HSE 2013 *HSIS7 Health and Safety (Sharp Instruments in Healthcare) Regulations 2013 Guidance for Employers and Employees* (Norwich: Health and Safety Executive) https://hse.gov.uk/pubns/hsis7.pdf (accessed 25 February 2023)

HSE 2015 *INDG334 Working Safely With Ionising Radiation. Guidelines for Expectant or Breastfeeding Mothers* http://hse.gov.uk/pubns/indg334.htm (accessed 31 January 2023)

HSE 2016 *L23 Manual handling—Manual Handling Operations Regulations 1992—Guidance on Regulations* 4th edn (Health and Safety Executive) www.hse.gov.uk/pubns/books/l23.htm (accessed 25 February 2023)

HSE 2018a *Ionising Radiations Regulations 2017—Guidance for Notifications, Registrations and Consents* (Norwich: Health and Safety Executive)

HSE 2018b L113 Safe use of lifting equipment Lifting Operations and Lifting Equipment Regulations *1998 Approved Code of Practice and Guidance* 2nd edn (Health and Safety Executive) https://hse.gov.uk/pubns/books/l113.htm (accessed 25 February 2023)

HSE 2018c L121 Work with ionising radiation Ionising Radiations Regulations *2017 - Approved Code of Practice and guidance* 2nd (Norwich: Health and Safety Executive) https://hse.gov.uk/pubns/books/l121.htm (accessed 31 January 2023)

IAEA 2005 *Safety Reports Series No. 40: Applying Radiation Safety Standards in Nuclear Medicine, Safety Reports Series.* (Vienna, Austria: International Atomic Energy Agency) https://iaea.org/publications/7116/applying-radiation-safety-standards-in-nuclear-medicine (accessed 25 February 2023)

IAEA 2014 *Nuclear Medicine Physics: A Handbook for Teachers and Students* ed D Bailey, J Humm, A Todd-Pokropek and A van Aswege (Vienna, Austria: IAEA) https://iaea.org/publications/10368/nuclear-medicine-physics (accessed 25 February 2023)

IAEA 2018 *Radiation Protection and Safety in Medical Uses of Ionizing Radiation. SSG-46, Safety Standards Series* SSG-46. (Vienna, Austria) https://iaea.org/publications/11102/radiation-protection-and-safety-in-medical-uses-of-ionizing-radiation (accessed 10 March 2023)

Iball G R, Bebbington N A, Burniston M, Edyvean S, Fraser L, Julyan P, Parkar N and Wood T 2017 A national survey of computed tomography doses in hybrid PET-CT and SPECT-CT examinations in the UK *Nucl. Med. Commun.* **38** 459–70

ICRP 1987 Protection of the Patient in Nuclear Medicine (and Statement from the 1987 Como Meeting of ICRP) *Ann. ICRP* **17** Publication 52 https://www.icrp.org/publication.asp?id=ICRP%20Publication%2052 (accessed 16 April 2024)

ICRP 1989 Radiological protection of the worker in medicine and dentistry *Ann. ICRP* **20** Publication 57 https://www.icrp.org/publication.asp?id=ICRP%20Publication%2057 (accessed 16 April 2024)

ICRP 2015 Radiation dose to patients from radiopharmaceuticals: a compendium of current information related to frequently used substances *Ann. ICRP* **44** Publication 128 https://www.icrp.org/publication.asp?id=ICRP%20Publication%20128 (accessed 16 April 2024)

IPEM 2015 *Report 111: Quality Control of Gamma Cameras and Nuclear Medicine Computer Systems* (York: Institute of Physics and Engineering in Medicine)

IR(ME)R(NI) 2018 *The Ionising Radiation (Medical Exposure) Regulations (Northern Ireland)* https://legislation.gov.uk/nisr/2018/17 (accessed 31 January 2023)

IR(ME)R 2017 *The Ionising Radiation (Medical Exposure) Regulations* www.legislation.gov.uk/uksi/2017/1322 (accessed 31 January 2023)

IRR(NI) 2017 *The Ionising Radiations Regulations (Northern Ireland)* http://legislation.gov.uk/nisr/2017/229 (accessed 31 January 2023)

IRR 2017 *The Ionising Radiations Regulations* (Great Britain) https://legislation.gov.uk/uksi/2017/1075) (accessed 31 January 2023

Klausen T L, Chakera A, Friis E, Rank F, Hesse B and Holm S 2005 Radiation doses to staff involved in sentinel node operations for breast cancer *Clin. Physiol. Funct. Imaging* **25** 196–202

Martin J C, Temperton D H D, Hughes A and Jupp T 2018 *Guidance on the Personal Monitoring Requirements for Personnel Working in Healthcare* (Bristol: IOP Publishing Ltd)

MHRA 2022 *Guidance: Regulating Medical Devices in the UK, GOV.UK Website* (Medicines and Healthcare Products Regulatory Agency) https://gov.uk/guidance/regulating-medical-devices-in-the-uk (accessed 23 November 2022)

Mountford P J 1997 Risk assessment of the nuclear medicine patient *Br. J. Radiol.* **70** 671–84

National Radiological Protection Board 2001 *Doses to the Embryo/Fetus and Neonate From Intakes of Radionuclides by the Mother. Part 1: Doses Recieved In Utero and From Activity Present At Birth* (Sudbury: National Radiological Protection Board)

NHS England no date *DH Health Building Notes, GOV.UK Website* (London) www.gov.uk/government/collections/health-building-notes-core-elements (accessed 10 March 2023)

NHS Estates 2001 *Facilities for Diagnostic Imaging and Interventional Radiology. HBN 6* (Norwich: HMSO) http://gov.uk/government/publications/facilities-for-diagnostic-imaging-and-interventional-radiology (accessed 9 February 2023)

NPL 2006 *Protocol for Establishing and Maintaining the Calibration of Medical Radionuclide Calibrators and their Quality Control NPL Good Practice Guide* ed R Gadd, M Baker, K Nijran, S Owens, W Thomson, M Woods and F Zananiri (Teddington: National Physical Laboratory) https://npl.co.uk/special-pages/guides/establishing-maintaining-calibration-radionuclide (accessed 25 February 2023)

NPL 2014 *Good Practice Guide No. 30 Practical Radiation Monitoring*. Issue 2 *NPL Good Practice Guide* (Teddington: National Physical Laboratory) https://npl.co.uk/special-pages/guides/gpg30_radiation (accessed 25 February 2023)

NRPB 1979 *NRPB-DL2: Derived Limits for Surface Contamination* ed A Wrixon, G Linsley, K Binns and D White (Chilton: National Radiological Protection Board)

RAMP 2022 *VARSKIN Dose Calculation for Skin Contamination, US National Regulatory Commission Website* https://ramp.nrc-gateway.gov/codes/varskin (accessed 2 June 2022)

RCR 2017 *iRefer Guidelines: Making the Best Use of Clinical Radiology (8th edition) Royal College of Radiologists Website* https://www.rcr.ac.uk/our-services/irefer/ (accessed 16 April 2024)

RSA 1993 *Radioactive Substances Act 1993* http://legislation.gov.uk/ukpga/1993/12 (accessed 11 February 2023)

SEPA 2022 *Guidance on Decommissioning of Nonnuclear Facilities for Radioactive Substances Activities, Version 4.0. 4.0* (Scottish Environmental Protection Agency) https://sepa.org.uk/media/594319/guidance_on_decommissioning_of-non-nuclear_facilities.pdf (accessed 25 February 2023)

SEPA, NRW, EANI and EA 2020 *Environment Agencies' Statement on Radioactive Waste Advisers (RWA-S-1 Version 2.1)* https://sepa.org.uk/media/520368/radioactive-waste-advisers-statement.pdf (accessed 25 February 2023)

SNMMI no date *Society of Nuclear Medicine & Molecular Imaging—Clinical Guidelines Society of Nuclear Medicine & Molecular Imaging Website* https://snmmi.org/ClinicalPractice/content.aspx?ItemNumber=10817 (accessed 26 May 2022)

Temperton D H 2009 *Pregnancy and Work in Diagnostic Imaging Departments. 2nd Edition* https://www.rcr.ac.uk/our-services/all-our-publications/clinical-radiology-publications/pregnancy-and-work-in-diagnostic-imaging-departments-second-edition/ (accessed 16 April 2024)

The Health and Safety (Sharp Instruments in Healthcare) Regulations (Northern Ireland) 2013 Northern Ireland https://legislation.gov.uk/nisr/2013/108) (accessed 25 February 2023

The Health and Social Care Act 2008 (Regulated Activities) Regulations 2014 2014 https://legislation.gov.uk/uksi/2014/2936/ (accessed 8 February 2023)

The Human Medicines Regulations 2012 https://legislation.gov.uk/uksi/2012/1916 (accessed 25 February 2023)

The Lifting Operations and Lifting Equipment Regulations (Northern Ireland) 1999 https://legislation.gov.uk/nisr/1999/304 (accessed 25 February 2023)

The Manual Handling Operations Regulations (Northern Ireland) 1992 www.legislation.gov.uk/nisr/1992/535 (accessed 25 February 2023)

UKHSA 2022 *Guidance: National Diagnostic Reference Levels (NDRLs) from 13 October 2022, GOV.UK website* https://www.gov.uk/government/publications/diagnostic-radiology-national-diagnostic-reference-levels-ndrls/ndrl (accessed 7 November 2022)

UK Radiopharmacy Group, IPEM NM SIG and BNMS 2017 *Advice for the Safe Drawing up and Administration of 223 Ra Radium—Chloride (Xofigo)* https://cdn.ymaws.com/www.bnms.org.uk/resource/resmgr/guidelines/addendum_to_safe_drawing_up_.pdf (accessed 25 February 2023)

van der Pol J, Vöö S, Bucerius J and Mottaghy F M 2017 Consequences of radiopharmaceutical extravasation and therapeutic interventions: a systematic review *Eur. J. Nucl. Med. Mol.r Imaging* **44** 1234–43

Waddington W, Keshtgar M R, Taylor I, Lakhani S R, Short M and Ell P 2000 Radiation safety of the sentinel lymph node technique in breast cancer *Eur. J. Nucl. Med.* **27** 377–91

Wall B F, Haylock R, Jansen J T M, Hillier M C and Hart D 2011 *Radiation Risks from Medical X-ray Examinations as a Function of the Age and Sex of the Patient* https://gov.uk/government/publications/medical-x-rays-radiation-risks-by-age-and-sex-of-patient (accessed 25 February 2023)

John Saunderson, Mohamed Metwaly, William Mairs, Philip Mayles, Lisa Rowley and Mark Worrall

Chapter 11

Positron emission tomography

Scope

11.1 This chapter covers the requirements for the production and use of positron emitting radionuclides for positron emission tomography-computed tomography (PET-CT). Much of the content is also applicable to PET/MR (see paragraphs 11.83–11.88 for *PET/MR*); however, as this is still an emerging technology, its use is not as widespread in the UK.

Hazards from PET sources

11.2 Due to the high energy annihilation photons emitted from PET radionuclides, there is the potential for high doses associated with the use of PET-CT and while the guidance in chapter 10, *Diagnostic use of unsealed radioactive substances*, applies, extra requirements are necessary to ensure doses to staff and the public are kept as low as reasonably practicable (ALARP).

11.3 A list of some radionuclides used in PET imaging is shown in table 11.1, along with their positron energy, range in tissue, half-life and production method.

Design of a PET-CT department

11.4 Many of the requirements for the design of a nuclear medicine department, outlined in chapter 10, *Diagnostic use of unsealed radioactive substances*, paragraphs 10.32–10.49, will apply for PET-CT as well. Specific guidance for PET-CT is available (Madsen *et al* 2006, IAEA 2010b, Peet *et al* 2012, Sutton *et al* 2012, Peet and Edyvean 2017), with the key points detailed below. Consideration must be given to the estimated workload and potential future developments, both in terms of physical space (e.g. uptake bays) and potential radiation doses. In particular, PET has historically often been used purely for

doi:10.1088/978-0-7503-2332-1ch11

Table 11.1. Radionuclides used in PET-CT imaging (Levin and Hoffman 1999, Delacroix *et al* 2002, Kim *et al* 2010, Velikyan 2014, Conti and Eriksson 2016).

Radionuclide	Positron energy (keV)	Positron range in tissue (mean, mm)	Half-life	Production
^{11}C	960	1.2	20.4 min	Cyclotron
^{13}N	1190	1.8	9.97 min	Cyclotron
^{15}O	1732	3.0	2.04 min	Cyclotron
^{18}F	634	0.6	1.83 h	Cyclotron
^{62}Cu	2926	6.1	9.7 min	Generator (^{62}Zn, $t_{1/2}$ = 9.2 h)
^{64}Cu	656	0.7	12.8 h	Cyclotron
^{68}Ga	1899	3.5	1.13 h	Generator (^{68}Ge, $t_{1/2}$ = 271 days)
^{82}Rb	3378	7.1	76 s	Generator (^{82}Sr, $t_{1/2}$ = 25 days)
^{89}Zr	902	1.3	78.4 h	Cyclotron
^{90}Y[a]	523(β-2284)	2.4	2.7 days	Reactor
^{124}I	2138	4.4	4.17 d	Cyclotron

[a] Primarily used for molecular radiotherapy.

fluorodeoxyglucose (FDG) imaging in oncology but is also used in other areas (e.g. in paediatrics or neurology, where patient considerations will be different) and with more than one tracer (where the organisation of the dispensing area will be affected). Future use and workload must therefore be considered in the design.

11.5 Shielding is required around the scanning room and uptake bays. Preparation areas may have specially designed dispensing, calibrator and waste cupboards, with lead shielding included. Shielding against PET radionuclides can normally stop at 2–2.2 m above the finished floor level. CT shielding needs to go to the underside of the soffit, with consideration of service ducts in and out of the scanner room. Consideration should also be given to the use of automatic dispensers (which should be used in addition to traditional manual systems in case of breakdown) and the space required to store them (Sutton *et al* 2012).

11.6 As per general nuclear medicine, the patient flow and departmental layout require consideration. Separate toilet facilities are required for patients who have been administered radiopharmaceuticals. A separate area for initial consultation, consenting, and cannulation of patients prior to injection to minimise doses to staff, patients, and carers is also preferred, either in shielded uptake rooms or preferably apart from the 'hot' areas. A discharge area for inpatients or patients being collected should also be available.

11.7 Consideration should be given to engineering and design solutions that reduce staff dose by minimising contact prior to and after scanning. For example, clear signage for 'hot' and 'cold' areas and, if possible, coloured lines on the floor for patients to follow to the toilet, scanner room, and exit. An intercom provides a two-way communication with patients to instruct them to change, go to the toilet, and make their way to the scan room, minimising any direct contact. 'Toilet in use' lights are also helpful when directing patients to prevent grouping in one area. Drinking water should be available in all individual hot uptake areas for patients to help themselves with minimal movement during the uptake phase. CCTV may be used; however, consideration should be given to data protection and to patient privacy and dignity, with separate areas for changing or turning off cameras while patients change.

11.8 Hand and foot monitor stations (where used) could require some degree of shielding, depending on positioning, to avoid false positive readings when injected patients are nearby. Safes and drawing up areas can also provide shielding in hot labs in addition to the wall shielding. Shielding above and below drawing up stations in hot labs should be considered, e.g. where a stock vial containing several GBq of activity may be exposed with the top of the tungsten/lead pot left off: in this situation, scatter from soffit calculations may indicate that additional shielding is required to reduce extraneous gamma photon count rates in a nearby hand and foot monitor. This shielding could be achieved by extending a thinner layer of shielding to the soffit above the primary shield, which may only extend to 2.2 m above floor level. Local environmental agency inspectors and the Counter Terrorism Security Adviser, CTSA (where appropriate), should be consulted for the level of security required for any hot labs due to the storage of tracers and sealed sources.

11.9 An assessment of drainage will be required for hot toilets and any sinks used for radioactive disposal to ensure these do not run through any populated areas and any radiation risk is appropriately managed. A risk assessment will be required to deal with any toilet blockages, covering the use of alternative temporary facilities and access by estates or external contractors.

Cyclotrons for PET radiopharmacy production

11.10 Cyclotrons should be housed in a dedicated vault within the radiochemistry facility. The shielding around the vault should be designed to effectively attenuate both neutron and gamma radiation that will be generated during cyclotron operation. Concrete is a popular shielding material, and for an unshielded cyclotron, a thickness of approximately 2 m is likely to be required. The positions of service penetrations, for instance, for ventilation or transport of radioactive products out of the vault, should be carefully considered to avoid significant defects in the shielding design. For 'unshielded' cyclotrons, where the vault itself forms the main shielding, outside dose rates arising from duct penetrations may be reduced by using a series of angled bends (or 'legs'). These significantly reduce neutron transmission and the expected dose rates as part of the shielding design (NCRP 1977, 2003).

11.11 The design of a new facility should take account of the strategy for eventual decommissioning under the principle of 'Best Available Techniques' (BAT) or 'Best Practical Means' (BPM). It is common, for instance, to install a sacrificial layer on the inside of the vault that can be easily removed and disposed of if it becomes activated during the lifetime of the facility.

11.12 Interlocks and safety systems must be in place to mitigate the risk of unintended exposures by staff that might be present in either the cyclotron bunker or the laboratory to which the automated delivery of radioactive products is made. Appropriate electronic security must also be in place to prevent malicious remote access to the cyclotron systems.

11.13 Consideration should be given to the potential exposure from gaseous by-products during the production of tracers, e.g. $^{11}CO_2$ gas during the production of ^{11}C tracers, $^{15}O_2$ gas in the production of ^{15}O water, and $^{18}F_2$ in the production of some ^{18}F tracers. The increased risk of inhalation of any radioactive gas associated with such procedures should be fully considered in the prior risk assessment, which should take into account the air change rates inside the laboratory facilities during a release scenario.

11.14 Further guidance on cyclotrons is available in the Institute of Physics and Engineering in Medicine (IPEM) Report 105 (IPEM 2011).

Generators

11.15 ^{68}Ga and ^{82}Rb can be produced in a hospital environment by the elution of a $^{68}Ge/^{68}Ga$ or $^{82}Sr/^{82}Rb$ generator, and the production of ^{68}Ga labelled radiopharmaceuticals is similar to the synthesis of ^{18}F-FDG (UK Radiopharmacy Group 2014). The life of the generator depends on several factors, not only the half-life but also the frequency of use, volume of elution, and purity of the HCl solution used. As such, a $^{68}Ge/^{68}Ga$ generator in routine clinical use may need to be replaced every 9–12 months (U.S. Nuclear Regulatory Commission 2017) and every 4–6 weeks for a $^{82}Sr/^{82}Rb$, and it is therefore important to have a programme and funds in place to return the old generator to the manufacturer and replace it with a new one.

11.16 The elution of the generator and production of ^{68}Ga labelled radiopharmaceuticals for patient use should be carried out in a GMP grade C environment or better (UK Radiopharmacy Group 2014), and is therefore likely to be performed in a dedicated PET shielded hot cell or isolator. Sufficient lead shielding in the hot cell or isolator and use of ancillary equipment to reduce operator dose will be identified in the radiation risk assessment (see also paragraph 11.23).

11.17 Due to the short half-life of ^{82}Rb elution, providing adequate security is in place, storage of the generator will generally be within the imaging room. The generators are stored and run through an infusion cart. Again, the radiation risk assessment must identify the necessary shielding and storage requirements. Sterility is maintained by sterility filters in the tubing used in the generator. The infusion cart is calibrated to ensure the correct activities are infused.

11.18 Extremity (finger) doses of operators eluting non-automated generators for PET imaging will not be insignificant, and this work alongside other routine radiopharmacy work may require operators to be designated as classified persons under the Ionising Radiations Regulations (IRR 2017, IRR(NI) 2017). To minimise operator doses it is recommended that a routine dispensing and labelling service is shared by multiple trained operators. The prior risk assessment should highlight if there is a sufficient number of staff and appropriate facilities to run the service prior to setting up.

11.19 When eluting both generators, there is a possibility of breakthrough of the long-lived ^{68}Ge radionuclide into the ^{68}Ga eluate, and ^{82}Sr with ^{85}Sr into the ^{82}Rb eluate, which will potentially cause an increase radiation exposure to patients. Quality control (QC) for breakthrough should be tested on a daily basis for ^{82}Sr/^{82}Rb generators. The amount of ^{68}Ge breakthrough is dependent on the generator (Sudbrock *et al* 2014), and the proportion of this will increase over time. For clinical use it is important to choose a generator with a low ^{68}Ge breakthrough, typically less than 0.001% (U.S. Nuclear Regulatory Commission 2017). Further information will be available in the product SPC (Summaries of Product Characteristics). A ^{68}Ga certificate of calibration from a traceable standard will be required if ^{68}Ga is manufactured and shipped to other customers.

11.20 Given the low percentage of ^{68}Ge and ^{82}Sr breakthrough, the amount of ^{68}Ge and ^{82}Sr in any eluate or solid waste may be under 40 kBq per item (Very Low Level Waste (VLLW)). While the limit applies in Scotland, it is not termed VLLW (RSA 1993, EPR 2016, BEIS *et al* 2018, EA(S)R 2018, SEPA 2021b, 2021a) and eligible for disposal in normal refuse. However, depending on the site circumstances, it may be more appropriate to store the ^{68}Ga and ^{82}Rb waste and dispose of it as longer lived waste under the conditions of the site permit/authorisation. In practice, the activity of ^{68}Ge or ^{82}Sr in the waste can be determined by multiplying the percentage breakthrough factor for the generator by the eluate or waste activity. If the ^{68}Ga labelled peptide or eluate is to be sent off site, the transport process is similar to shipping ^{18}F-FDG, and a suitably shielded transport case must be used (see chapter 18 *Keeping, accounting for, and moving radioactive substances (including transport)*).

Deliveries

11.21 Transport and delivery of radioactive materials are covered in chapter 18, *Keeping, accounting for, and moving radioactive substances (including transport)*.

Due to the short half-life of many PET tracers, delivery will generally be within a few hours of use.

11.22 For transport across hospital sites (i.e. from cyclotrons/PET radiopharmacy), an appropriate radiation risk assessment must be performed to ensure the dose to the public and staff is minimal and adequate shielding and protective measures are in place to ensure the security and integrity of the radiopharmaceutical during transport.

11.23 Pneumatic systems may be utilised for tracer delivery. There should be written instructions on the operation of the pneumatic system, and staff should be aware of tracer delivery and acceptance procedures. Access to the pneumatic system should be strictly controlled at both ends. The carriers used in the pneumatic system should contain adequate shielding to reduce the dose rates in public areas adjacent to the route of the pneumatic system to below 7.5 microsieverts per hour. Pipes must be labelled at 1 m intervals, and estates must be aware of the use of the pipe and access points installed at regular intervals. Suggested labelling includes the radioactive trefoil, the potential hazard, e.g. ^{18}F, and contact details of the appropriate person in case access is required. A radiation risk assessment should cover the expected dose rates depending on the amount to be delivered and the route of the pipes. Contingency plans must be in place in the event of a blockage. This should include a plan of the pneumatic system route and an immediate effort to identify the position of the source and restrict access to the area around it until the source has decayed. The carriers should be designed to contain any liquid that might leak from the tracer vial.

Gaseous delivery

11.24 Gaseous delivery of PET tracers will require a suitable extractor system and measurement of any vented product using stacks. Two techniques for measuring the stack emissions from PET tracer radiopharmaceutical facilities are commonly used; continuous measurement from pairs of probes inside the exhaust stack of the concentration of radioactivity present near the detectors, or sampling the contaminated air stream at a certain volumetric rate and measuring the concentration in the sample. Both of these methods require scaling to the total flow rate ($m^3\ s^{-1}$), which is commonly measured using a pitot sampling flow grid, to obtain the total release rate ($Bq\ s^{-1}$).

11.25 Emissions from the facility must be monitored and recorded to demonstrate discharges are compliant with the limits and conditions specified in the relevant environmental agency site Permit or Authorisation. In order to obtain a Permit/Authorisation for the site covering the releases to the atmosphere, it is necessary to carry out a radiological assessment of the impact of the discharges. Since hospitals consist of a complex array of buildings and are commonly situated in built-up areas, the dispersion of emitted radioactive plume gases is a complex problem, as noted in IPEM

Report 105 (IPEM 2011). Furthermore, simple assessment models may not give sufficient accuracy when there are nearby air intakes into buildings that could result in radioactive exposure from positron emitting gases. The results of a detailed study at a hospital site showing the results from wind tunnel measurements and a comparison with simpler models, have been published (Gallacher *et al* 2016a, 2016b), along with methods for converting radioactive plume concentrations into effective doses for people in the surrounding locations.

Activity preparation

11.26 PET tracers are usually delivered in vials from a reasonably close cyclotron in multiple deliveries due to their short half-life. As such, high activities will normally be received on delivery. All permit conditions under the Radioactive Substance Regulations (RSR) (RSA 1993, EPR 2016, SEPA 2021b) must be complied with for the storage and receipt of deliveries (see chapter 18 *Keeping, accounting for and moving radioactive substances (including transport)*).

11.27 Aseptic techniques should be carried out when dispensing from multi-dose vials. The UK Radiopharmacy group have issued advice on multi-dose dispensing (UK Radiopharmacy Group 2012). Further information is available in chapter 12 *Preparation of radiopharmaceuticals.*

11.28 There are automatic dose dispensers and delivery systems available, both as standalone units or as an all-in-one delivery kit. These offer the potential for reducing finger doses, which are often an issue in PET, particularly if staff work purely in that area as there is a limited option for rotation. This should be assessed against the costs of running the automated system (capital costs, consumables, and maintenance) as well as the accuracy of the system. Small volumes, such as those required if high concentrations are delivered, or small activities, such as those required for paediatrics, can be difficult to reproduce accurately in a consistent manner. All such equipment will require commissioning and regular QC checks to ensure it is performing within expected parameters across the concentration of activity used. Care should be taken when considering the dose from the unit itself and an accurate measurement of the radionuclide before and after administration. An external dose calibrator is still necessary as a backup unit for manual dispensing and for QC purposes.

11.29 Where automatic units are not available, a shielded pot-turner may be used (Hogg *et al* 2010) to draw up the appropriate volume, which should be calculated beforehand to take account of decay. In a contingency situation, a long needle (e.g. spinal needle (20 G $\times$ 90 mm)) may be used as a backup to reduce dose. Dispensing pots (lead shielding $\geqslant$ 30 mm, tungsten shielding $\geqslant$ 20 mm (Hogg *et al* 2010)) with an aperture for the needle should be used with all manipulations carried out behind appropriate L-shields. Operator technique and training are essential to reducing finger doses as much as possible, and PET specified syringe shields should be used at all times when handling syringes containing PET radionuclides. Tungsten syringes are believed to be more effective at dose reduction than Perspex (Peet and Edyvean 2017). PET finger doses must always be measured for staff dispensing and injecting patients, and the use of PPE must be considered in the risk assessment to prevent contamination directly on the skin, e.g. gloves and sleeve covers. Appropriate contamination monitoring should be performed following any manipulation of PET radiopharmaceuticals.

Administration

11.30 To minimise staff doses, prior to administration of the radiopharmaceutical, the patient should be given all the relevant information (i.e. what happens during the procedure) and advice following administration (i.e. wait times, restriction advice, toilet advice). See paragraph 10.81 for further details.

11.31 Prior to administering the radiopharmaceutical intravenously, trained staff should cannulate the patient in accordance with local radiation employer procedures and test the patency with saline. The radionuclide should only be brought through and administered to the patient when the operator is satisfied that everything is lined up properly. For slow injections, a shield pump should be considered. Portable lead screens may be utilised to reduce doses to staff where practical.

Extravasation

11.32 Extravasation of diagnostic nuclear medicine radiopharmaceuticals, including PET tracers, will not require interventions by the patient (van der Pol *et al* 2017); however, a compress and gentle massage of the area by the patient may help disperse any activity. If possible, the site of extravasation should be imaged with PET and low dose CT for attenuation correction (AC) to make a rough assessment of tissue activity. Any residual activity in the syringe can also be used to assess tissuing. Consideration should be given to any ‘top up’ activity to be administered to the patient at a different site in discussion with the medical physics expert (MPE), the practitioner, and the scanning team. Dose considerations, the amount of activity tissued, and the importance of Standardised Uptake Value (SUV) to the clinical condition should all be assessed, as should the possibility of bringing the patient back on a different day.

Radionuclide calibrators

11.33 Rather than rely on the calibrator manufacturer’s standard settings, users should determine calibration factors experimentally for each radionuclide in each geometry, i.e. syringe size and type, for each calibrator used (Zimmerman *et al* 2001). Measurements should also be traceable to a primary standard through an inter-comparison procedure such as that provided by the National Physical Laboratory (NPL). Calibrators used in a large manufacturing unit will need to be able to measure much higher activities than a standard nuclear medicine chamber.

11.34 Residual activity in the syringe and tubing should be measured, decay corrected, and subtracted from the final injected activity used for SUV calculation. With some automated systems, there is no significant residue. If this is established within centres, this may negate the need for re-measuring; however, there should always be the option of doing so (i.e. a dose calibrator) if delivery is suspected to be incomplete.

General procedures in PET departments

11.35 Evidence based indications for clinical PET-CT are published by the Royal College of Physicians (RCP) and the Royal College of Radiologists (RCR) (RCR and RCP 2022). The International Atomic Energy Agency (IAEA) has also published Appropriate Use of FDG for Cancer Patients (IAEA 2010a). Clinical imaging guidelines are available from the European Association of Nuclear Medicine (EANM) website (EANM 2017); however, all exposures and studies should still be optimised using the relevant equipment available, as laid out in chapter 10 *Diagnostic uses of open radioactive substances.*

11.36 Recommendations for clinical scientist support are available (IPEM 2005a, Evans *et al* 2014, EFOMP *et al* 2016, Greaves and BNMS 2016, IAEA 2018, IPEM *et al* 2021). The level of clinical scientist support must be sufficient to cover the workload and the range of studies performed. This can increase rapidly where non-FDG tracers are used for more complex studies and on-site radiopharmaceutical production. Additional support is also required for research studies. MPE input is required as described in chapter 2 *Radiation protection of persons undergoing medical exposures* and with specific roles laid out in IR(ME)R (IR(ME)R 2017, IR(ME)R(NI) 2018).

11.37 The Administration of Radioactive Substances Advisory Committee (ARSAC) Notes for Guidance (ARSAC 2024) publish diagnostic reference levels (DRLs) for PET tracers. Weight-based administration is usually applied in PET imaging and optimised for local procedures, e.g. time per bed position and overlap.

11.38 The CT in hybrid PET-CT systems is primarily used for attenuation correction of PET data and for anatomical localisation of the functional information provided by the PET images. The CT scanner is also capable of performing standard diagnostic CT protocols either as part of the PET study or standalone. CT protocols should therefore be consistent with radiology, where applicable, appropriate to the clinical question, and the dose justified. Advice on setting up CT scanning protocols should be sought from the CT MPE and the manufacturer, and where appropriate, optimisation should involve the multidisciplinary team. CT dosimetry can be performed using the method(s) outlined by Shrimpton *et al* (2016) (see chapter 3 *Diagnostic (other than dental) and interventional radiology* for further details on CT). Iterative reconstruction techniques offer the potential for significant dose reduction, although must be introduced carefully.

11.39 For dosimetry using PET-CT, e.g. ^{90}Y, and ^{18}F/^{68}Ga-PSMA for ^{177}Lu-PSMA therapy, see Appendix 20, *Dosimetry in radiomolecular therapy*.

11.40 National Diagnostic Reference Levels (NDRLs) are available online (UKHSA 2022) for hybrid PET-CT 'half'-body (base-of-brain to mid-thigh, the most common examination) for attenuation correction (AC) and localisation. The source paper (Iball *et al* 2017) also provides CT doses from PET cardiac and brain imaging that may be used for reference.

11.41 Local CT DRLs should be set for each exam type and audited on a regular basis. Typical doses for CT used for attenuation correction and localisation for half body (mid-brain to thighs) PET-CT are 6.5 mSv (Iball *et al* 2017). If using CT for attenuation correction only, such as for cardiac studies, the doses can be reduced to a much lower level of approximately 0.9 mSv (Iball *et al* 2017).

11.42 The ARSAC Notes for Guidance (ARSAC 2024) list effective doses for some commonly used PET tracers. These doses are enough for most dose calculations for clinical and research applications using the Integrated Research Application System (IRAS) applications. Many tracer types not covered by the ARSAC guidance notes can be found in ICRP Report 128 (ICRP 2015). For tracers that have been through phase 1 trials, information may be available in the literature. For pre-phase 1 work, models may be used, such as mouse-to-man estimations, if no human data exists.

11.43 General radiation protection considerations discussed in chapter 10, *Diagnostic use of unsealed radioactive substances*, will still apply. Due to the higher gamma energies involved in PET, increased shielding is required when compared to conventional nuclear medicine, with the half-value layer in lead for 511 keV photons being 6 mm (2.7 mm for tungsten) and the tenth value layer being 17 mm (Madsen *et al* 2006). Further information is available from the IAEA report Radiation Protection in PET (IAEA 2008). Consideration should also be given to the potentially high skin doses from ^{18}F exposure, which arises from positron exposure as well as annihilation gammas, with a dose rate of 7.88×10^{-1} mSv h^{-1} for a 0.05 ml droplet of 1 kBq (Delacroix *et al* 2002), leading to a skin dose of 2 mSv (when integrated over infinity).

11.44 Due to the increased shielding requirements for PET imaging, a full risk assessment should be performed, in compliance with the radiation employer's procedures, for moving and lifting all shielded items. For staff with existing manual handling limitations, an assessment should be made with occupational health for individual staff to determine what duties they may undertake. The use of automatic shielded doors (with appropriate security such as swipe or keycode) should also be considered.

11.45 The following measures should also be considered to limit the occupational exposure of staff working in PET:

(a) layout of the work area should be well thought out to ensure that the operator will not require to reach across or cross areas of high dose rates;
(b) long handled tongs (25–40 cm (Hogg *et al* 2010) for moving unshielded vials and syringes. This can provide significant dose savings when measuring syringes (IPEM 2014);
(c) syringe shields should be specific to PET (tungsten $\geqslant$ 5 mm (Hogg *et al* 2010));
(d) carriers—when transporting activity a sufficiently shielded carry case and a trolley should be used;
(e) L-shield with the appropriate thickness (at least 3 cm lead equivalent glass and 3 cm lead barrier) should be employed for dose measurement and manipulation;
(f) mobile whole-body lead screens in the scanning and injection rooms should also be considered, particularly if setting up large numbers of radiotherapy patients or performing dynamic imaging with slow injections;
(g) for slow injections for dynamic imaging, the use of a shielded pump should be considered;
(h) calibrators should be sunk into the unit with 5 cm lead shielding around the chamber to reduce doses during tracer measurement and to reduce background;
(i) there should be an easily accessible waste bin, again sunk into a shielded compartment on the bench, with a shielded, sliding lid;
(j) storage—a risk assessment of the quantities to be stored should determine the lead thickness. Source security and storage requirements, outlined in chapter 18 *Keeping, accounting for & moving radioactive substances (including transport)* or new chapter 19 (old 18) *Disposal of radioactive waste*, will apply;
(k) weighing scales will require annual calibration;
(l) weighing scales should be situated in an area away from hot patients;
(m) blood glucose monitors will also require calibrations, in accordance with the radiation employer's procedures.

11.46 Contingency plans should include information for emergency situations such as cardiac arrest or anaphylaxis, for example, relevant information and advice to attending staff, estimation of doses, advice to pregnant staff etc. If the patient is transferred to the emergency department or a ward, the relevant information should be passed along as well, along with an assessment of potential staff doses.

Waste

11.47 The management of radioactive waste is discussed in chapter 19, *Disposal of radioactive waste*, and the majority of radioactive waste arising from PET activity can be decayed and disposed of as VLLW (<40 kBq per item, < 400 kBq per 0.1 m^3, <2 MBq $year^{-1}$) (RSA 1993, EPR 2016, EA(S)R 2018). All waste procedures should be discussed with the Radioactive Waste Adviser (RWA) for the Employer. For radionuclides with a half-life less than 2 h, waste can generally be stored for 36 h before disposal.

11.48 Consideration should be given to contaminants, for example, breakthrough products from generators, as discussed in paragraphs 11.19 and 11.20. A calculation of the activity should be performed to ensure disposals are compliant with VLLW if this route is chosen. This could be a one-off calculation for the worst case scenario for normal conditions (i.e. maximum permitted breakthrough for clinical use). An RWA should be consulted. Waste contaminated with longer lived radionuclides will need to be stored for a longer period of time.

11.49 The target material in gas and liquid cyclotron targets is enclosed by metallic foils that become activated during the operation of the cyclotron. When a target is removed from the cyclotron, these foils become radioactive waste. The radionuclides present in the foils and the activities of these radionuclides will depend on the type of foil and the duration of bombardment; possible activation products include cobalt, zinc, manganese, and copper isotopes, and a single foil might have a total activity of several hundred MBq at the time of removal from the cyclotron. Foils should therefore be stored for decay in a designated, shielded safe, ideally for a period of several months or years, under the conditions specified in the environmental agency site permit/authorisation. After this period, the activity of the foils should be quantified by gamma spectroscopy. If the activity present in a foil is below 40 kBq, it can be disposed of as VLLW (RSA 1993, EPR 2016, EA(S)R 2018); otherwise, it should be consigned off-site to a facility permitted to receive radioactive waste.

Staff doses and monitoring

11.50 Staff doses can be high in PET-CT. The basic principle of minimal contact with the patient following administration is key, with consideration to distance and shielding, e.g. the gantry of the scanner. Monitoring doses with electronic personal dosimeters (EPDs) is recommended, especially for new staff or staff with high recorded doses. These should be plotted over time for all staff members to ensure optimisation of doses and identify any further training issues (Peet and Edyvean 2017). EPDs may not be suitable for monitoring in PET/MRI however. Monthly whole-body monitoring is advised for all PET workers (Martin *et al* 2018); however, this is based on local practice and risk assessment.

11.51 Eye doses are unlikely to exceed any statutory limits or require the classification of PET workers. Work has been carried out to suggest $H_p(3)$ values for eye dose (with a dosimeter worn at that level) were approximately 1.5 times the $H_p(10)$ value worn at the waist (Walsh, O'Connor, and O'Reilly 2014, Martin *et al* 2018). A preliminary calculation based on these values will determine if eye monitoring is required, as documented in the radiation risk assessment.

11.52 For staff involved in PET imaging, ring dosimeters are recommended for both hands, worn at the base of the index finger. If scaling factors have not been assessed at the centre to calculate fingertip dose, a factor of 6 should be applied (Martin *et al* 2018). This may require the classification of staff based on finger doses.

Paediatric patients

11.53 The use of PET-CT in paediatric imaging is not as well established as in adult imaging (RCR 2014), and far fewer studies are performed in children. Imaging children is more complex and requires careful planning and preparation to ensure success. The centre should employ staff experienced in the care and imaging of children and provide an appropriate environment. The UK PET-CT Advisory Board has produced a report that outlines the essential criteria for the establishment of a paediatric PET-CT centre (UK PET-CT Advisory Board 2009), and the RCR has published Guidelines for the Use of PET-CT in Children, which give a useful overview of the practicalities of scanning children. Before embarking on imaging paediatric patients, sites should work with the local paediatric services to ensure the local facilities and expertise are suitable so any deficiencies can be addressed or the child should be referred to a specialist centre (ARSAC 2024).

11.54 Any procedures involving radiation must be clinically justified and appropriate to answer the clinical question, this is even more important in children, where the detrimental risks of radiation are increased (ICRP 2007). Where studies are considered justified, the imaging procedures should be optimised to ensure the dose is kept as low as possible while still providing the clinical information required. This includes the use of weight-based injected activities as recommended by the ARSAC (ARSAC 2024) and the reduction of CT doses by the use of age and indication based protocols (SoR no date). For example, in cardiac and brain imaging where the CT aspect of the PET-CT study is only required for attenuation correction, tube current can often be reduced to the minimum settings available on the scanner to achieve this (Alessio and Kinahan 2012). Liaison with paediatric radiology and an appropriate MPE may be helpful.

11.55 The use of sedation and anaesthesia may be required for paediatric PET imaging, which must employ additional paediatric medical and nursing staff and specialist equipment. In this case, radiation risk assessments must be in place to cover the additional patient care required, and contingency plans should be in place to minimise staff doses while ensuring the safety of the child. Ideally, additional segregated areas for uptake and recovery are required with access to personnel and equipment for advanced paediatric life support.

Reconstruction algorithms

11.56 A PET reconstruction algorithm is required to convert the detected counts into an image. The choice of algorithm will directly impact the final image and its quantitative accuracy. An MPE, working with the IR(ME)R practitioner, must be directly involved in optimising the reconstruction settings of the algorithm (Lamare *et al* 2014).

11.57 The most common algorithm used is ordered subset expectation maximisation (OSEM), this requires the number of iterations, number of subsets, and amount of filtration to be optimised (Hudson and Larkin 1994). The reconstruction settings may be different for each radionuclide used and may also vary depending on the intended use. For example, different reconstruction settings are generally used for whole-body ^{18}F-FDG imaging compared to brain only ^{18}F-FDG imaging.

11.58 For multi-centre work, it is essential that reconstruction settings are matched across sites to allow data to be comparable. Particular care should be taken when using newer reconstructions such as those including point spread function modelling as the performance of these varies across scanner manufacturers (Rahmim *et al* 2013). In addition, if new algorithms are introduced clinically, results may not be directly comparable to older style algorithms (Teoh *et al* 2015).

PET quantification

11.59 PET is a quantitative technique, and the number of PET coincidence events measured in an image is directly proportional to the tracer uptake in a volume of tissue. The scanner needs to be calibrated against a known amount of radioactivity in a known volume to determine the conversion from coincidence counts per second to radioactivity concentration in kBq ml^{-1}. Any error in this cross calibration will produce a systematic error in PET uptake measurements.

11.60 Absolute physiological quantification is complex, and generally requires dynamic imaging as well as blood sampling, so more simple semi-quantitative measures have been introduced and are now widely used clinically. In particular for FDG-PET, the SUV is utilised; however, comparison of uptake with background in the mediastinum or liver is gaining popularity and avoids some of the issues with the SUV. SUV is defined below and is a simple semi-quantitative measure that allows comparison between patient groups by normalising uptake measured in a region of interest using patient habitus and injected activity. Assume 1 ml = 1 g, where the activity concentration is decay corrected to the administration time.

$$\mathrm{SUV}\left(\frac{\mathrm{g}}{\mathrm{ml}}\right)=\frac{\text{activity concentration}\left[\frac{\mathrm{kBq}}{\mathrm{ml}}\right]}{\text{injected activity[kBq] / patient weight [g]}}.$$

11.61 SUV normalises the measured activity concentration to the patient weight, Lean Body Mass (Wahl *et al* 2009) or Body Surface Area (analogous to the distribution volume), and the injected activity. This allows the rate of glucose uptake to be compared across time points and across patients.

11.62 There are a number of factors that can affect the SUV, which have been summarised by Boellaard *et al* (Boellaard 2009). These can be minimised through rigorous QC procedures for the scanner and ancillary equipment (IPEM 2013) and standardisation of patient preparation and scanning technique (Boellaard *et al* 2014).

11.63 The calibration factor to convert counts per second (cps) to kBq ml^{-1} is determined using a standard phantom filled with a known activity concentration. Any error in the determination of the calibration factor directly impacts the accuracy of the kBq ml^{-1} and thus the SUV measured; therefore, it is important that the procedure is followed carefully, and where possible, a daily check with a long-lived ^{68}Ge uniform cylinder is recommended (Lockhart *et al* 2011). The calibration should be performed using the same radionuclide calibrator used for measuring patient injected activities, and ideally, this should be traceable to a primary standard (NPL 2006).

11.64 It is important that patient weight and injected activity be measured as accurately as possible for each patient. Patient weight should be measured on each visit to have a PET scan, and the patient weighing scales should be subject to appropriate QC testing (Office for Product Safety and Standards 2022). With the short-lived radiopharmaceuticals used in PET, inaccuracies in timing will lead to errors in the determination of injected activity. Therefore, scanner clocks should be synchronised to the clocks used for recording injection and assay times, and residual activities should be measured for the determination of the final injected activity.

Acceptance and quality control

11.65 As outlined in paragraphs 1.245–1.259, *Critical examination*, and 2.88–2.94, *Acceptance, commissioning and routine testing*, a critical examination of all new equipment is required under IRR legislation (IRR 2017, IRR(NI) 2017), and acceptance testing is required under IR(ME)R legislation (IR(ME)R 2017, IR(ME)R(NI) 2018). Acceptance testing must be carried out prior to use for both the PET and CT components and all ancillary equipment. Guidance on appropriate testing is readily available for PET (IAEA 2009, Busemann Sokole *et al* 2010, Busemann Sokole *et al* 2010, IPEM 2013, NEMA 2018), CT (IPEM 2003, 2005b) (see chapter 4, *X-ray equipment for diagnostic radiology (excluding dental) and fluoroscopically guided interventions)*, and for Radiotherapy related equipment (e.g. alignment lasers) (IPEM 2018).

Comparison of scanners

11.66 Corrections to PET imaging include decay, scatter, randomness, attenuation, deadtime, normalisation and scanner geometry. For clinical applications, PET data is reconstructed using iterative reconstruction techniques, and these corrections may be applied within the system model. The reconstruction algorithms and implementation of these corrections vary between manufacturers and scanner models. In addition, PET-CT scanner hardware and software have advanced significantly with the introduction of fully 3D iterative reconstruction, time-of-flight, point spread function modelling, continuous motion, and digital detectors, and this continues to evolve. The resultant PET image quality and uptake measurements, such as SUV, are influenced by variations in available scanner technology as well as the acquisition and reconstruction parameters chosen at each centre. Radiologists reporting the PET scans should be notified of potential differences in SUV and whether matching with other systems has been achieved.

11.67 Given the variation in the capabilities of PET-CT systems and clinical practice, it is not feasible to be very prescriptive in determining how PET studies should be performed and images reconstructed. There are, however, guidelines for ^{18}F-FDG-PET for tumour imaging that give general recommendations for patient preparation, image acquisition, and reconstruction to improve consistency in ^{18}F-FDG-PET reporting and the use of semi-quantitative measures across platforms and institutes. To allow comparison of PET images and quantitative measures across different scanners and sites, standard phantoms are now widely used. In the UK and Europe, matching of recovery coefficients determined using the NEMA image quality phantom is the accepted method for inter-comparison between systems (Makris *et al* 2013, Boellaard *et al* 2014). The use of this technique has been successfully applied for multi-centre trials in the UK (Barrington *et al* 2011) and is currently used by the UK PET Core Lab (UK PET Core Lab no date) and EARL (EAMN no date) for PET-CT scanner accreditation.

11.68 The SUVs are often reported clinically and are used to determine malignant versus benign tumours or response to treatment. However, there are several sources of bias and variation in the determination of the SUV (Boellaard 2009), which can impact clinical diagnosis or response assessment (Boellaard 2011).

Ideally, to minimise bias, sites would want to implement all the latest technology and software corrections to produce the best image quality and SUV accuracy available. Minimising variation requires identical patient preparation, acquisition, and reconstruction protocols, which would require patients to have follow-up PET studies acquired on the same scanner with the same software as their initial baseline scan. In a clinical setting, this is not always feasible, and patients can often be scanned on different PET-CT scanners or at different sites throughout their clinical pathway. It is therefore recommended to use matching of recovery curves to develop harmonised PET protocols across scanners and centres, allowing clinical PET images and SUVs to be compared for patients undergoing repeat PET studies either on the same PET-CT system or on different systems located at different sites. The NEMA image quality phantom can also be used as a useful tool for protocol optimisation or for assessing the impact of new technologies, either through comparison to baseline scans or to the accreditation specifications provided by EARL (see paragraph 11.67).

PET-CT for diagnostic radiology

11.69 Please see chapters 3 and 4 for general CT advice. Most commonly, nuclear medicine/PET technologists do not have training in CT for diagnostic radiology procedures, and staff will require additional training (see guidance produced by the SCOR (SoR 2016)). This does not cover the administration of contrast, as this requires personnel to be registered healthcare professionals and to work under Patient Group Directions (PGDs). Technologists may administer contrast as part of a PET-CT procedure under Patient Specific Directions (PSDs) (BMA 2016).

11.70 Training should cover radiation safety issues for CT, optimisation of imaging, use of national and local DRLs, importance of patient positioning for modulated mA procedures. Evidence of operator training, both practical and theoretical, must be kept. Training can be done in-house if a suitable training programme is available, and rotation through diagnostic CT may be of value. Details of the National Occupational Standards for the production of CT images for diagnostic purposes can be found online (Skills for Health 2019).

11.71 In general, it is recommended to perform CT with IV contrast after the PET acquisition to avoid PET artefacts and inaccuracies in SUV introduced by the CT attenuation correction in the presence of contrast.

11.72 The cannula used for contrast administration is likely to be unused for over an hour during the uptake phase and initial imaging, so it is important to flush the cannula before contrast administration. In this case, use of the mobile shield is recommended, as is minimising the time spent near the patient. Typical diagnostic procedures would involve a member of staff in the room during the initial contrast injection, and mobile shields should be used where required. A risk assessment for the case of an allergic reaction must be undertaken, and a contingency plan must be put in place. Reaction to contrast takes priority over radiation safety issues, but where possible, note contact times and names of staff involved for any dose calculations.

PET-CT for radiotherapy treatment planning

11.73 There is increasing use of PET-CT for radiotherapy planning, either by indirect planning, where a separate PET-CT is acquired in the radiotherapy position and registered with a planning CT, or by direct planning, where the planning CT is entirely replaced by a planning PET-CT. Guidance is available from EANM (Thorwarth *et al* 2012). Additional equipment and staff training are required for the use of the scanner for radiotherapy planning, and a radiotherapy MPE should be involved in the specification of the QA programme and tolerances on QC tests (Somer *et al* 2012, Thomas *et al* 2014).

11.74 In general, the scanner room will require external radiotherapy lasers, a flat scanning bed, and immobilisation devices that can be shared with the radiotherapy department. If PET-CT is to be used for direct planning, the scanner needs to have a full QA programme in place, which is akin to that performed on a radiotherapy CT simulator (Gregoire and Chiti 2010). In addition, tests should cover PET to CT alignment and the integrity of transfer to the planning system.

11.75 A risk assessment should be performed for staff setting up the patient in their immobilisation devices. If the workflow allows, doses given to staff can be reduced by coaching the patient at previous appointments. For direct planning, it is recommended to use trained radiotherapy radiographers to position the patients to ensure accurate and reproducible positioning. In this case, the radiotherapy radiographers should read the local rules and have radiation protection training for working with open sources. Doses should be monitored with personal dosimeters and EPDs. Staff should be reminded to limit their close proximity to injected patients, where possible. There should be clear responsibilities attributed to the radiotherapy and nuclear medicine radiographers/technologists, e.g. patient administration, protocol selection, positioning etc.

Mobile PET

11.76 Mobile PET-CT may be provided in a variety of models, ranging from a host site simply renting and running a mobile scanner, a 'modular unit' in place for a fixed time, to a fully managed service by a third party provider. This section considers a managed service where a mobile scanner is brought on-site periodically and might include some or all of the following functions; management of bookings, ordering of radiopharmaceuticals, clerking of patients, administration of the radiopharmaceutical, imaging, and reporting.

As for a fixed site, an Employer license covering all procedures must be held by the Employer responsible for the administration (ARSAC 2024). All roles and responsibilities must be clearly defined between all involved parties, for example, the Employer for IRR (IRR 2017, IRR(NI) 2017) and IR(ME)R (IR(ME)R 2017, IR(ME)R(NI) 2018) purposes, any radiation protection advisers (RPAs), RWAs, MPEs and duty holders under IRMER. There should also be a clear means to communicate between employers and the relevant MPEs and RPAs.

11.77 An RPA will be appointed by the provider for the mobile trailer, and this arrangement would normally include a service level agreement for the provision of radiation protection advice for each third party site where mobile PET-CT services are provided. The RPA will advise on the radiation risk assessment, which addresses radiation exposure to others from work within the mobile unit. The provider will work with the host site to establish safe processes and responsibilities. This is paramount as a variety of configurations of the service are possible, with variations in who manages hot toilets, radioactive waste, advice to those who come into contact with hot patients, and other parameters. In some cases, a mobile service may be provided on a site where other radiation work is not carried out, and so the RPA for the mobile PET-CT provider will be the sole RPA at that site. A radiation protection supervisor (RPS) will also be appointed by the provider, who will supervise the arrangements in the Local Rules wherever the mobile scanner is operating.

11.78 It is the responsibility of the mobile PET-CT provider to appoint an MPE for each scanner. Depending on how responsibilities for the service are defined, it may be the responsibility of the mobile PET-CT provider and/or the host site to provide MPE support for the service. There may be multiple MPEs who have responsibilities for different aspects of service. An MPE may support multiple sites and practitioners in this respect and be required by the provider to optimise different site specific protocols.

11.79 The radiation work planned, managed, and carried out by the provider must be consistent with the circumstances at the host site and in co-operation with the host site. To this end, RWAs working for the host and provider must both agree on all aspects of environmental agency permits/authorisations and share appropriate information. A trailer specific permit/authorisation is required for sealed sources that remain on the mobile scanner. A separate permit/authorisation to allow open sources, receipt, accumulation, and disposal is also required for each trailer, which contains a list of sites to which it applies. As radioactive waste has to remain on the site at which it is generated, the provider must supply sufficient information to the host site for them to manage the radioactive waste. See chapter 19 for the management of radioactive waste.

11.80 The location and arrangement of the PET-CT mobile scanner with associated services may not be as optimal as with a fixed site facility. Particular consideration is needed for risk assessment throughout the patient pathway (including toilets) and the radioactive materials pathway (including delivery and radioactive waste store). Consideration should also be given to area designation outside the unit, as determined by the prior risk assessment, the occupancy of the unit, and how demarcation is enforced. The dose rates from the CT scanning and PET tracer (before and after administration) through the roof of the mobile scanner should be taken into account where the location of the mobile may be viewed from positions overlooking the mobile, e.g. through higher floor windows.

11.81 A mobile service can, in theory, provide much of the same PET services as a fixed site scanner, provided that it is operated with the same rigor as a fixed unit. The range and type of scanning carried out will depend on accessibility and the availability of support services. The mobile scanner will not usually be connected directly to a main hospital building, so the transport of inpatients on beds or chairs may be restricted. Unless there is a designated hot waiting area, it may not be possible to include patients who need to wait for transport. Mobile scanning may not be appropriate for some research trials that rely on the same scanner being used for each session. Emergency hospital services such as the resuscitation and fire safety teams must be notified and consulted before the beginning of a mobile service. Security must also be considered, as providers may need access to areas containing radioactive substances that are managed by the host organisation.

11.82 Further considerations and details can be found in the BNMS Report—Mobile PET: What Are the Issues? (BNMS 2007). For mobile PET units run by an external provider, the RSR (RSA 1993, EPR 2016, EA(S)R 2018) permit may be held by the provider, who will account for radioactive waste generated from the unit. Further details and responsibilities are outlined in the IPEM newsletter Regulation of Mobile PET under the Environmental Permitting Regulations (EPR) in England and Wales (Griffiths 2017).

PET-MR

11.83 PET-MR scanners are based on the integration of the PET detector ring between the intrinsic MR radio-frequency (RF) body coil and the gradient coils and are designed in such a way as to minimise the effect of high magnetic static fields and rapidly changing magnetic fields on the PET counting circuitry. Both detection methods are very temperature sensitive and require cooling circuits with very sensitive feedback loops. This requires that chiller cabinet facilities, provided by the estate management team, be set to provide temperature fluctuations well within the allowable specifications set by the manufacturer. The consequence of not doing this is that the PET scanner may shut itself down, with the potential for corruption or loss of data. The MRI scanner technology is based on the commercially available clinical 3T systems and operates in the same way as a standalone MRI scanner with no detriment to performance (Herzog and Lerche 2016).

11.84 The MRI element of the PET-MRI scanner does not pose a risk in terms of ionising radiation, reducing the overall radiation burden on patients compared to PET-CT scanning. However, MRI poses a different set of risks to both staff and patients. These include hazards arising from the large static magnetic field (e.g., projectiles, movement of implants), switching magnetic field gradients (e.g. peripheral nerve stimulation), and RF fields (e.g. RF burns from inductive loops, heating effects etc). There are statutory provisions under the Control of Electromagnetic Fields at Work Regulations 2016 (CEMFAW 2016) for regulating the use of MRI, but there are exemptions for the development, testing, installation, use, and maintenance of, or research related to, magnetic resonance imaging equipment for patients in the health sector, where: the exposure of employees to electromagnetic fields is as low as reasonably practicable; and employees are protected against the health effects and safety risks arising from their exposure to electromagnetic fields. Under the Health and Safety at Work Act 1974, there is also an overall duty on the employer to ensure staff safety. Guidance on MRI safety has been published by the MHRA (MHRA 2021), and a suitable Magnetic Resonance Safety Expert (MRSE) should be consulted during the facility planning and design stages as well as in the drawing up of risk assessments and safety procedures.

11.85 The walls of the room housing a PET-MRI scanner must be sufficiently shielded to keep the radiation dose to appropriate levels for staff and members of the public in surrounding areas (see 'BIR shielding book' (Sutton *et al* 2012)). An RF cage must also be located around the MR scanner room. All services, i.e. tubing, optical fibers, and cables, can only pass into the scanning room through waveguides or filter panels. It may be beneficial to slant the waveguides to minimise defects in the gamma ray shielding provided by the walls. Consideration of the likely procedures that may take place in the PET-MRI room must be taken when designing the layout of the facility. All equipment used within the vicinity of the MRI scanner needs to be MR-compatible and safe to use in the MRI room with appropriate labelling. For example lead shielded holders for PET tracer administration, fluid analysers for continuous arterial blood sampling etc. Equipment must be appropriately labelled. Also, conventional radiation monitors do not work in the high magnetic fields in an MR scanning room, so new processes for environmental monitoring have to be established, such as wipe testing at the end of each day.

11.86 Dose rates from PET patients are typically 14 $\mu Sv\ h^{-1}$ at 2 m following administration of FDG (Delacroix *et al* 2002). This can give rise to high staff doses when setting up patients in the coils of the MRI scanner, e.g. 13 μSv for 10 min spent near the patient. Setting the patient up should be rehearsed, and more than one person should assist to make the process as efficient as possible. It is difficult to monitor doses in real time as most electronic body dosimeters do not work in high magnetic fields. Currently, due to long scanning slots for clinical PET-MRI examinations (ranging from 30–60 min) and the fact that PET-MRI is often used for research lasting up to 4 h, the number of patients scanned per day is low, keeping staff dose at a similar or lower level than PET-CT.

11.87 The RF coils used for MRI imaging are either included in the attenuation map or manufactured from low-density materials to reduce the attenuation of the PET gamma rays. Any additional objects located within the field of view of the scanner (e.g. immobilisation devices for radiotherapy planning or headphones) can affect the quantitative accuracy of the PET images if not accounted for in the attenuation map. For example, an 11% decrease in measured PET activity concentration occurs when conventional headphones are used to reduce noise exposure in patients undergoing a brain scan (Büther *et al* 2016). Complex methods for registering and correcting for attenuation by these objects with a CT image are required. In the case of the headphones, it is safe to scan without them, provided a risk assessment is performed. This should include measurement of sound levels experienced during the scan to compare with MHRA occupational exposure limits (MHRA 2021, appendix A2.2.2).

11.88 The combination of PET and MRI imaging modalities presents a challenge to the training of radiographers and nuclear medicine technicians and the planning of efficient workflows. MRI radiographers will require formal radiation protection training. It is recommended that MRI radiographers not familiar with ionising radiation work alongside PET-CT colleagues for a significant period to become familiar with the local rules governing radiation protection for staff and patients before working unsupervised. There is also an opportunity during this time to appreciate the difficulties of balancing a scanning list where timing is of the essence when it comes to achieving good quality imaging using radiopharmaceuticals. Similarly, nuclear medicine technicians/radiographers need to appreciate the dangers presented by ferrous objects in the vicinity of high magnetic fields, which if not policed rigorously can result in fatalities. Consequently, training a workforce with the necessary skill mix remains challenging. A hybrid unit sited in a PET-CT department will benefit from training MR radiographers to understand the management of a PET radiopharmaceutical list. Alternatively, a hybrid unit sited in an MR department should consider the recruitment of a PET-CT technician or radiographer to provide assistance in managing the worklist and radiation protection procedures. It is difficult for a PET-CT technologist/technician to gain sufficient MR planning skills if they are not working full time on the MR scanner, i.e. rotating between PET-CT and SPECT scanners week by week. Ideally, a PET-MRI unit sited near both a PET-CT and MRI unit could be comfortably staffed by a pool of a minimum number of rotating MR and PET-CT radiographers.

References

Alessio A M and Kinahan P E 2012 *CT Protocol Selection in PET-CT Imaging, Image Wisely* https://www.imagewisely.org/~/media/ImageWisely Files/NucMed/CT Protocol Selection in PETCT Imaging.pdf (accessed 20 April 2024)

ARSAC 2024 *Notes for guidance on the clinical administration of radiopharmaceuticals and use of sealed radioactive sources* (March 2024. Administration of Radioactive Substances Advisory Committee) https://www.gov.uk/government/publications/arsac-notes-for-guidance (accessed 18 April 2024)

Barrington S F *et al* 2011 Establishment of a UK-wide network to facilitate the acquisition of quality assured FDG-PET data for clinical trials in lymphoma *Ann. Oncol.* **22** 739–45

BMA 2016 *Patient Group and Patient Specific Directions, British Medical Association Website* https://bma.org.uk/advice-and-support/gp-practices/prescribing/patient-group-and-patient-specific-directions (accessed 4 March 2023)

BNMS 2007 *Mobile PET: What Are the Issues* ed P Hinton, C Englefield, S Ebdon-Jackson and P Julyan (British Nuclear Medicine Society) https://cdn.ymaws.com/www.bnms.org.uk/resource/resmgr/guidelines/mobile_pet_what_are_the_issu.pdf (accessed 9 March 2023)

Boellaard R 2009 Standards for PET image acquistion and quantitative data analysis *J. Nucl. Med.* **50** 11S–20S

Boellaard R 2011 Need for standardization of 18F-FDG PET/CT for treatment response assessments *J. Nucl. Med.* **52** 93S–100S

Boellaard R *et al* 2014 FDG PET/CT: EANM procedure guidelines for tumour imaging: version 2.0 *Eur. J. Nucl. Med. Mol. Imaging* **42** 328–54

Busemann Sokole E, Płachcínska A and Britten A 2010 Acceptance testing for nuclear medicine instrumentation *Eur. J. Nucl. Med. Mol. Imaging* **37** 672–81

Busemann Sokole E, Płachcínska A, Britten A, Lyra Georgosopoulou M, Tindale W and Klett R 2010 Routine quality control recommendations for nuclear medicine instrumentation *Eur. J. Nucl. Med. Mol. Imaging* **37** 662–71

Büther F, Vrachimis A, Becker A and Stegger L 2016 Impact of MR-safe headphones on PET attenuation in combined PET/MRI scans *EJNMMI Res.* **6** 20

CEMFAW 2016 *The Control of Electromagnetic Fields at Work Regulations* https://legislation.gov.uk/uksi/2016/588 (accessed 9 March 2023)

Conti M and Eriksson L 2016 Physics of pure and non-pure positron emitters for PET: a review and a discussion *EJNMMI Phys.* **3** 1–17

Delacroix D, Guerre J P, Leblanc P and Hickman C 2002 Radionuclide and radiation protection data handbook 2nd edition (2002) *Radiat. Prot. Dosim.* **98** 1–168

EAMN no date *EARL F-18 PET/CT Accreditation, EANM Website* https://earl.eanm.org/18f-pet-ct_pet-mr/ (accessed 20 May 2022)

EANM 2017 *European Association of Nuclear Medicine Guidelines, European Association of Nuclear Medicine Website* https://eanm.org/publications/guidelines/ (accessed 19 April 2022)

EA(S)R 2018 The Environmental Authorisations (Scotland) Regulations 2018 http://legislation.gov.uk/ssi/2018/219 (accessed 31 January 2023)

EFOMP Evans, Christofides S, Brambilla S and M 2016 European Federation of Organisations for Medical Physics. Policy Statement No. 7.1: *The Roles Responsibilities and Status of the Medical Physicist Including the Criteria for the Staffing Levels in a Medical Physics Department Approved by EFOMP Council, Physica Medica* (Associazione Italiana di Fisica Medica) https://www.physicamedica.com/article/S1120-1797(16)00520-2/fulltext (accessed 20 April 2024)

EPR 2016 *Environmental Permitting (England and Wales) Regulations* https://legislation.gov.uk/uksi/2016/1154 (accessed 31 January 2023)

Evans S, Guerra A, Malone J and Bunton R 2014 *European Commission Radiation Protection N° 174 European Guidelines on Medical Physics Expert Annex 2 Medical Physics Expert Staffing Levels in Europe* https://efomp.org/uploads/rp_174_full.pdf (accessed 6 February 2023)

Gallacher D, Robins A, Burt A, Chadwick S, Hayden P and Williams M 2016a Dispersion of positron emitting radioactive gases in a complex urban building array: a comparison of dose modelling approaches *J. Radiol. Prot* **36** 746–84

Gallacher D, Robins A and Hayden P 2016b Conversion of simulated radioactive pollutant gas concentrations for a complex building array into radiation dose *J. Radiol. Prot* **36** 785–818

Greaves CBNMS 2016 *Scientific Support for Nuclear Medicine* https://cdn.ymaws.com/www.bnms.org.uk/resource/resmgr/guidelines/scientific_support_for_nucle.pdf (accessed 9 March 2023)

Gregoire V and Chiti A 2010 PET in radiotherapy planning: particularly exquisite test or pending and experimental tool? *Radiother. Oncol.* **96** 275–6

Griffiths K 2017 *Regulation of Mobile PET under the Environmental Permitting Regulations (EPR) in England and Wales* (York: Institute of Physics and Engineering in Medicine)

Herzog and Lerche 2016 Advances in clinical PET/MRI instrumentation *PET Clin.* **11** 95–103

Hogg P, Meadows A and Heathcote A 2010 *Principles and Practice of PET/CT: Part 1: A Technologists Guide* https://eanm.org/publications/technologists-guide/principles-practice-petct-part-1 (accessed 9 March 2023)

Hudson H M and Larkin R S 1994 Accelerated image reconstruction using ordered subsets of projection data *IEEE Trans. Med. Imaging* **13** 601–9

IAEA 2008 *Radiation Protection in Newer Medical Imaging Techniques: PET/CT, Safety Reports Series No. 58* (Vienna, Austria: International Atomic Energy Agency) https://iaea.org/publications/7955/radiation-protection-in-newer-medical-imaging-techniques-pet/ct (accessed 9 March 2023)

IAEA 2009 *Quality Assurance for PET and PET/CT Systems, IAEA Human Health Series* https://iaea.org/publications/8002/quality-assurance-for-pet-and-pet/ct-systems (accessed 9 March 2023)

IAEA 2010a *Appropriate Use of FDG-PET for the Management of Cancer Patients, IAEA Human Health Series* https://iaea.org/publications/8367/appropriate-use-of-fdg-pet-for-the-management-of-cancer-patients (accessed 9 March 2023)

IAEA 2010b *Planning a Clinical PET Centre* https://iaea.org/publications/8368/planning-a-clinical-pet-centre (accessed 9 March 2023)

IAEA 2018 *Human Health Reports No. 15. Medical Physics Staffing Needs in Diagnostic Imaging and Radionuclide Therapy: An Activity Based Approach, Human Health Reports* (Vienna, Austria: International Atomic Energy Agency) https://www-pub.iaea.org/MTCD/Publications/PDF/PUB1797_web.pdf (accessed 10 March 2023)

Iball G R, Bebbington N A, Burniston M, Edyvean S, Fraser L, Julyan P, Parkar N and Wood T 2017 A national survey of computed tomography doses in hybrid PET-CT and SPECT-CT examinations in the UK *Nucl. Med. Commun.* **38** 459–70

ICRP 2007 The 2007 recommendations of the international commission on radiological protection. ICRP Publication 103 *Ann. ICRP* **37** 1–337 https://www.icrp.org/publication.asp?id=ICRP%20Publication%20103 (accessed 20 April 2024) ed S Mattsson

ICRP 2015 Radiation dose to patients from radiopharmaceuticals: a compendium of current information related to frequently used substances. ICRP Publication 128 *Annals of the ICRP* ed S Mattsson, L Johansson, S Leide Svegborn, J Liniecki, D Noßke, K Riklund, M Stabin, D Taylor, W Bolch, S Carlsson, K Eckerman, A Giussani, L Söderberg, and S Valind S 44(2S) https://www.icrp.org/publication.asp?id=ICRP Publication 128 (accessed 20 April 2024)

IPEM 2003 *Report 32: Measurement of the Performance Characteristics of Diagnostic X-Ray Systems Part III Computed Tomography X-ray Scanners* 2nd edn S Endyvean, M A Lewis, N Keat and A P Jones (York: Institute of Physics and Engineering in Medicine)

IPEM 2005a *IPEM Recommendations for Clinical Scientist Support for PET-CT: Support Required for Fixed Site Performing FDG Oncology Studies* (York: IPEM) https://cdn.ymaws.com/www.bnms.org.uk/resource/resmgr/guidelines/ipem_clinical_scientist_supp.pdf (accessed 20 April 2024)

IPEM 2005b *Report 91: Recommended Standards for the Routine Performance Testing of Diagnostic X-ray Imaging Systems* (York: Institute of Physics and Engineering in Medicine)

IPEM 2011 *Report 105: Medical Cyclotrons (Including PET Radiopharmaceutical Production)* ed S Evans (York: Institute of Physics and Engineering in Medicine)

IPEM 2013 *Report 108: Quality Assurance of PET and PET/CT Systems* (York: Institute of Physics and Engineering in Medicine)

IPEM 2014 *Report 109: Radiation Protection in Nuclear Medicine* ed M McJury and C Tonge (York: Institute of Physics and Engineering in Medicine)

IPEM 2018 Section 3.7 PET-CT *Report 81: Physical Aspects of Quality Control in Radiotherapy* 2nd edn I Patel, S Weston, L A Palmer, W P M Mayles, P Whittard, R Clements, A Reilly, T J Jordan and S Wright (York: Institute of Physics and Engineering in Medicine) pp 89–90

IPEM, ARSAC, BNMS and BIR 2021 *IPEM Policy Statement Medical Physics Expert Support for Nuclear Medicine* ed F McKiddie, A Fletcher, C Kalirai, D McGowan, L Fraser, N Parkar, K Adamson and P Julyan (York: Institute of Physics and Engineering in Medicine, Administration of Radioactive Substances Advisory Committee, British Nuclear Medicine Society, British Institute of Radiology) https://ipem.ac.uk/media/lbblkxyn/mpe-support-for-nuclear-medicine.pdf (accessed 4 March 2023)

IR(ME)R 2017 *The Ionising Radiation (Medical Exposure) Regulations* www.legislation.gov.uk/uksi/2017/1322 (accessed 31 January 2023)

IRR(NI) 2017 *The Ionising Radiations Regulations (Northern Ireland)* http://legislation.gov.uk/nisr/2017/229 (accessed 31 January 2023)

IR(ME)R(NI) 2018 *The Ionising Radiation (Medical Exposure) Regulations (Northern Ireland)* https://legislation.gov.uk/nisr/2018/17 (accessed 31 January 2023)

IRR 2017 *The Ionising Radiations Regulations* (Great Britain) https://legislation.gov.uk/uksi/2017/1075 (accessed 31 January 2023)

Kim Y C, Kim Y H, Uhm S H, Seo Y S, Park E K, Oh S Y, Jeong E, Lee S and Choe J G 2010 Radiation safety issues in Y-90 microsphere selective hepatic radioembolization therapy: possible radiation exposure from the patients *Nucl. Med. Mol. Imaging* **44** 252–60

Lamare F, Le Maitre A, Dawood M, Schäfers K P, Fernandez P, Rimoldi O E and Visvikis D 2014 Evaluation of respiratory and cardiac motion correction schemes in dual gated PET/CT cardiac imaging *Med. Phys.* **41** 072504

Levin C S and Hoffman E J 1999 Calculation of positron range and its effect on the fundamental limit of positron emission tomography system spatial resolution *Phys. Med. Biol.* **44** 781

Lockhart C M, MacDonald L R, Alessio A M, McDougald W, Doot R K and Kinahan P E 2011 Quantifying and reducing the effect of calibration error on variability of PET/CT standardized uptake value measurements *J. Nucl. Med.* **52** 218–24

Madsen M T, Anderson J, Halama J R, Kleck J, Simpkin D J, Votaw J R, Wendt R E, Williams L E and Yester M V 2006 AAPM Task Group 108: PET and PET/CT shielding requirements *Med. Phys.* **33** 4–15 (accessed 31 January 2023)

Makris N E, Huisman M C, Kinahan P E, Lammertsma A A, Boellaard R, Lammertsma A A and Boellaard R 2013 Evaluation of strategies towards harmonization of FDG PET/CT studies in multicentre trials: comparison of scanner validation phantoms and data analysis procedures *Eur. J. Nucl. Med. Mol. Imaging* **40** 1507–15

Martin J C, Temperton D H D, Hughes A and Jupp T 2018 *Guidance on the Personal Monitoring Requirements for Personnel Working in Healthcare* (Bristol: IOP Publishing Ltd)

MHRA 2021 *Safety Guidelines for Magnetic Resonance Equipment in Clinical Use, GOV.UK* (London: Medicines and Healthcare Products Regulatory Agency) https://gov.uk/government/publications/safety-guidelines-for-magnetic-resonance-imaging-equipment-in-clinical-use (accessed 15 February 2023)

NCRP 1977 *NCRP Report 51, Radiation Protection Design Guidelines for 0.1–100 MeV Particle Accelerator Facilities* (Washington, DC: National Council of Radiation Protection and Measurements)

NCRP 2003 *Report No. 144: Radiation Protection for Particle Accelerator Facilities* (Bethesda, MD: National Council on Radiation Protection & Measurements)

NEMA 2018 NEMA Standards Publication NU 2- 2018 Performance Measurements of Positron Emission Tomographs (PET) *National Electrical Manufactuers Association* (Rosslyn, VA: National Electrical Manufactuers Association)

NPL 2006 *NPL Good Practise Guide 93-Protocol for Establishing and Maintaining the Calibration of Medical Radionuclide Calibrators and their Quality Control* (Teddington: National Physical Laboratory) https://npl.co.uk/special-pages/guides/establishing-maintaining-calibration-radionuclide (accessed 9 March 2023)

Office for Product Safety and Standards 2022 *Non-Automatic Weighing Instruments Regulations 2016—GOV.UK, GOV.UK Website* https://gov.uk/government/publications/non-automatic-weighing-instruments (accessed 30 April 2022)

Peet D and Edyvean S 2017 Radiation safety and CT dosimetry in PET/CT imaging ed M M Kahlil *Basic Science of PET Imaging* (Berlin: Springer)

Peet D J, Morton R, Hussein M, Alsafi K and Spyrou N 2012 Radiation protection in fixed PET/CT facilities—design and operation *Br. J. Radiol.* **85** 643–6

Rahmim A, Qi J and Sossi V 2013 Resolution modeling in PET imaging: theory, practice, benefits, and pitfalls *Med. Phys.* **40** 064301

RCR 2014 *Guidelines for the Use of PET-CT in Children* 2nd edn (London: The Royal College of Radiologists)

RCR and RCP 2022 *Evidence-based indications for the use of PET-CT in the United Kingdom 2022* (London: Royal College of Radiologists) https://www.rcr.ac.uk/our-services/all-our-publications/clinical-radiology-publications/evidence-based-indications-for-the-use-of-pet-ct-in-the-united-kingdom-2022/ (accessed 20 April 2024)

RSA 1993 *Radioactive Substances Act 1993* http://legislation.gov.uk/ukpga/1993/12 (accessed 11 February 2023)

SEPA 2021a *Environmental Authorisations (Scotland) Regulations 2018 Guide to Standard Conditions for Radioactive Substances Activities v2* https://sepa.org.uk/media/591433/guide-to-standard-conditions-v2.pdf (accessed 9 March 2023)

SEPA 2021b *Environmental Authorisations (Scotland) Regulations 2018 Standard Conditions for Radioactive Substances Activities V2.0* 2nd edn (Stirling) https://sepa.org.uk/media/593756/standard-conditions-for-radioactive-substances-activities-v2.pdf (accessed 9 March 2023)

Shrimpton P C, Jansen J T M and Harrison J D 2016 Updated estimates of typical effective doses for common CT examinations in the UK following the 2011 national review *Br. J. Radiol.* **89** 20150346

Skills for Health 2019 *CI.D.2019 Produce Computed Tomography (CT) Scanning Images for Diagnostic Purposes* (Skills for Health) https://tools.skillsforhealth.org.uk/competence-details/html/4303 (accessed 9 March 2023)

Somer E J, Pike L C and Marsden P K 2012 Recommendations for the use of PET and PET-CT for radiotherapy planning in research projects *Br. J. Radiol.* **85** e544–8

SoR 2016 *Computerised Tomography (CT) Scanners in Nuclear Medicine Facilities; Use By Nuclear Medicine Practitioners From Both Radiographic and Technologist Backgrounds* ed S Johnson (Society of Radiographers) https://sor.org/getmedia/c1a99485-1a80-4cb2-b88a-f41b15ae848d/Computerised (accessed 9 March 2023)

SoR no date *Radiation Protection (Minimising Dose in Imaging Examinations) in Children* (Society of Radiographers) https://studylib.net/doc/5849223/scor-advice-and-guidelines-in-relation-to-children-and-young- (accessed 23 March 2023)

Sudbrock F, Fischer T, Zimmermanns B, Guliyev M, Dietlein M, Drzezga A and Schomäcker K 2014 Characterization of SnO2-based 68Ge/68Ga generators and 68Ga-DOTATATE preparations: radionuclide purity, radiochemical yield and long-term constancy *EJNMMI Res.* **4** 36

Sutton D G, Martin C J, Williams J R and Peet D J 2012 *Radiation Shielding for Diagnostic Radiology* 2nd edn (London: The British Institute of Radiology) (accessed 31 January 2023)

Teoh E J, McGowan D R, Macpherson R E, Bradley K M and Gleeson F V 2015 Phantom and clinical evaluation of the Bayesian penalized likelihood reconstruction algorithm Q.Clear on an LYSO PET/CT system *J. Nucl. Med.* **56** 1447–53

Thomas C M, Pike L C, Hartill C E, Baker S, Woods E, Convery D J and Greener A G 2014 Specific recommendations for accurate and direct use of PET-CT in PET guided radiotherapy for head and neck sites *Med. Phys.* **41** 041710

Thorwarth D *et al* 2012 Integration of FDG-PET/CT into external beam radiation therapy planning: technical aspects and recommendations on methodological approaches *Nucl. Med.* **51** 140–53

UK PET-CT Advisory Board 2009 *Criteria for the Establishment of a Paediatric PET-CT Centre* (UK) https://cdn.ymaws.com/www.bnms.org.uk/resource/resmgr/guidelines/criteria_for_-the_establishme.pdf (accessed 9 March 2023)

UK PET Core Lab no date *UK PET Core Lab, UK PET Core Lab Website* http://ncri-pet.org.uk (accessed 20 May 2022)

UK Radiopharmacy Group 2012 *Safe Drawing Up of Radiopharmaceuticals in Nuclear Medicine Departments* https://cdn.ymaws.com/www.bnms.org.uk/resource/resmgr/guidelines/ukrg_-drawing_up_feb-12.pdf (accessed 9 March 2023)

UK Radiopharmacy Group 2014 *Guidance for Introduction of a 68 Ge/ 68 Ga Generator and Labelling Service into Routine Clinical Practice* https://cdn.ymaws.com/www.bnms.org.uk/resource/resmgr/ukrg/gmp_and_practical_requiremen.pdf (accessed 9 March 2023)

UKHSA 2022 National Diagnostic Reference Levels (NDRLs) from 19 August 2019 *GOV.UK websitev.Uk Website.* 24/11/2022 https://gov.uk/government/publications/diagnostic-radiology-national-diagnostic-reference-levels-ndrls/ndrl (accessed 5 December 2022)

U.S. Nuclear Regulatory Commission 2017 ‘Eckert and Ziegler GalliaPharm TM Germanium-68 / Gallium-68 Pharmacy Grade Generator Licensing Guidance’, (301) https://www.nrc.gov/docs/ML1707/ML17075A488.pdf (accessed 20 April 2024)

van der Pol J, Vöö S, Bucerius J and Mottaghy F M 2017 Consequences of radiopharmaceutical extravasation and therapeutic interventions: a systematic review *Eur. J. Nucl. Med. Mol. Imaging* **44** 1234–43

Velikyan I 2014 Prospective of 68Ga-radiopharmaceutical development *Theranostics* **4** 47–80

Wahl R L, Jacene H, Kasamon Y and Lodge M A 2009 From RECIST to PERCIST: evolving considerations for PET response criteria in solid tumor *J. Nucl. Med.* **50** 122–50

Walsh C, O’Connor U and O’Reilly G 2014 Eye dose monitoring of PET/CT workers *Br. J. Radiol.* **87** 1–4

Zimmerman B, Kubicek G, Cessna J, Plascjak P and Eckelman W 2001 Radioassays and experimental evaluation of dose calibrator settings for 18F *Appl. Radiat. Isot.* **54** 113–22

IOP Publishing

Medical and Dental Guidance Notes (Second Edition)
A good practice guide on all aspects of ionising radiation protection in the clinical environment:
IPEM Report 113
John Saunderson, Mohamed Metwaly, William Mairs, Philip Mayles, Lisa Rowley and Mark Worrall

Chapter 12

Preparation of radiopharmaceuticals

Scope

12.1 This chapter applies to the production and preparation of radiopharmaceuticals for subsequent administration to human or animal subjects for the purpose of diagnosis, treatment, or research. This includes the labelling of commercially available kits or other pharmaceuticals and labelling of autologous blood components. Also included are those radiopharmaceuticals for human or animal use that may be purchased as ready-to-use products or products requiring dilution or sub-division. The topics covered in this chapter may go well beyond radiation protection and the use of ionising radiation since there is often a conflict between those topics and the requirements of the Medicines and Healthcare Products Regulatory Agency (MHRA) for safe radiopharmaceutical manufacture, so some information about Good Manufacturing Practice (GMP) has been included.

Hazards from open radioactive substances and principles for their control

12.2 The hazards from open sources and the principles for their control are the same as for diagnostic (see paragraphs 10.8–10.13, *Hazards from open radioactive substances and principles for their control*) and therapeutic uses (see paragraphs 13.3 and 13.4, *Hazards from open radioactive substances for therapeutic use*). However, the activities handled by radiopharmacy staff are often much greater than those encountered by staff in areas where the products will be administered or used. Additional radiation protection measures are almost always required.

doi:10.1088/978-0-7503-2332-1ch12

12.3 MHRA guidance (MHRA 2020b) quotes the EEC Directive 2001/83 (European Commission 2001) definition of a medicine as 'any substance or combination of substances which may be administered to human beings or animals with a view to making a medical diagnosis or to restoring, correcting, or modifying physiological functions in human beings or animals'. Therefore, applying this definition, radiopharmaceuticals are medicines.

12.4 In most NHS organisations, the purchase of medicines falls under the remit of the Chief Pharmacist (or similar title, e.g. Clinical Director of Pharmacy) as the person responsible for the safe use and custody of medicines within that organisation. Good communication with the Chief Pharmacist about arrangements for the purchase, receipt, storage, supply, and disposal of radiopharmaceuticals is essential (Croasdale and Lowe 2020). It is recommended that a Technical (Quality) Agreement be in place to clarify the responsibilities and agreed practice of key individuals (Croasdale and Lowe 2020), such as the Chief Pharmacist, Radiation Protection Adviser (RPA), Radiation Protection Supervisor (RPS), Radioactive Waste Adviser (RWA), Medical Physics Expert (MPE), and radiopharmacy lead.

12.5 Manufacturing needs to comply with the principles of GMP (European Commission 2010b, MHRA 2020a). For sterile medicines, manufacturing needs to take place in a cleanroom environment.

12.6 Radiopharmaceuticals may be presented as single-dose vials, multidose vials, or syringes. Multidose vials of radiopharmaceuticals manufactured and supplied for same-day use should be drawn up in a way that minimises the risk of maladministration or microbial contamination, as described in Safe Drawing Up of Radiopharmaceuticals (UK Radiopharmacy Group 2012). This guidance advises, among other things, that drawing up should be done in an area supplied with Grade A air. This can be done at a suitable workstation or via the use of a retro-fitted high-efficiency particulate absorbing filter over the bench.

12.7 Guidance on safe drawing up of radium-223 is available as an addendum to the document. The radiopharmaceutical is contained within a single-dose vial, and the risk of microbial contamination is low. However, since it is an alpha emitter, there is special emphasis on radiation protection, and it is recommended that this be drawn up at a suitable workstation in order to maximise containment of any potential contamination.

12.8 EU Directive 2010/32/EU (European Commission 2010a) on preventing sharps injuries in the hospital and healthcare sector is implemented into UK legislation through The Health and Safety (Sharp Instruments in Healthcare) Regulations 2013 (*The Health and Safety (Sharp Instruments in Healthcare) Regulations (Northern Ireland*) 2013, HSE 2013). This states that the re-sheathing of needles is not permitted. However, it has been recognised that in some specialist areas, such as radiopharmacy, re-capping may be appropriate so long as certain conditions have been met. A risk assessment must be in place, and suitable re-sheathing devices must be used (HSE 2013, UK Radiopharmacy Group 2013).

Staff doses issues

12.9 Careful vigilance is required to ensure that doses are kept as low as reasonably practicable (ALARP). Optimisation of procedures, rigorous attention to technique, and the use of remote handling equipment are required. However, in some cases, the classification may still be necessary (IRR regulation 21 (IRR 2017, IRR(NI) 2017) and Paragraphs 432–446 in L121 (HSE 2018)). This is usually a result of high finger doses or accident scenarios rather than high whole-body doses. It is recommended that all radiopharmacy staff preparing ^{99m}Tc radiopharmaceuticals be monitored using finger stalls on the tips of both index fingers, provided this does not interfere with staff duties. Ring dosimeters may also be worn on the index fingers of both hands, with the active element on the palmer side but with a correction factor applied. In the absence of local data (i.e. finger stall tip and ring measurements), a multiplying factor of 6 should be applied to ring dosimeters, or a factor of 2 if the rings are worn on the middle finger, close to the tip (Martin *et al* 2018). Classified staff must wear the dosimetry as directed by the Approved Dosimetry Service and record the results as reported. Appropriate syringe shields should always be used. Finger stalls should be used for staff dispensing beta emitters for molecular radiotherapy (Martin *et al* 2018).

12.10 Measurement of eye dose has been assessed, and the evidence does not indicate that routine monitoring is necessary. However, this should be based on local risk assessment, using measurements to demonstrate that all necessary precautions are being taken. Studies published on eye dose in nuclear medicine indicate that the dose recorded on whole-body dosimeters will be greater than the eye dose.

12.11 The frequency of dosemeter change should be determined by risk assessment, taking into account the activities handled, facilities available, and previous dose measurements. The wearing of electronic personal dosimeters is particularly useful in the training of new staff and the assessment of the impact of new procedures and equipment. Using an external fixed or hand-held monitor to assess internal contamination of the thyroid may be indicated for staff performing radioiodinations. If any of the relevant annual dose classification levels (i.e., 6 mSv effective dose, 150 mSv skin dose, 6 mSv eye dose) are likely to be exceeded in normal operation or as a result of a reasonably foreseeable incident, in spite of full radiation protection measures, the staff must be classified (see also paragraph 1.188).

12.12 The risks to the foetus or breastfeeding infant must be assessed where radiopharmaceuticals are prepared by pregnant or breastfeeding staff

(Temperton 2009). Paragraphs 10.14 and 10.15, *Pregnant or breastfeeding staff*, apply. It may be necessary to adjust systems of work or reallocate duties for such staff, particularly if therapies are being made or radioiodine is dispensed in liquid form.

Protection of the patient

12.13 The use of radioactive substances for medical exposures is covered by the IR(ME)R regulations (IR(ME)R 2017, IR(ME)R(NI) 2018). Further details are given in paragraphs 10.16–10.31, *Roles and responsibilities of duty holders in nuclear medicine*. Staff who prepare radiopharmaceuticals for clinical use are defined as operators under IR(ME)R (see chapter 2 *Radiation protection of persons undergoing medical exposures* for their general duties).

Design and inspection of radiopharmacies

12.14 The two over-riding, and often conflicting principles applied to the design of a radiopharmacy are protection of the operator from ionising radiation and protection of the radiopharmaceutical, and hence the subject, from microbial contamination and pyrogens.

12.15 The design process should be primarily concerned with the safe production of radiopharmaceuticals in terms of Good Manufacturing Practice (GMP) (European Commission 2010b, MHRA 2020a), aseptic dispensing, and radiation protection practices. The pharmacist providing specialist advice, the MHRA, the external Quality Assurance Pharmacist/Specialist and the RPA should be involved from an early stage.

12.16 The accommodations required by a radiopharmacy will depend principally on the range and type of procedures undertaken, hence the likely designation of the area. The need for the following areas should be considered:

(a) cleanroom suite or facility, including appropriately graded changing area (s);
(b) appropriate support area(s) for the cleanroom suite;
(c) separate blood labelling facility, including changing areas;
(d) laboratory—e.g. for quality control;
(e) radionuclide store;
(f) reception and despatch area for radioactive materials; and
(g) storage and disposal facilities for radioactive waste, as well as office space and staff facilities such as a cloakroom, shower, and toilet.

12.17 The general provisions of paragraphs 10.32–10.49, *Design of nuclear medicine departments*, apply to all areas where radionuclides are handled although, as indicated, where aseptic procedures take place, there are some further considerations:

(a) Whilst sinks or drains should not be sited in the aseptic area or in the immediate support or changing areas (GMP), there must be adequate cleaning and decontamination facilities for persons and equipment, taking into account contingency arrangements; in particular, hand-washing and drying facilities should be available as close as aseptic requirements will allow and should be accessible before exit from the controlled radiation area.

(b) Classification of areas will be required in terms of GMP (grade of air required with regard to particle concentration) as well as the associated radiation hazard, and

(c) The aseptic area will require a positive pressure differential with respect to adjacent areas. Radiation protection for the operator must be provided either through the use of a negative pressure isolator or through the use of a microbiological safety cabinet, which protects the operator through the use of airflow patterns to prevent the release of radioactive materials to the operator and the environment. Radiation protection measures should also account for the consequences of any radioactivity that may be released remotely via air ducts (see paragraph 12.25). This should be detailed in the radiation risk assessment.

12.18 Radiopharmacies with a Manufacturer's 'Specials' Licence must conform to several standards (NHS Executive 1997) to ensure product pharmaceutical quality that may conflict with radiation protection measures; for example, radiation protection requirements for vial shields are in discord with the GMP requirement to minimise the number of objects/materials in the aseptic area. Radiopharmacies must be designed to conform to the requirements of the Human Medicines Regulations (2012). Units must either be licensed under the Human Medicines Regulations (2012) or claim a section 10 exemption from the Medicines Act (1968) as non-licensed units. In the latter case, they should comply with equivalent standards. Licensed Radiopharmacies are inspected against 'the Orange Guide' (MHRA 2022) by the Medicine and Healthcare Products Regulatory Agency (MHRA). Non-licensed units are inspected by NHS Regional Quality Assurance Pharmacists against the Quality Assurance of Aseptic Preparation Services (Beaney 2016), which contains references to Health Building Note 14–01 (DH 2013), particularly in terms of facilities. Standards for radiopharmaceuticals specifically are described in Quality Assurance of Radiopharmaceuticals (UK Radiopharmacy Group and NHS Pharmaceutical Quality Assurance Committee 2016). The design features required for a radiopharmacy are covered in these documents and in several others.

12.19 Radiopharmacies manufacturing under the terms of a Manufacturers' Specials Licence must have named individuals acting in the roles of Production Manager and Quality Controller. Those that operate under the section 10 exemption must identify an Accountable Pharmacist, along with a suitable Supervising Pharmacist(s) (NHS Executive 1997).

12.20 Radiopharmacies involved in the manufacture of medicines for clinical trials must hold a manufacturing authorisation for Investigational Medicinal

Products (IMP) (MHRA 2011). Manufacturing must comply with the requirements of GMP, and this is inspected by the MHRA. It should be noted that radiopharmaceuticals that have to be manufactured in the radiopharmacy for clinical trials, as opposed to being supplied as finished products, often involve handling radionuclides other than ^{99m}Tc. A radiological risk assessment must be carried out prior to the commencement of the trial in order to assure operator and product protection.

12.21 Blood component labelling is classified as a clinical procedure. It is not subject to the requirements of the Human Medicines Regulations (2012) although many of the design features are similar. Further information is given in paragraphs 12.38–12.42. The MHRA Inspectorate will only scrutinise the procedure if the labelling has the potential to impact the non-blood work of the radiopharmacy. Protection from the inherent biohazards is usually consistent with good radiation protection practice.

12.22 Work involving volatile radioactive materials such as iodination may be undertaken without the product being administered to humans or animals. In this case, the requirements of the following section should be met. Iodinated products for administration to humans or animals should be prepared in a facility conforming to the standards referenced in paragraph 12.18.

Enclosures and extract systems

12.23 All aseptic manipulations, including generator elution, must be performed in a Grade A area (as classified according to airborne particulate and micro-organism contamination levels). This is achieved through the use of a workstation or isolator sited in a room conforming to the specifications referenced in paragraph 12.18 (although also see paragraph 18.18), which also protects the operator from released radioactivity.

12.24 A workstation for non-aseptic work involving volatile radioactive materials should conform to the requirements of BS 5726 Microbiological Safety Cabinets (BSI 2005) for the general protection of the operator. The workstation may be a microbiological safety cabinet, an isolator, or a fume cupboard, which also inherently protects the operator from released radioactivity, but, as noted above, radiation protection measures should account for the consequences of radioactivity that may be released remotely via air ducts.

12.25 Where the exhaust from contained workstations cannot be safely discharged within the laboratory because of a risk of radioactive or other contamination, it should be separated from the normal ventilation system and discharged through ducting that is at an adequate height and not in close proximity to windows or air intakes. A discharge point 2 or 3 m above the roof and well away from any point of re-entry to the building is usually sufficient. Provided that each outlet point is well selected, a risk assessment should show that filters to trap radioactive contamination will not be necessary for the amounts normally handled in hospitals and other medical establishments (although iodination may be an exception, and note that filters may be expected in aseptic facilities). The extractor fan should be positioned close to the outlet in order to maintain negative pressure throughout the duct, and the motor should be outside the duct.

12.26 The local exhaust ventilation system should be inspected and tested at least annually by a competent person under the Provision and Use of Work Equipment Regulations (PUWER) (PUWER 1998, PUWER(NI) 1999) and for compliance with BS 5726 (BSI 2005), to ensure that it is adequate in adverse circumstances, e.g. when windows and doors are open and the wind conditions are unfavourable. It should prevent radioactive dust and vapours from escaping the enclosure into the work area or into the air of neighbouring rooms through the external vent. Smoke tests are one useful method for checking airflows; they are usually carried out annually for GMP purposes, obviating the need for a specific RP-based test.

12.27 Exhaust filters will be periodically changed, and it may become necessary to decontaminate the exhaust ducting, e.g. if maintenance work is required or the unit is being decommissioned. Ducting and filter enclosures should be labelled if any hazards are likely. The need for decontamination or labelling is most likely if long-lived radioactive substances are used in the enclosure. Sections of the ducting should be detachable, made of a non-absorbent material, and monitored by staff members who are knowledgeable about handling radioactivity before disposal.

Equipment and clothing

12.28 Reference should be made to paragraphs 10.51–10.65 (*Equipment: installation, maintenance and quality assurance* and *Provision of protective equipment and clothing* sections) and 12.18. There are usually more stringent clothing requirements (e.g. for non-shedding material) to comply with the Human Medicines Regulations (2012).

General procedures in radiopharmacies

12.29 The advice in paragraphs 10.66–10.80, *General procedures in nuclear medicine departments*, is also applicable to the handling of open radioactive substances in radiopharmacies and in blood labelling and radioiodination facilities except that:

(a) paper tissues must not be brought into an aseptic area because of particulate contamination from them. However sterile, low-lint cloths should be available; and

(b) contamination of isolator gloves will increase finger dose to staff; training of operators should ensure vigilance for this eventuality and awareness of appropriate corrective actions.

12.30 The person(s) responsible for procuring radionuclides need(s) to be aware of the limitations imposed by the permits or certificates of registration and authorisation granted under the Radioactive Substances Regulation (RSR) (RSA 1993, EPR 2016, EA(S)R 2018) or the conditions of an exemption or general binding rules to these regulations, so as not to exceed the limits on holdings or radioactive waste (see chapter 19 *Disposal of radioactive waste*). If radionuclides are to be transferred off-site, this should also be verified for the receiving establishment. Advice may be sought from the RPA and/or RWA.

12.31 Only the minimum number of personnel required should be present in areas set aside for aseptic work. Wristwatches and jewellery should not be worn. Entry to the room must be restricted to essential materials and trained personnel (or those undergoing training). Entry of materials or people, staff changing, and washing must follow written procedures designed to minimise the entry of particulate contamination and the spread of radioactive contamination in or out of the area.

12.32 If radiopharmaceuticals are dispatched to other sites, the Employer becomes a consignor, and if in-house vehicles and/or drivers are used, the employer also acts as a carrier. As such, the employer is required to meet all relevant transport obligations contained in the Carriage of Dangerous Goods and Use of Transportable Pressure Equipment Regulations (CDG 2009, CDG(NI) 2010) and in the IRR (IRR 2017, IRR(NI) 2017). A Dangerous Goods Safety Advisor (DGSA) should be appointed to provide advice on transport obligations where the employer undertakes regular transport of Type A packages (typically more than once a month). The transport of radioactive substances is covered in more detail in chapter 18, *Keeping, accounting for, and moving radioactive substances (including transport)*. Consideration should be given to the validation of arrangements for transport in terms of maintaining the cold chain.

Quality assurance of radiopharmaceuticals

12.33 Radiopharmaceuticals are considered medicinal products under the Human Medicines Regulations (Medicines Act 1968, The Human Medicines Regulations 2012) (see paragraph 12.18). A key requirement of the manufacture of medicinal products is a system of quality management covering all matters concerned with the quality of the product.

12.34 As many of the tests are not completed before the product is released for use, a greater emphasis is placed on in-process controls to ensure the continued quality of the radiopharmaceuticals produced. This is achieved by:

(a) all operating procedures being documented and strictly observed;
(b) the keeping of accurate and up-to-date records;
(c) a system of in-process checks;

(d) routine monitoring of the production environment with respect to microbiological, particulate and radioactive contamination;
(e) a fully validated method of product recall if QA testing does not meet acceptable limits;
(f) planned preventative maintenance on equipment and instruments routinely used, including a full QA programme on, e.g. radionuclide calibrators (see paragraphs 10.51–10.65, *Equipment: installation, maintenance, and quality assurance*, and *Provision of protective equipment and clothing* sections).

12.35 Acceptable radiochemical and radionuclide purity of the radiopharmaceutical is essential to the minimisation of the radiation dose to the patient and the optimisation of the quality of the diagnostic test or efficacy of therapeutic treatment. Radiochemical purity tests may be conducted as defined in the UKRG Quality Assurance of Radiopharmaceuticals (UK Radiopharmacy Group and NHS Pharmaceutical Quality Assurance Committee 2016) and Sampson's Textbook of Radiopharmacy (Theobold 2010) and are required when preparing unlicensed products. Individual methods can be found in the manufacturer's Summary of Product Characteristics (SmPC) for the product in question. Radionuclide purity is unlikely to be an issue in terms of purpose and subject radiation dose, but it can be significant in terms of radioactive waste disposal.

12.36 QC samples of product usually require sterility testing outside the radiopharmacy (UK Radiopharmacy Group and NHS Pharmaceutical Quality Assurance Committee 2016). Samples must be stored to allow the radioactivity to decay before being transferred to the analysis laboratory. Processes must be in place to ensure the materials used to test the aseptic environment, e.g. settle plates, are not radioactive before dispatch to external laboratories.

12.37 Unlicensed products should only be used if there is no alternative licensed product. Their use must be carefully considered by the responsible clinician and by the person responsible for supply in terms of clinical use, efficacy, and safety. Written disclaimers must be in place, and appropriate end-product testing must be carried out (MHRA 2014). Detailed guidance on the level of quality management deemed acceptable is given in Quality Assurance of Aseptic Preparation Services (Beaney 2016) and in the Quality Assurance of Radiopharmaceuticals guidance document (UK Radiopharmacy Group and NHS Pharmaceutical Quality Assurance Committee 2016), published jointly by the UK radiopharmacy group and the NHS Pharmaceutical Quality Assurance Committee. Sampson's Textbook of Radiopharmacy (Theobold 2010) also describes in some detail the parameters by which the quality of radiopharmaceuticals may be checked. A quality management system such as BS EN ISO 9001 Quality Management Systems (BSI 2015) may be used to demonstrate compliance with the various statutory requirements.

Blood cell labelling

12.38 When handling blood components, there is an additional risk of viral cross-contamination (for example, hepatitis B and HIV) of other products as well as the potential for accidental infection of the Operator. The radiation hazard is generally relatively small because of the limited amount of radioactivity involved.

12.39 Generally, closed procedures should be used wherever practicable when making radiopharmaceuticals. These are procedures whereby the container has a rubber bung and is not open to the environment, and transfers are made using syringes and needles, although opening a vial of saline for immediate use is acceptable. Open procedures, which can be defined as those whereby an ingredient or semi-finished product is at some stage open to the environment, are usually preferred for blood labelling, particularly for white cells and platelets. This is because it is often not appropriate to transfer blood products using a syringe and needle, for example, if pipetting is necessary or if the use of the needle may damage the cells. Open procedures can save time, thereby reducing radiation dose, and can also eliminate the risk of needle-stick injury. However, note that stringent monitoring for contamination is required.

12.40 It is preferable, especially when cell labelling is carried out frequently, to have a dedicated facility. Previously, it was considered safe to use common facilities/support areas for aseptic work and blood labelling, and in some older units, this practice is being permitted to continue, although there are now expectations that it be phased out because of the potential risk of cross-contamination. A comprehensive risk assessment, reviewing the potential for cross-contamination, must be in place, and mitigating strategies must be acted upon. In new builds and where an existing unit is being refurbished, dedicated facilities are required. Separate rooms and garments should be used to prevent the operator from moving from the room in which blood is handled to a room in which other radiopharmaceuticals are prepared without a change of garment (UK Radiopharmacy Group 2009).

12.41 A dedicated workstation of the type referenced in paragraph 12.17(c) must be reserved for all stages of blood labelling. Simultaneous labelling of blood from more than one person must not be carried out under any circumstances. All work should be carried out on a drip tray large enough to contain any spillage. Centrifuges with sealable buckets are required to contain blood if the primary container leaks or breaks.

12.42 The surfaces of the workstation, benches, and all equipment used in the process must be disinfected after use to prevent cross-contamination. It is essential to use a disinfectant that is active against spores and against viruses such as hepatitis B and HIV, taking due notice of the COSHH Regulations (COSHH 2002, COSHH(NI) 2003), as such agents can be both toxic and corrosive. Special attention should be paid to the potential for disinfecting agents to reduce the visibility through leaded-acrylic vision panels and to cause rusting or degradation of the cleanroom facility and equipment that may compromise containment and/or shielding.

PET radiopharmaceutical production

12.43 Development and clinical production of PET radiopharmaceuticals should be carried out in a dedicated radiochemistry facility or a dedicated area within an existing radiopharmacy, comprising purpose-built laboratories as well as, ideally, an on-site cyclotron to supply short-lived radionuclides. The design of such facilities should be carefully considered from the points of view of both radiation protection and the requirements of good manufacturing practices. Much of the other information in chapter 12, *Preparation of radiopharmaceuticals*, is also relevant here.

12.44 Given the high dose rates from PET radionuclides, it is necessary that tracer production be carried out where possible using automated synthesis modules housed in hot cells containing shielding equivalent to 5–10 cm of lead. In routine use, interlocks should prevent access to hot cells that contain high activities. Research and development work that is carried out on the benchtop or in open hot cells should be risk assessed and proceed according to written systems of work, with careful monitoring of the extremity and whole-body doses received by staff. The dose to the eyes, given recent more stringent legal requirements on eye dose limits, and head doses should be taken into account in radiation risk assessments and monitoring in scenarios where staff may be viewing over benchtop shielding. See the IPEM report: *Guidance on the personal monitoring requirements for personnel working in healthcare* (Martin *et al* 2018).

12.45 Laboratories in which radioactive gases are present or potentially present should be monitored for dose rate or radioactivity in air concentration with automated alarms to alert staff in the event of a gaseous release into the laboratory. Air extracted from hot cells and from the facility heating, ventilation, and air conditioning (HVAC) system should be extracted to a stack that is permitted by the Environment Agency for radioactive gaseous releases. Appropriate abatement techniques, such as active gas compression systems, active charcoal filters, or delay systems/tanks, should be used to minimise gaseous releases to the environment. The release of radioactive gas from the stack should be monitored using calibrated radiation detectors to ensure that permitted limits are not exceeded.

12.46 Further guidance is available from the International Atomic Energy Agency (IAEA) for production (IAEA 2012a), gas and liquid targets (IAEA 2012b), and IAEA facility design (IAEA 2012a).

Monitoring of persons

12.47 Routine measurement of whole-body and extremity radiation doses by radiopharmacy staff is necessary (see paragraphs 12.9–12.12, *Staff dose issues*). Frequent monitoring for personal contamination is also required, especially after leaving the controlled radiation area(s) of the radiopharmacy. External monitoring of the thyroid may be indicated for staff performing radioiodination (see paragraphs 12.9–12.12). Monitoring schedules should be established as part of the prior risk assessment, taking activity levels, working conditions, and procedures into account in conjunction with the RPA.

Monitoring of work areas

12.48 The advice in paragraphs 10.97–10.100, *Monitoring of work areas and persons*, applies to all rooms in the radiopharmacy in which work with open radioactive substances is undertaken.

12.49 The frequency of monitoring should be decided on the basis of the prior risk assessment, taking into account the advice of the RPA. The monitoring programme should be able to detect significant changes in radiation and contamination levels and ensure that these levels are within as low as reasonably practicable (ALARP). Monitoring will also be required for particulate and microbiological contamination of the air in the aseptic suite. Radiation dose levels in adjacent rooms and corridors should be monitored at appropriate intervals.

12.50 Transport containers should be checked for damage and/or radioactive contamination prior to each use (see chapter 18 *Keeping, accounting for, and moving radioactive substances (including transport)*). On-going contamination monitoring of the containers once they have been packed is required using a wipe test. If not done prior to each consignment, this must be justified, and the potential for radioactive contamination must be controlled through a series of in-process contamination checks. Wipe testing rather than using a contamination monitor directly will be necessary to determine the presence of contamination, unless radiation dose rates are negligible, which is unlikely.

12.51 Before returning radionuclide generators to the manufacturer for recycling, external surfaces should be swabbed to check for contamination. The site permit or open-source registration (issued under RSR (RSA 1993, EPR 2016, EA(S)R 2018))) should have sufficient capacity to cover ^{99}Mo/^{99m}Tc generators if they are awaiting collection. If the generators are being held for decay to recover the lead before subsequent disposal as waste, provision for this should be included in the site permit/authorisation for accumulation and disposal. If generator shielding contains depleted uranium, this should be included in the site-sealed source permit or registration.

12.52 It may be necessary to monitor filters removed from the ventilation system and the ventilation ducting before disposal (see paragraph 12.27). Personal protective equipment, such as gloves, overalls, and disposable masks, should be used as dictated by risk assessment.

Decontamination

12.53 Decontamination procedures detailed in paragraphs 10.101–10.112, *Decontamination of areas, surfaces and equipment*, are applicable, although it can be difficult to cope with radioactive spills in aseptic areas because of the general prohibition on materials that might shed particles. However, it is possible to find absorbent materials that are non-shedding. Contingency plans for potentially serious incidents such as cracked generator columns, broken eluate vials, etc may have to include instructions to bring large quantities of absorbent material into the area. In this case, aseptic work cannot restart until appropriate radioactive decontamination and sanitation procedures are performed and the resulting particulate count is below the appropriate level.

References

Beaney A 2016 *Quality Assurance of Aseptic Preparation Services* 5th edn (London: Royal Pharmaceutical Society) https://rpharms.com/recognition/setting-professional-standards/quality-assurance-of-aseptic-preparation-services

BSI 2005 BS *5726:2005* Microbiological Safety Cabinets *Information to be Supplied by the Purchaser to the Vendor and to the Installer, and Siting and use of Cabinets—Recommendations and Guidance* (London: British Standards Institute)

BSI 2015 *BS EN ISO 9001:2015 Quality Management Systems—Requirements* (London: British Standards Institute)

CDG 2009 *The Carriage of Dangerous Goods and Use of Transportable Pressure Equipment Regulations 2009* https://legislation.gov.uk/uksi/2009/1348 (accessed 20 April 2024)

CDG(NI) 2010 *The Carriage of Dangerous Goods and Use of Transportable Pressure Equipment Regulations (Northern Ireland)* 2010 https://legislation.gov.uk/nisr/2010/160 (accessed 20 April 2024)

COSHH 2002 *The Control of Substances Hazardous to Health Regulations 2002* https://legislation.gov.uk/uksi/2002/2677 (accessed 20 April 2024)

COSHH(NI) 2003 *Control of Substances Hazardous to Health Regulations (Northern Ireland)* https://legislation.gov.uk/nisr/2003/34 (accessed 20 April 2024)

Croasdale J and Lowe R 2020 *The Responsibilities of Chief Pharmacists for Radiopharmaceuticals* (UK Radiopharmacy Group and the NHS Pharmaceutical Quality Assurance Committee) 3rd edn https://sps.nhs.uk/wp-content/uploads/2020/07/Responsibilities-of-Chief-Pharmacists-for-Radiopharmaceuticals-Edn.3.pdf (accessed 20 April 2024)

DH 2013 *Medicines management. Health Building Note 14-01: Pharmacy and radiopharmacy facilities* (London: Department of Health) https://england.nhs.uk/publication/designing-pharmacy-and-radiopharmacy-facilities-hbn-14-01/ (accessed 20 April 2024)

EA(S)R 2018 *The Environmental Authorisations (Scotland) Regulations 2018* http://legislation.gov.uk/ssi/2018/219 (accessed 20 April 2024)

EPR 2016 *Environmental Permitting (England and Wales) Regulations* https://legislation.gov.uk/uksi/2016/1154 (accessed 20 April 2024)

European Commission 2001 *Directive 2001/83/EC of the European Parliament and of the Council of 6 November 2001 on the Community Code Relating to Medicinal Products for Human Use* https://eur-lex.europa.eu/legal-content/en/ALL/?uri=CELEX%3A32001L0083 (accessed 20 April 2024)

European Commission 2010a *EU Directive 2010/32/EU—Prevention from Sharp Injuries in the Hospital and Healthcare Sector* https://eur-lex.europa.eu/LexUriServ/LexUriServ.do?uri=OJ:L:2010:134:0066:0072:EN:PDF (accessed 20 April 2024)

European Commission 2010b *EudraLex—Volume 4—Good Manufacturing Practice (GMP) Guidelines* https://ec.europa.eu/health/medicinal-products/eudralex/eudralex-volume-4_en (accessed 20 April 2024)

HSE 2013 *HSIS7 Health and Safety (Sharp Instruments in Healthcare) Regulations 2013 Guidance for Employers and Employees* (Norwich: Health and Safety Executive) https://hse.gov.uk/pubns/hsis7.pdf (accessed 20 April 2024)

HSE 2018 *L121 Work with Ionising Radiation Ionising Radiations Regulations 2017—Approved Code of Practice and Guidance* 2nd edn (Norwich: Health and Safety Executive) https://hse.gov.uk/pubns/books/l121.htm (accessed 20 April 2024)

IAEA 2012a *Cyclotron Produced Radionuclides: Guidance on Facility Design and Production of [18F]Fluorodeoxyglucose (FDG), IAEA Radioisotopes and Radiopharmaceuticals Series* (Vienna, Austria: International Atomic Energy Agency) https://iaea.org/publications/8529/cyclotron-produced-radionuclides-guidance-on-facility-design-and-production-of-fluoro-deoxyglucose-fdg (accessed 20 April 2024)

IAEA 2012b *Cyclotron Produced Radionuclides: Operation and Maintenance of Gas and Liquid Targets, IAEA Radioisotopes and Radiopharmaceuticals Series* (Vienna, Austria: International Atomic Energy Agency) https://iaea.org/publications/8783/cyclotron-produced-radionuclides-operation-and-maintenance-of-gas-and-liquid-targets (accessed 20 April 2024)

IR(ME)R 2017 *The Ionising Radiation (Medical Exposure) Regulations* www.legislation.gov.uk/uksi/2017/1322 (accessed 20 April 2024)

IR(ME)R(NI) 2018 *The Ionising Radiation (Medical Exposure) Regulations (Northern Ireland)* https://www.legislation.gov.uk/nisr/2018/17 (accessed 20 April 2024)

IRR 2017 *The Ionising Radiations Regulations.* Great Britain https://legislation.gov.uk/uksi/2017/1075 (accessed 20 April 2024)

IRR(NI) 2017 *The Ionising Radiations Regulations (Northern Ireland)* http://legislation.gov.uk/nisr/2017/229 (accessed 20 April 2024)

Martin J C, Temperton D H D, Hughes A and Jupp T 2018 *Guidance on the Personal Monitoring Requirements for Personnel Working in Healthcare* (Bristol: IOP Publishing Ltd)

Medicines Act 1968 https://legislation.gov.uk/ukpga/1968/67 (accessed 20 April 2024)

MHRA 2011 *FAQ for Investigational Medicinal Products (IMP), MHRA Website* (Medicines and Healthcare Products Regulatory Agency) https://forums.mhra.gov.uk/showthread.php?32-MHRA-produced-FAQs-for-Investigational-Medicinal-Product (accessed 20 April 2024)

MHRA 2014 *Guidance Note 14: The Supply of Unlicensed Medicinal Products ('specials')* (London: Medicines and Healthcare Products Regulatory Agency) https://gov.uk/government/publications/supply-unlicensed-medicinal-products-specials (accessed 20 April 2024)

MHRA 2020a *Guidance: Good Manufacturing Practice and Good Distribution Practice, GOV.UK Website* (UK: Medicines and Healthcare products Regulatory Agency) https://gov.uk/guidance/good-manufacturing-practice-and-good-distribution-practice (accessed 20 April 2024)

MHRA 2020b *Guidance Note 8: A Guide to What Is a Medicinal Product* (London: Medicines & Healthcare products Regulatory Agency) https://assets.publishing.service.gov.uk/government/uploads/system/uploads/attachment_data/file/872742/GN8_FINAL_10_03_2020__combined_.pdf (accessed 20 April 2024)

MHRA 2022 *Rules and Guidance for Pharmaceutical Manufacturers and Distributors 2022 (The MHRA Orange Guide)* 11th edn (London: Pharmaceutical Press)

NHS Executive 1997 *Aseptic Dispensing in NHS Hospitals: EL(97)52*

PUWER 1998 *Provision and Use of Work Equipment Regulations.* https://legislation.gov.uk/uksi/1998/2306 (accessed 20 April 2024)

PUWER(NI) 1999 *Provision and Use of Work Equipment Regulations (Northern Ireland)* 1999 http://legislation.gov.uk/nisr/1999/305 (accessed 20 April 2024)

RSA 1993 *Radioactive Substances Act 1993* http://legislation.gov.uk/ukpga/1993/12 (accessed 20 April 2024)

Temperton D H 2009 *Pregnancy and Work in Diagnostic Imaging Departments. 2nd Edition* https://www.rcr.ac.uk/our-services/all-our-publications/clinical-radiology-publications/pregnancy-and-work-in-diagnostic-imaging-departments-second-edition/ (accessed 20 April 2024)

The Health and Safety (Sharp Instruments in Healthcare) Regulations (Northern Ireland) 2013 (Northern Ireland) https://legislation.gov.uk/nisr/2013/108 (accessed 20 April 2024)

The Human Medicines Regulations 2012 https://legislation.gov.uk/uksi/2012/1916 (accessed 20 April 2024)

Theobold A 2010 *Sampson's Textbook of Radiopharmacy* 4th edn (London: Pharmaceutical Press)

UK Radiopharmacy Group 2009 *Guidelines for the Safe Preparation of Radiolabelled Blood Cells, BNMS Website* https://cdn.ymaws.com/www.bnms.org.uk/resource/resmgr/guidelines/ukrg_blood_labelling_2009.pdf (accessed 20 April 2024)

UK Radiopharmacy Group 2012 *Safe Drawing Up of Radiopharmaceuticals in Nuclear Medicine Departments* https://cdn.ymaws.com/www.bnms.org.uk/resource/resmgr/guidelines/ukrg_-drawing_up_feb-12.pdf (accessed 20 April 2024)

UK Radiopharmacy Group 2013 *Guidance on the Recapping of Needles in Radiopharmacy and Nuclear Medicine, BNMS Website* https://cdn.ymaws.com/www.bnms.org.uk/resource/resmgr/guidelines/recapping_needles_ukrg_guida.pdf (accessed 20 April 2024)

UK Radiopharmacy Group and NHS Pharmaceutical Quality Assurance Committee 2016 *Quality Assurance of Radiopharmaceuticals. BNMS Website* 4th edn (UK: NHS Pharmaceutical Quality Assurance Committee & UK Radiopharmacy Group) https://cdn.ymaws.com/www.bnms.org.uk/resource/resmgr/guidelines/qa_of_radiopharmaceuticals_e.pdf (accessed 20 April 2024)

**John Saunderson, Mohamed Metwaly, William Mairs, Philip Mayles,
Lisa Rowley and Mark Worrall**

Chapter 13

Therapeutic uses of open radioactive substances

Scope

13.1 This chapter applies to the use of open (dispersible) radioactive substances that are administered to people as patients for treatment or for research into treatment.

13.2 Most of the information on the diagnostic uses of open radioactive substances given in chapter 10, *Diagnostic uses of unsealed radioactive substances*, is also applicable to therapeutic uses, including that in relation to the designation of areas (paragraphs 1.104–1.115, *Controlled areas and supervised areas*, and 10.36). Additional information specific to therapy administration will be given in this chapter.

Hazards of the therapeutic use of open radioactive substances

13.3 The radiation emitted, the administered activity, and the physical half-lives of the radionuclides used for molecular radiotherapy (mRT) will be different from those usually encountered in diagnostic work and will require additional radiation protection measures.

13.4 The advice of the radiation protection adviser (RPA), radioactive waste adviser (RWA), and medical physics expert (MPE) should be sought before new therapeutic procedures are introduced. A radiation risk assessment must be made, and that should also include consideration of the administration and design of facilities, as detailed further in paragraphs 13.6–13.10, *Design of the treatment area*.

doi:10.1088/978-0-7503-2332-1ch13

Pregnant or breastfeeding staff

13.5 Special care should be taken when involving a pregnant or breastfeeding member of staff in a therapeutic procedure with open sources. A radiological risk assessment of likely doses to the foetus or breastfeeding infant should be made (HSE, 2015). Guidance is given in paragraphs 10.14 and 10.15, *Pregnant or breastfeeding staff*. Reallocation of some duties may be indicated.

Design of the treatment area

13.6 Patients treated with open radioactive substances may require inpatient facilities, for example, for higher activities of ^{131}I for thyroid cancer, ^{131}I-MIBG, ^{90}Y, and ^{177}Lu peptide therapies. The design of treatment areas and wards for radionuclide therapy, whether new-build or adapted from existing accommodation, should conform to the relevant requirements for the design of laboratories and nuclear medicine departments for diagnostic uses of open sources (see paragraphs 10.32–10.49, *Design of nuclear medicine department*) and be subject to a radiation risk assessment.

13.7 The RPA, RWA and/or MPE, as appropriate, should be contacted for advice on:
- (a) whether special precautions are required to ensure that relevant dose constraints are not exceeded;
- (b) the appropriate designation of the areas required;
- (c) precautions for the protection of staff and visitors, including carers and comforters (IR(ME)R 2017; IR(ME)R(NI) 2018);
- (d) whether extra shielding is required in the walls, doors, windows, ceiling or floors to protect persons in adjacent rooms from external irradiation;
- (e) monitoring arrangements (patient and area);
- (f) a suitable waste management system, including aqueous waste drainage and its route off-site (e.g. ideally this should be direct to the main waste drain and avoid passing through other wards); and
- (g) any other special precautions or procedures required.

13.8 A risk assessment should be performed to determine whether special precautions are necessary to ensure that relevant dose constraints are not likely to be exceeded based on estimates of radiation exposure (of staff, other patients, or visitors) from the patient or from any associated contamination risk.

13.9 Patients kept in the hospital due to high dose rates should be accommodated in single rooms designed or adapted for the purpose, ideally with their own toilet, washing facilities, and, perhaps, a food preparation area. The need for each of these dedicated facilities should be assessed against the level of risk. For example, en-suite toilet facilities are essential wherever significant amounts of radioactivity will be excreted in urine or faeces. The design of safe and comfortable accommodations for carers and comforters is important, particularly for paediatric facilities. If prisoners may be treated, means of appropriate supervision by prison officers might also be considered.

13.10 Secure areas should be provided for bins for the temporary storage of linen and waste contaminated with radioactive substances. The storage areas and bins should be in or near the treatment area or ward and clearly marked using the radiation warning sign (appendix 7, *Warning signs and notices*). The design of some pedal bins is such that a lock can be placed to secure the bag and prevent/remind domestic staff not to remove it. These features should be considered, particularly in areas where there is high domestic staff turnover or the use of external contractors.

Administration of therapeutic radioactive substances

13.11 The general provisions of paragraphs 10.16–10.31, *Roles and responsibilities of duty holders in nuclear medicine*, apply and, where diagnostic procedures are referred to, the provisions should generally be taken to include therapeutic procedures as well, except that:

(a) All therapeutic administrations must be planned individually to the patient (IR(ME)R 12(2) (IR(ME)R 2017, IR(ME)R(NI) 2018)). This may be based on risk, i.e. staging, disease progression, as well as the dose to radiosensitive organs.

(b) Diagnostic reference levels (DRLs) are not applicable to therapeutic administrations, however, departments should set up local limits for the permitted tolerance for the prescribed activity that can be administered, taking into account the limitations of the measuring equipment and radiation dose to the operator. Different tolerances may be set for pre-dispensed therapeutic radiopharmaceuticals delivered to a centre, such as ^{131}I capsules or ^{90}Y Theraspheres, and locally dispensed items, such as ^{223}Ra-chloride and ^{90}Y-microspheres. The tolerance of the prescribed activity chosen should be agreed upon with the practitioner and recorded in the therapeutic protocol.

(c) Significant, Accidental, and Unintended Exposure (SAUE) Guidance (CQC 2023) sets criteria for notification to the CQC of mRT overexposures and underexposures based on prescribed activity or dose to the treatment volume and the overexposure of organs at risk (CQC 2023).

(d) Advice to patients concerning the avoidance of conception after therapeutic administrations was provided in the ARSAC Notes for Guidance (ARSAC 2024) and chapter 16 *Patients leaving the hospital after administration of radioactive substances*, table 16.4. Any advice should take into account the clinical condition in conjunction with the referring medical team.

(e) Female patients should be advised as appropriate that pregnancy and breastfeeding are contraindicated (after almost all therapeutic administrations) (ARSAC 2024) (chapter 16 *Patients leaving hospital after administration of radioactive substances*, table 16.4)

(f) An MPE must be closely involved in all therapeutic administrations of open radioactive sources, other than those for which a standard protocol is followed; in the case of standard treatments, the MPE must be 'involved' (BIR *et al* 2020)—available on-site or, at least, contactable (IR(ME)R 2017, IR(ME)R(NI) 2018), and

(g) Verification of treatment delivery should be performed for molecular radiotherapy (IR(ME)R 2017, IR(ME)R(NI) 2018) (see paragraphs 13.20 and 13.50).

13.12 All arrangements should be discussed as fully as possible with the patient prior to radioactive substance administration, ideally well in advance of the appointment for treatment. If the patient is unable to understand fully, the most appropriate person acting on their behalf should also be consulted. Issues to be discussed include:

(a) the manner and place of administration;
(b) whether an inpatient stay will be required;
(c) whether the patient and any necessary visitors while in the hospital are willing and able to cooperate with any necessary restrictions on behaviour (e.g. is the patient incontinent or catheterised?);
(d) Is the patient able to follow simple instructions? (Does the patient have religious views that may conflict with the requirements?);
(e) whether the administration will require a carer and comforter;
(f) whether there are particular difficulties in restricting exposure to other persons once the patient leaves the hospital, such as children, pregnant colleagues, carers or people the patient cares for (see chapter 16 *Patients leaving the hospital after administration of radioactive substances*);
(g) whether the patient is returning to a place where people are employed to care for them, such as a hospice and including community nurses visiting the patient's home (see paragraph 16.16–16.21, *Discharge of patients to outside organisations etc*, for guidance on whether notification to HSE under IRR17 (IRR 2017, IRR(NI) 2017) and/or arrangements to comply with radioactive substances regulation (RSR) (RSA 1993, EPR 2016, EA (S)R 2018) will be required); and
(h) the arrangements for transport home or to the receiving institution;
(i) if there is advice/restrictions for medical emergencies/surgery/burial/ cremation, etc following discharge (see paragraphs 13.27, 13.40–13.41 *Procedures in operating theatres* and chapter 17 *Precautions after the death of a patient to whom radioactive substances have been administered*).

Any potential issues arising from the above should be discussed with the MPE and the Practitioner in advance of administration, as well as with the RPA regarding IRR (IRR 2017, IRR(NI) 2017) and the RWA for RSR (RSA 1993, EPR 2016, EA(S)R 2018).

13.13 Persons receiving treatment with open radioactive substances and their carers and comforters, as appropriate, must receive instructions and information on precautions to be taken before and after they leave the treatment facility and any hazards involved (see also chapter 16 *Patients leaving the hospital after administration of radioactive substances*). The instructions and information must be given in writing (IR(ME)R 12 (6) (IR(ME)R 2017, IR(ME)R(NI) 2018) and may also be given verbally, in accordance with the employer's procedure.

13.14 Some treatments may exceptionally result in the need for a patient to be buried rather than cremated for a period following treatment, should they die. Where applicable, this should be discussed with the patient (and their next of kin) prior to treatment. Written consent from the patient should be obtained, and any precautions should be recorded in the patient's notes (also see paragraph 17.11).

13.15 Treatments should be administered in the designated treatment area. The treatment area chosen must be suitable for the purpose and meet the design provisions set out in paragraphs 13.6–13.10, *Design of the treatment area.*

13.16 The identity of the patient must be checked immediately prior to administration, as detailed in the employer's procedures under IR(ME)R (IR(ME)R 2017, IR(ME)R(NI) 2018).

13.17 Local shielding should be used as appropriate against external radiation (see paragraphs 10.75 and 10.77). Automated methods of activity administration and vital signs monitoring (e.g. syringe pumps and blood pressure monitors) should be used, where possible when extended administration times are required for therapeutic procedures (e.g. ^{131}I labelled mIBG).

13.18 Staff should wear protective clothing—reference should be made to paragraphs 10.60–10.65, *Provision of protective equipment and clothing*. It is likely that an assessment of risk will lead to the conclusion that staff must wear gloves, single-use sleeves, an apron or a long-sleeved apron, and overshoes in treatment areas and when caring for patients undergoing mRT. These should be removed when leaving the area, stored if appropriate, and monitored as necessary.

13.19 The Operator administering the treatment must be appropriately trained and entitled to perform the administration. They must be familiar with all contingency plans designed to cope with such situations that may arise, such as extravasation or contamination, and have the appropriate equipment, e.g. a spill kit or extravasation kit, on hand if required. Some protection from possible external contamination should be provided for patients who receive oral administration of radioactive liquids for therapy. An absorbent paper or cotton apron/towel with a plastic backing should be sufficient to protect the patient and their clothing. If the patient is feeling nauseous prior to an oral administration, the Operator should seek advice from the Practitioner (and RPA, if appropriate) as to the advisability of proceeding. A prophylactic anti-emetic may be considered. Patients who feel nauseous after oral administration should not be allowed to leave the area until this contamination hazard passes. The use of visors for the administration of oral mRT should be considered in case the patient coughs or discharges the capsule.

13.20 Following an oral administration, the patient should be asked to wait in the department to verify that the activity has been swallowed and retained—comparing readings obtained with a hand-held dose rate meter over the patient's neck and stomach can confirm this.

13.21 Following intravenous administrations, it is advisable to assess for extravasation of the therapy agent by comparing measurements from contamination/dose rate monitors taken directly over the injection site with those taken over the same area of the contralateral arm. This may require the use of a lead cuff over the injection site when taking measurements over the contralateral arm. There should be a documented extravasation policy and kit readily available, with all staff involved in administration aware of the procedures to follow in such an event.

13.22 When radioiodinated compounds are to be administered for the treatment of non-thyroid conditions, the use of a thyroid-blocking agent should be considered in order to reduce the radiation dose to the patient's thyroid. Further advice is available in the ARSAC Notes for Guidance (ARSAC 2024).

13.23 The radioactive substance should be clearly labelled, indicating the radionuclide, chemical form, activity at a given date and time, and batch number if appropriate. Detailed records should be kept of all administrations of radioactive substances (see paragraph 10.79).

Extravasation

13.24 Extravasation of therapy radionuclides can lead to radiation ulcers and radionecrosis (van der Pol *et al* 2017). A contingency for this event should be in place. Advice can be sought from the radiopharmaceutical manufacturer with regard to appropriate immediate steps to mitigate any damage and may include the use of cold/warm compresses, withdrawal of as much radiopharmaceutical as possible and elevation of the arm. An assessment of the quantity of extravasated radiopharmaceutical may be made using monitors or gamma camera imaging. In extreme cases, surgical intervention, and a corresponding radiation risk assessment, may be required. Similar procedures within the treating facility may already be in place for chemotherapy extravasation and may be used as a starting point. The MPE and Practitioner should be contacted for advice. Follow-up of the patient may be required following an intervention. An example flow chart of actions to follow in the event of extravasation is shown in figure 13.1.

General procedures in wards

13.25 The advice in paragraphs 10.44, 10.66–10.80 (*General procedures in nuclear medicine departments*) and 10.90–10.95 (*Procedures in wards and clinics*) is also applicable to wards used by patients treated with open radioactive substances.

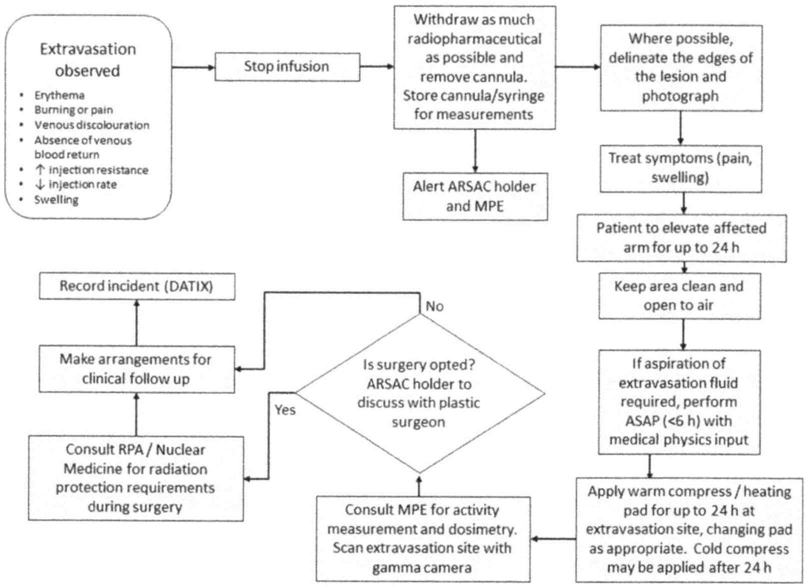

Figure 13.1. An example flow chart for actions to be undertaken in the event of mRT extravasation, based on (van der Pol *et al* 2017).

13.26 Appropriate local rules and radiation protection training must be made available to all relevant staff for inpatient treatments, including nurses, doctors, domestic staff, porters, night staff, and escorts. Local rules should include or refer to any specific radiation safety precautions considered necessary for each technique involving radioactive substances. This will include situations where there is a risk of significant contamination from bodily fluids or on account of external radiation from a patient.

13.27 For some inpatient therapies, only essential nursing procedures should be carried out within the patient room, as detailed in the risk assessment. These should be done as rapidly as is consistent with good nursing practice. In a medical emergency, the care of the patient is the priority, which may result in these restrictions not being fully complied with. In the event of such an emergency, the dose received by the staff involved and any potential contamination may need to be assessed. If the medical condition of the patient deteriorates such that intensive nursing care becomes necessary, the advice of the RPA should be sought immediately and the contingency plans in the local rules enacted. This should include the maximum time that individual healthcare professionals should spend with the patient and any necessary PPE. This information should be clearly visible and available within the treatment room. An emergency PPE kit for this use should be close to hand. Urgent medical care is the priority, so this should not be delayed while advice is sought. In the event of the death of the patient, special precautions may need to be taken (see chapter 17 *Precautions after the death of a patient to whom radioactive substances have been administered*).

13.28 Any nursing or other procedures that are not urgent should be postponed for as long as possible after administration to take full advantage of the reduction of activity by decay and excretion. During the initial period, there should be only the minimum handling of contaminated bed linen, clothing, towels, crockery, etc The documentation should state for how long the precautions should be maintained.

13.29 Beds or rooms used for inpatient treatments should have a notice indicating the designation of the area and include a radiation warning sign (see appendix 7, *Warning signs and notices*). The nursing staff should be made familiar with the implications of the notice and be given details of the nature and activity of the radioactive substances, the time and date of administration, and any relevant instructions for nurses and visitors.

13.30 The maximum dose rates at suitable distances from the patients should be determined as part of a risk assessment. This information will assist in identifying controlled areas and determining the appropriate arrangements for entry to the areas by visitors and staff. These arrangements should be written into the local rules.

13.31 Family or friends visiting or caring for patients who have received therapeutic administrations of radionuclides may incur a significant radiation dose themselves. Restrictions and advice should be given to keep all such doses as low as reasonably practicable (ALARP). Those acting as carers and comforters for the patient, as defined in IR(ME)R (IR(ME)R 2017, IR (ME)R(NI) 2018) and outlined in paragraph 2.156, are not subject to any dose limit; however, a dose constraint for this group must be established (IR (ME)R 6 (d)(ii) (IR(ME)R 2017)). All other visitors are regarded as members of the public for radiation protection purposes and subject to the appropriate dose limit of 5 mSv in 5 years for persons exposed resulting from another's medical exposure (IRR17 Schedule 3 Paragraph 6 (IRR 2017, IRR(NI) 2017)). Procedures for identifying and advising potentially pregnant visitors and carers should be in place following consultation with the RPA

13.32 All visitors should be informed of any necessary precautions to be taken when entering the room of a patient treated with an open radioactive source. This may include wearing appropriate protective clothing, visiting time restrictions, not using the patient's designated toilet, wearing an electronic personal dosimeter (EPD), and not eating or drinking in the room. The radiation risk assessment will inform the level of precautions required.

13.33 Where required by the radiation risk assessment, therapy patients should not leave their ward suite or treatment room without the approval of the radiation protection supervisor (RPS) and clinical lead for the area. A person who has been instructed on the necessary precautions should escort them inside the hospital. Guidance on procedures to be followed when patients leave the hospital after administration of radioactive substances is given in chapter 16 *Pleaving the leaving hospital after administration of radioactive substances.*

13.34 Whenever possible, therapy patients should use designated toilets (see paragraph 13.9) in accordance with the instructions provided. Simple precautions such as placing plastic-backed absorbent paper/Packexe® securely on the floor around the toilet bowl and instructing the patient to flush the toilet twice after each use will help to minimise the external radiation and contamination hazards. The patient may also be advised to sit down when using the toilet, to avoid the risk of splashing (European Commission 1998). Incontinent or catheterised patients will pose additional risks that should have been identified and assessed as described in paragraph 13.12(c). Where a bedpan or urine bottle is provided for such a patient, it should be kept for that patient's exclusive use, preferably in the toilet, and should not be used by another patient until it has been checked and decontaminated as appropriate. Disposable bedpans and urine bottles may be preferable, and consideration may be given to sealable bottles depending on local procedures. An authorised route, under RSR (RSA 1993, EPR 2016, EA(S)R 2018), must be used for their disposal and the disposal of incontinence pads, if required, in consultation with the RWA. Instructions may be required for patients on managing incontinence pads when leaving the hospital, as determined by risk assessment. Flushable stoma bags that dissolve over time are available; however, advice should be sought from the appropriate clinical team on their suitability for the patient to be treated and the frequency at which they should be changed to prevent leakage.

13.35 Crockery and cutlery may become contaminated by patients undergoing mRT. The local rules (see paragraph 13.26) should specify washing-up procedures (e.g. by the patient where possible or with a dedicated dishwasher) and any necessary segregation of utensils for such patients. Any contaminated disposable items should be treated as radioactive waste.

13.36 Particular care should be taken to avoid contamination and limit its spread in the case of an incontinent patient and, following oral administration, if there are clinical reasons for the belief that the patient may vomit. Bedding and personal clothing should be changed promptly if contaminated (see paragraphs 10.104–10.106) and retained for monitoring.

13.37 If a patient is returned or transferred to a different ward, to another hospital, or to a nursing home, any necessary information on appropriate precautions and the person(s) to contact in case of difficulty, should accompany the patient (see chapter 16 *Patients leaving the hospital after administration of radioactive substances*). The receiving institution may have to notify HSE as discussed in paragraphs 16.16–16.21, *Discharging patients to outside organisation etc.* This should be done as soon as the need for notification is identified, and certainly, before the patient is transferred. Any radioactive waste generated must be disposed of appropriately in accordance with RSR (RSA 1993, EPR 2016, EA(S)R 2018), either through an existing authorisation/ permit if there is one, by invoking an exemption or general binding rules under RSR, or by transferring the waste back to the administering hospital, if they are authorised to receive it, for disposal by an authorised route there. The transport of any radioactive waste must comply with the Carriage of Dangerous Goods Act 2009 (as amended) (CDG 2009, CDG(NI) 2010). If the waste is classed as clinical, there may also be extra transport requirements (see chapter 18 *Keeping, accounting for & moving radioactive substances including transport* paragraph 18.155, *Transport of clinical waste*). If the receiving institution has an appointed RPA and/or RWA, they must be informed.

Decontamination procedures

13.38 Decontamination procedures used for diagnostic uses of open sources are also applicable to therapeutic uses (see paragraphs 10.101–10.119, *Decontamination of areas, surfaces and equipment* and *Decontamination of persons*).

13.39 The use of a dedicated washing machine for the decontamination of bedding is advised when possible to reduce storage requirements. One or more cycles of washing may be required, as per paragraph 10.105. Upon sufficient decontamination, below the limits (see paragraph 10.102), the items should be sent to the main laundry system of the hospital. Contaminated linen in storage or waste should not be in alginate bags because these are designed to disintegrate during washing and are likely to degrade over time. If washing facilities are not available, consider disposal as radioactive waste to avoid long-term storage of biologically contaminated items.

Procedures in operation theatres

13.40 Before surgery (other than in an emergency) on a patient that has recently received mRT, information should be sought on the activity remaining in the body. Any precautions against external radiation and possible contamination from body fluids should be determined in collaboration with the RPA. If surgery is not urgent, it should be postponed until the radioactivity in the body has fallen to a suitable level, as determined by risk assessment. The wearing of two pairs of surgical gloves will give some protection to the hands against beta radiation (see Varskin or similar for calculations (RAMP 2022)). If the gloves are cut or torn during the operation and the surgeon's hands are injured, the advice given in paragraph 10.117 should be followed. Face visors should be considered for surgical staff to reduce the risk of contamination. If practicable, two separate areas should be defined—a radiation area, encompassing the scope of the patient and any area that may become contaminated, and a non-radioactive area, ideally with clear demarcation between the two. Any staff and equipment leaving the radiation area must be monitored for contamination, including staff hands and feet. It is helpful to have a trained staff member physically standing by the demarcated area with a contamination monitor and a large container for contaminated items. After the operation has been completed the operating theatre, surgical instruments and equipment, and protective clothing should be thoroughly checked for contamination and, if necessary, decontaminated (see paragraphs 10.113–10.119, *Decontamination of persons*). The use of Packexe® (Packexe Ltd, Exeter, UK) or Benchkote (Whatman Ltd Chalfont St Giles, UK), if permitted in the theatre, prior to commencing the surgery may aid in the clean-up.

13.41 The advice in paragraph 13.40 may be used for the administration of radiopharmaceuticals under fluoroscopic guidance, for example, e.g. ^{90}Y microspheres. However, in these cases, there will be two staff groups familiar with a different set of radiation hazards, and each staff group will require training in the other's local rules, with special emphasis on the separate contingency arrangements.

Monitoring of work areas

13.42 The advice concerning monitoring in paragraphs 10.97–10.100, *Monitoring of work areas and persons*, should be applied with particular rigour in laboratories and work areas used for therapeutic applications of open radioactive substances.

13.43 Contamination and dose rate monitors appropriate to the type and energy of the radiation concerned should be readily available and used at all stages of the therapy procedure. Monitoring should be carried out at a frequency decided upon with advice from the RPA to detect changes in radiation and contamination levels. Radiation dose rate levels in adjacent rooms and corridors will need to be assessed and monitored, as appropriate. The findings should be recorded and detailed in the risk assessment.

13.44 Prior to the discharge of patients undergoing mRT (see paragraphs 13.8 and 13.12) from a ward, clothing and other personal property should be monitored for contamination. Depending on the levels of contamination found, the items may need to be stored or decontaminated (see paragraphs 10.101–10.119, *Decontamination of areas, surfaces and equipment*, *Decontamination of persons*). In some cases, items of personal property may be sealed and given to the patient to store at home. The RPA should be consulted on determining appropriate guidelines for acceptable levels of contamination.

13.45 Subsequent to the discharge of a patient who has undergone mRT, the area of the ward used by the patient should be monitored and, if necessary, decontaminated before further use. Where the area was designated as controlled, it cannot be undesignated until monitoring for contamination and any necessary decontamination have taken place.

Specific precautions for alpha emitters

13.46 Alpha particles travel a few centimetres in air and only micrometres in tissue and are unable to penetrate the cornified epithelium (dead skin layer) (IAEA 2004). The risk of internal exposure through needlesticks, ingestion, or inhalation is the dominant consideration for work with alpha-emitting radionuclides and should be specifically addressed in local protocols and radiation risk assessments (IAEA 1999a, 1999b).

13.47 The decay chains of alpha-emitting radionuclides include additional beta or gamma emissions, which should be taken into account when assessing the shielding requirements for mRT (e.g. use of syringe shields) (Dauer *et al* 2014, UK Radiopharmacy Group, IPEM NM SIG and BNMS 2017).

13.48 Most centres are unlikely to be using pure alpha emitters and will probably have contamination monitors that will readily detect beta and gamma radiation. Radium-223, for example, is easily detectable with conventional contamination monitors. For pure alpha emitters, suitably calibrated contamination monitors with thin window probes should be readily available. Alpha monitoring should be performed thoroughly at a short distance (a few millimetres) from the surface monitored and taking into account that uneven surfaces and self-absorption can mask alpha contamination (IAEA 2000, 2004, NPL 2014). Consideration should be given to swabbing surfaces in the radiation risk assessment.

13.49 The need for additional training of staff above and beyond conventional handling of beta and gamma emitters and training for alpha monitoring should be assessed.

Dosimetry in molecular radiotherapy

13.50 The legislative requirement for dosimetry in radiation therapies is laid out in IR(ME)R regulation 12(2) (IR(ME)R 2017, IR(ME)R(NI) 2018). '*In relation to all radiotherapeutic exposures, the practitioner must ensure that exposures of target volumes are individually planned and their delivery appropriately verified, taking into account that doses to non-target volumes and tissues must be as low as reasonably practicable and consistent with the intended radiotherapeutic purpose of the exposure.*' Interpretation of this requirement for nuclear medicine therapies is provided by the health department and the professional bodies (Department of Health and Social Care 2018 page 23, Radiotherapy Board 2020 page 69).

13.51 The rationale for dosimetry within radionuclide therapies is central to key areas of patient care:

(a) Patient protection: planning treatment to minimise toxicities to critical organs and the probabilities of secondary malignancies.

(b) Treatment optimisation and planning: to maximise dose and control toxicities, thereby optimising therapeutic ratios.

(c) Standardisation: in both the planning and administration of molecular radiotherapy and the practice of dosimetry for cross-comparisons in multicentre studies.

13.52 Standard-administered activities are often employed for common molecular radiotherapies, consistent with clinical guidelines (Perros *et al* 2014). Verification imaging is also routinely performed for most therapies, following administration. The ARSAC guidance notes (ARSAC 2024) recommend the measurement of absorbed doses to tumours and non-target volumes and tissues for cancer treatments and the calculation or estimation of absorbed doses for the treatment of benign conditions. Applications for ARSAC Employer's licenses that include therapy administrations should specify what dosimetry will be performed. Methodology and details for dosimetry are outlined in appendix 20 *Dosimetry in radiomolecular therapy*.

References

ARSAC 2024 *Notes for guidance on the clinical administration of radiopharmaceuticals and use of sealed radioactive sources* (Administration of Radioactive Substances Advisory Committee) https://www.gov.uk/government/publications/arsac-notes-for-guidance (accessed 18 April 2024)

BIR, RCR, IPEM, SCoR and PHE 2020 *IR(ME)R Implications for Clinical Practice in Diagnostic Imaging Interventional Radiology and Diagnostic Nuclear Medicine* https://www.rcr.ac.uk/our-services/all-our-publications/clinical-radiology-publications/ir-me-r-implications-for-clinical-practice-in-diagnostic-imaging-interventional-radiology-and-diagnostic-nuclear-medicine/ (accessed 23 April 2024)

CDG 2009 *The Carriage of Dangerous Goods and Use of Transportable Pressure Equipment Regulations 2009* https://legislation.gov.uk/uksi/2009/1348 (accessed 23 April 2024)

CDG(NI) 2010 *The Carriage of Dangerous Goods and Use of Transportable Pressure Equipment Regulations (Northern Ireland)* 2010 https://legislation.gov.uk/nisr/2010/160 (accessed 23 April 2024)

CQC 2023 *Ionising Radiation (Medical Exposure) Regulations (IR(ME)R)*. 19/4/23 *CQC Website. 19/4/23* (London: Care Quality Commission) https://cqc.org.uk/guidance-providers/ionising-radiation/ionising-radiation-medical-exposure-regulations-irmer (accessed 23 April 2024)

Dauer L T *et al* 2014 Radiation safety considerations for the Use of 223RaCl2 DE in men with castration-resistant prostate cancer *Health Phys.* **106** 494–504

DHSC 2018 Guidance to the Ionising Radiation (Medical Exposure) Regulations 2017 *Gov.UK Website* (Department of Health and Social Care) https://gov.uk/government/publications/ionising-radiation-medical-exposure-regulations-2017-guidance (accessed 31 January 2023)

EA(S)R 2018 *The Environmental Authorisations (Scotland) Regulations 2018* http://legislation.gov.uk/ssi/2018/219 (accessed 23 April 2024)

EPR 2016 *Environmental Permitting (England and Wales) Regulations* https://legislation.gov.uk/uksi/2016/1154 (accessed 23 April 2024)

European Commission 1998 *Radiation Protection 97: Radiation Protection Following Iodine*-131 *Therapy (Exposures Due to Out-Patients or Discharged In-Patients)*, *Radiation Protection* 97 https://energy.ec.europa.eu/system/files/2014-11/097_en_1.pdf (accessed 23 April 2024)

HSE 2015 *INDG334 Working Safely with Ionising Radiation. Guidelines for Expectant or Breastfeeding Mothers* http://hse.gov.uk/pubns/indg334.htm (accessed 23 April 2024)

IAEA 1999a *Assessment of Occupational Exposure Due to External Sources of Radiation, Safety Standards Series* (Vienna, Austria: International Atomic Energy Agency) https://iaea.org/publications/5742/assessment-of-occupational-exposure-due-to-external-sources-of-radiation (accessed 23 April 2024)

IAEA 1999b Assessment of occupational exposure due to intakes of radionuclides *Safety Standards Series No. RS-G-1.2 [Preprint]. RS-G-1.2* https://iaea.org/publications/5743/assessment-of-occupational-exposure-due-to-intakes-of-radionuclides (accessed 23 April 2024)

IAEA 2000 Calibration of radiation protection monitoring instruments *SRS, Safety Reports Series No.16. SRS* (Vienna, Austria: International Atomic Energy Agency) https://iaea.org/publications/5149/calibration-of-radiation-protection-monitoring-instruments (accessed 23 April 2024)

IAEA 2004 *Workplace monitoring for radiation and contamination.* PRTM-1 *(Re, Practical Radiation Technical Manual No. 1 (Rev. 1). PRTM-1* (Vienna, Austria: International Atomic Energy Agency) https://iaea.org/publications/6757/workplace-monitoring-for-radiation-and-contamination (accessed 23 April 2024)

IR(ME)R 2017 *The Ionising Radiation (Medical Exposure) Regulations* www.legislation.gov.uk/uksi/2017/1322 (accessed 23 April 2024)

IR(ME)R(NI) 2018 *The Ionising Radiation (Medical Exposure) Regulations (Northern Ireland)* https://legislation.gov.uk/nisr/2018/17 (accessed 23 April 2024)

IRR 2017 *The Ionising Radiations Regulations* https://legislation.gov.uk/uksi/2017/1075 (accessed 23 April 2024)

IRR(NI) 2017 *The Ionising Radiations Regulations (Northern Ireland)* http://legislation.gov.uk/nisr/2017/229 (accessed 23 April 2024)

NPL 2014 *Good Practice Guide No. 30 Practical Radiation Monitoring. Issue 2, NPL Good Practice Guide. Issue 2* (Teddington: National Physical Laboratory) https://npl.co.uk/special-pages/guides/gpg30_radiation (accessed 23 April 2024)

Perros P *et al* 2014 British Thyroid Association Guidelines for the management of thyroid cancer *Clin. Endocrinol.* **81** 1–122

Radiotherapy Board 2020 *Ionising Radiation (Medical Exposure) Regulations: Implications for Clinical Practice in Radiotherapy Guidance from the Radiotherapy Board* (London: Royal College of Radiologists) https://www.rcr.ac.uk/media/smmkkrsa/ionising-radiation-medical-exposure-regulations-implications-for-clinical-practice-in-radiotherapy.pdf (accessed 23 April 2024)

RAMP 2022 VARSKIN Dose Calculation for Skin Contamination *US National Regulatory Commission Website* https://ramp.nrc-gateway.gov/codes/varskin (accessed 23 April 2024)

RSA 1993 *Radioactive Substances Act 1993* http://legislation.gov.uk/ukpga/1993/12 (accessed 23 April 2024)

UK Radiopharmacy Group, IPEM NM SIG and BNMS 2017 *Advice for the Safe Drawing up and Administration of 223 Ra Radium-Chloride (Xofigo)* https://cdn.ymaws.com/www.bnms.org.uk/resource/resmgr/guidelines/addendum_to_safe_drawing_up_.pdf (accessed 23 April 2024)

van der Pol J, Vöö S, Bucerius J and Mottaghy F M 2017 Consequences of radiopharmaceutical extravasation and therapeutic interventions: a systematic review *Eur. J. Nucl. Med. Mol. Imaging* **44** 1234–43

IOP Publishing

Medical and Dental Guidance Notes (Second Edition)
A good practice guide on all aspects of ionising radiation protection in the clinical environment: IPEM Report 113
John Saunderson, Mohamed Metwaly, William Mairs, Philip Mayles, Lisa Rowley and Mark Worrall

Chapter 14

Ionising radiation protection in general laboratories

Scope

14.1 This chapter applies to the use of open radioactive substances in medical diagnosis or research where the activities of radionuclides employed in the procedure are typically less than those required for the quantity for notification to the HSE or HSENI under the Ionising Radiations Regulations, IRR (IRR 2017, IRR(NI) 2017) (tabulated in column 3 of Table 10.1). It includes the use of small quantities of ^{35}S and ^{32}P in cytogenetics and may include work on tissue samples from the theatre, such as those from sentinel node localisation procedures (see paragraph 10.44). Equipment that might be used in laboratories, for example, blood irradiators, x-ray specimen cabinets, and electron microscopy, is also mentioned. Workers using higher activities should refer to chapter 10, *Diagnostic uses of open radioactive substances*. Specific advice relating to all work with tritium is provided in this chapter.

14.2 This chapter does not apply to work with volatile radioactive materials, such as iodination, which is discussed in chapter 12, *Preparation of radiopharmaceuticals*.

14.3 The Association of University Radiation Protection Officers (AURPO) produces guidance notes appropriate to work in research and teaching establishments (AURPO Scientific & Technical Committee 2010).

Patient samples

14.4 Radioactive patient samples may be analysed by staff who are not radiation workers (as determined by the radiation risk assessment), for example, blood

doi:10.1088/978-0-7503-2332-1ch14

from patients having undergone nuclear medicine procedures or biopsy samples from sentinel lymph node procedures. Where possible, the analysis should be delayed until samples are no longer radioactive to reduce doses to staff.

14.5 In the case of sentinel lymph node biopsies, finger doses to lab staff will be low (Waddington *et al* 2000). The radiation risk assessment should include the workload, the time of analysis following surgery (and therefore administration), and the nature of the samples being measured, for example, the tumour site, nodes, etc, to determine whether staff require monitoring.

14.6 Depending on local procedures, the radiation risk assessment will also inform any requirements under radiation legislation for transport, restriction of exposure, contingency planning, monitoring, disposal, and staff training. All radioactive specimens must be appropriately labelled and written procedures must be provided to the lab staff involved.

Blood irradiators, equipment with internal sources and similar devices

14.7 The work of departments and laboratories falling within the scope of this chapter may include the irradiation of clinical samples. Blood irradiators are well-shielded devices incorporating large, sealed radioactive sources. Many of these irradiator sources may be High Activity Sealed Sources (HASS) with reference to the definition thereof in the European directive (European Commission 2014), Schedule 23 Part 5 of the Environmental Permitting Regulations (EPR 2016), Schedule 8 of the Environmental Authorisations (Scotland) Regulations 2018 (EASR 2018, EA(S)R 2018), and in Northern Ireland, the High-activity Sealed Radioactive Sources and Orphan Sources Regulations 2005 (HASS(NI) 2005). Thus, they will need particular consideration with respect to security facilities and consent from the HSE or HSENI under the IRR (IRR 2017, IRR(NI) 2017). Chapter 15, *Diagnostic uses of sealed or other solid radioactive sources*, gives further information on similar devices that may be helpful.

14.8 Cabinets for x-ray of tissue samples (e.g. Faxitron™ units) may also be in use and may require the Employer to gain registration with the HSE or HSENI under IRR (IRR 2017, IRR(NI) 2017).

14.9 Advice should be obtained from the radiation protection adviser (RPA) on the radiation risk assessment, purchase, installation, and maintenance of devices of either kind and also from a radioactive waste adviser (RWA) on those devices containing radioactive sources. Advice should include the need for the designation of controlled or supervised areas, the content of local rules, and the testing of any engineering controls, safety features, and warning devices. The RPA may also advise on the calibration of the devices. The devices must undergo a critical examination before being brought into use (see paragraphs 1.246–1.260 *Critical examination* and paragraphs 2.95–2.98 if the equipment is

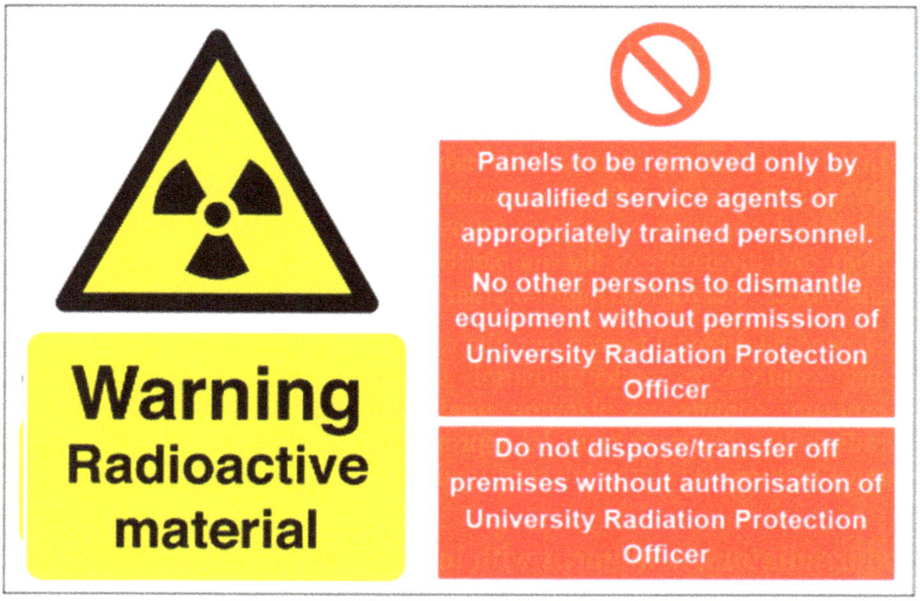

Figure 14.1 An example sign for the external surface of a device containing a sealed source.

on loan). Only staff instructed in their safe and proper operation may use them. The RPA or medical physics expert (MPE), as appropriate, may provide guidance on a system of quality assurance.

14.10 Guidance on leak testing of sealed sources is provided in L121 (HSE 2018), with performance tests detailed in the ISO document ISO 2919:2012 (ISO 2012). A liquid scintillation counter effectively carries out its own leakage test when performing a standard count using the internal source. This may alert the user to a leaking source if there is an unusually high background count.

14.11 For devices containing sealed sources, appropriate warning signs should be placed on clearly visible, external surfaces, to prevent disposal or dismantling without consideration for the source within. An example sign is shown in figure 14.1.

Uranium staining in electron microscopy

14.12 In electron microscopy, uranium compounds are commonly used for staining samples for investigation. Stains are prepared from stock bottles of uranium salts by dissolving a small quantity of uranium powder in an appropriate medium. Once prepared, stains can be stored for long periods in darkness for gradual use. Depending upon the activity and activity concentration of the material held, it may be necessary to register with or notify HSE or HSENI under IRR (IRR 2017, IRR (NI) 2017).

Laboratory protection measures

14.13 Facilities for working with uranium compounds should be designed with containment and ease of decontamination in mind. Safety features common to most laboratories should include non-absorbent, easily cleaned surfaces, and dedicated disposal sinks.

14.14 The dispensing of powders and preparation of stains should be carried out within a fume cupboard or contained workstation.

14.15 Most protection from exposure during work with open radioactive materials is achieved by following systems of work to prevent the spread of contamination; for example, the use of drip trays and the performance of regular contamination monitoring (see chapter 10, *Diagnostic uses of open radioactive substances*, for more information).

14.16 Good laboratory practice and worker training should minimise the potential for spreading radioactive contamination outside the working area. Written arrangements, informed by the radiation risk assessment, should be prepared to minimise the potential for such a spread. Suitable and appropriate personal protective equipment should be worn, including lab coats (fastened to the neck), disposable gloves, and protective eyewear.

14.17 A suitable store for stocks of radioactive materials, prepared stains, and any radioactive experimental samples and waste should be available nearby. The store should be fire resistant and secured with a key retained for issue only to authorised persons.

Contamination monitoring

14.18 Contamination monitoring, using suitable, calibrated instruments, should be performed before, during, and after dispensing or experiments involving open sources. Contamination monitors must be adequately tested and examined at appropriate intervals (HSE 2018). At the very least, a record should be made to confirm that the work area is free from radioactive contamination at the end of the session and that the sink is clear of contamination following the disposal of aqueous radioactive waste. Clear action levels should be defined. Records should be quantitative, detailing the instrument used, the level of contamination measured, and the name of the person carrying out the monitoring. The use of expressions such as 'OK' or the simple use of ticks is open to interpretation and should therefore be avoided.

14.19 If contamination is identified, it should be promptly dealt with. Significant events, such as the spread of contamination outside the working area or personal contamination, should be reported to the radiation protection supervisor (RPS) and, if relevant, the URPO (University Radiation Protection Officer), as well as local management, with the RPA consulted as appropriate.

14.20 A suitable instrument for contamination monitoring during work with uranium/thorium compounds would be a large area end-window Geiger–Muller detector or scintillation detector (e.g. zinc sulphide or dual phosphor type). Hand-held monitoring can be supplemented by wipes for liquid scintillation analysis.

Radioactive stock and radioactive waste disposal conditions

14.21 Work with uranium compounds must be carried out in accordance with any relevant radioactive substances regulation (RSR) (RSA 1993, EPR 2016, EA (S)R 2018) or operate under an exemption (BEIS *et al* 2018) or binding rules.

14.22 Uranium, together with plutonium and thorium, is classified as a 'nuclear material' under the International Safeguard Rules (IAEA 2001, ONR 2022). In accordance with those rules, employers holding uranium compounds are required to keep accurate records of the number of uranium holdings (including stocks, prepared stains, and waste) and to make routine returns to the Office for Nuclear Regulation of their entire inventory of nuclear materials (see chapter 18, *Keeping, Accounting for, and moving radioactive substances including transport*, for further details).

Hazards from radioactive substances and principles for their control

14.23 The radiation hazards associated with open radioactive substances and principles for their control are as detailed in paragraphs 10.8–10.13, *Hazards from open radioactive substances and principles for their control.* Biohazards and chemical hazards, such as solvent or drain disposals, must also be considered.

14.24 Where work with open radioactive materials involves more than one employer, it is the duty of all such employers to liaise and co-operate in order that all legislative requirements are met (see IRR regulation 16 (IRR 2017, IRR(NI) 2017)). See chapter 1, *General measures for radiation protection*, for more detail.

Protection of the patient

14.25 If the work involves the administration of radioactive materials to humans, the general provisions of paragraphs 10.16–10.31, *Roles and responsibilities of duty holders in nuclear medicine*, apply. There is no exemption for the small activities involved. The Practitioner must individually justify any radiation exposure to patients. The Practitioner and the Employer must hold appropriate ARSAC licences (ARSAC 2024).

14.26 Staff who carry out such procedures on patients, calibrate or check any of the associated equipment, or prepare the radiopharmaceutical to be given must be given suitable training and comply with the responsibilities of Operators as defined in the Employer's procedures required under IR(ME)R (2017,

IR(ME)R(NI) 2018)) (see chapter 2, *Radiation Protection of persons undergoing medical exposures*).

14.27 The appointment of an appropriate MPE will also be required, with the duties of the MPE listed in appendix 15.

Design of work areas

14.28 The requirements for work with open radioactive substances vary with the type of procedures being undertaken. An RPA and RWA need to be consulted about the design of laboratories and other work areas, and a formal appointment of such experts is likely to be required by the Employer or RSR Permit holder. Detailed plans should be drawn up in collaboration with the person(s) responsible for the work. Changes that may become necessary from time to time should also be discussed with the RPA and RWA.

14.29 Special procedures to restrict radiation exposure are very unlikely to be needed for work coming within the scope of this chapter, and hence the work areas will normally only require supervision if a designation is required at all. A radiation risk assessment should be undertaken to inform the decision. Chapter 10, *Diagnostic uses of open radioactive substances*, should be consulted where larger amounts of radionuclides are used.

14.30 The general provisions of paragraphs 10.32–10.49, *Design of nuclear medicine departments*, apply to these areas. A risk assessment will normally show that good modern laboratory facilities need little or no upgrading to conform to the requirements of radioactive work within the scope of this chapter.

14.31 It may be sufficient and indeed desirable to confine space for work involving radioactive materials to a specified area within a laboratory. Particularly in a research environment, radioactive work may be spasmodic, and the area may have to be released for other purposes. This should only be done after all radioactive sources have been removed, monitoring has confirmed that there is no radioactive contamination remaining, and, finally, all radioactive warning signs have been concealed or removed.

14.32 Secure storage of radioactive materials must be provided in compliance with IRR (IRR 2017, IRR(NI) 2017) NaCTSO guidelines (for sealed sources) (IAEA 2019, NaCTSO 2011), and with consultation from the local environmental protection agency inspector. The store should be sited in a suitable place away from flammable materials, taking into account the dose rate from it and the convenience and reduced potential for accidents in transit if it is close to the work area. Facilities for the receipt of radioactive materials and the storage and disposal of radioactive waste should be agreed upon with the RPA and RWA. This is discussed further in chapter 18, *Keeping, accounting for, and moving radioactive substances (including transport)*.

14.33 An area for the preparation and counting of radioactive samples may be required. It may need to be separate from other areas used for radioactive work.

14.34 Guidance on the design of laboratories for work with open radioactive material can be found in the Association of University Radiation Protection Officers (AURPO) *Guidance Notes on Working with Ionising Radiations in Research and Teaching* (Appendix 4 of which contains the Environment Agency Guidance on Standards for Radiochemical Laboratories in Non-Nuclear Premises) (AURPO Scientific & Technical Committee 2010).

14.35 If radioactive materials are to be administered to patients, then space must be allocated for the taking of biological samples and any clinical measurements involved. Patients receiving the small tracer amounts covered by this chapter may wait in general waiting rooms and do not need separate toilets.

Equipment and clothing

14.36 The general provisions of paragraphs 10.51–10.65, *Equipment: installation, maintenance and quality assurance* and *Provision of protective equipment and clothing*, apply. Radioactive sources should be kept in suitable shielding, handled with tongs, and worked within drip trays to contain spills. Separate laboratory coats should be worn solely for radioactive work, and disposable impermeable gloves should be worn whenever unsealed sources are handled. Suitable monitoring equipment should be readily available.

General procedures

14.37 The general provisions of paragraphs 10.66–10.80, *General procedures in nuclear medicine departments*, apply.

Administration of radioactive materials to patients

14.38 The general provisions of paragraphs 10.81–10.88, *Administration of radiopharmaceuticals*, apply.

Monitoring of staff

14.39 TLDs, OSLs, EPDs, and other common personal monitoring devices do not detect radionuclides of very low penetrating power and energy of emission, such as ^{3}H and ^{14}C. The advice of the RPA should be sought. Personal extremity monitoring may be appropriate in some circumstances, but, in general, rigorous laboratory technique and good housekeeping are the best safeguards for staff working with these levels and types of radioactive materials.

14.40 Contamination monitoring for tritium is particularly difficult and can be expensive because of the very low energy of the emitted beta particles. Wipe testing and liquid scintillation counting (to determine the amount of removable contamination on working surfaces) after each use are normally sufficient to confirm that arrangements are adequate. The surfaces should be wiped in suitable sections with wetted filter paper or other suitable material, and then the activity on each swab can be measured by liquid scintillation counting. This will provide a qualitative method of confirming whether contamination is occurring and, if so, whether further actions may be needed to reduce it.

14.41 Where a risk assessment identifies that there is a risk of significant intakes of radioactivity, advice should be sought from the RPA on the appropriate method of monitoring potential internal exposures. For example, tritium-in-urine monitoring and *in vivo* thyroid monitoring (for radioiodine) may be advisable in specific situations.

14.42 It may not be necessary to vary the duties of members of staff who are pregnant or breastfeeding on radiation protection grounds, as long as procedures to avoid contamination keep doses as low as reasonably practicable (ALARP). A properly prepared risk assessment should consider this and take into account that some radionuclides (such as ^{32}P) are preferentially taken up by the unborn child.

Monitoring of areas

14.43 Monitoring of surfaces used for radioactive work should be carried out after every procedure or regularly, at a frequency dependent upon use, and with a monitor suitable for the purpose. Advice should be sought from the RPA and RWA if necessary. The comments made with regard to monitoring for tritium in paragraph 14.40 are applicable.

Decontamination

14.44 Advice on decontamination of persons, equipment, and surfaces is given in paragraphs 10.101–10.119, *Decontamination of areas, surfaces and equipment* and *Decontamination of persons.*

Keeping, using and transporting radioactive materials, and accumulating and disposal of radioactive waste

14.45 Requirements under the legislation for the keeping, use, and transport of radioactive materials are outlined in chapter 18, *Keeping, accounting for, and moving radioactive substances (including transport).* Chapter 19, *Accumulation and disposal of radioactive waste*, details the requirements for the disposal of radioactive waste.

References

ARSAC 2024 *Notes for guidance on the clinical administration of radiopharmaceuticals and use of sealed radioactive sources* (Administration of Radioactive Substances Advisory Committee) https://www.gov.uk/government/publications/arsac-notes-for-guidance (accessed 20 April 2024)

AURPO Scientific & Technical Committee 2010 *AURPO Guidance Notes on Working with Ionising Radiations in Research and Teaching*

BEIS, DEFRA, Welsh Government and DAERA 2018 *Scope of and Exemptions from the Radioactive Substances Legislation in England Wales and Northern Ireland Guidance Document* https://gov.uk/government/publications/guidance-on-the-scope-of-and-exemptions-from-the-radioactive-substances-legislation-in-the-uk (accessed 20 April 2024)

EA(S)R 2018 The Environmental Authorisations (Scotland) Regulations 2018 http://legislation.gov.uk/ssi/2018/219 (accessed 20 April 2024)

EPR 2016 *Environmental Permitting (England and Wales) Regulations* https://www.legislation.gov.uk/uksi/2016/1154 (accessed 20 April 2024)

European Commission 2014 1.123, and repealing Directives 89/618/E89/618/Euratom, 90/641/Euratom, 96/29/Euratom, 97/43/E *Off J Eur Commun.* **L13** 1–73 https://eur-lex.europa.eu/LexUriServ/LexUriServ.do?uri=OJ:L:2014:013:0001:0073:EN:PDF (accessed 20 April 2024)

HASS(NI) 2005 The High-activity Sealed Radioactive Sources and Orphan Sources Regulations 2005 http://legislation.gov.uk/en/uksi/2005/2686 (accessed 20 April 2024)

HSE 2018 *L121 Work with Ionising Radiation Ionising Radiations Regulations 2017—Approved Code of Practice and Guidance* 2nd (Norwich: Health and Safety Executive) https://hse.gov.uk/pubns/books/l121.htm (accessed 20 April 2024)

IAEA 2001 *IAEA Safeguards Glossary International Nuclear Verification Series* (Vienna, Austria: International Atomoc Energy Authority) https://iaea.org/sites/default/files/iaea_safeguards_-glossary.pdf (accessed 20 April 2024)

IAEA 2019 *Security of Radioactive Material in Use and Storage and of Associated Facilities, Nuclear Security Series No. 11-G (Rev.1).* Vienna, Austria: International Atomic Energy Agency https://www.iaea.org/publications/12360/security-of-radioactive-material-in-use-and-storage-and-of-associated-facilities (accessed 20 April 2024)

IR(ME)R 2017 *The Ionising Radiation (Medical Exposure) Regulations* www.legislation.gov.uk/uksi/2017/1322 (accessed 20 April 2024)

IR(ME)R(NI) 2018 *The Ionising Radiation (Medical Exposure) Regulations (Northern Ireland)* https://www.legislation.gov.uk/nisr/2018/17 (accessed 20 April 2024)

IRR 2017 *The Ionising Radiations Regulations* (Great Britain) https://legislation.gov.uk/uksi/2017/1075 (accessed 20 April 2024)

IRR(NI) 2017 *The Ionising Radiations Regulations (Northern Ireland)* http://legislation.gov.uk/nisr/2017/229 (accessed 20 April 2024)

ISO 2012 *2919:2012 Radiological Protection—Sealed Radioactive Sources—General Requirements and Classification* (Geneva: International Organization for Standardization)

NaCTSO 2011 *Security Requirements for Radioactive Sources* (National Counter Terrorism Security Office)

ONR 2022 *Nuclear safeguards, Office for Nuclear Regulation Website* https://onr.org.uk/safeguards/ (accessed 20 April 2024)

RSA 1993 *Radioactive Substances Act 1993* http://legislation.gov.uk/ukpga/1993/12 (accessed 20 April 2024)

Waddington W, Keshtgar M R, Taylor I, Lakhani S R, Short M and Ell P 2000 Radiation safety of the sentinel lymph node technique in breast cancer *Eur. J. Nucl. Med.* **27** 377–91

IOP Publishing

Medical and Dental Guidance Notes (Second Edition)
A good practice guide on all aspects of ionising radiation protection in the clinical environment: IPEM Report 113
John Saunderson, Mohamed Metwaly, William Mairs, Philip Mayles, Lisa Rowley and Mark Worrall

Chapter 15

Diagnostic uses of sealed or other solid radioactive sources

Scope

15.1 This chapter contains guidance on the use of sealed or other solid sources (see paragraph 9.2) for:

(a) transmission scanning or as anatomical markers used in nuclear medicine investigations; or

(b) diagnostic or clinical research purposes in the fields of bone mineral measurements, x-ray fluorescence scanning of the thyroid and neutron activation analysis.

15.2 The guidance in this chapter also applies when the equipment is being used for experimental purposes, is being tested or calibrated at its place of use, or when sources are being changed.

15.3 Guidance on leakage testing of sources is given in paragraph 15.8 and chapter 18, *Keeping, accounting for, and moving radioactive substances (including transport)*.

15.4 Some clinical research projects may require diagnostic reference measurements on healthy subjects, e.g. bone mineral density, using sealed sources. Guidance on the medical exposure of volunteers, who are included in the term 'patients', is given in appendix 19, *Guidance on medical research exposures.*

Protection of the patient

15.5 If the radioactive source irradiates the patient or causes induced radioactivity within the patient, then the provisions of IR(ME)R (IR(ME)R 2017, IR(ME)

doi:10.1088/978-0-7503-2332-1ch15

R(NI) 2018) apply, and the relevant guidance in paragraphs 10.16–10.31, *Roles and responsibilities of duty holders in nuclear medicine*, should be followed.

Sources and equipment—general

15.6 Whenever reasonably practicable, sources used in diagnostic and analytical equipment should be sealed sources conforming to BS ISO 2919:2012 (ISO 2012). Attention should be paid to the supplier's recommendations on working life and the environment of use (see chapter 18, *Keeping, accounting for, and moving radioactive substances (including transport)*). Sources with the lowest activity consistent with satisfactory clinical results throughout the useful life of the source should be selected for these applications.

15.7 Prior to receipt of the source, consideration must be given as to whether the source requires notification or registration under the relevant Radioactive Substances Regulation (RSR) (RSA 1993, EPR 2016, EA(S)R 2018). The type of registration, whether for sealed or mobile sources, and the security and financial provisions (including for disposal) must also be considered. Registration/notification of the use of radioactive material with HSE or HSENI under the Ionising Radiations Regulations, IRR (IRR 2017, IRR (NI) 2017), may also be required.

15.8 No new source should be used until a leakage test has been carried out unless a leakage test certificate issued by the manufacturer has been obtained. An assessment of the potential ways in which containment could be lost and their likelihood of occurring should be carried out initially as part of the radiation risk assessment required under the IRR (IRR 2017, IRR(NI) 2017). The potential radiation exposure of the person carrying out the leak test should also be estimated. The findings should be used to determine a suitable leak test method, a pass-fail criteria, and the frequency of such testing (IRR Reg. 28 (IRR 2017, IRR(NI) 2017)). For a wet/dry wipe test, BS ISO 9978:2020 states that if activity under 200 Bq is detected, the source is considered leak-tight (BSI 2020). Where testing is appropriate under normal operating conditions, the interval between tests should not exceed two years. The frequency of testing should be increased if sources are stored in harsh environments or when the source is used beyond its recommended life. Records should be kept of the results of leakage testing. Fuller guidance is given in L121 (HSE 2018) concerning IRR regulation 28 (IRR 2017, IRR(NI) 2017).

15.9 Records of each sealed/solid source must be kept indicating the source details, i.e. radionuclide(s), activity and reference date, source identification number/mark, date of receipt, location, and when appropriate, the date and manner of disposal (see paragraphs 18.39–18.51, *Stock records and control procedures*). The presence of each sealed source should be confirmed on a regular basis, depending on the frequency of use. The loss or theft of a sealed source must be notified to the appropriate environment agency and may also require notification to HSE depending on the source activity (see chapter 20 *Contingency planning and emergency procedures for radioactive substances activity*). Any theft must also be reported to the police. Fuller information on accounting is given in L121 (HSE 2018) (IRR regulation 29 (IRR 2017, IRR(NI) 2017). Consider whether the intended use of the source requires its details to be included in the IR(ME)R equipment inventory (see paragraphs 2.99–2.102, *Equipment inventory*).

15.10 If a sealed source is to be used outside the manufacturer's recommended working life (RWL) of the source, an assessment must be carried out to ensure the continued safe use of the source, and leak tests should be carried out at a frequency determined by the risk assessment. If a source exceeds twice its RWL or has not been used in a year with no future plans for use, it should be disposed of (EA 2018).

Anatomical marker sources

15.11 Anatomical marker sources used in nuclear medicine investigations should be placed on or attached to the patient's skin or clothing for the minimum time necessary. Special care must be exercised to ensure that any such sources are not left attached to the patient. The frequency of auditing these sources should be chosen to reflect this possibility. The rest of this chapter does not apply to anatomical marker sources.

Sources and equipment—diagnostic investigations

15.12 Except where paragraphs 15.17 or 15.18 apply, sources should be mounted, either permanently or when the equipment is in use, in a housing that has an aperture for the radiation beam. The aperture should be such that the radiation beam will not irradiate a greater part of the patient's body than is necessary. A high degree of collimation will be needed for most scanning techniques.

15.13 Shutters should be interlocked and operated, for example, by means of a patient-presence sensor. For a scanning device, the opening and closing of the shutter may be linked with the scanning movement. Equipment with a manually operated deadman-type shutter should be provided with a lock so that the shutter cannot be opened accidentally when the equipment is not in use.

15.14 The automatic interlocks/shutters of equipment with such devices must be tested at a suitable frequency dependent on the dose implications of their failure. Records of these tests should be kept.

15.15 The equipment should be clearly marked to indicate:

(a) that it contains a radioactive source; and

(b) whether any shutter or cover is open or shut.

The marking should include a radiation warning sign (see appendix 7 *Warning signs and notices*).

15.16 In the case of equipment where the radiation beam is transmitted through the patient, the detection system should fully intercept the emerging beam and be effective as a beam stop. Preferably, the source housing and the detection system should be mechanically linked so that this condition is met whenever the shutter is open.

15.17 A different relationship between source and patient may be needed occasionally, e.g. a technique in which the patient grips a ^{252}Cf source within a moderator for neutron activation studies. Here, the source should be in a shielded housing with an aperture large enough only for the hand, and there should be a lockable cover for the aperture when the equipment is not in use.

15.18 If the equipment is to be used exclusively for examining pathological specimens, e.g., bones, it should be completely enclosed and provided with a loading drawer mechanism, or, if it is a cabinet, it should have an interlocked door and external shutter mechanism.

Operating procedures

15.19 If the source, its bonding, or its immediate containment will be in contact with the body or need to be placed in the mouth or other cavity, then IR(ME)R (IR(ME)R 2017, IR(ME)R(NI) 2018) will apply and Employer & Practitioner licenses will be required. These Regulations also apply to neutron activation analysis (see paragraphs 10.16–10.31, *Roles and responsibilities of duty holders in nuclear medicine*).

15.20 The patient should be properly positioned before the shutter is opened, the cover removed, or the source brought into position, and should not be exposed to the radiation beam for longer than is necessary to carry out the diagnostic test.

15.21 No one other than the patient should be exposed to the radiation beam. Particular care should be taken that the fingers of the staff are neither exposed to the beam nor placed close to a fluorescence or back-scattering device. This applies not only during a diagnostic test but also during any calibration or experimental work when no patient is present.

15.22 If the shutter or cover does not operate automatically, the aperture should be closed immediately after the test. Complete closure of the shutter should be confirmed using an appropriate radiation monitor after each procedure.

15.23 When the equipment is not in use, it should be made safe, e.g. by means of a secure locking mechanism or kept in a suitably shielded and locked room or store.

Source handling

15.24 Where it is necessary to remove or replace sources, e.g. for storage purposes, they should never be handled directly. A handling tool should be used. It may be necessary to use shielded handling tools where dose rates warrant it. Specialist installers are required for the loading and unloading of high-activity sealed sources (HASS).

15.25 Storage and movement of sources are dealt with in chapter 18, *Keeping, accounting for, and moving radioactive substances (including transport)*.

Localisation of tissue with Iodine-125 seeds

15.26 Radioactive ^{125}I seeds may be used for localisation of tissue prior to surgery. The requirements of IR(ME)R (IR(ME)R 2017, IR(ME)R(NI) 2018) will apply (see 15.5 and 15.19). Typically, this is used as an alternative to guidewire localisation in breast surgery. The tissues most commonly localised are breast tumours, but may also include axillary lymph nodes in the neoadjuvant chemotherapy setting. Seeds are typically inserted 7–14 days prior to surgery and excised using a sentinel node probe. Seed activity should be less than 10 MBq on insertion. Multiple seeds may be inserted for multi-site tumours or large tumours.

15.27 The clinical applications defined in the manufacturer's marketing authorisation must be reviewed by the Practitioner and Referrers. Typically, ^{125}I seeds are only licenced for use in prostate brachytherapy but not for localisation prior to surgery. The detail for this application is described by Goudreau *et al* (2015). The seeds should be CE-marked. The period of time that the marketing authorisation states that the seed is sterile must not be breached.

15.28 The main radiation safety concern is source security. There are multiple manipulations of the seed at insertion, excision, and dissection. Contingency plans for the loss of radioactive material are covered in chapters 18, *Keeping, accounting for, and moving radioactive substances (including transport)*, and chapter 20, *Contingency planning and emergency procedures—radioactive substances*, and a full risk assessment must be performed to assess and mitigate the risks from any foreseeable contingencies. A check sheet to record the presence of seeds at the expected location and the exclusion of seeds from where they originated is recommended. The presence or absence may be determined by x-ray or a radiation detector at points where there is a manipulation of the seed or a transfer of responsibility to another member of staff.

15.29 In-house leak testing and activity measurement of ^{125}I seed sources are not practicable, therefore, the manufacturer's certificate must be inspected to ensure it has been tested and passed and is the anticipated activity.

15.30 The extremity dose to staff loading seeds into the needle, inserting the loaded needle into the patient, excising the specimen, and dissecting the specimen is minimal, provided that the seed is not handled directly. Forceps or something similar should be used where the seed is not shielded by tissue to keep exposure as low as reasonably practicable.

15.31 Contact restrictions to minimise the radiation exposure to family and friends of patients with a ^{125}I seed inserted are unlikely to be required unless multiple seeds are inserted. No more than four breast seeds or one seed per nodal bed are likely to be required clinically.

15.32 Appropriate radiation detection probes must be selected for use in the theatre and other units handling the seeds. Within theatres, the energy discrimination between ^{125}I seeds and ^{99m}Tc radiopharmaceuticals for sentinel node biopsies should be evaluated by an MPE. Where energy discrimination is not required, then a collimator fitted to a routine contamination monitor will improve seed localisation.

15.33 The radiation safety training of staff falls into three categories.

(a) IR(ME)R (IR(ME)R 2017, IR(ME)R(NI) 2018) training is required for those operators inserting and excising the seed, as both are influencing the dose to the patient. This should comprise theory consistent with the syllabus in IR(ME)R (IR(ME)R 2017, IR(ME)R(NI) 2018) as well as supervised practice.

(b) Initial and continuing theoretical and practical training is required for key members of staff under IR(ME)R (IR(ME)R 2017, IR(ME)R(NI) 2018). Continuing education may include the support of a mentoring site where the service is established and regular team meetings for feedback, service development, and incident actions.

(c) Radiation protection training, briefly covering IRR (IRR 2017, IRR(NI) 2017), IR(ME)R (IR(ME)R 2017, IR(ME)R(NI) 2018) and RSR (RSA 1993, EPR 2016, EA(S)R 2018), is recommended for all staff in a supporting role of the ^{125}I seeds service where they take responsibility for the seed.

References

BSI 2020 BS ISO *9978:2020* Radiation Protection: Sealed Sources . *Leakage Test Methods* (London: BSI)

EA 2018 *How to Comply With Your EPR RSR Environmental Permit—Sealed Sources. 2.0* (Bristol: Environment Agency) https://gov.uk/government/publications/rsr-environmental-permit-how-to-compy-sealed-sources (accessed 20 April 2024)

EA(S)R 2018 *The Environmental Authorisations (Scotland) Regulations 2018* http://legislation.gov.uk/ssi/2018/219 (accessed 20 April 2024)

EPR 2016 *Environmental Permitting (England and Wales) Regulations* https://legislation.gov.uk/uksi/2016/1154 (accessed 20 April 2024)

Goudreau S H, Joseph J P and Seiler S J 2015 Preoperative radioactive seed localization for nonpalpable breast lesions: technique, pitfalls, and solutions *RadioGraphics* **35** 1319–34

HSE 2018 L121 Work with ionising radiation Ionising Radiations Regulations 2017 *Approved Code of Practice and Guidance* 2nd edn (Norwich: Health and Safety Executive) https://hse.gov.uk/pubns/books/l121.htm (accessed 20 April 2024)

IR(ME)R 2017 *The Ionising Radiation (Medical Exposure) Regulations* www.legislation.gov.uk/uksi/2017/1322 (accessed 20 April 2024)

IR(ME)R(NI) 2018 *The Ionising Radiation (Medical Exposure) Regulations (Northern Ireland)* https://legislation.gov.uk/nisr/2018/17 (accessed 20 April 2024)

IRR 2017 *The Ionising Radiations Regulations* https://legislation.gov.uk/uksi/2017/1075 (accessed 20 April 2024)

IRR(NI) 2017 *The Ionising Radiations Regulations (Northern Ireland)* http://legislation.gov.uk/nisr/2017/229 (accessed 20 April 2024)

ISO 2012 *2919:2012 Radiological Protection—Sealed Radioactive Sources—General Requirements and Classification* (Geneva: International Organization for Standardization)

RSA 1993 *Radioactive Substances Act 1993* http://legislation.gov.uk/ukpga/1993/12 (accessed 20 April 2024)

IOP Publishing

Medical and Dental Guidance Notes (Second Edition)

A good practice guide on all aspects of ionising radiation protection in the clinical environment: IPEM Report 113

John Saunderson, Mohamed Metwaly, William Mairs, Philip Mayles, Lisa Rowley and Mark Worrall

Chapter 16

Patients leaving hospital after administration of radioactive substances

Scope

16.1 This chapter gives advice on the conditions under which patients may be allowed to leave the hospital following the administration of open or sealed radioactive sources. Its purpose is to minimise the risk of radioactive contamination and the risk from external radiation to family and friends (including those who may be classed as Carers and Comforters under IR(ME)R (IR(ME)R 2017, IR(ME)R(NI) 2018), the general public, or non-radiation workers who may be classed as members of the public for radiation protection purposes).

Principles

16.2 Care should be taken to ensure that patients only leave the hospital if they are unlikely to cause a significant radiation or contamination hazard to other people. Appropriate scenarios and dose constraints should be used in planning discharge dates (see paragraphs 16.6–16.9, *Risk assessment*). Arrangements should be in place to restrict the exposure of those who come into contact with the discharged patient as far as is reasonably practicable.

16.3 Arrangements for giving advice on the radiation protection of patients and their personal contacts after the patient has left the hospital must be documented. For Carers and Comforters this will be detailed in the IR(ME)R Employer's Procedure (Schedule 2 (i) (IR(ME)R 2017, IR(ME)R(NI) 2018). Responsibility lies with the Employer who may delegate it to the medical physics expert (MPE) to oversee such arrangements. The form and content of

doi:10.1088/978-0-7503-2332-1ch16

the advice should be agreed upon with the MPE, and the task of communicating it to the patient may be delegated as detailed in the employer's procedure. Personal contacts who are not Carers and Comforters are members of the public and subject to public dose limits (IRR 2017, IRR(NI) 2017). As such, the radiation protection adviser (RPA) must be consulted on the radiation protection advice provided to those contacts.

16.4 Advice should be based on generic and individual risk assessments and should be developed for each clinical procedure. Specific details given here are based on IAEA guidance entitled *Release of Patients After Radionuclide Therapy* (IAEA 2009).

Potential hazards

16.5 Radioactive substances present an external irradiation hazard and, if open, a contamination hazard as well. The risks posed to others by discharging a radioactive patient from the hospital will depend on the personal circumstances of the patient as well as the type, activity, and biokinetics of the radioactive substance administered.

Risk assessment

16.6 Potential risks to other persons should be assessed by reviewing published data, existing risk assessments, or in-house measurements that have been validated by the MPE and/or RPA. Calculation of the effective dose to other persons from first principles, making realistic or, if that is not possible, conservative assumptions about the behaviour of the patient and other persons, may be necessary if no other suitable data are available.

16.7 The risk of contamination and any hazards arising from the administration must also be assessed. For example, the possibility of a patient causing contamination due to incontinence (urinal or faecal), vomiting or the use of stoma bags should be considered. The risk assessment must include any ongoing medical treatments if they necessitate the storage, handling or disposal of radioactive material (e.g. frequent blood sampling to monitor blood sugars or anti-coagulant therapy; see paragraph 10.44). Risk assessments should also cover transport (see chapter 18, *Keeping, accounting for and moving radioactive substances (including transport)*) and the appropriate regulations must be followed (CDG 2009, CDG(NI) 2010, IRR 2017, IRR(NI) 2017, DfT 2020).

16.8 Any advice to reduce radiation exposure resulting from the risk assessment should be discussed with the patient (and/or the patient's representative as appropriate) well before the radioactive substance is administered (see also paragraphs 10.25, 10.81, 10.87 and 13.12). It is imperative to check with the patient or the patient's representative that the assumptions on which the advice is based are reasonable in the individual circumstances. Any problems should be discussed with the IR(ME)R Practitioner and the MPE for further dose assessment, or with the person specified in the Employer's procedure. The

Referrer must also be involved if postponement of the procedure is thought necessary.

16.9 There must be an Employer's procedure concerning the provision of any required written or verbal advice on radiation protection to patients (see paragraphs 2.155–2.168, *Information and instructions for nuclear medicine patients* and *Carers and comforters*, and also 10.87), relatives, or anyone, within reason, likely to be exposed as a result of the clinical procedure undertaken.

Departure from hospital and conduct at home

16.10 The radiation doses to members of the patient's household and to other members of the public that result from the discharge of a radioactive patient from the hospital must be kept as low as reasonably practicable (ALARP). Estimates of potential doses to other people should be compared with a relevant dose constraint (e.g. see table 16.1) set by the Employer. If the potential doses are significant compared to the constraint, then the advice (as indicated in paragraphs 16.6–16.9, *Risk assessment*) must reflect this, with further input provided from the RPA. Consideration must also be given to delaying discharge if doses cannot be reduced. They must not exceed the dose limit (where applicable) indicated in table 16.1 (and see paragraph 13.31). Consideration should also be given to the need for repeat exposures within a calendar year (e.g. multiple cycles of molecular radiotherapy).

16.11 If someone caring for the patient is pregnant, the risks that they, and the unborn child, may incur as a result of exposure to ionising radiation must be explained. Advice should be given to limit the dose to the unborn child to below 1 mSv over the term of the pregnancy from all routes of exposure, i.e. external and contamination hazards.

16.12 Carers and Comforters (as defined in IR(ME)R (IR(ME)R 2017, IR(ME)R (NI) 2018)), see paragraphs 2.156–2.168, *Carers and comforters*) must be capable of giving informed consent to the person outlined in the employer's procedures for any exposure they may incur. This should be 'knowingly and willingly' and therefore they should be provided with written instructions that specify how doses to them may be restricted as far as reasonably practicable and set out the risks associated with ionising radiation.

16.13 Regulation 6(5) of IR(ME)R requires the employer to establish dose constraints for Carers and Comforters. Although a dose constraint of 5 mSv per procedure should be appropriate for most cases, it is intended that the Employer set a value according to local circumstances. In some situations it may even be necessary to adopt a higher dose constraint, but this should only be done on a case-by-case basis with full consent of the Carer and Comforter following a one-to-one discussion from an appropriately trained employee, such as a physicist, radiographer or technologist.

Table 16.1. Recommended dose constraints and legal dose limits for classes of persons exposed as a result of discharge of radioactive patients from hospital (IR(ME)R 2017, IRR 2017).

	Effective dose constraint per procedure	Effective dose limit
Carers and comforters	5 mSv[a]	None
Other members of the household	1 mSv	5 mSv in 5 consecutive years
Members of the general public	0.3 mSv	1 mSv year^{-1}

[a] From IAEA safety guide No. SSG-46 (section 2.49) (IAEA 2018).

16.14 If, on the basis of the assumptions made about the behaviour of the patient and other persons (paragraphs 16.6–16.9, *Risk assessment*), it is estimated that the doses received by persons with whom the patient comes into contact may exceed the recommended dose constraints, it may be necessary to keep the patient in the hospital. Difficulties are most likely if the patient is the sole or main carer of a small child (Barrington *et al* 1999) or the child acts as the patient's carer (see paragraphs 2.156–2.168, *Carers and comforters*). In this case, the following options are also available;

(a) delay the treatment until the child is older;
(b) use of the 5 mSv in 5 consecutive years dose limit; or
(c) send the child to stay with relatives or friends for a suitable period following the administration and consider admitting the patient to the hospital or helping to facilitate community care.

The likelihood of repeat administrations should be discussed with the IR(ME)R Practitioner and MPE and systems should be put in place to detect such events and ensure that annual dose limits are not thereby exceeded.

16.15 Where it is necessary to restrict the extent of contact between the patient and others, verbal and written instructions should be given and discussed with the patient or their representative (see paragraph 16.8). An instruction card, to be carried while restrictions apply, giving key information and summarising the advice, is a suitable means of reminding patients and alerting those who may deal with them in the event of unforeseen illness or death (see chapter 17, *Precautions after the death of a patient to whom radioactive substances have been administered*). The card, or any other written information provided (see paragraph 16.22 for diagnostic administrations), may also be used to identify patients who may set off radiation detectors at large public events or border crossings. Timescales for this aspect will generally significantly exceed the standard restriction times. The card may need to include the following information:

(a) name and address of the patient;
(b) name, address and telephone number of the administering hospital;

(c) contact name in case of difficulty for radiation-related advice;
(d) type and activity of radionuclide administered;
(e) date of administration;
(f) restrictions to be followed, especially concerning contact with young children; and
(g) the period of time for which the restrictions apply.

Discharging patients to outside organisation etc

16.16 For patients administered with radioactive substances and returning to care or nursing homes or other outside organisations where the patient is resident in a workplace, such as a prison, the Ionising Radiations Regulations (IRR 2017, IRR(NI) 2017) require the organisation to register with the HSE or HSENI. See chapter 10, *Diagnostic uses of open radioactive substances*, for registration activities and chapter 1 starting at paragraph 1.127 for *Cooperation between employers and others*. The organisation must also consult/appoint an RPA. Sufficient cooperation must take place between the discharging unit and the outside organisation to fully comply with IRR before the patient is transferred. Registration requirements include a radiation risk assessment and a measurement/estimate of the employee's exposure. Both of these should be covered in the administering nuclear medicine radiation risk assessment. Training, information and instruction can be provided to the care home by a nursing leaflet provided by the administering nuclear medicine department. The risk assessment is likely to show that contingency plans, designation of areas and local rules will not be required.

16.17 The administering hospital should provide the organisation caring for the patient with:

1. A cover letter.
2. A care leaflet, customised to the patient (i.e. activity administered, radionuclide, considerations on dealing with radioactive waste, contact times etc).
3. Appropriate radiation risk assessment.
4. Guide to registering with the HSE.
5. Details of appointing a suitable RPA (including a template appointment letter for the episode or a permanent contract, if multiple patients are expected and the administering hospital is taking on the RPA role).

16.18 The administering department must also ensure compliance with data protection, depending on who is organising the patient's medical care, when contacting outside organisations. For example, information can be provided directly to staff if the organisation has a duty of care to the patient, e.g. care home or another hospital. However, if the patient lives in a residence and is arranging their own care, e.g. warden controlled residence, information must not be sent directly to the home, but can be supplied to the patient to pass on to staff who may be involved in their care.

16.19 Suitable arrangements for radioactive waste disposal (see chapter 19, *Accumulation and disposal of radioactive waste*) should be considered alongside the need for the receiving Employer to appoint a radioactive waste adviser (RWA) if required, however, Radioactive Substances Regulations (RSR) (RSA 1993, EPR 2016, EA(S)R 2018) exemptions and general binding rules are likely to cover the majority of radioactive waste arising from the patient once they have been discharged from hospital. Advice from the RWA should be sought if existing permits are held by the receiving Employer under RSR.

16.20 After the administration of a radioactive substance, a patient's condition may deteriorate requiring admission to a care establishment that is not familiar with their administration and so will not have suitably trained staff or relevant procedures. Ideally, the establishment will be made aware that the patient is radioactive by the patient themselves or their partner/carer by presenting the information card issued at the time of treatment. The care establishment should contact the administering hospital to get advice on patient management.

16.21 A record of estimated aqueous disposals must be kept for future inspection by the regulator or retrospective calculation of discharges.

Diagnostic procedures with open sources

16.22 In general, the quantities of radioactive material currently administered for diagnostic procedures (as in the ARSAC notes for guidance (ARSAC 2024)) do not necessitate any special precautions or restrictions to be placed on the patient for one-off exposures within a calendar year. Known exceptions fall into the following categories:

(a) patients who provide the majority of close care to babies and have been administered with greater than the activities specified for the following (Harding *et al* 1985, Mountford and Coakley 1989, O'Doherty *et al* 1993, Greaves and Tindale 1999):
   i. 10 MBq ^{111}In leucocyte;
   ii. 150 MBq ^{201}Tl chloride;
   iii. 800 MBq ^{99m}Tc myocardial perfusion agent
(b) breastfeeding mothers (ARSAC 2024);
(c) patients administered with 30 MBq or more of ^{131}I, which should be treated as a therapeutic administration for the purpose of radiation protection;
(d) patients administered with 20 MBq or more of ^{111}In leucocyte or 220MBq or more of ^{111}In octreotide (Singleton *et al* 2003).

Restrictions may also be necessary after administration of positron-emitting radionuclides (see paragraphs 16.25–16.28, *Patient restrictions following administration of positron-emitting radionuclides*). If the patient works with radioactive materials, or if his or her work is radiation-sensitive, the caution in paragraph 16.43 may be relevant.

16.23 Repeat exposures in a calendar year for some procedures, such as MUGA studies where the estimated maximum dose to an adult partner is 0.6 mSv for an 800 MBq administration (Singleton *et al* 2003), or ^{111}In studies as in paragraph 16.22(d), may require simple advice, such as a note in the appointment letter to avoid close contact (< 1 m) with family members until the following morning and a verbal reminder on exit from the department.

16.24 Sensitive dose monitoring devices are commonly installed in airports, border crossings, rail terminals and at large events. Patients may be advised to carry a separate letter describing details of their administration, or their appointment letter (ARSAC 2007). A sign in the waiting areas asking patients to ask staff for more information, if applicable, should be sufficient. This should be carried for 1 week for ^{99m}Tc, and at least three months for longer-lived diagnostic tracers (ARSAC 2007).

Patient restrictions following administration of positron-emitting radionuclides

16.25 Information on patient restrictions should be provided to the patient prior to administration to reduce the radiation dose to staff. Children and pregnant persons should not accompany the patient to the PET centre (Cronin *et al* 1999), and this should be explained to the patient prior to the appointment.

16.26 For ^{18}F administrations under 400 MBq, the dose to a member of the public 90 min post injection from the external dose rate at close contact will not exceed any dose limits for a one-off exposure (e.g. public transport). However, it is advised that close contact (< 1 m) with pregnant people and children be avoided for 4 h post-administration and an advice slip is provided to the patient as such. This should be doubled if the patient has a catheter bag, due to the risk of contamination. For patients undergoing repeat exposures in a year, these restrictions should be applied to adults as well.

16.27 For other tracers with a half-life greater than 120 min, such as ^{64}Cu, advice may be required to ensure public doses are within a suitable constraint (Pereira *et al* 2015). For tracers with a half-life below 1 h, restrictions are not required.

16.28 There is no restriction on conception from a radiation point of view following an ^{18}F-FDG-PET scan for males or females (ARSAC 2024).

Pregnant or breastfeeding patients

16.29 The advice given in the ARSAC notes for guidance (ARSAC 2024) regarding performing clinical procedures on pregnant or breastfeeding patients should be followed.

16.30 There must be an Employer's procedure for making enquiries of individuals of childbearing potential to establish whether the individual is or may be pregnant or breastfeeding. Only those investigations that are imperative should be conducted during pregnancy (ARSAC 2024).

16.31 Specific instructions must be given to breastfeeding patients to minimise the radiation dose to the infant through the ingestion of milk, with additional consideration given to the external dose rate from the mother. These should be discussed with the mother as soon as practicable and prior to the test. Interruption times are listed for commonly used radiopharmaceuticals in the ARSAC guidance notes (ARSAC 2024) and include consideration of the external dose from the mother. A risk assessment must be made for procedures where a specific recommendation is not available, employing the dose limit of 1 mSv for the breastfeeding baby. Measurements of breast milk following administration may be used to determine if and when the mother may resume feeding in these cases, including consideration of external dose, as described in the ARSAC guidance notes (ARSAC 2024).

16.32 Consideration should also be given to repeat exposures while the patient is pregnant or breastfeeding. Reduced activity with longer imaging times may be considered for this group of patients, who are generally fit enough to tolerate increased scanning times.

16.33 On occasion, the patient may discover at some stage after administration that they are pregnant. For administrations involving radioiodine, it should be noted that radioiodine freely crosses the placenta. It is important to determine the point of gestation at the time of administration. The foetal thyroid may be present after week 8 and will accumulate radioiodine which crosses the placenta. Administration of radioiodine at or later than that time can lead to very high foetal thyroid doses as there is such a small amount of thyroid tissue present. Equally, administration before that time can lead to relatively smaller doses of the foetal thyroid, as there is significantly less free iodine in the mother at the time the foetal thyroid is formed. Stable potassium iodide (KI) could be given to partially block both the maternal and foetal thyroids, mitigating the large thyroid dose (IAEA 2009), ideally within 12 h of administration of radioiodine.

16.34 Table 16.2 shows the foetal thyroid dose for administration to the mother at various timescales in relation to the gestational age for treatment of thyrotoxicosis (Pauwels *et al* 1999, ICRP 2001). Although such high doses from ^{131}I administration for thyrotoxicosis at about 12 weeks of gestation or greater would not warrant consideration of termination, they are at a deterministic level that could lead to the destruction of the foetal thyroid tissue and result in a hypothyroid baby. Further information can be found online (HPS no date, IAEA no date)

Table 16.2. Foetal thyroid dose for 600 MBq I-131 administered at different times (in weeks) relative to conception.

Weeks from conception	Foetal thyroid dose
−26	<0.0006 mSv
0	47 mSv
5	140 mSv
10	1.9 Sv
15	140 Sv
25	410 Sv
35	660 Sv

Breastfeeding and therapeutic procedures

16.35 Breastfeeding is contra-indicated for all therapeutic radiopharmaceuticals that are expressed in breast milk (ARSAC 2024). If the patient is breastfeeding, then they should be given advice on the appropriate cessation of breastfeeding. This advice, obtained from a lactation consultant, should allow the mother's breast status to return to normal prior to any therapeutic molecular radiotherapy administration (ARSAC 2024). This aspect is important, since the breast dose to the mother can otherwise be substantially increased. The typical breast dose (no breastfeeding) following 600 MBq ^{131}I administration is 48 mSv. If the patient is still lactating at normal levels, the increased dose to the breast from this can be 1.25 Sv, about 25 times greater (ICRP 2004, Thomson and Lewis 2015). This must be checked and planned well in advance of administration.

Therapeutic procedures with open sources

16.36 A radiation risk assessment must be carried out prior to any therapeutic administration, as described in paragraphs 16.6–16.9, *Risk assessment*. For many therapeutic administrations, this may be based directly on a generic risk assessment carried out for the 'typical' patient. However, each patient must be assessed to ensure that the assumptions made are applicable to the individual concerned.

16.37 The radionuclides most commonly used for molecular radiotherapy are shown in table 16.3 (IPEM 2014).

Pregnancy after therapeutic procedures

16.38 There are guidelines available for the avoidance of pregnancy following molecular radiotherapy from the IAEA (IAEA 2009) and ARSAC (ARSAC 2024), shown in table 16.4. Any advice on the avoidance of pregnancy must be agreed with the IR(ME)R Practitioner and the referring clinical team, to take account of both radiation risk and the future clinical management of the

Table 16.3. Commonly used Radionuclides for molecular radiotherapy.

Radionuclide	Radiation type	Physical half-life	Max (mean) particulate energy	Max (mean) particulate range in tissue
^{32}P	Beta	14.3 days	1.71 (0.70) MeV	8.0 (3.0) mm
^{89}Sr	Beta	50.5 days	1.46 (0.58) MeV	7.0 (2.4) mm
^{90}Y	Beta	2.7 days	2.27 (0.94) MeV	11.9 (5.3) mm
^{111}In	Auger, gamma	2.8 days	25 keV	10 μm
^{131}I	Beta, gamma	8.0 days	0.61 (0.19) MeV	2.4 (0.8) mm
^{153}Sm	Beta, gamma	1.9 days	0.81 (0.23) MeV	0.6 mm
^{177}Lu	Beta, gamma	6.7 days	0.50 (0.15) MeV	2.0 (0.5) mm
^{186}Re	Beta, gamma	3.7 days	1.07 (0.35) MeV	1.1 mm
^{223}Ra	Alpha (95%), beta (<4%), gamma (<2%)	11.4 days	(5.78) MeV	<100 μm

Table 16.4. Time to avoid pregnancy following treatment based on IAEA (IAEA, 2009), ARSAC (ARSAC 2024) and advice from the NET patient foundation (NET Patient Foundation, 2019). The life cycle of a sperm cell is under 4 months (ARSAC 2024).

Radionuclide and form	Disease treated	Activity (upper limit) MBq	Pregnancy avoidance period (months)	Fathering a child (months)
^{131}I Iodine	Hyperthyroidism	800	6 to 12	4
^{131}I Iodine	Thyroid cancer	6000	6 to 12	4
^{131}I mIBG	Neuroendocrine tumours	7500	3	4
^{32}P phosphate	Myeloprliferative disease	200	3	4
^{89}Sr chloride	Bone metastases	150	24	4
^{90}Y colloid	Synovectomy	400	0	*
^{90}Y colloid	Malignancies	4000	1	*
^{198}Au colloid	Malignancies	10 000	2	*
^{169}Er	Synovectomy	400	0	*
^{153}Sm-colloid	Bone metastases	400	0	*
$^{177}Lu/^{90}Y$ peptide	NET	7400	6	*

patient, i.e. the need for repeat molecular radiotherapy, chemotherapy etc. In cases where possible pregnancy cannot be excluded, it is advised that a pregnancy test be performed before any molecular radiotherapy (mRT) administration, with the results recorded in the patient notes. A negative pregnancy test does not necessarily exclude pregnancy, depending on timing, and clear guidelines will be required on criteria, agreed by the Practitioner.

Travel

16.39 There can be several travel situations that require restriction advice for patients. The patient needs to know if they can travel home by public transport after administration. Also, after what time could they consider further travel such as holiday flights etc. These situations tend to be 'one-off' events, where the principal dose consideration is to an adjacent member of the public. Therefore, the dose constraint value of 0.3 mSv should apply. This would also cover the situation of the adjacent passenger being pregnant. If the travel is with a partner or family member, then this additional dose should be considered in conjunction with the dose received at home.

16.40 Travel restriction times have been calculated in paragraphs 16.54–16.57 for thyrotoxicosis patients, based on their dose rates. These might therefore be applied to patients having undergone ^{131}I ablation therapy in some circumstances, i.e. on discharge, but will not be accurate for all scenarios.

16.41 As highlighted in paragraph 16.24, sensitive dose monitoring devices are now commonly installed in airports, ports, rail terminals etc. Restriction periods for therapeutic administrations are normally only for periods of a few weeks, however, it is important that the patient is aware that there continues to be a very small remnant of longer-lived radionuclides in them for periods of several months. For example, even at four months there may still be approximately 1 kBq of ^{131}I present after thyrotoxicosis treatment, and it is possible that this could trigger an alarm.

Return to work following therapeutic doses of radioactive substances

16.42 Restrictions may be necessary for patients returning to work if they are likely to spend substantial periods at distances of less than 2 m from other workers or if their work involves close contact with an individual, particularly a child or pregnant person, for more than 15 min in a day (Radiation Protection Committee of the BIR 1999). Any restrictions provided for the patient should take this into account. The risk of contamination for those handling and preparing food for others must also be considered. Patient-specific modelling could be used, such as the methodology outlined by Cormack and Shearer (1998).

16.43 If radiation-sensitive work is undertaken at the patient's place of work or the patient works with radioactive substances, practical difficulties may arise. Although the patient's Employer may have procedures in place to detect the situation, the patient should be advised to inform their employer in order to prevent inconvenience and possible alarm and to discuss whether they must remain away from work.

Iodine-131 administered for hyperthyroidism

16.44 The restrictions given in tables 16.5–16.11 are based on recommendations that make assumptions about the normal behaviours of patients and those who live with them and compliance with the restrictions outlined. If the validity of the assumptions is in doubt for any particular patient, the original referenced texts should be consulted or individual assessments made. The main general assumptions are that the patient:
(a) poses no significant risk of causing a contamination hazard;
(b) is being discharged to a private dwelling (not a nursing home, hospital or other workplace);
(c) will not be in prolonged contact with nurses or other persons employed to care for them at the private dwelling;
(d) is only given one administration of radioiodine in a year;
(e) is able to make suitable arrangements for the alternative care of children under three years of age if they are the primary carer.

16.45 Under these conditions, the restrictions set out are sufficient to control the external dose hazard in the majority of cases of treatment of thyrotoxicosis with up to 800 MBq of ^{131}I. Periods of restriction for other administered activities can generally be estimated by linear interpolation between the tabulated range for each row (although the restriction calculations are not strictly linear processes). However, it is important to note that restriction times should not be estimated by extrapolation beyond the tabulated values.

16.46 Restrictions for the patient's partner and family members within the home generally relate to a 1 mSv per year dose constraint for members of the public, in agreement with table 16.1.

16.47 In exceptional cases, it may be necessary to consider the use of the 5 mSv in five years dose limit for members of the public. However, since a repeat

Table 16.5. Restriction Period (days) for the patient and partner to refrain from sleeping together for a total dose constraint of 5 mSv or 3 mSv or 1 mSv. An assumption of average restrictions during the daytime is used, with a total daytime dose figure of 0.5 mSv/600 MBq assumed (Thomson *et al* 1996). The restrictions periods are not linear for interpolation in relation to dose, and so restriction periods are also given for 3 mSv. Night time dose figure is based on measured data (Thomson *et al* 1993, Thomson and Harding 1995) and uses the 95 percentile figure for the total night time dose of 5.4 mSv/600 MBq that would result at night without any sleeping restrictions.

	Administered ^{131}I activity			
	200 MBq	400 MBq	600 MBq	800 MBq
5 mSv	0	0	2	5
3 mSv	0	4	8	10
1 mSv	1	15	22	28

treatment cannot be predicted, it is not appropriate to apply the dose limit of 5 mSv for one episode. Advice might be based on dose values between 1–3 mSv, for example. For adults providing care to the patient in the family home, where restrictions to reduce the dose cannot be complied with, the designation and consent of an individual as a 'carer and comforter' will likely be applicable (see paragraphs 2.156–2.168, *Carers and comforters*, and 16.12–16.13).

16.48 It should be noted that the general assumptions normally used to calculate the dose to others involve the measured dose rates at a distance from the patient. In practice, this effectively provides the individual's entrance surface dose. In cases of lateral exposure to ^{131}I, e.g. sitting together, the true effective dose to the person is one half the lateral entrance dose (ICRP 1996). This factor has been incorporated into the travel restrictions only, where this aspect is generally true.

Restriction times for the patients sharing a bed

16.49 These restriction times given in table 16.5 are calculated for the patient and their adult partner normally being together seven nights a week and sharing a double bed. There is also an assumption that there will be standard contact restrictions observed during the daytime to reduce the daytime exposure values. Such daytime exposure restrictions will have separate restriction times. It should be noted that twin beds with a separation of about 1 m (i.e. about 2 m between bed centres) can substantially reduce the night time dose.

Restriction times for other members of the family

16.50 There are derived models for close contact that assume the patient is normally the main carer for a child in three age groups; under 2 years, 2–5 and 5–11 years (Rose *et al* 1990, O'Doherty *et al* 1993). The dose value for the 0.1 m distance accounts for more than 92% of the total calculated dose for all three models, and so is a critical component for any dose reduction restrictions. Please see table 16.6.

16.51 Restrictions for children >11 years and for adults can generally be more easily applied and observed. The restriction periods are given for an assumed typical daily contact model of 0.5 h at 0.3 m, 1 h at 0.5 m and 4 h at 1 m.

16.52 However, it is important to note that the restrictions in table 16.6 are for the 1 mSv dose limit, not the 0.3 mSv dose constraint. Also, these restriction calculations for children were made using median dose rate values measured by O'Doherty *et al* (1993). This does indicate the importance of conveying to patients that the restriction periods for children in table 16.6 are the minimum allowed and are important to observe. If a patient is the sole carer for a child under 2 years, the options in paragraph 16.14 should be considered to ensure the dose to the child is under 1 mSv.

Table 16.6. Restriction periods (days) for close contact with children in three age groups and other family members. These are intended to ensure doses are below the 1 mSv dose limit.

	Administered ^{131}I activity			
	200 MBq	400 MBq	600 MBq	800 MBq
<2 years old	15	21	25	27
2–5 years old	11	16	20	22
5–11 years old	5	11	14	16
Adolescents and adults	0	6	10	12

16.53 Contamination in the home has also been assessed by direct measurement (Barrington *et al* 2008), with the effective dose calculated for 93% of children in the study being under 0.2 mSv. If good hygiene is followed, contamination in the home is therefore not considered an additional radiation hazard and does not form any component of the restriction periods above. However, it is noted that these results are for thyrotoxicosis patients only and may not apply to ablation patients.

Immediate travel from the hospital after administration

16.54 Maximum travel times immediately following administration are listed in table 16.7. This is based on direct measurements made immediately after ^{131}I administration and relates to the lateral entrance dose given to someone sitting next to the patient (Thomson and Harding 1995, Gunasekara *et al* 1996).

16.55 It should be stressed to the patient that if travelling by car, then doses are substantially reduced if the patient sits in the rear of the vehicle, diagonally opposite the driver, with no other passengers. This increased distance, typically from 40 cm sitting adjacent to 1 m sitting in the rear, can reduce doses by approximately a factor of 6. If a driver is provided (employed or volunteer) who may drive a number of radioactive patients in the same year then a risk assessment must be undertaken to determine what precautions are appropriate.

Travel (or other) restrictions for future 'one-off' journeys or events

16.56 For travel as a one-off journey, e.g. holiday, air travel etc, the dose for an adjacent member of the public is considered in table 16.8. This may also be used for advice on the restriction period for 'one off' attendance at events where the patient may sit beside a member of the public for a long period e.g. at cinemas and theatres.

Table 16.7. Maximum travel times immediately after administration for a single journey to limit the radiation dose to an adjacent passenger to 0.3 mSv dose constraint and 1 mSv dose limit. Standard seating is assumed—closer contact (e.g. crowded tube) may shorten permitted travel times.

	Administered ^{131}I activity			
	200 MBq	400 MBq	600 MBq	800 MBq
Travel time for 0.3 mSv	2.8 h	1.4 h	1 h	0.8 h
Travel time for 1 mSv	9.6 h	4.8 h	3 h	2.4 h

Table 16.8. Restriction (days) before a single travel journey or event of the following duration, given that the restrictions are given for the 0.3 mSv dose constraint, with the value for the 1 mSv dose limit in brackets.

	Administered activity of ^{131}I			
Travel time	200 MBq	400 MBq	600 MBq	800 MBq
1 h	0 (0)	0 (0)	0 (0)	1 (0)
2 h	0 (0)	1 (0)	3 (0)	5 (0)
3 h	0 (0)	3 (0)	6 (0)	9 (1)
6 h	3 (0)	9 (1)	12 (2)	15 (4)
12 h	5 (0)	15 (4)	18 (8)	21 (10)

Repeated journey with the same person

16.57 Another travel restriction to note is if the patient travels routinely with the same person each day, e.g. to work. This requires the cumulative dose of such repeated travel to be considered. See table 16.9. Again, if an arrangement of sitting diagonally opposite is possible (see paragraph 16.55), then the restriction periods would be considerably less.

Restrictions for work

16.58 Some restriction period before returning to work is very likely. This will often need to be individually assessed for the patient's work arrangement and circumstances. Therefore, consideration of this should happen prior to the patient attending for the administration. The restriction should generally be considered to reduce the dose to fellow employees to a dose constraint of 0.3 mSv. This will also cover the possible pregnancy of any employee.

16.59 Although every patient's circumstance will be different, some example situations are presented in the following sections. The most important factor to ascertain is the general distance and time model that the patient has with other employees. The importance of the distance factor can be seen from

Table 16.9. Restriction (days) for a total journey time per day (hours) routinely travelled on five days a week sitting adjacent to the same person on each journey. Restriction values (days) are shown for 0.3 mSv dose constraint, with the value for 1 mSv given in brackets.

	Administered activity of ^{131}I			
	200 MBq	400 MBq	600 MBq	800 MBq
0.5 h day^{-1}	0 (0)	3 (0)	8 (0)	10 (0)
1 h day^{-1}	3 (0)	9 (0)	14 (3)	16 (4)
1.5 h day^{-1}	8 (0)	13 (2)	16 (7)	18 (9)
2 h day^{-1}	9 (0)	15 (4)	18 (9)	22 (11)

Table 16.10. Restriction period (days) for the patient before returning to work based on three different fixed distances from colleagues for 8 h a day (assumed for five days a week). Based on a 0.3 mSv dose constraint to a fellow worker.

	Administered ^{131}I activity			
	200 MBq	400 MBq	600 MBq	800 MBq
8 h at 2 m	0	0	1	3
8 h at 1 m	3	10	14	16
8 h at 0.5 m	14	20	23	27

table 16.10, where the restriction period for 0.3 mSv for the three models is compared.

^{131}Iodine administered for carcinoma of the thyroid

16.60 The decision to discharge a patient who retains more than 800 MBq of ^{131}I should be based on an individual risk assessment of the external dose rate and contamination hazard, taking into account the home circumstances, particularly the ability and willingness of all potentially exposed individuals to comply with the advice given to them to restrict their exposure. It would also be possible to discharge such patients on compassionate grounds, e.g. to visit a dying friend or relative, as long as other potential exposures have been identified and assessed as detailed in paragraphs 16.6–16.9, *Risk assessment*.

16.61 For patients receiving 1.1 GBq ^{131}I, administration may take place in the morning at a centre, with discharge later in the day, to allow for some excretion and a drop in activity if the risk assessment in paragraph 16.60 indicates this is required.

16.62 The time variation of the external dose rate following a first ablative administration of ^{131}I for thyroid cancer is known to differ substantially from that associated with subsequent administrations for residual or recurrent disease (Barrington *et al* 1996). In both cases, the dose rate decreases more rapidly than is seen in the treatment of hyperthyroidism. A dose rate measurement of the patient may be made during their inpatient stay, and restrictions may be provided based on modelling. Studies have found ^{131}I ablation administration follows a bi-exponential model, with an initial 'fast' phase followed by a slower phase (Barrington *et al* 1996, Tabei *et al* 2012). It has also been shown that recombinant thyroid-stimulating hormone (rTSH) does not alter clearance behaviour, and measurements at time points later than 24 h post administration are needed to accurately predict clearance (Agius 2015). Contact restrictions will differ for patients and should be based on modelling of the dose rate data, incorporating patterns of close contact (Rose *et al* 1990, O'Doherty *et al* 1993). The contact restrictions above for thyrotoxicosis patients are not appropriate for Iodine ablation patients and should not be applied (IPEM 2014). Consideration must also be given to the need for repeat treatment within the calendar year when calculating restrictions.

16.63 A survey of centres within the UK supported the following common considerations as standard practice (Driver and Atkinson 2017):

(a) patients should take their personal clothing home and wash separately from other items before use. No clothes should be stored by the hospital;
(b) personal electronic items may be used uncovered (and without gel-type phone protectors) and decontaminated prior to discharge. These should be kept for the personal use of the patient only. Alternatively, they can be covered in protective material, such as Packexe®, to be removed on discharge;
(c) measurement of dose rate/residual activity can be used to confirm discharge and to give individualised radiation protection advice;
(d) centres choosing to carry out 1.1 GBq treatments as out-patient procedures should risk assess each case individually, e.g. use of shared bathrooms, etc;
(e) contact restrictions should be based on measurements and individual risk assessment—only in exceptional circumstances are these likely to exceed seven days post administration.

Other therapeutic administrations using open gamma ray emitters

16.64 The time for which a patient receiving a therapeutic dose of radioactivity will pose a radiation hazard will depend on the amount retained at the time of discharge and its effective half-life. These should be determined before discharge as part of a risk assessment in order to determine the appropriate restrictions to be applied. Where there are sufficient and reliable data concerning dose rates, an effective half-life and patterns of close contact with others, the modelling approach should be employed.

Therapeutic administrations using radionuclides that only emit beta particles

16.65 Up to 200 MBq of ^{32}P, ^{90}Y or ^{89}Sr may generally be administered without placing any restrictions on the patient (ARSAC 2024), although additional radiation protection advice may still be needed in the event of patient death while radioactive (see chapter 17, *Precautions after the death of a patient to whom radioactive substances have been administered*). The possibility of contamination and the generation of radioactive waste must be considered and may require precautions to be taken if the patient is discharged to a nursing home or other workplace or needs other invasive medical intervention or tests for a period after treatment.

16.66 Patients receiving intra-articular, intra-peritoneal or intra-pleural administrations of radioactive substances should remain at the treatment centre until monitoring has shown that the insertion site is unlikely to leak.

^{177}Lu peptide receptor radionuclide therapy (PRRT)

16.67 The usual practice is to hospitalize patients for 24 h to permit medical observation (e.g. monitoring of side effects following amino acid infusion) and optimal post-therapy imaging, as well as significant physiological clearance. The relatively low external dose rate, even a few hours post-administration has enabled some UK centres to treat and discharge ^{177}Lu dotatate patients from hospital on the same day. However, this should be balanced against the increased contamination risk (particularly if there are continence issues) and home circumstances, and an appropriate radiation risk assessment should be performed. It should also be noted that due to current UK provision, although external dose rates measured following therapy administration are low, the duration of the patient's journey home may be much greater than for other, more routinely available therapies. Additionally, four (or more) cycles of PRRT are typically prescribed, which means that the *cumulative* radiation dose in a calendar year of accompanying persons may also need to be considered.

16.68 Typical radiation dose rates for 7.4 GBq ^{177}Lu dotatate (measured at 2m) immediately post-administration are approximately 10 μSv h^{-1}, which decrease rapidly to 2–3 μSv h^{-1} by 24 h post-administration (Fernandez *et al* 2015).

16.69 Generic contact restrictions for patients undergoing ^{177}Lu PRRT for neuroendocrine tumours (NETs), taking into account dose rate and contamination, are shown in table 16.11. Behavioural models by O'Doherty *et al* (1993) were applied, assuming a four-cycle treatment in a calendar year with 7.4 GBq administered. Given the age of the majority of these patients, the following assumptions have been made:

Table 16.11. Contact and general restrictions for patients undergoing ^{177}Lu peptide receptor radionuclide therapy (PRRT), assuming four cycles in a calendar year and 7.4 GBq administration.

	Minimum distance or restriction for the time indicated		
	7 days	15 days	20 days
Children (<18 years) and pregnant woman	Limit contact to 2 h per day > 1 m	1 m	No restrictions
Adults (>18 years)	1 m	1 m	–
General	Flush the toilet twice	–	Sleep in separate beds

(a) the patient is not the primary carer for children under 2 years (behavioural models for children aged 2–5 have been applied);
(b) the patient is retired/at home and spends 8 h at 1 m from their partner.

^{90}Y selective internal radiotherapy

16.70 For patients undergoing Selective Internal Radiotherapy (SIRT) with ^{90}Y microspheres, patients are advised to wash their hands thoroughly after using the toilet and clean up any bodily fluids/spills by disposing of them in the toilet for 24 h following administration. Close contact restrictions are not required (Sirtex Medical 2013, 2020).

Therapeutic administrations using ^{223}Ra

16.71 No contact restrictions are required for ^{223}Ra treatment; however, patients are advised to flush the toilet twice for the first seven days following administration. It is also advised that no blood samples be taken within 4 h of administration. In cases of vomit or faecal contamination in areas, patients are advised to wear gloves when cleaning, isolate contamination rubbish in an outside bin, and wash any clothes separately from other items. The patient must be issued an information card, and it is recommended that cards be issued to both the patient and their next of kin/partner/carer and carried for the whole duration of the course of treatments and at least four weeks following the last treatment.

16.72 If patients are admitted for surgery more than four weeks after treatment, then no radiation protection precautions are necessary. If surgery is required less than four weeks post therapy, then contact with the surgery team should be made. The minimum involvement should be contamination monitoring and segregation of any radioactive waste generated during the procedure, with advice, as above, for the management of solid radioactive waste depending on the type of site. Reassurance from the theatre staff may also be required. An individual assessment may be required where surgery involves areas of bone close to metastases or the intestines. Wards should be contacted for reassurance that routine hygiene procedures are all that is required for the management of these patients and to request that they feed back to the advising physics team if there were any incidents involving bodily fluids.

Therapeutic administrations using sealed sources

16.73 Patients should not be discharged from the hospital with temporary therapeutic implants or surface moulds containing sealed sources. Patients receiving permanent implants should remain in the hospital long enough to ascertain that the sources are unlikely to migrate from the implant site and be expelled from the body.

16.74 For patients receiving permanent implants of ^{125}I seeds for prostate cancer, the advice in IPEM Report 106 should be followed (IPEM 2012), including:

1. No restrictions for contact with adults following the implant.
2. For two months following the implant, reduced close contact with children, i.e. contact <0.5 m is permitted for only a few minutes per day.
3. For two months following the implant, reduced close contact with pregnant people, i.e. <0.5 m for prolonged periods.
4. No travel restrictions.
5. Condoms should be used for the first five ejaculations.

16.75 The precautions that are necessary following other types of permanent implants should be determined for each individual patient on the basis of the advice of the appropriate MPE and RPA.

16.76 Medical Devices Regulation (EU) 2017/745 (MDR) (European Commission 2017) has requirements on implant cards provided to patients and the information required on them (MDCG 2020).

References

Agius S 2015 Predictive modelling of external dose rate from radionuclide therapy patients following discharge from hospital *Master's Thesis* The University of Malta

ARSAC 2007 *Newsletter* Issue 2 Administration of Radioactive Substances Advisory Committee

ARSAC 2024 *Notes for guidance on the clinical administration of radiopharmaceuticals and use of sealed radioactive sources.* March 2024. Administration of Radioactive Substances Advisory Committee https://www.gov.uk/government/publications/arsac-notes-for-guidance (accessed 20 April 2024)

Barrington S, O'Doherty M, Kettle A, Thomson W, Mountford P, Burrell D, Farrell R, Batchelor S, Seed P and Harding L 1999 Radiation exposure of the families of outpatients treated with radioiodine (iodine-131) for hyperthyroidism *Eur. J. Nucl. Med.* **26** 686–92

Barrington S F, Anderson P, Kettle A G, Gadd R, Thomson W H, Batchelor S, Mountford P J, Harding L K and O'Doherty M J 2008 Measurement of the internal dose to families of outpatients treated with 131I for hyperthyroidism *Eur. J. Nucl. Med. Mol. Imaging* **35** 2097–104

Barrington S F, Kettle A G, O'Doherty M J, Wells C P, Somer E J R and Coakley A J 1996 Radiation dose rates from patients receiving iodine-131 therapy for carcinoma of the thyroid *Eur. J. Nucl. Med.* **23** 123–30

CDG 2009 *The Carriage of Dangerous Goods and Use of Transportable Pressure Equipment Regulations 2009* https://legislation.gov.uk/uksi/2009/1348 (accessed 31 January 2023)

CDG(NI) 2010 *The Carriage of Dangerous Goods and Use of Transportable Pressure Equipment Regulations (Northern Ireland) 2010* https://legislation.gov.uk/nisr/2010/160 (accessed 31 January 2023)

Cormack J and Shearer J 1998 Calculation of radiation exposures from patients to whom radioactive materials have been administered *Phys. Med. Biol.* **43** 501–16

Cronin B, Marsden P K and O'Doherty M J 1999 Are restrictions to behaviour of patients required following fluorine-18 fluorodeoxyglucose positron emission tomographic studies? *Eur. J. Nucl. Med.* **26** 121–8

DfT 2020 *Carriage of Dangerous Goods: Approved Derogations and Transitional Provisions, GOV. UK Website* https://gov.uk/government/publications/the-carriage-of-dangerous-goods-approved-derogations-and-transitional-provisions/carriage-of-dangerous-goods-approved-derogations-and-transitional-provisions (accessed 28 June 2022)

Driver I and Atkinson S 2017 Standardisation of patient restrictions following the treatment of thyroid cancer with radioiodine: a collaborative project involving 15 Nuclear Medicine Departments in England and Scotland and Butterfly Thyroid Cancer Trust *IPEM Radiation Protection in Nuclear Medicine Meeting* (Birmingham: Institute of Physics and Engineering in Medicine)

EA(S)R 2018 *The Environmental Authorisations (Scotland) Regulations 2018* http://legislation.gov.uk/ssi/2018/219 (accessed 31 January 2023)

EPR 2016 *Environmental Permitting (England and Wales) Regulations* https://legislation.gov.uk/uksi/2016/1154 (accessed 31 January 2023)

European Commission 2017 Regulation (EU) 2017/745 of the European Parliament and of the Council of 5 April 2017 on medical devices *Official Journal of the European Union* Document 02017R0745-20200424 L 117 https://eur-lex.europa.eu/eli/reg/2017/745/2020-04-24 (accessed 10 March 2023)

Fernandez R, Allen S and Lewington V 2015 Ensuring safe & effective delivery of Lutetium-177 dotatate therapy *J. Nucl. Med.* **56** 479

Greaves C and Tindale W 1999 Dose rate measurements from radiopharmaceuticals: implications for nuclear medicine staff and for children with radioactive parents *Nucl. Med. Commun.* **20** 179–87

Gunasekara R, Thomson W H and Harding L K 1996 Use of public transport by I131 therapy outpatients *Nucl. Med. Commun.* **17** 275

Harding L, Mostafa A, Roden L and Williams N 1985 Dose rates from patients having nuclear medicine investigations *Nucl. Med. Commun.* **6** 191–4

HPS no date *Nuclear Medicine and the Pregnant Patient Q&A, Health Physics Society website.* https://hps.org/physicians/nuclear_medicine_pregnant_patient_qa.html (accessed 28 June 2022)

IAEA 2009 Release of Patients After Radionuclide Therapy. SRS No 63 *Safety Reports Series No. 63. SRS No 63* (Vienna: International Atomic Energy Agency) https://iaea.org/publications/8179/release-of-patients-after-radionuclide-therapy (accessed 31 January 2023)

IAEA 2018 Radiation Protection and Safety in Medical Uses of Ionizing Radiation *SSG-46, Safety Standards Series. SSG-46* (Vienna, Austria) https://iaea.org/publications/11102/radiation-protection-and-safety-in-medical-uses-of-ionizing-radiation (accessed 10 March 2023)

IAEA no date Radiation Protection of Pregnant Women in Nuclear Medicine *International Atomic Energy Authority Website* (Vienna, Austria) https://iaea.org/resources/rpop/health-professionals/nuclear-medicine/pregnant-women (accessed 28 June 2022)

ICRP 1996 Conversion coefficients for use in radiological protection against external radiation *Ann. ICRP* **36** Publication 74 https://www.icrp.org/publication.asp?id=ICRP%20Publication%2074 (accessed 25 April 2024)

ICRP 2001 Doses to the embryo and fetus from intakes of radionuclides by the mother. ICRP Publication 88 *Ann. ICRP* https://www.icrp.org/publication.asp?id=ICRP%20Publication%2088 (accessed 25 April 2024)

ICRP 2004 Dose to infants from ingestion of radionuclides in mother's milk. ICRP Report 95 *Ann. ICRP* **34** Publication 95 https://www.icrp.org/publication.asp?id=ICRP%20Publication%2095 (accessed 25 April 2025)

IPEM 2012 *Report 106: UK Guidance on Radiation Protection Issues following Permenant Iodine-125 Seed Prostate Brachytherapy* ed P Bownes (York: Institute of Physics and Engineering in Medicine)

IPEM 2014 *Report 109: Radiation Protection in Nuclear Medicine* ed M McJury and C Tonge (York: Institute of Physics and Engineering in Medicine)

IR(ME)R 2017 *The Ionising Radiation (Medical Exposure) Regulations* www.legislation.gov.uk/uksi/2017/1322 (accessed 31 January 2023)

IR(ME)R(NI) 2018 *The Ionising Radiation (Medical Exposure) Regulations (Northern Ireland)* https://legislation.gov.uk/nisr/2018/17 (accessed 31 January 2023)

IRR 2017 *The Ionising Radiations Regulations* https://legislation.gov.uk/uksi/2017/1075 (accessed 31 January 2023)

IRR(NI) 2017 *The Ionising Radiations Regulations (Northern Ireland)* http://legislation.gov.uk/nisr/2017/229 (accessed 31 January 2023)

MDCG 2020 Medical Devices: Guidance Document *Implant Card relating to the application of Article 18 Regulation (EU) 2017/745 of the European Parliament and of the Council of 5 April 2017 on Medical Devices* (Medical Device Coordination Group) 2019th–8 v2 edn https://health.ec.europa.eu/system/files/2020-09/md_mdcg_2019_8_implant_guidance_card_en_0.pdf (accessed 10 March 2023)

Mountford P J and Coakley A J 1989 Body surface dosimetry following re-injection of 111In-leucocytes *Nucl. Med. Commun.* **10** 497–501

NET Patient Foundation 2019 *Your Guide to Living with Neuroendocrine Cancer* (Neuroendocrine Cancer UK) https://neuroendocrinecancer.org.uk/wp-content/uploads/2020/03/NET-Patient-Foundation-Handbook.pdf (accessed 16 March 2023)

O'Doherty M, Kettle A, Eustance C, Mountford P and Coakley A 1993 Radiation dose rates from adult patients receiving 131I therapy for thyrotoxicosis *Nucl. Med. Commun.* **14** 160–8

Pauwels E K, Thomson W H, Blokland J A, Schmidt M E, Bourguignon M, El-Maghraby T A, Broerse J J and Harding L K 1999 Aspects of fetal thyroid dose following iodine-131 administration during early stages of pregnancy in patients suffering from benign thyroid disorders *Eur. J. Nucl. Med.* **26** 1453–7

Pereira S, Woods E, John J, Jacob A, Abreu C, Alves L and Pike L 2015 64Cu-ATSM PET studies: are radiation protection restrictions required for patients after scan? *Eur. J. Nucl. Med. Mol. Imaging* **42** S836

Radiation Protection Committee of the BIR 1999 Patients leaving hospital after administration of radioactive substances *Br. J. Radiol.* **72** 121–5

Rose M R, Prescott M C and Herman K 1990 Excretion of iodine-123-hippuran, technetium-99m-red blood cells, and technetium-99m-maccroaggregated albumin into breast milk *J. Nucl. Med.* **31** 978–84 https://jnm.snmjournals.org/content/jnumed/31/6/978.full.pdf (accessed 10 March 2023)

RSA 1993 *Radioactive Substances Act 1993* http://legislation.gov.uk/ukpga/1993/12 (accessed 11 February 2023)

Singleton M, Griffiths C, Morrison G and Soanes T 2003 Dose Constraints for Comforters and Carers. Prepared by Royal Hallamshire Hospital Dose Constraints for Comforters and Carers *HSE Research Report 155* https://hse.gov.uk/research/rrpdf/rr155.pdf (accessed 31 January 2023)

Sirtex Medical 2013 *A Patient's Guide Selective Internal Radiation Therapy (SIRT) for Liver Tumours Using SIR-Spheres® Microspheres* (Bonn, Germany: Sirtex Medical Limited) https://sirtex.com/eu/patients/resources/educational-materials/ (accessed 10 March 2023)

Sirtex Medical 2020 *SIR-Spheres ® Y-90 Resin Microspheres, Training Manual, Physicians and Institutions (TRN-RW–06)* (Bonn, Germany: Sirtex Medical)

Tabei F, Asli I N, Azizmohammadi Z, Javadi H and Assadi M 2012 Assessment of radioiodine clearance in patients with differentiated thyroid cancer *Radiat. Prot. Dosim.* **152** 323–7

Thomson W, Mills A, Smith N, Mostafa A, Notghi A and Harding L 1993 Day and night radiation doses to patients' relatives implications of ICRP 60 *Nucl. Med. Commun.* **14** 275

Thomson W H, Bray D, Mills A and Harding L K 1996 Reducing the radiation dose to relatives of 131I therapy thyrotoxic patients *Nucl. Med. Commun.* **17** 300

Thomson W H and Harding L K 1995 Radiation protection issues associated with nuclear medicine out-patients *Nucl. Med. Commun.* **16** 879–92

Thomson W H and Lewis C 2015 Breast dose from lactation following I131 treatment *Nucl. Med. Commun.* **36** 536

IOP Publishing

Medical and Dental Guidance Notes (Second Edition)
A good practice guide on all aspects of ionising radiation protection in the clinical environment:
IPEM Report 113
John Saunderson, Mohamed Metwaly, William Mairs, Philip Mayles,
Lisa Rowley and Mark Worrall

Chapter 17

Precautions after the death of a patient to whom radioactive substances have been administered

Scope

17.1 This chapter applies to:
- (a) the carrying out of post-mortems/autopsies on corpses which contain radioactive substances; and
- (b) the preparation of such corpses for burial or cremation.

17.2 For the purposes of this chapter, all administrations of ^{131}I greater than 30 MBq should be considered therapy procedures.

Principles

17.3 The procedures used when handling or preparing a radioactive corpse must comply with the requirements of the Ionising Radiations Regulations (IRR) (IRR 2017, IRR(NI) 2017) and Radioactive Substances Regulations (RSR) (RSA 1993, EPR 2016, EA(S)R 2018), particularly in relation to dose restriction, the handling of radioactive materials, and the management of radioactive waste (noting that while under UK law a corpse is not 'waste' itself, waste may arise from a corpse). There may also be associated transport obligations to consider, particularly with the latter (CDG 2009, CDG(NI) 2010).

General advice

17.4 The administering department should prepare generic radiation risk assessments for each of its clinical procedures to be applied in the event of the death of the patient following the administration of a radioactive substance. This

doi:10.1088/978-0-7503-2332-1ch17

should identify the groups likely to be exposed to any significant risk and the control measures necessary to ensure that exposures to ionising radiation are as low as reasonably practicable (ALARP). Reference Greaves and Tindale (2001) illustrates a practical example. Where precautions are indicated, the period for which they are necessary needs to be determined in the radiation risk assessment. Written instructions should be available to send to funeral directors and mortuaries so that they can make arrangements to comply with the requirements of IRR (IRR 2017, IRR(NI) 2017) before receiving a radioactive body. The means by which they will be informed that the corpse is radioactive should also be considered (see paragraph 16.15). As this information may be supplied by the family, when consenting patients for molecular radiotherapy, the next of kin should be made aware of the need to contact the administering medical physics department following death within a specified time frame following administration. This information should also be recorded in the patient's notes. HSE guidance, *Managing infection risks when handling the deceased* (HSE 2018), provides a template 'Hazard notification sheet' as one way of providing the necessary information required to safely handle the deceased where there are infection risks or hazards from implantable devices and radioactive sources.

17.5 Instructions to mortuaries and funeral directors should encompass all possible work with radioactive corpses and should consider:

(a) the requirement for undertakers and mortuaries for notification (Regulation 5) or registration (Regulation 6) to the HSE or HSENI that they are working with ionising radiation if they have never previously done so (IRR 2017, IRR(NI) 2017), see chapter 10 table 10.1) and any requirements under RSR (RSA 1993, EPR 2016, EA(S)R 2018) (see chapter 19, *Accumulation and disposal of radioactive waste*);
(b) the requirement for the undertakers and mortuaries to consult with a Radiation Protection Adviser (RPA), undertake their own radiation risk assessment and implement the findings before receiving a radioactive corpse (they may appoint the hospital RPA);
(c) precautions necessary to keep doses ALARP;
(d) contact telephone numbers for further advice;
(e) any need for whole body and/or extremity monitoring of personnel and how this should be arranged;
(f) any need for monitoring of the premises and how this should be arranged.

17.6 The external radiation hazard associated with the handling of a corpse after most diagnostic administrations of radioactive substances will be small, and special precautions are unlikely to be required.

17.7 The generic radiation risk assessments for therapy procedures should include a radiological impact assessment for burial and cremation.

17.8 There will be a potential for contamination and the production of radioactive waste if a post-mortem and/or embalming are carried out on any corpse containing radioactive substances. Appropriate precautions on the handling of radioactive materials, as detailed in chapter 10, *Diagnostic uses of open radioactive substances*, should be followed. Paragraphs 13.40 and 13.41 on work in operating theatres may be helpful.

17.9 Skin doses may be significant if the hands are likely to come into contact with any body tissues that may have concentrated levels of radioactivity. Finger monitoring may be necessary in such cases.

17.10 For patients in hospitals, nursing homes, hospices, etc, information about any radioactive substances administered, the time for which the patient will be radioactive, and who to contact if anything untoward happens to the patient should always be available (see paragraphs 10.90 to 10.95, *Procedures in wards and clinics*, paragraph 10.96, *Procedures in operating theatres*, and 16.16 to 16.21, *Discharging patients to outside organisation etc*). If a patient who is radioactive dies in such establishments, the corpse should be labelled as being radioactive, and anyone who may subsequently come into contact with it should be notified of any precautions to be taken during post-mortem. Instructions on the precautions to be taken should be available from the department that administered the radioactive substance, as detailed in paragraph 17.5.

17.11 When a patient is sent home during the precautionary period (see paragraph 17.4), the presence of the card described in paragraph 16.15 should be enough to alert any attending healthcare professional to the need to ask for advice from the administering department in the event of the death of the patient, including seeking advice on behalf of relatives who may wish to carry out ritual washing of the corpse for religious reasons. For most diagnostic administrations, the hazards to those coming into contact with the corpse are very low, and the risk of contravention of IRR (IRR 2017, IRR(NI) 2017) or RSR (RSA 1993, EPR 2016, EA(S)R 2018) is also low. Risk assessments for those returning home after therapeutic administrations should consider the possibility of the death of the patient, and suitable systems should be put in place to deal with this situation should it arise. A rota system for the provision of expert advice by telephone may be sufficient.

Death following administration of diagnostic quantities of radioactive substances

17.12 For most diagnostic administrations of radioactive substances, no extra precautions beyond those normally employed when carrying out post-mortems/embalming will be necessary unless death occurs within the precautionary period determined by the risk assessment (see also paragraph 17.4). The maximum precautionary period for any diagnostic administration is 48 h (Singleton *et al* 2007).

17.13 There is no need to place any restrictions on the method of disposal of the body.

Death following administration of therapeutic quantities of radioactive substances

17.14 The medical physics expert (MPE) involved in the administration should be contacted for any further information required on the level of hazard remaining in all cases where the patient dies while still under restrictions applied after a therapeutic administration of a radioactive substance. It may be helpful to establish the following information:

(a) where the body of the patient is located;
(b) the exact date of the death;
(c) the date of therapy treatment;
(d) the administered activity and radiopharmaceutical;
(e) if a post-mortem examination is planned, provide contact details of who will carry out the post-mortem and where it will be performed;
(f) whether it is intended for the body to be embalmed;
(g) whether it is intended for the body to be buried or cremated;
(h) who the RPA/radioactive waste adviser (RWA) for the receiving establishment is.

17.15 For patient death during inpatient molecular radiotherapy, staff should be made aware that normal procedures of releasing waste by pressing on the abdomen of the deceased patient must not be followed and orifices should not be blocked (see Singleton *et al* 2007 for details). Friends and family should be advised appropriately by medical physics staff on time and contact restrictions for viewing the body. The body should be placed in a bag as soon as possible to prevent leakage and the spread of contamination.

17.16 Temporary implants of radionuclides should be removed from corpses as soon as possible after death and before the body is released for post-mortem or disposal.

17.17 Nuclear-powered cardiac pacemakers must never be left in a corpse. At the time these were brought into use (the early 1970s to late 1980s), it was arranged for patients fitted with pacemakers of this kind to wear an identity bracelet and carry an identity (ID) card. These patients also agreed at the time of the implant to the removal of their pacemakers on death. RPAs may wish to discuss the following points with local cardiology departments to minimise the likelihood of a potentially serious problem:

(a) check whether there are any unaccounted devices;
(b) check whether any patients fitted with such pacemakers are still alive and confirm that they still carry a bracelet and ID card; and
(c) advise cardiac and mortuary services of the potential hazards when removing pacemakers. Disposal of such pacemakers must be in full

compliance with RSR (RSA 1993, EPR 2016, EA(S)R 2018) (see chapter 19, *Accumulation and disposal of radioactive waste*).

Post-mortem examinations

17.18 If a post-mortem examination is to be performed on a patient who has recently received radioactive substances, the post-mortem room and any equipment used may become contaminated. Radioactive waste, particularly aqueous liquid waste, may be produced. Procedures are described in some detail in Singleton *et al* (2007).

17.19 Mortuaries in hospitals that have nuclear medicine departments should be provided with written guidance for staff carrying out post-mortem examinations on patients who are radioactive. The guidance should include the information and instructions described in paragraph 17.5. Table 17.1 gives

Table 17.1. Guideline maximum remaining activities of radionuclides, or time periods from administration, for disposal of corpses without special precautions for radiation during the procedure (activities in MBq).

Radionuclide	Post-mortem/embalming	Burial	Cremation
^{131}I	<10 MBq[e]	<400 MBq[a]	<400 MBq[a]
^{125}I seeds for prostate brachytherapy	Leave prostate *in situ* if <20 months post administration[f]	<4,000 MBq[b]	After 20 months (After 22 months for scattering ashes, see paragraph 17.35)[f]
^{125}I seeds for breast tumour localisation	Leave breast tissue *in situ* if <20 months post administration[f]	<4,000 MBq[b] (i.e. no restriction)	Remove seeds prior to cremation or > 6 months if seeds still *in situ*
^{103}Pd seeds	–	<15,000 MBq[a]	see paragraphs 17.30 to 17.32
^{90}Y colloid	<200 MBq[e]	<2,000 MBq[b]	<70 MBq[c]
^{198}Au seeds	–	<4,000 MBq[a]	see paragraphs 17.30 to 17.32
^{198}Au colloid	–	<400 MBq[a]	<100 MBq[c]
^{32}P	<100 MBq[e]	<2,000 MBq[b]	<30 MBq[c]
^{89}Sr	<50 MBq[e]	<2,000 MBq[b]	<200 MBq[c], [d]
^{177}Lu Dotatate	<200 MBq[g]	<2000 MBq[g]	<90 MBq[g]
^{223}Ra Xofigo	<0.2 MBq[h] (4 weeks post treatment)	No restriction[h]	1 month[h]

The values in columns 2, 3, and 4 relate to the greatest risk to those involved in the procedures.
[a] Based on the dose rate external to the body (NRPB *et al* 1988).
[b] Based on the bremsstrahlung dose at 0.5 m (NRPB *et al* 1988).
[c] Based on contamination hazards, assuming that these radionuclides remain in the ash (NRPB *et al* 1988).
[d] Relaxed from earlier guidance, HSE Advice 1989 (HSE 1989).
[e] From IAEA guidance (IAEA 2009).
[f] IPEM Report 106 (IPEM 2012).
[g] From ^{90}Y and the ratio of ALI to ^{90}Y for cremation.
[h] Work carried out by Paul Hinton (Hinton 2018)]

the indicative maximum remaining activities of radionuclides for conducting post-mortems without any special radiation precautions.

17.20 When post-mortem examinations are performed at establishments not covered in paragraph 17.18, written procedures and advice should be requested from the department that administered the radioactive substance. Further advice may be obtained from the RPA for either the administering or receiving establishment.

17.21 The MPE should be consulted about radiation levels likely to be encountered and the hazards involved whenever a post-mortem needs to be carried out before the end of the precautionary period (see paragraph 17.4). The appropriate RPA and RWA may also need to be contacted for advice. The mortuary and staff involved in the procedure should be monitored for contamination and the results recorded if still within the restriction period in table 17.1. Any solid waste must be bagged, monitored, and labelled, and, if necessary, decay stored.

17.22 If an isolation facility is available, the post-mortem should be conducted there for ease of decontamination afterwards. Similar precautions should be taken to those for operating theatres (see paragraphs 10.96, 13.40, and 13.41, *Procedures in operating theatres*). Those carrying out the post-mortem should wear heavy-duty, impermeable gloves, waterproof coveralls and face visors. Leakage from the body should be carefully contained to prevent the spread of radioactive contamination.

17.23 If samples of tissues are required for histology or forensic assessment, the receiving laboratory should be notified if the samples may be radioactive. They should also be told about any precautions that need to be taken, including dealing with any radioactive waste (see paragraph 10.44). Dispatch of the samples should be in accordance with the requirements of IRR (IRR 2017, IRR(NI) 2017) and road transport regulations (CDG 2009, CDG(NI) 2010), as appropriate (see chapter 18, *Keeping, accounting for, and moving radioactive substances (including transport)*).

Embalming

17.24 Embalming should not be carried out if death occurs within the period covered by an instruction card (see paragraph 16.15) or within 48 h of the administration of radioactive substances unless a risk assessment indicates otherwise. The embalmer should be advised by the RPA/RWA/Dangerous Goods Safety Adviser (DGSA) appointed to the hospital where the radioactive material was administered as to the necessary precautions to ensure that doses are kept ALARP and relevant legislation is complied with, including transport associated with wastes arising. Sites may use body bags, and if required, families and friends should be encouraged to view the body prior to placing it in the bag. Some religions may wish for the body to be washed. For treatments such as ^{223}Ra or ^{90}Y-SIRT, there is no reason why this should not go ahead, given consideration of the external dose rate. Orifices may be packed to prevent leakage. Table 17.1 gives the indicative maximum remaining activities of radionuclides for embalming without any special precautions for radiation during the procedure.

Burial

17.25 No special precautions are necessary during the burial of corpses that have residual activities less than those indicated in column 3 of table 17.1.

17.26 For higher activities and before the corpse is released, the RPA should be consulted to identify control measures, such as lead-lined coffins, that may need to be applied to those who approach the corpse to ensure that doses are ALARP.

Cremation

17.27 No special precautions need to be taken during cremation where the residual activity is less than that in column 4 of table 17.1, except where the source is encapsulated or contained, e.g. ^{125}I or ^{103}Pd seeds in prostate brachytherapy or ^{32}P stents in interventional techniques; see paragraphs 17.30 to 17.32, *Encapsulated-sealed sources*.

17.28 The advice given in paragraph 17.24 is appropriate for the handling of corpses during preparation for cremation.

17.29 For higher activities, and before the corpse is released, the RPA should be consulted to identify control measures that may need to be applied to ensure that doses are ALARP for those who approach the corpse.

Encapsulated-sealed sources

17.30 The generic radiation risk assessment (and environmental impact assessment) for the burial or cremation of a patient who has undergone radiotherapy treatment where the source is encapsulated or contained (e.g. ^{125}I or ^{103}Pd seeds in prostate brachytherapy or ^{32}P stents in interventional techniques) should include consideration of the fact that the source is most likely a sealed source manufactured to BS EN ISO 2919:2014 (BSI 2014) or ISO 2919:2012 Sealed Radioactive Sources (ISO 2012) and, as such, should not be dispersed by cremation and might not be contained by burial.

17.31 It is recommended that in the event of death within a given number of half-lives of the radionuclide after the administration (the actual number to be determined by risk assessment and administered activity), cremation be avoided. If this is not possible, the excision of the sources should be considered (see paragraph 9.10).

17.32 For the cremation of corpses where the source is encapsulated or contained, the advice and control measures in the generic radiation risk assessment should be communicated to the Employer at the crematorium and followed to ensure that doses are ALARP. This should include consideration of whether restrictions on the use of a cremulator (used to grind the bone fragments into a fine powder) are required. It is recommended that such generic radiation risk and environmental impact assessments be documented for discussion with the statutory inspectors at the next inspection.

Death following administration of ^{125}I seeds for the treatment of prostate cancer

17.33 Guidance on procedures to follow in the event of death following administration of ^{125}I seeds for the treatment of prostate cancer is detailed in IPEM Report 106 (IPEM 2012). There are restrictions if death occurs within 20 months of administration, highlighted in the following paragraphs.

17.34 A post-mortem may be performed; however, the prostate and surrounding tissue should remain *in situ*. There are no restrictions after 20 months post-administration. Embalming procedures may be performed, provided they are not invasive around the prostate.

17.35 If the body is to be cremated within 20 months after administration, the prostate should be removed beforehand. Ashes must be scattered over a large enough area to ensure an activity concentration of less than 30 $Bq.cm^{-2}$. After 22 months, the ashes should be scattered over at least 1 m^2. Burial may take place at any time following administration.

17.36 If the prostate is removed, sectioning must not be performed before 20 months post-administration. The tissue should be stored or dissolved in accordance with the family's wishes and the applicable Human Tissue Act (Human Tissue Act 2004, Human Tissue (Scotland) Act 2006). All centres performing ^{125}I treatment must ensure sufficient scope on their permit/authorisation in the event of patient death or prostate removal. If this occurs at a site where treatment is not performed, the prostate should be returned to the largest local licensed implanting centre, under the guidance of the relevant environment agency inspector. Guidance should be provided by the implanting hospital. Financial provision must also be in place prior to the commencement of a ^{125}I prostate service.

17.37 The above issues should be discussed with the patient and next of kin prior to treatment.

Death following administration of ^{223}Ra

17.38 ^{223}Ra-dichloride (Xofigo) is a targeted radionuclide therapy agent for patients with prostate cancer to treat metastatic disease in their bones. This treatment may also be extended to patients with breast cancer. The standard treatment protocol is to administer six doses via intravenous injection at four weekly intervals. The prescribed activity is scaled by the patient's body weight at 55 kBq kg^{-1} which corresponds to an injected activity of 3.9 MBq for a 70 kg patient.

17.39 The physical half-life of ^{223}Ra is 11.43 days. At 4 h post-injection 44%–77% of the injected activity is in the bones. At seven days post-injection a median of 76% of the administered activity is cleared from the body, and this clearance is mostly dependent on the intestinal transit rate, with only 5% being eliminated via the urine (Bayer PLC 2017). If death occurs at least one week after treatment, a simple estimate of the radioactivity remaining in the corpse is to assume 25% retention of the administered activity and to calculate the decay of this amount correctly, use the number of days between the administration date and the date of death.

17.40 The death of a patient should not result in any additional precautions above those normally carried out on the basis of hygiene for simple handling of the body. Ideally, advice should be sought in the first instance from the administering site's RPA.

17.41 With reference to table 17.1, the external dose rate from the body will present no restrictions to post-mortem examinations, and routine practices should avoid any ingestion of small levels of radionuclide contamination. At four weeks post-administration, the estimated activity remaining will be below 200 kBq. Routine clinical practice regarding hygiene will provide sufficient protection, and special precautions for contamination and radioactive solid waste will not need to be implemented.

17.42 Additional precautions will be required within the four weeks following administration for post-mortem, embalming, and cremation, with the support of the physics department from the administering site or other local medical physicists. The RPA and RWA should offer practical advice if it is not feasible for someone to conduct monitoring in person. It should be remembered that faecal excretion is the major route of elimination from the body in the first week and therefore the bowels and any contents should be handled with particular caution and may need to be treated as radioactive material/waste. Precautions as stated in paragraphs 17.18 to 17.23, *Post-mortem examinations*, above should be followed.

17.43 Advice for embalmers is the same as for post-mortem staff. The advice in paragraph 17.24, *Embalming*, should be followed. During the embalming process, bone fragments may be generated and will end up in the sluice. These are unlikely to present any hazard although it may be prudent to avoid embalming if death is within two weeks of treatment.

17.44 If the death occurs at least one week after a radium treatment, it is assumed that any remaining radioactivity is contained within the bones and is retained in the bony remains after cremation. Crematoria staff should wear gloves,

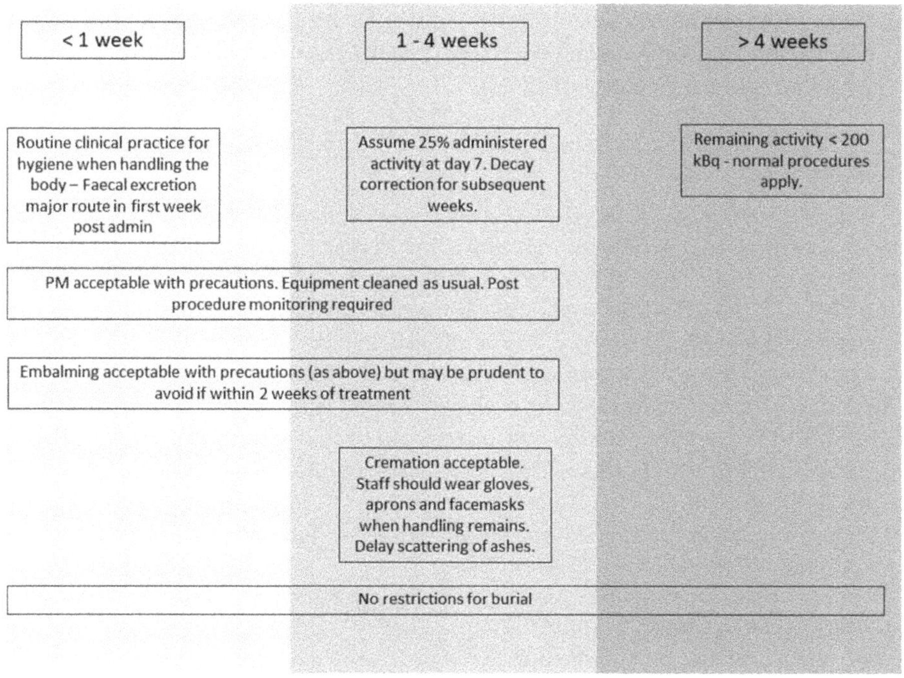

Figure 17.1. Flow chart following the death of a patient treated with ^{223}Ra-dichloride (Xofigo).

aprons, and face masks when handling any remains from patients who have undergone treatment with ^{223}Ra. With protective equipment, there is little risk to the staff when transferring remains from the cremator to the cremulator. The biggest risk to staff occurs after the cremation, during the transfer of the remains to the urn, when fine dust is released from the radioactive bones. This activity must take place inside a cabinet with air extraction capability and filtered exhaust. Note that in the calculations used for the advice provided, the inhalation dose coefficient for ^{223}Ra from ICRP publication 119 (ICRP 2012) has been applied. ICRP publication 137 (ICRP 2017) states that for ^{223}Ra-dichloride, for 'radiation protection purposes, it is normally sufficient to control exposures on the basis of the intake of ^{211}Pb', which is a lower inhalation dose coefficient; however, this does not affect the advice provided (figure 17.1).

Repatriation

17.45 If a corpse is to be transported within Great Britain or overseas, advice on transport requirements should be obtained from a DGSA in consultation with the RPA/RWA appointed to the hospital where the radioactive material was administered. The Office for Nuclear Regulation can be contacted for further information as required thereafter. The relevant Consulate may also need to be contacted before transportation, and the relevant modal transport legislation and any specific requirements of the receiving country should be checked and met before dispatch.

References

Bayer PLC 2017 Summary of Product Characteristic: Xofigo *EMA Website* (Committee for Medicinal Products for Human Use) https://ema.europa.eu/en/documents/product-information/xofigo-epar-product-information_en.pdf (accessed 22 April 2024)

BSI 2014 BS EN ISO 2919:2014 Radiological Protection: Sealed Radioactive Sources *General Requirements and Classifications* (London: British Standards Institute)

CDG 2009 *The Carriage of Dangerous Goods and Use of Transportable Pressure Equipment Regulations 2009* https://legislation.gov.uk/uksi/2009/1348 (accessed 22 April 2024)

CDG(NI) 2010 *The Carriage of Dangerous Goods and Use of Transportable Pressure Equipment Regulations (Northern Ireland) 2010* https://legislation.gov.uk/nisr/2010/160 (accessed 22 April 2024)

EA(S)R 2018 *The Environmental Authorisations (Scotland) Regulations 2018* http://legislation.gov.uk/ssi/2018/219 (accessed 22 April 2024)

EPR 2016 *Environmental Permitting (England and Wales) Regulations* https://legislation.gov.uk/uksi/2016/1154 (accessed 22 April 2024)

Greaves C and Tindale W 2001 Radioiodine therapy: care of the helpless patient and handling of the radioactive corpse *J. Radiol. Prot.* **21** 381–92

Hinton P 2018 *Personal Correspondance: from Paul Hinton to Lisa Rowley. Adivce and RA—Ra223 Patient Death (April 2018)*

HSE 1989 Advice of February 1989 *Radiation Protection Topic Group News Sheet No. 30* (York: Institute of Physics and Engineering in Medicine)

HSE 2018 *HSG283 Managing Infection Risks When Handling the Deceased* (Norwich: Health and Safety Executive) https://hse.gov.uk/pubns/priced/hsg283.pdf (accessed 22 April 2024)

Human Tissue Act 2004 https://legislation.gov.uk/ukpga/2004/30 (accessed 22 April 2024)

Human Tissue (Scotland) Act 2006 https://legislation.gov.uk/asp/2006/4 (accessed 22 April 2024)

IAEA 2009 Release of patients after radionuclide therapy. SRS No 63 *Safety Reports Series No. 63. SRS No 63* (Vienna: International Atomic Energy Agency) https://iaea.org/publications/8179/release-of-patients-after-radionuclide-therapy (accessed 22 April 2024)

ICRP 2012 Compendium of dose coefficients based on ICRP Publication 60 *Ann. ICRP* **41** Publication 119 ed C Eckerman, K Harrison, J Menzel, H-G Clement https://icrp.org/publication.asp?id=ICRP%20Publication%20119 (accessed 22 April 2024)

ICRP 2017 Occupational intakes of radionuclides: part 3. ICRP Publication 137 *Ann. ICRP* **46** Publication 137 ed J D H F Paquet https://icrp.org/publication.asp?id=ICRP%20Publication%20137 (accessed 22 April 2024)

IPEM 2012 *Report 106: UK Guidance on Radiation Protection Issues following Permenant Iodine-125 Seed Prostate Brachytherapy* ed P Bownes (York: Institute of Physics and Engineering in Medicine)

IRR 2017 *The Ionising Radiations Regulations* https://legislation.gov.uk/uksi/2017/1075 (accessed 22 April 2024)

IRR(NI) 2017 *The Ionising Radiations Regulations (Northern Ireland)* http://legislation.gov.uk/nisr/2017/229 (accessed 22 April 2024)

ISO 2012 *2919:2012 Radiological Protection—Sealed Radioactive Sources—General Requirements and Classification* (Geneva: International Organization for Standardization)

NRPB, HSE, Department of Health and Social Security, Department of Health and Social Services (Northern Ireland), Scottish Home and Health Department and Welsh Office 1988

RSA 1993 *Radioactive Substances Act 1993* http://legislation.gov.uk/ukpga/1993/12 (accessed 22 April 2024)

Singleton M, Start R D, Tindale W, Richardson C and Conway M 2007 The radioactive autopsy: safe working practices *Histopathology* **51** 289–304

John Saunderson, Mohamed Metwaly, William Mairs, Philip Mayles, Lisa Rowley and Mark Worrall

Chapter 18

Keeping, accounting for, and moving radioactive substances (including transport)

Introduction

18.1 Keeping and accounting for radioactive substances under the Ionising Radiations Regulations (IRR 2017, IRR(NI) 2017) are legal requirements for Employers, in addition to the fact that they must be registered or have obtained a permit under the Environmental Authorisations (Scotland) Regulations 2018 (Scotland) (EA(S)R 2018), The Radioactive Substances Act 1993 (Amendment) Regulations (Northern Ireland) 2011 (Northern Ireland) (RSA 1993, RSA(A)R(NI) 2011) or the Environmental Permitting Regulations 2016 (England and Wales) as amended (EPR 2016) (referred to hereafter as Radioactive Substances Regulations (RSR)) or operate under a relevant exemption (BEIS *et al* 2018) or binding rules. Employers are legally required to comply with the conditions for keeping and accounting of radioactive materials specified in their Permit, Certificate of Registration (referred to hereafter as permits), or Exemption Order, including a suitable management system. Guidance for compliance with radioactive substances regulations for non-nuclear sites is available online (DEFRA, DECC and Welsh Assembly Government 2011, Department of Agriculture and Rural Affairs 2020, EA 2021, SEPA 2021). Authorisations/Permits (referred to hereafter as permits) to dispose of radioactive waste must be obtained prior to keeping, use, and disposal (see chapter 19—*Accumulation and Disposal of Radioactive Waste*). The implementation of a single system for keeping and accounting for radioactive materials that satisfies both sets of legal requirements should normally be readily achievable (see IRPS8 Control of Radioactive Substances (HSE *et al* 2010)).

doi:10.1088/978-0-7503-2332-1ch18

Regulatory aims

18.2 Radioactive substances regulation has three main aims: (i) to establish and maintain control over the keeping, use, and security of radioactive sources; (ii) to ensure the impact on the environment and public is minimised and limited; and (iii) consideration of costs for the disposal of sealed sources (financial provision for HASS sources) (DEFRA, DECC, and Welsh Assembly Government 2011). To this end, the use of Best Available Techniques (BAT) or Best Practical Means (BPM) is employed.

General practical principles

18.3 Radioactive substances that are not in use should be kept in a properly designed store, unless their removal from the work area is likely to result in a greater hazard to the persons concerned. The arrangements made should provide security (as appropriate to their security category) against loss or theft and damage by fire and flood, as well as protection against radiation hazards. Details of security categories are available from local Police Counter Terrorism Security Advisers (CTSA). The IAEA has also published guidance on source security (IAEA 2019). The UK security requirements for the storage of high activity sealed sources and sealed sources exceeding the category 5 limits are classed as security sensitive information and are therefore not detailed in this publication. They can be obtained from the local CTSA, in the first instance via the local environment agency inspector, on a need-to-know basis. These security requirements cover information security as well as physical security.

18.4 Appropriate designation of the store as a Controlled or Supervised area under IRR (IRR 2017, IRR(NI) 2017) will depend on the contents, contamination risk, and external dose rates (see paragraphs 1.104 to 1.115, *Controlled areas and Supervised areas*, and appendix 6, *Designation of controlled and supervised areas*). Consideration should also be given to radionuclides with gaseous decay products, with all such sources being stored in an appropriately ventilated area as determined by risk assessment. Temperature and humidity should also be taken into account for package integrity (IAEA 2006).

18.5 Radioactive substances should not be retained unnecessarily once they are no longer required. Arrangements should be made for their transfer or disposal as radioactive waste (where appropriate, after a period to allow radioactive decay; see chapter 19, *Accumulation and disposal of radioactive waste*). The need to retain substances should be reviewed during the annual audit (see paragraph 18.51).

Responsibility

18.6 A condition of permits with authorised disposals under the radioactive substances regulations (RSA 1993, RSA(A)R(NI) 2011, EPR 2016, EA(S)R 2018) is that the Employer must appoint a Radioactive Waste Adviser (RWA) with the appropriate experience and competence in the relevant industry to advise on radioactive waste management and environmental radiological protection (EA 2012, SEPA 2020b). Where medical physics is the only user of radioactive substances on a hospital site, the RPS in the department may also be the custodian of radioactive substances for the entire site or department, with responsibility for the keeping and use of all radioactive substances at the establishment and for all the necessary records. However, the roles are distinct in law. The RPS supervises the arrangements for work with ionising radiation as set out in the local rules. The custodial duties described in this chapter should be clearly documented, e.g. in the BAT compliance document. There should be appropriate management systems in place that detail the responsibilities of the named individuals and highlight the chain of command.

18.7 Radioactive substances issued from a store should be in the care of responsible individuals at all times until their return or disposal. The responsibilities of individuals for the care of radioactive substances must be clearly set down in writing.

Design of stores

18.8 A store for radioactive substances may be a room or separate space outside the work area, or it may be a locked cupboard, locked refrigerator, locked freezer, or safe, often situated in the work area. The suitability of a store for radioactive materials should be the subject of a prior risk assessment, taking into account the following:

1. whether open and/or sealed sources are to be stored;
2. the length of time storage is required;
3. the adequacy of any shielding, given the emission type and activity of the radionuclides, and the security, designation and occupancy of all surrounding areas, including above the ceiling and below the floor;
4. the need for reduced temperatures or other forms of control of the storage environment (e.g. humidity);
5. ease of decontamination of the materials from which the store is made;
6. security requirements (see paragraph 18.3); and
7. if human tissue is being stored, the requirements of related regulation (Human Tissue Act 2004, *Human Tissue (Scotland) Act* 2006).

18.9 The appropriate environmental agency inspector should be consulted for the design of new stores, in particular the security requirements. The inspector can advise on whether the local CTSA should also be consulted, which is not generally required if only category 5 sealed sources and open sources are in use, unless there are particular security concerns.

18.10 Radioactive materials should be separated from non-radioactive to avoid cross-contamination and to aid in waste minimisation (but see chapter 19,

Accumulation and disposal of radioactive waste, when the disposal route is with domestic refuse).

18.11 The store should be sited in a suitable place, taking into account the dose rate from it and the convenience and reduced potential for accidents in transit if it is close to the work area. The store should be designed so that all persons are adequately protected during both storage and the transfer of substances to and from the store.

18.12 Shielding for radioactive substances should take account of scattered radiation and source distribution. For example, it is not sufficient to store large activities behind a thick barrier that protects the torso if the radiation scattered above the barrier presents a significant hazard to the eyes. Sometimes advantage can be taken of self-shielding, e.g. by arranging a group of spent radionuclide generators so that the one with the highest activity is at the centre.

18.13 The store should be ventilated by mechanical means when radioactive gas, dust, or vapour is likely to be present. The mechanical means may be an extractor fan that should operate for a sufficient time to dispel any airborne radioactivity before the room is entered or the cupboard opened and while persons are present in the storeroom or near an open store cupboard. Stack height and position should be sufficient to ensure adequate dispersal of the extracted air into the environment. Filters are not usually required for the quantities and types of radioactive materials normally stored in hospitals. If fitted, they may cause additional maintenance and disposal considerations.

18.14 Surfaces, including workbenches, floors, and ceilings, where radioactive materials may be handled or could be spilt, must be smooth and easy to decontaminate, with minimal joins. Consideration should be given to the strength of workbenches if they are used to support lead shielding.

18.15 The store should be sited and constructed with due regard for the need for security and the radiation hazards that may arise from stored radioactive substances in the event of fire or flood. Advice on these matters may be sought from both the HSE/HSENI and the Chief Fire Officer. Flammable materials, e.g. solvents, should not be stored near radioactive substances. However, paraffin wax used as a neutron moderator may be kept in the store provided that it is in a closed metal container.

18.16 A fireproof warning notice should be displayed outside the store if in a secure area or placed inside the store to be visible on entry if in a public area: a suitable design for unsealed sources is shown in figure A7.6 in appendix 7, *Warning signs and notices*. This notice should list the principal contents of the store: this can be helpful in cases of emergency. Alternatively, a 'grab pack', containing maps outlining the location of radioactive stores in the area and the maximum activity of radionuclides in that store can be prepared for emergency situations and stored in a secure area. Advice should be sought from the Chief Fire Officer and the CTSA, where appropriate.

18.17 Stores (including refrigerators and freezers) should be kept locked for security.

Keeping of radionuclide generators

18.18 All generators should be kept secure, whether in use or not. While in current use, generators, such as ^{99}Mo/^{99m}Tc and ^{68}Ge/^{68}Ga, used in the production of radiopharmaceuticals should remain in the radiopharmacy and be stored as referenced in paragraph 12.23. Suitable instructions should be made available to assure the adequate security of generators, such as ^{81}Rb/^{81m}Kr and ^{82}Sr/^{82}Rb, which must be sited in the imaging room while in use and possibly stored there, where appropriate, when not in use. These instructions should be developed from a risk assessment that takes account of security considerations such as department layout and availability of staff, as well as advice from the relevant environment agency and local CTSA. The instructions would normally be referred to or included in the local rules. Spent generators awaiting recycling or disposal should be moved to a store. Where the radioactive substances are liable to become airborne, see paragraphs 10.40, 10.80, and 10.86. The shielding of some makes of ^{99}Mo/^{99m}Tc generators may contain depleted uranium, which should be included on the site's sealed source Permit by weight (e.g. 80 kg) if the amount exceeds exemption levels (5 kg on premises) (BEIS *et al* 2018, SEPA 2020a). If exempted, exemption requirements for records, etc must be adhered to.

18.19 All civil nuclear material (thorium, uranium, and plutonium) in the UK is subject to the safeguards provisions of the Agreement between the IAEA and UK for the Application of Safeguards in the UK in Connection with the Treaty on the Non-Proliferation of Nuclear Weapons (IAEA and United Kingdom 2018). The Nuclear Safeguards (EU Exit) Regulations (2019) are primarily aimed at the nuclear industry but may also apply to any civil nuclear material held by small holders, e.g. this may include depleted uranium and compounds such as uranyl acetate. The Office for Nuclear Regulation (ONR) made a policy decision in 2019 not to include hospital inventories within reporting criteria. The ONR safeguard team can be contacted at UKSO@onr.gov.uk for confirmation of these arrangements.

Keeping of mobile and portable equipment

18.20 Mobile and portable equipment containing sealed sources, e.g. low activity after-loading equipment, should be moved to a store or to a secure place when not in use. The store must conform to the security requirements of the source type.

Special precautions for keeping open radioactive substances

18.21 Breakable containers, such as glass vials, should be stored inside robust, leak-proof containers with absorbent packing. If not for immediate use, the containers should be kept in a store (see paragraphs 18.7 to 18.8, *Design of stores*).

18.22 Some chemically unstable solutions containing radioactive substances can be hazardous and should be stored in vented containers and kept in ventilated storage areas. Examples include solutions that are by-products of HPLC analysis or purification, nitric acid or other oxidising solutions containing traces of organic material, peroxides, and chlorates.

18.23 Special care should be taken when retrieving or opening containers of radioactive substances that have been kept in store for more than a few months and where there may be a risk of bursting or frothing. They should be opened only over a drip tray in a safety cabinet.

18.24 Solutions stored in refrigerators or cold-storage units should be in suitable containers that will not break at low temperatures.

Spillage, lost or damaged sources

18.25 Loss of containment of sources (open and sealed) must be notified to the HSE if the quantities exceed those given in IRR Schedule 7 Part 1 Column 5 (IRR 2017, IRR(NI) 2017) (see chapter 10, table 10.1). Notification to the appropriate environment agency might also be required, unless the release is permitted under the relevant RSR (RSA 1993, EPR 2016, EA(S)R 2018) and within specially designed facilities, i.e. nuclear medicine holding tanks (HSE 2018).

18.26 Lost or stolen sources (open and sealed) will have to be notified to the HSE as well as the appropriate environment agency if the quantities exceed those given in IRR Schedule 7 Part 1 Column 6 (IRR 2017, IRR(NI) 2017) (see chapter 10, table 10.1).

18.27 Lost or stolen sources below the quantities listed in the IRR (IRR 2017, IRR (NI) 2017) but not out of scope of the radioactive substances regulation must be reported to the appropriate Environment Agency, the police, and, if in relation to transport related loss, the ONR.

Identification, inspection and leakage testing of radioactive substances

18.28 Every radioactive source or quantity of open radioactive substance should be identified uniquely by number, mark, label or other appropriate method, unless the activity is so low that the source is not considered radioactive (see Paragraph 21 of L121 (HSE 2018)) or out of scope for practices, as defined in the guidance document '*Scope of and exemptions from the Radioactive*

Substances Legislation in England, Wales, and Northern Ireland' (BEIS *et al* 2018) for Great Britain and Northern Ireland and *Scope of Radioactive Substances Legislation in Scotland—Guidance document* for Scotland (Scottish Government 2020). For very small sources (i.e. no dimension greater than 5 mm) and dispersible radioactive substances, the identification should be on the container, e.g. source applicator, magazine containing ^{125}I grains, or vial of solution. Identification of the contents should appear on containers. If lengths of foil or wire, e.g. ^{192}Ir, are cut off for immediate use, they need not be identified individually, but a container holding cut lengths should be labelled.

18.29 Some sealed sources, such as ^{125}I seeds, may be self-shielding and will have two associated activities; the 'actual' activity and the 'apparent' activity. The higher of the two should be accounted for within the site permit and records.

18.30 Labels on containers such as bottles of radioactive solution or attached to lengths of wire or foil sources should bear the radiation symbol.

18.31 Where two or more types of solid source have a similar appearance, a method of discrimination should be adopted; a check should be made at appropriate intervals depending upon the frequency of use to verify that the sources have retained their individual identification.

18.32 Sources should be inspected after use for evidence of damage before being returned to store. A competent person should also examine them with sufficient frequency to permit the early detection of progressive damage that might lead to the leakage of radioactive substances.

18.33 Leakage tests should be made thoroughly but quickly, using forceps, and in a manner that minimises the radiation dose to the operator. Guidance on the determination of methods and frequency of testing is given in paragraph 15.8.

18.34 Bottles of radiochemicals in store should be closed tightly to prevent leakage, and confirmatory leakage tests may be needed. Areas where vials of radiopharmaceuticals are kept should be monitored for contamination.

18.35 If a sealed source is to be used outside the manufacturer's recommended working life (RWL) of the source, a risk assessment must be carried out to ensure the continued safe use of the source, and leak tests should be carried out at a frequency determined by the risk assessment. If a source exceeds twice its RWL or has not been used in a year with no future plans for use, it should be disposed of (EA 2018). If the source is an approved 'special form' design, then consideration should be given as to whether it remains 'special form' beyond its RWL. This may affect classification under the Radiation (Emergency Preparedness and Public Information) Regulations (REPPIR 2019, REPPIR(NI) 2019) and transport requirements (ADR no date).

Maintenance and checking of stores

18.36 The custodian (paragraph 18.49) or another responsible person (see also paragraph 18.51) should maintain the stores in an orderly manner and conduct routine inspections.

18.37 Monitoring should be undertaken and recorded in and around stores to ensure that contamination has not occurred, with care taken to ensure contamination monitors are not responding to localised dose rates rather than waste held in the stores. Wipes may be used, if appropriate.

18.38 Monitoring of dose rates in and outside stores to confirm their appropriate designation should also be undertaken. The frequency of the monitoring will depend upon the use, quantity, and form of the substances; monthly monitoring should suffice for a typical store; for stable use, annual monitoring may suffice (see paragraphs 10.97 to 10.100, *Monitoring of work areas and persons*). Records of the monitoring should be kept.

Stock records and control procedures

18.39 See appendix 10 for information on *record keeping* and how long records must be retained.

18.40 A central register of sealed and solid sources should be kept in a secure and restricted area, which gives particulars of all sources. This register should include the following relevant information:

1. the radionuclide;
2. the activity on a given date;
3. the serial number or other distinguishing mark;
4. the date of receipt;
5. the normal location of the source; and
6. the date and manner of disposal.

It may be useful to add further information, such as a more detailed description or photograph of the source. Entries should be made relating to examinations, leakage tests, and repairs; alternatively, leakage test certificates may be attached to the register. Where lengths are cut from a foil or wire source (e.g. of ^{192}Ir) for an individual patient, the total amount should be accounted for, but records of the activity of each individual piece need not be kept for IRR (IRR 2017, IRR(NI) 2017) (also see paragraph 9.18).

18.41 Transfers of radioactive substances to and from departments or areas must be recorded (IRR regulations 29, 30 (IRR 2017, IRR(NI) 2017).

18.42 Records should be kept in each store of radioactive substances received, issued, and returned.

18.43 The records for sealed or other solid sources should contain sufficient detail and be arranged so that information concerning the whereabouts of every solid source is immediately available. This will also provide a means for showing at any time the number of sources of each type actually in each store and available for issue. The total quantity of each radionuclide held on the premises should also be available.

18.44 Details of sealed sources must not be passed over to unsecured email accounts.

18.45 Records of the receipt, issue, and disposal of open sources should be kept in such a way as to allow stock control, enable loss or theft to be noticed quickly, and estimate the total quantity of each radionuclide at any particular time. These should all be kept in a secure and restricted area.

18.46 When ordering open sources, a check should be made that the activity ordered will not breach any permit or authorisation upon delivery, taking into account the reference time and delivery time.

18.47 Where radionuclide generators are used, there should be:

1. a record showing the receipt of new generators with their activity and reference date and the return or disposal of spent ones; and
2. a record of use for each generator showing all relevant information, e.g. the date, time, and activity of each elution for a ^{99}Mo/^{99m}Tc or ^{68}Ge/^{68}Ga generator, and the date and list of patients administered with activity from a ^{81}Rb/^{81m}Kr or ^{82}Sr/^{82}Rb generator.

18.48 When radioactive substances are administered to patients, entries should be made in the hospital's stock records and in the patients' individual records (see paragraph 10.82). Entries will also be needed when temporary implants are removed.

18.49 Where sources are readily accessible, the custodian or the person responsible for the store must carry out an audit at regular intervals to account for every solid source in storage or in use. The frequency of the audit should be established in a radiation risk assessment. Sources in the store should be counted. A check of serial numbers must be included if the risk assessment indicates it. An entry in the record should cover sources that are no longer available in the store.

18.50 The stock of open radioactive substances should be inspected to ensure that it agrees with receipts, issues, and disposals, taking into account radioactive decay.

18.51 The Employer should nominate a senior staff member to conduct an annual check audit.

Movement of radioactive substances (in areas wholly within the employer's control)

18.52 Containers used to store radioactive substances may not be suitable for movement and may need additional physical protection when taken outside a building. Consideration should be given to the need for total containment in the event of spills and any requirement to reduce the external dose rate. Further advice is provided in L121 (HSE 2018) for guidance on IRR regulation 30 (2) (IRR 2017, IRR(NI) 2017).

18.53 The container should be marked clearly to indicate that its contents are radioactive, and the label should carry the warning sign shown in appendix 7, *Warning signs and notices.*

18.54 Only members of staff who are adequately trained should move radioactive substances between rooms or buildings, taking into account the manual handling requirements for shielding (see paragraphs 10.42, 10.44, 10.75–10.77).

18.55 If radioactive goods are moved within an establishment and there is no use of public roads or railways, the transport is subject to the safety regulations within the establishment, and the requirements of transport legislation do not apply. However, if the site is open to the public, IRR and the Carriage of Dangerous Goods and Use of Transportable Pressure Equipment Regulations (CDG) as amended (CDG 2009, CDG(NI) 2010, CDG(A) 2019, CDG(A)(NI) 2019) will apply (L121 paragraph 597) (HSE 2018).

Transport of radioactive materials

18.56 Most obviously, the transport of radioactive materials by road occurs when the materials are placed in a vehicle and taken outside the confines of the establishment or on any road running through the establishment to which the public has access. However, transport is more than the actual journey by road. Legislation places obligations on those designing and manufacturing packages for use, as well as those maintaining and repairing them, those preparing a package for transport to consign radioactive material, and a number of other participants in the transport chain. All transport of radioactive materials should conform to the relevant national and international regulations and codes of practice relating to the various modes of transport used (Department for Transport 2012), including road (CDG 2009, CDG (NI) 2010, CDG(A) 2019, CDG(A)(NI) 2019), rail (COTIF 2011), sea (The Merchant Shipping (Dangerous Goods and Marine Pollutants) Regulations 1997, The Dangerous Goods in Harbour Areas Regulations 2016, IMO 2016) or air (ICAO 2015). IRR (IRR 2017, IRR(NI) 2017) also applies in relation to the transport of radioactive material, and a radiation risk assessment will be required. The transport of radioactive material by road within Great Britain is governed by the Carriage of Dangerous Goods and Use of

Transportable Pressure Equipment Regulations 2009 (CDG 2009) as amended (CDG(A) 2019), and in Northern Ireland by the Carriage of Dangerous Goods and Use of Transportable Pressure Equipment Regulations (Northern Ireland) 2010 (CDG(NI) 2010) as amended (CDG (A)(NI) 2019). CDG reads out the current European agreement, '*Accord européen relatif au transport international des marchandises dangereuses par route*' (known as ADR). The ADR is updated biennially on the 1st of January, with a six-month grace period. The following refers to the 1st January 2019 version (ADR 2018), with references to the relevant paragraph/ table. It is unusual for major changes in text or paragraph numbers to be made for the transportation of radioactive substances between versions, but procedures should always be checked against the latest version (ADR no date). There are also approved derogations (DfT 2020a) relating to national carriage in particular circumstances, and ONR has issued various authorisations in line with CDG Regulation 12 for use in particular situations.

18.57 CDG does not apply to radioactive material implanted or incorporated into someone for medical diagnosis or treatment (ADR 1.7.1.4 (c)) or to a person subject to contamination or accidental/deliberate intake of radioactive material ([ADR 1.7.1.4 (d)). They do apply to all those involved in the transport operation, not just the movement. This includes packers, loaders, drivers, and those unloading consignments.

18.58 The ONR is the enforcing body within Great Britain for CDG in relation to civil transport, and the Department of Agriculture, Environment, and Rural Affairs (DAERA) is the enforcing authority for the equivalent legislation in Northern Ireland. The ONR also enforces the IRR (IRR 2017, IRR(NI) 2017) in relation to the civil transport of radioactive materials by road. The ONR has published guidance to assist non-nuclear duty holders in their compliance with transport regulations (ONR 2022b).

18.59 The radiation protection adviser (RPA) should be consulted for advice before the establishment first transports any radioactive material to ensure compliance with the IRR (IRR 2017, IRR(NI) 2017).

18.60 Where radioactive materials are to be transported other than excepted packages only, then as a consignor or carrier, there is a need to appoint a Dangerous Goods Safety Adviser (DGSA) (ONR 2021a) (Note that this was a change from the previous wider interpretation of exemption from DGSA requirements). The DGSA being considered for appointment should be suitably qualified and experienced in providing advice relating to class 7 dangerous goods. The DGSA is required to monitor compliance, provide advice, and prepare an annual report. More detail is given in ADR 1.8.3.3. Note: Exemptions from the need to appoint a DGSA relate to national carriage only. Further guidance on DGSAs is available on the ONR website.

18.61 The information below was written specifically for small users in the medical sector transporting radioactive materials on the road within Great Britain, generally between hospital sites and suppliers of medical radioactive material, in standard cars, assuming the transport of radioactive material is the sole purpose of the journey (i.e. there are no other deliveries of different categories of dangerous goods en route or delays in public areas). There may be additional requirements for other journeys, such as the transport of Type B packages and transfrontier shipments, and users should seek further advice from their Dangerous Goods Safety Advisor (DGSA).

18.62 If duty holders are involved in the transport of High Consequence Radioactive Material (HCRM), as consignors or carriers, additional security requirements are required. The HCRMs are those that have the potential for misuse in a terrorist event. A package containing material with an activity greater than 3000 × A_2 will generally be considered HCRM, with the ADR listing different threshold activities for 25 specific radionuclides, including Cs-137 and Ir-192 (ADR 1.10.1.3).

18.63 Transport duty holders are required to make a suitable and sufficient assessment of the risks to employees and others associated with the transport of radioactive material (IRR, Reg 8). This is required before transport commences. To be suitable and sufficient, all relevant matters in the associated Approved Code of Practice (ACoP) paragraphs 70 and 71 (HSE 2018) should have been duly considered and appropriate decisions made on measures appropriate to restricting exposure to ionising radiation.

18.64 Carriers of radioactive material should ensure they have the necessary registration and/or consent from HSE in line with IRR ahead of this practice commencing.

Management system

18.65 There is a requirement for transport duty holders to establish and implement a management system for all activities within the scope of CDG/ADR to ensure compliance with all relevant provisions. All procedures should be reviewed periodically and updated to reflect the latest legislation and guidance. This could be set as a standing item on the Organisational Radiation Safety Committee meeting agenda for review. A review of suppliers related to transport, e.g. suppliers of package containers, labels, etc should be included within this.

Radiation protection programme

18.66 CDG/ADR refers to the need for a radiation protection programme, with specific requirements set out in ADR 1.7.2. Any Employer involved in the transportation of non-exempt radioactive materials must have a radiation protection programme in place for compliance with IRR (IRR 2017, IRR

(NI) 2017), which outlines the systematic arrangements needed to protect staff and the general public while sources are in transit. As IRR is more developed in this regard, duty holders may make reference to the fact that compliance with a radiation protection programme is achieved via IRR to demonstrate the requirements in ADR 1.7.2 have been considered.

Receipt and collection of packages containing radioactive material

18.67 When radioactive generators and other radiopharmaceuticals or sources are delivered, there are certain security arrangements that need to be in place (HSE 2003). This particularly applies when deliveries are made 'out of hours' when departments are closed. A suitable risk assessment is required, and this should consider the appropriate levels of security arrangements for all radioactive materials on the premises at all times. In this section, 'Employer' generally refers to the hospital employer receiving radioactive materials for use or returning spent generators, etc.

18.68 From HSE reports (HSE 2003), the most common time for any loss of radioactive materials to occur is at the time of transfer of responsibility, i.e. when deliveries or collections are being made. When radioactive materials are delivered to a hospital, the receiving employer is immediately responsible for all aspects of security, storage, and appropriate record keeping.

18.69 The hospital must have sufficient arrangements in place to ensure that any new generators or other radioactive items are received knowingly. The hospital's source records need to accurately reflect the radioactive materials and their location on the premises. Similarly, the return of spent generators must be considered as to when and who removed them from the premises if collection is outside of normal working hours.

18.70 It is not acceptable for the hospital to be unaware of the presence, security, or condition of any radioactive delivery until the next morning or after the weekend. Also, if deliveries are made outside of the normal working hours of the nuclear medicine department or radiopharmacy, it is not acceptable to rely on the delivery driver of a separate employer for this security.

18.71 Given the above framework, there are five principles to consider for any arrangements (Lawson *et al* 2004):

1. Responsibility—the point of transfer of responsibility for the radioactive package from the delivery company to the Employer (and vice versa) must be clear.
2. Training—Appropriate procedures for the receipt and dispatch of radioactive packages should be produced by the Employer, and all staff involved should be instructed in them.
3. Accounting—The Employer must ensure that appropriate records are kept of the receipt or collection of all radioactive packages.

4. Timeliness—Such accounting procedures should be capable of quickly identifying delivery issues, such as the wrong source delivery, the loss of any source, or breach of permit. In particular, the receipt or collection of a package needs to have a mechanism for prompt acknowledgement.
5. Security—There must be a suitable store for all radioactive packages when they are delivered or collected. This store must offer physical security at a level consistent with the Employer's permit in consultation with the local environment agency inspector. A suitable security risk assessment should be in place to cover relevant areas where radioactive packages are delivered. Templates are available from the Department for Transport (DfT 2020b).

18.72 The simplest mechanism for the receipt (or dispatch) of a radioactive package is that a member of staff employed directly by the Employer can accept (or hand over) the package and sign for the receipt (or transfer). Whoever carries this out must be trained in the procedure (including what to do with the package on receipt to ensure its security) and should be aware of their delegated responsibilities in this regard, including any record keeping. This should also include instructions in the event that the package appears to be damaged.

18.73 Note that some staff working for the Employer may be employed by a third party, e.g. agency staff or porters and security staff employed by an outside agency. Such staff can sign for receipt (or dispatch) provided that the Employer has a specific agreement with the agency staff to carry out such duties on behalf of the Employer.

18.74 Generally, a delivery driver working for a carrier company would not be able to sign for the delivery or receipt on behalf of the Employer. However, the Employer could also have a specific written agreement with the carrier company authorising named drivers to carry out such procedures on behalf of the Employer. This would need to include authorisation to sign the Employer's records for receipt or delivery of packages, which would allow such procedures to be carried out by the driver alone. Such drivers would need to have appropriate instruction in the relevant Employer procedures. This solution is only practical for situations where only a small number of regular drivers are used for such deliveries or pickups by the carrier company. Also, access for such drivers should only be to the store, which is used solely for the receipt or pickup of such radioactive packages.

18.75 All packages on delivery must be placed immediately in the designated secure store. It is not acceptable for packages to be left unattended in corridors or at reception, e.g. under desks. If a delivery or receipt is made outside of the normal working hours of the radiopharmacy/nuclear medicine department, then a radioactive store may be needed close to the point of delivery. The procedure will detail the arrangements for appropriate staff to unlock the store and record the delivery (or pick up).

18.76 A risk assessment must be carried out to determine if the radioactive store requires designation under IRR (HSE 2018). If the store is only used for receipt and collection of radioactive packages and the packages are stored in their transport packages, then it may not require designation or only be designated as a supervised area in consultation with the RPA.

18.77 Another factor to consider is whether access to the radioactive store is through a controlled area. If this is the case, or the radioactive store itself is designated a controlled area, then arrangements for out of hours access as required by Employer staff (e.g. security or porter staff) should be included in the department's local rules. Such staff will then access the radioactive store under a written system of work and will require appropriate training. Such training only needs to be very specific to the tasks performed by these staff. However, this does mean that only named individuals with the appropriate training can carry out the task. The same system of work could be written to permit a delivery driver to enter the controlled areas if they are accompanied by a member of staff trained to work in that controlled area. However, for a driver to be able to access the areas unaccompanied, they would have to fall into the category of being a named driver working for a company with which the Employer has a written agreement that the driver can act on behalf of the Employer with regard to the security of the radioactive packages. Such named drivers would then also need the specific training required to access the controlled area. Since the driver is categorised as an outside worker, a suitable liaison between the Employer and the driver's employer is needed to establish the driver's radiation exposure within the Employer's premises.

18.78 Some possible scenarios for consideration are given below. These are only examples, and an individual Employer is free to consider any different scenario. However, the security aspects of this should be discussed with the relevant Environment Agency (as advised by the NaCTSO) prior to implementation.

Scenario 1

The simplest scenario is that deliveries and collections are made to the radiopharmacy/nuclear medicine departments only during working hours. Employer staff can then take responsibility at the point of handover, either signing for the receipt of packages or getting the delivery driver to sign for goods going out. The member of staff should then check the contents of the packages for any damage and ensure that they are the appropriate goods. The contents are then placed in the radioactive store, and appropriate records are made.

Scenario 2

If packages are delivered outside of normal working hours, a trained member of Employer staff opens the radiopharmacy/nuclear medicine department and oversees the driver placing the packages in an agreed secure storage area according to protocol. A member of staff signs for the packages, and a copy of the transport document is placed in a designated location in the store. For collection, this process would essentially operate in reverse, with the driver signing the collection record. Staff should verify that the collection has taken place and that records have been completed at the next practical opportunity.

Scenario 3

Where it is not practical for staff to access the radiopharmacy/nuclear medicine department out of hours, an alternative secure radioactive store needs to be available, ideally near the agreed drop off/collection point. The driver reports to the drop off/collection point. A member of the Employer staff trained in the protocol places them in the secure store and signs the local record. A member of the radiopharmacy/nuclear medicine department then retrieves the packages as soon as is practical, signs the local record, takes the packages to the department where they are opened and inspected, and places them in store with details entered in departmental records.

Scenario 4

The transport company has a formal written agreement with the Employer allowing their employees to act on behalf of the Employer with regard to the secure delivery and collection of radioactive packages outside normal working hours. Such specified and trained drivers will have direct access to a secure radioactive store (e.g. they have been issued with a key or a coded digital lock). The driver places the packages in the store, completes and signs the local record, and locks the store. At the next opportunity, radiopharmacy/nuclear medicine staff retrieve the packages, complete the local record, take the packages to the department for inspection and storage, and also complete the departmental records.

Scenario 5

There may be a scenario where there is no staffed delivery point or formal written agreement, but the delivery drivers have received training in the protocol for delivery/pick up. An acceptable procedure could have a security controller able to monitor the radioactive store with CCTV. The driver contacts the security monitor, who observes the storage of the packages in the radioactive store, and the driver signs the local record. The security controller enters the delivery into the security log, which in effect acts as the point of transfer of responsibility to the Employer.

18.79 Incoming packages of radioactive materials should be opened only by persons having the necessary knowledge and in an area set aside for the purpose. While waiting to be opened, packages should be stored securely. For multiple out of working hours deliveries by different drivers, a locker system may be employed to secure packages.

Use of external companies

18.80 If third party carriers are used for the transport of radioactive goods, the consignor must take appropriate measures to ensure the consignment is compliant with the ADR (1.4.2.1.2). Some details, such as a company's DGSA, training certificates, audit reports, etc can be checked on contract renewal/award through a simple questionnaire. If a company is regularly used to transport a consignment from a department, e.g., radiopharmaceuticals produced in the radiopharmacy on a daily basis, spot checks may be performed and recorded for audit purposes on the vehicles and equipment carried to ensure compliance. It is helpful for consignors to hold copies of the photo ID of all drivers, alongside copies of their training certificates, to check before handing over any packages for consignment. Consignors may also wish to check that the third party being used has the necessary registration and/or consent from the HSE in relation to IRR Regulations 6/7 (IRR 2017, IRR(NI) 2017) in place.

Packages

18.81 The types of radioactive packages generally transported for medical use and covered within this chapter are excepted and type A.

Exempt consignments

18.82 Exempt consignments (i.e. the total activity in any package(s) or load presented by a consignor for carriage) contain minimal quantities or concentrations of radioactive materials (ADR 2.2.7.1.1) and are exempt from all the requirements of the transport regulations (see columns 4 and 5 of table 18.1). (Note that 'exempt' in the CDG regulations is different from 'exempt' in the radioactive substances regulations and more analogous to 'out of scope'.) This would generally include all empty Type A packages that previously contained sources, provided the conditions in table 18.1 were met and there had been no contamination during transport (ADR 2.2.7.1.2). To meet this requirement, procedures must be in place to prove there is no demonstrable contamination of the packages if contamination monitoring and leak testing are not sensitive enough to measure the required activity. This can include procedures where cases are unpacked away from areas where radioactive manipulations take place. If contamination of empty Type A packages cannot be discounted, the package must be transported as an excepted package.

Excepted packages

18.83 Excepted packages typically contain limited quantities of radioactive substances and are subject to exemption from some of the requirements of the ADR (1.1.3.4; see columns 6 and 7 of table 18.1).

Table 18.1. Activity limits for some radionuclides of particular interest in healthcare for exempt (ADR table 2.2.7.2.2.1), excepted (ADR table 2.2.7.2.4.1.2), and Type A (ADR 2.2.7.2.4.4) packages.

					Activity limits (MBq)		
					Excepted material package[a] limits (non-special form)		
Radionuclide	A_1(TBq)	A_2(TBq)	Activity concentration for exempt material (Bq/g)	Activity limit for an exempt consignment (Bq)	Solid ($10^{-3}A_2$)	Liquid ($10^{-4}A_2$)	Type A material package[b] limits (non-special form) (A_2)
Am-241	1×10^{1}	1×10^{-3}	1×10^{0}	1×10^{4}	1	0.1	1000
Co-57	1×10^{1}	1×10^{1}	1×10^{2}	1×10^{6}	10 000	1000	10 000 000
Co-60	4×10^{-1}	4×10^{-1}	1×10^{1}	1×10^{5}	400	40	400 000
Cr-51	3×10^{1}	3×10^{1}	1×10^{3}	1×10^{7}	30 000	3000	30 000 000
Cs-137	2×10^{0}	6×10^{-1}	1×10^{1}	1×10^{4}	600	60	600 000
F-18	1×10^{0}	6×10^{-1}	1×10^{1}	1×10^{6}	600	60	600 000
Ga-67	7×10^{0}	3×10^{0}	1×10^{2}	1×10^{6}	3000	300	3 000 000
Gd-153	1×10^{1}	9×10^{0}	1×10^{2}	1×10^{7}	9000	900	9 000 000
Ge-68	5×10^{-1}	5×10^{-1}	1×10^{1}	1×10^{5}	500	50	500 000
I-123	6×10^{0}	3×10^{0}	1×10^{2}	1×10^{7}	3000	300	3 000 000
I-125	2×10^{1}	3×10^{0}	1×10^{3}	1×10^{6}	3000	300	3 000 000
I-129	Unlimited	Unlimited	1×10^{2}	1×10^{5}	Unlimited	Unlimited	Unlimited
I-131	3×10^{0}	7×10^{-1}	1×10^{2}	1×10^{6}	700	70	700 000
In-111	3×10^{0}	3×10^{0}	1×10^{2}	1×10^{6}	3000	300	3 000 000
Ir-192	1×10^{0}	6×10^{-1}	1×10^{1}	1×10^{4}	600	60	600 000
Kr-81	4×10^{1}	4×10^{1}	1×10^{4}	1×10^{7}	40 000	4000	40 000 000
Lu-177	3×10^{1}	7×10^{-1}	1×10^{3}	1×10^{7}	700	70	700 000
Mo-99	1×10^{0}	6×10^{-1}	1×10^{2}	1×10^{6}	600	60	600 000
P-32	5×10^{-1}	5×10^{-1}	1×10^{3}	1×10^{5}	500	50	500 000
Ra-223	4×10^{-1}	7×10^{-3}	1×10^{2}	1×10^{5}	7	0.7	7000

Table 18.1. (*Continued*)

Radionuclide	A_1(TBq)	A_2(TBq)	Activity concentration for exempt material (Bq/g)	Activity limit for an exempt consignment (Bq)	Activity limits (MBq): Excepted material package[a] limits (non-special form)		Activity limits (MBq): Type A material package[b] limits (non-special form) (A_2)
					Solid ($10^{-3}A_2$)	Liquid ($10^{-4}A_2$)	
Rb-81	2×10^{0}	8×10^{-1}	1×10^{1}	1×10^{6}	800	80	800 000
Se-75	3×10^{0}	3×10^{0}	1×10^{2}	1×10^{6}	3000	300	3 000 000
Sm-153	9×10^{0}	6×10^{-1}	1×10^{2}	1×10^{6}	600	60	600 000
Sr-89	6×10^{-1}	6×10^{-1}	1×10^{3}	1×10^{6}	600	60	600 000
Tc-99m	1×10^{1}	4×10^{0}	1×10^{2}	1×10^{7}	4000	400	4 000 000
Tl-201	1×10^{1}	4×10^{0}	1×10^{2}	1×10^{6}	4000	400	4 000 000
Y-90	3×10^{-1}	3×10^{-1}	1×10^{3}	1×10^{5}	300	30	300 000

[a] (for gases, special form solids, and instruments or articles in excepted packages, see table 2.2.7.2.4.1.2).
[b] (for special form, use A_1; otherwise, use A_2—see ADR 2.2.7.2.4.4).

Type A packages

18.84 Type A Packages contain larger quantities of radioactive substances (see column 8 of table 18.1). Activity limits for these packages are calculated as follows in table 18.1 for specific radionuclides, commonly used in the medical sector. For radionuclides not listed, refer to table 2.2.7.2.2.1 in the ADR.

18.85 For mixtures of radionuclides in a Type *A* package, the following will apply (ADR 2.2.7.2.4.4):

$$\sum_i \frac{B(i)}{A_1(i)} + \sum_j \frac{C(j)}{A_2(j)} \leqslant 1$$

where $B(i)$ is the activity of radionuclide i as special form radioactive material;
$A_1(i)$ is the A_1 value for radionuclide i;
$C(j)$ is the activity of radionuclide i as other than special form radioactive material; and
$A_2(j)$ is the A_2 value for radionuclide j.
For example, for a non-special form consignment (i.e. A_2 values used) of 50GBq of ^{99m}Tc, 20GBq ^{131}I and 10GBq ^{123}I:

$$\sum \frac{0.05}{4} + \frac{0.02}{0.7} + \frac{0.01}{3} = 0.044$$

Thus satisfying the conditions for consignment as a Type A package.

18.86 The labels required for Type A packages are classified into three subcategories by the maximum dose rate on the external surface of the package and the Transport Index (TI) (the dose rate in mSv.h^{-1} at 1 m from the external surface, multiplied by 100 (ADR 5.1.5.3.1) (or measured in μSv.h^{-1} and divided by 10)). The different categories and their corresponding surface dose rates and TI are shown in table 18.2. Guidance on radiation and contamination monitoring requirements when transporting radioactive material and how to determine a TI is available from the ONR website (ONR 2016).

18.87 If the maximum surface dose rate and TI fall into different categories, the one corresponding to the higher category should be chosen, with WHITE I being the lowest category and YELLOW III the highest.

Consignment requirements

18.88 The requirements for the transportation of excepted and Type A packages in the UK are outlined below.

UN number and shipping name

Excepted

18.89 UN 2908 Radioactive Material, Excepted Package, Empty Package (ADR 3.2 table A).

Table 18.2. The label requirements for type A package category (ADR table 5.1.5.3.4). Note: there is scope to exceed these TI/dose rates, but this is conditional and further advice from the DGSA would be appropriate.

Category		Dose rate on external surface	Transport index (TI)
I (White)		Not more than 5 μSv.h^{-1}	0[a]
II	(Yellow) 	5–500 μSv.h^{-1}	0–1
III	(Yellow) 	500–2000 μSv.h^{-1}	1–10

[a] Where the TI is not greater than 0.05, a figure of 0 may be entered

18.90 UN 2910 Radioactive Material, Excepted package, Limited Quantity Of Material (ADR 3.2 table A).

18.91 UN 2911 Radioactive Material, Excepted Package, Instruments or Articles (ADR 3.2 table A). For instruments and articles, see ADR 2.2.7.2.4.1.3 and ADR 2.2.7.2.4.1.6. Note that the activity limits set out in table 18.1 do not apply.

Type A

18.92 UN 2915 Radioactive Material, Type A Package, non-special form, non-fissile or fissile-excepted (ADR 3.2 table A).

18.93 UN 3332 Radioactive Material, Type A Package, special form, non-fissile or fissile-excepted (ADR 3.2 table A). (For special form (ADR 1.6.6.4) material, note that the activity limits set out in table 18.1 do not apply—see ADR 2.2.7.2.4.4).

Maximum dose rate on the external surface of the package

Excepted

18.94 Does not exceed 5 $\mu Sv.h^{-1}$ (ADR 2.2.7.2.4.1.2).

Type A

18.95 Does not exceed 2 $mSv.h^{-1}$ (ADR 4.1.9.1.11) **unless transported under conditions for exclusive use.**

Package requirements

18.96 All packaging exterior containment, such as boxes, drums, or bags, and internal components, such as lead pots and inserts/spacers should be inspected periodically to ensure they remain in good condition. Inspection may consist of daily checks prior to packing and a more detailed inspection of hinges for dents or scratches that could compromise the integrity of the shielding. These checks should be done in accordance with any manufacturer's information, with a local protocol developed and results recorded.

Excepted

18.97 Excepted packages should be designed to meet the requirements in ADR 6.4.4. Note: Not all the requirements are covered in detail here.

18.98 Excepted packages consigned under UN2910 (limited quantity of material) must retain the radioactive contents under routine conditions of carriage (ADR 2.2.7.2.4.1.4(a)).

18.99 Excepted packages consigned under UN2910 must be marked as 'RADIOACTIVE' on an internal surface that is visible on opening the package or on the outside if it is impractical to mark an internal surface (ADR 2.2.7.2.4.1.4(b)).

Type A

18.100 Type A packages should be designed to meet the requirements in ADR 6.4.2 and ADR 6.4.7. Note: Not all the requirements are covered in detail here.

18.101 The package should only contain items necessary for the use of the radioactive material (ADR 4.1.9.1.3).

18.102 The outside of the package should include a seal, such as a padlock, to provide evidence the package has not been opened during transport (ADR 6.4.7.3).

18.103 If the package contains liquid radioactive material, there should be enough absorbent material to absorb twice the volume contained, or the package itself should incorporate a containment system to entirely enclose the liquid in the event of a leak (ADR 6.4.7.16).

18.104 Tests to prove compliance with the requirements for Type A and excepted packages are described in sections ADR 6.4.12 to ADR 6.4.17.4. Type A packages are available commercially, along with certificates of conformance. These certificates have an expiration date and should be updated with changes to the ADR to ensure the package used conforms to the requirements of the legislation. They can usually be obtained from the original supplier or are posted on the supplier's website.

18.105 If a centre designs and uses their own Type A packages, they should ensure testing is completed and the packages conform to the requirements. Retesting may be required with changes to the ADR, and an updated conformance certificate may be issued.

External package requirements

18.106 The level of non-fixed contamination on external surfaces of a package (averaged over 300 cm^2) must not exceed 4 $Bq.cm^{-2}$ for beta and gamma radiation and low toxicity alpha emitters and 0.4 $Bq.cm^{-2}$ for all other alpha emitters (ADR 4.1.9.1.2). This should be checked before consignment using swabs and a suitable radiation monitor.

Package marks

18.107 The outside of the package must:

(a) show the UN number of the package, preceded by 'UN';
(b) Identify the consignor, consignee or both;
(c) Show the permissible gross mass if this exceeds 50 kg.

(excepted—ADR 5.1.5.4.1, *type A—ADR* 5.2.1.7.1)

The marks must be legible and durable to withstand conditions in transport.

Type A

18.108 The package must be marked as 'Type A' (ADR 5.2.1.7.4(b)) with the proper shipping name (ADR 5.2.1.7.2, 5.2.2.1.6(a)).

At least two of the correct category labels (of dimensions defined in ADR 5.2.2.2.1.1 and design defined in ADR 5.2.2.2.2, see table 18.2) are on opposing sides (ADR 5.2.2.1.11.1), with the correct information (contents, total activity, and TI as appropriate) completed as required (ADR 5.2.2.1.11.2). Radionuclides should be written as in table ADR 2.2.7.2.2.1, e.g. ^{99m}Tc, ^{123}I (ADR 5.2.2.1.11.2 (a) (i)). The originating VRI code (international vehicle registration identification code, e.g. 'UK') and package sponsor identifier in Type A mark requirements are also needed ADR 5.2.1.7.4 (c).

Vehicle placarding

Excepted

18.109 Not required (ADR 1.7.1.5).

Type A

18.110 The placard will be designed as specified in ADR 5.3.1.7.2.

18.111 The placards should be affixed to both sides and the rear of the vehicle (ADR 5.3.1.5.2); see below. These must be removed when the Type A packages have been delivered. Typically, large placards are used, as below; however, there is scope for these to be reduced in size to 100 mm × 100 mm for smaller vehicles.

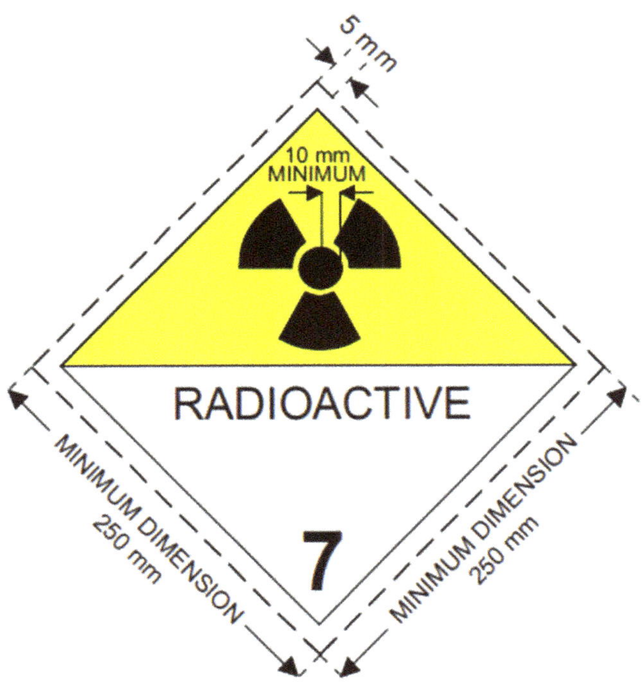

Orange plate markings

18.112 Two orange plates will be affixed to the front and rear of the vehicle, as specified in ADR 5.3.2.

Excepted

18.113 Not required (ADR 1.7.1.5).

Type A

18.114 Two orange plates conforming to ADR 5.3.2.2.1 at the front and rear of the vehicle (ADR 5.3.2.1.1).

Or

18.115 For small vehicles (that do not exceed 3.5 tonnes), where the number of Type A packages transported does not exceed 10 and the sum of the transport indices does not exceed 3, a fireproof sign with the following text should be securely posted in the vehicle where it is plainly visible to the driver but does not obscure their view of the road, as an alternative to the orange plates: Note: *Road derogation* 9 (DfT 2020a) contains conditions relating to the dimensions of the fireproof plate and text to be displayed that must be adhered to if this is used as an alternative to orange plates.

> This vehicle is carrying
> RADIOACTIVE MATERIALS
> In case of accident, get in touch at once with
> THE POLICE
> In an emergency, contact the 'JOB TITLE or DEPARTMENT'
> DEPARTMENT
> ADDRESS
> TEL: XXXXX XXXXXX or XXXXX XXXXXX (out of hours)

Vehicle requirements

18.116 Packages should be transported by cars (or vans) that are in good condition.

18.117 Public transport, motorbikes, taxis, and bicycles should not be used.

18.118 Insurance policies should cover the transport of radioactive materials. If staff use their own cars to transport radioactive material, they should check that their insurance will cover them in the event of an incident. Insurance companies will often cover the vehicle; however, the cost of any clean up in the event of an incident, e.g. a radioactive spill, is the responsibility of the individual. It may be necessary to have a written letter from the employer stating that they will cover the costs on such an occasion or to use a car insured by the employer.

18.119 The package itself must be stowed securely within the vehicle, as far from the drivers and vehicle crew as possible, to prevent free motion during transport and reduce radiation dose in accordance with the as low as reasonably achievable (ALARA) principle (ADR 7.5.7.1, 7.5.11, CV33). Minimum segregation distances outlined in CV33 (1.1) (ADR table A) must also be considered.

18.120 It is advised that the vehicle remain locked during the journey, and the driver should not leave the car unattended where possible. If overnight journeys are made or the vehicle must be left for long periods of time, arrangements should be made for the secure parking of the vehicle, as outlined in ADR 8.4. These requirements do not apply if the vehicle is locked and the dose rate outside the vehicle is less than 5 $\mu Sv.h^{-1}$ at any point [*S21*], therefore the vehicle may be left briefly if delivering a package or paying for petrol.

18.121 If a vehicle is routinely used for the transportation of excepted and Type A packages, or other equipment is regularly used, it should be monitored periodically for contamination at a time interval proportional to the likelihood and extent to which radioactive packages are transported (ADR 7.5.11, 5.3). In practice, for nuclear medicine consignments, the risk is quite low, as all packages should be wipe tested before consignment. The ONR recommends that for vehicles transporting liquid radiopharmaceuticals on a daily basis, a monthly check would be appropriate (ONR 2016).

Transport documents

18.122 A copy of the Emergency procedures should be readily available to the driver in case of an incident. See paragraphs 18.140 to 18.147, *Emergency arrangements*.

Excepted

18.123 The Transport Document for excepted packages must contain the United Nations (UN) number and the Name and address of the consignor and the consignee (ADR 5.1.5.4.2).

18.124 Instructions in writing not required (ADR 8.4, *S5*).

Type A

18.125 The transport document must contain the following information in any location and order, but with the first eight in that order and no other information interspersed (ADR 5.4.1.1.1, 5.4.1.2.5.1, 8.1.2.1):

1. UN number, preceded by 'UN' e.g. UN 2915;
2. Proper shipping name: e.g. 'RADIOACTIVE MATERIAL, TYPE A PACKAGE;
3. Class number: 7;
4. The tunnel code (only required if transport is through a tunnel or there are tunnels in the area);
5. Name or symbol of each radionuclide;
6. Description of the physical or chemical form;
7. Maximum activity of the radioactive contents in Bq;
8. Category of the package (e.g. White I, Yellow II, Yellow III);
9. Transport Index (Yellow II or Yellow III only);

10. For special form, low dispersible, or special arrangement consignments, the identification for each competent authority approval certificate applicable to the consignment;
11. Name and address of consignor;
12. Name and address of consignee.

18.126 A Declaration (ADR 5.4.1.1.1 (i)) is only required if transporting goods under a special agreement (as approved by the competent authorities) and, as such, does not apply for the arrangements described in this chapter. However, for accounting and audit purposes, signatures and names of persons acting as consignors, transporters, and consignees of the package should be recorded.

18.127 The instructions in writing (four pages specified in ADR 5.4.3.4) are available to download separately (must be in colour) from the full ADR online (ADR 2022). These are periodically updated, so the most current version must be carried. A photo ID (e.g. passport or driving licence) and driver training certificate should be kept in the vehicle's crew's cab (ADR 8.1.2). The training certificate is useful in case the vehicle is stopped by the police. It is recommended that the emergency procedures be kept with these documents in a plastic wallet.

18.128 Emergency arrangements appropriate to the consignment must also be provided by the consignor (ADR 5.4.1.2.5.2) (see paragraphs 18.140 to 18.147, *Emergency arrangements*).

Documentation

18.129 Training records should be held for at least a period determined by the competent authority (ADR 1.3.3). The ONR recommends four years of security training (ONR 2021b).

18.130 Copies of dispatched and received orders should be kept for at least three months (ADR 5.4.4.1). Copies of these are kept by the relevant departments receiving or dispatching packages. All documents should be signed and dated by the consignor, the driver, and the consignee on receipt of delivery.

Equipment

Type A

18.131 For each vehicle (8.1.5.2):

(a) a wheel chock of a size suited to the maximum mass of the vehicle and to the diameter of the wheel;
(b) two self-standing warning signs;
(c) eye rinsing liquid;
(d) Fire extinguisher (within reach in an emergency situation) (2 x 2 kg if vehicle 3.5 tonnes)(ADR 8.1.4.1). The extinguishers should be rated for inflammability classes A, B and C.

18.132 For each member of the vehicle crew:
(a) a warning vest (e.g. as described in the EN 471 standard);
(b) portable lighting apparatus;
(c) a pair of protective gloves;
(d) eye protection (e.g. protective goggles);
(e) photo identification (ADR 8.1.2.1).

(There are no equipment requirements for transporting only excepted packages.)

18.133 Equipment for the transportation of Type A packages should be checked on a regular basis and before use to ensure all equipment, e.g. eye rinse solution, is in date and useable. Fire extinguishers should be checked to ensure they are suitable for use (i.e. correct weights, etc) and seals are intact (ADR 8.1.4.4) with the date of the next inspection or maximum permissible period of use clearly marked. The results of the checks should be recorded.

Tunnel restrictions

18.134 Passage is forbidden through tunnels of category E (ADR table A, 8.6). The codes for tunnels in the UK can be found online and are currently the Blackwell, East India Dock, Limehouse, and Rotherhide tunnels.

Miscellaneous requirements to be complied with by the vehicle crew [ADR 8.3, 1.1.6.3.2, 8.5 *S5*]

18.135 A driver or driver's assistant should not open the package containing dangerous goods.

18.136 Smoking, including electronic cigarettes, should be prohibited during handling operations in the vicinity of and inside the vehicles.

(NB: 8.3.4 on portable lighting equipment does not apply if there is no subsidiary risk, i.e. another class of dangerous goods (ADR 8.5 *S6*)).

Type A

18.137 There should be no passengers besides the vehicle crew unless the package(s) are only category WHITE I (ADR 8.5 *S6*).

18.138 Members of the vehicle crew should know how to use firefighting equipment.

18.139 The engine should be shut off during loading and unloading with the handbrake engaged.

Training [ADR 1.3, ADR 8.2 (vehicle crew)]

18.140 Persons involved in the transport of dangerous goods by road should receive training appropriate to their duties, including the driver, consignor, and those who load or unload packages (ADR 8.2.3). This includes those who pack and unpack excepted or Type A packages and any security staff receiving deliveries out of hours. The training should cover radiation protection (ADR 1.7.2.5), general awareness (ADR 1.3.2.1), function specific training (ADR 1.3.2.2), safety training (ADR 1.3.2.3), including emergency situations, and security awareness (ADR 1.10.2). In addition, specific training may be required for those involved in transporting high activity sealed sources (HASS), as described in IRR (Regulation 15(2)) (IRR 2017, IRR(NI) 2017). All the training should be documented by the employer (ADR 1.3.3) and refresher training held (ADR 1.3.2.4). It is recommended that the training be repeated biennially. No person should perform any of the duties described without formal training unless supervised by someone who has completed the training (ADR 1.3.1).

Excepted

18.141 There is no requirement for specialised training (ADR 8.5 *S5*), but all drivers carrying excepted packages should be given basic radiation protection training, as specified in ADR 1.3.

Type A

18.142 If the total number of packages does not exceed 10 and the sum of the transport indices does not exceed 3, the requirement for specialised training (ADR driver training) in ADR 8.2.1.4 does not apply (ADR 8.5 S12). Drivers instead should receive '*appropriate training, commensurate with and appropriate to their duties, which provides them with an awareness of the radiation hazards involved in the carriage of radioactive material*'. A record of this training should be kept by the Employer. This training can be carried out in-house within a department, with a recommended frequency of two years. Suggested topics would include:

(a) awareness of Legislation;
(b) current ADR;
(c) CDG/IRR;
(d) packaging requirements;
(e) documentation;
(f) vehicle placarding and requirements;
(g) equipment;
(h) incidents and emergencies;
(i) general rules;
(j) security.

Compliance may be demonstrated with training timetables, signed attendance lists, certificates, and test results, where appropriate, available for inspection. It may be useful for the driver to carry a copy of the training certificate, which could include a statement referencing the current ADR and why the driver does not require full ADR training if stopped by the authorities.

18.143 For drivers exceeding the package limit stated above, specialised ADR training is required by a competent authority approved training provider.

Security

18.144 Guidance on security arrangements for the carriage of class 7 material is available from the ONR website (ONR 2021b) and includes details on recruitment, storage, and training. Appropriate security checks should be carried out on all staff involved in the transport of radioactive material, and the job description should contain a security element.

18.145 While on Employer property, the security of carriers should be considered. Parking should be readily available as close to the drop off point as possible. Drivers should either be met and escorted to the delivery point by organisational security outside of working hours, or a camera system may be enabled with the driver monitored by security.

18.146 Adequate security should be ensured wherever vehicles are parked, particularly if this is a place other than that set aside for unloading. The police may be able to advise in cases of particular difficulty.

Emergency arrangements

18.147 An assessment of the estimated dose an individual may receive from a transport emergency should be carried out in the radiation risk assessment for the transport of radioactive material, performed under IRR. If the estimated effective radiation dose to individuals exceeds 1 mSv, the event would be considered a radiation emergency. An emergency plan is then required (ONR 2023).

18.148 Detailed guidance on emergency arrangements for the non-nuclear sector is available from the ONR website (ONR 2023, 2022a) and should be used for the preparation of plans.

18.149 All emergency plans should be tested, rehearsed (including all relevant parties, e.g. third party carriers transporting the goods), and revised periodically to incorporate learning from testing. This should happen annually and at least every three years, and a report should be prepared, including lessons learned and recommendations for improvement, and sent to the ONR within the prescribed timelines (ONR 2023).

18.150 The following scenarios should be considered:

(a) mechanical breakdown;
(b) theft;
(c) road traffic accident;
(d) fire;
(e) loss of containment/shielding;
(f) illness.

18.151 The health and safety of the driver, crew, and anyone else involved in an emergency situation take priority. Radioactive contamination, the possible dose rate from a package, the dose to members of the public, and actions to reduce that dose (including the emergency services) should be taken into consideration.

18.152 The emergency procedures must include the contact details of the consignor (a 24 h phone number), as well as the contact details of the ONR, the police, and the fire services. There should be detail on where to obtain advice from, for example, the RPA, RWA, and/or DGSA, in relation to matters such as assessing the condition of the consignment and its suitability for onward transport.

18.153 There should also be an emergency procedure in place for responders, i.e. in the event of an emergency, the driver should contact the consigning medical physics department. These procedures should include details on contacting the relevant authorities (if necessary), alternative arrangements (e.g., In the event of breakdown, having an alternative vehicle, having a spill kit if required), and contacting the consignee to inform them of the situation.

18.154 Notification to other regulatory bodies, such as the Environment Agencies, may also be required in the event of an emergency (see chapter 20, *Contingency planning and emergency procedures- radioactive substances*).

Transport of clinical waste

18.155 If clinical waste to be collected and transported off site falls under the radioactive 'excepted' category, the clinical waste risk takes precedence, and the package must be categorised as UN3291, Regulated Medical Waste, N. O.S., Radioactive Material, excepted package, limited quantity of material. If the radioactive part falls under Type A, this takes precedence over the clinical waste requirements, which may be ignored.

Transport of radioactive blood and tissue samples

18.156 For ADR carriage purposes, UN3373 is the UN number for substances and patient samples (bloods/tissue) that are being transported and are considered to have a low probability of containing infectious substances/pathogens (except for those of the 'Category A' type, which are always allocated to UN2814). This requires the outer package of such samples to display the 'UN3373' marking and the words 'Biological Substances, Category B'.

If the above packaging and package marking conditions are met, then ADR Special Provision 319 allows such packages to be exempt from all other requirements of ADR.

18.157 When such blood samples contain an excepted level of radioactive material, the criteria of ADR's Special Provision (SP) 290 (ADR 2020) apply, and, as with clinical waste, it is expected that the blood hazard will take precedence over the excepted level of radioactive material. However, ADR Special Provision 290 includes the following clause:

'When the substance meets a special provision that exempts this substance from all dangerous goods provisions of the other classes, it shall be classified in accordance with the applicable UN number of Class 7 *and all requirements specified in ADR* 1.7.1.5 *shall apply.'*

Which suggests that for bloods that are exempted from ADR via SP319 (ADR 2020) but which contain excepted levels of radioactive material, unlike clinical wastes, the excepted levels of radioactive material take precedence, the UN3373 aspects can be totally disregarded, and the package and accompanying Dangerous Goods Note (DGN) need marking as 'UN2910'.

18.158 However, if the package containing the radioactive-coated blood sample is packed in package which only displays the 'UN2910' marking, then, by virtue, such a package does not display the 'UN3373' marking that is required by SP319 and, thus, cannot be exempted by SP319, which means the above statement no longer applies and means, like the clinical wastes, the blood 'UN3373' element now takes precedence over the excepted levels of radioactive material.

18.159 In conclusion, the package containing these radioactive-coated bloods needs only the 'UN3373' marking and the words 'Biological Substance, Category B' but, as with the clinical wastes, it is advised to issue the driver with a transport document that refers to these radioactive-coated bloods as:

UN3373, Biological Substance, Category B, Radioactive Material, Excepted Package—Limited Quantity of Material, Class 6.2.

References

ADR 2018 *ADR* 2019 *(files) European Agreement concerning the International Carriage of Dangerous Goods by Road UN Economic Commission for Europe Website* (United Nations) https://unece.org/adr-2019-files (accessed 1 July 2022)

ADR 2020 Special provisions applicable to certain articles or substances *Agreement Concerning the International Carriage of Dangerous Goods by Road* (United Nations) ch 3.3, 589–638 https://unece.org/sites/default/files/2021-01/ADR2021_Vol1e_0.pdf (accessed 10 March 2023)

ADR 2022 *ADR* 2023—*Agreement concerning the International Carriage of Dangerous Goods by Road UN Economic Commission for Europe Website Economic Comm* (United Nations) https://unece.org/transport/standards/transport/dangerous-goods/adr-2023-agreement-concerning-international-carriage (accessed 10 March 2023)

ADR no date About the ADR. Agreement concerning the International Carriage of Dangerous Goods by Road *UN Economic Commission for Europe Website* (United Nations) https://unece.org/about-adr (accessed 10 March 2023)

BEIS, DEFRA, Welsh Government and DAERA 2018 Scope of and Exemptions from the Radioactive Substances Legislation in England *Wales and Northern Ireland Guidance Document* https://gov.uk/government/publications/guidance-on-the-scope-of-and-exemptions-from-the-radioactive-substances-legislation-in-the-uk (accessed 11 February 2023)

CDG(A)(NI) 2019 *The Carriage of Dangerous Goods (Amendment) Regulations (Northern Ireland)* (Northern Ireland) https://legislation.gov.uk/nisr/2019/111 (accessed 18 February 2023)

CDG(A) 2019 *The Carriage of Dangerous Goods (Amendment) Regulations* https://legislation.gov.uk/uksi/2019/598 (accessed 18 February 2023).

CDG(NI) 2010 *The Carriage of Dangerous Goods and Use of Transportable Pressure Equipment Regulations (Northern Ireland) 2010* https://legislation.gov.uk/nisr/2010/160 (accessed 31 January 2023)

CDG 2009 *The Carriage of Dangerous Goods and Use of Transportable Pressure Equipment Regulations 2009* https://legislation.gov.uk/uksi/2009/1348 (accessed 31 January 2023)

COTIF 2011 Regulations concerning International Carriage of Dangerous Goods by Rail (RID—appendix C- to the Convention) *Convention Concerning International Carriage by Rail (COTIF)* (Berne: Intergovernmental Organisation for International Carriage by Rail (OTIF)) pp. 104–5 https://otif.org/fileadmin/new/3-Reference-Text/3A-COTIF99/06_Appendix_C.pdf (accessed 11 March 2023)

DEFRA, DECC and Welsh Assembly Government 2011 *Environmental Permitting Guidance Radioactive Substances Regulation* (London: Welsh Assembly Government, Deparment of Energy and Climate Change, Department forEvironment Food and Rural Affairs) https://gov.uk/government/publications/radioactive-substance-regulations-rsr-guidance (accessed 30 June 2022)

Department for Transport 2012 Moving dangerous goods *GOV.UK Website* 4/09/2012 https://gov.uk/guidance/moving-dangerous-goods (accessed 30 June 2022)

Department of Agriculture and Rural Affairs 2020 *Radiation—overview, Department of Agriculture and Rural Affairs website* https://daera-ni.gov.uk/articles/radiation-overview (accessed 30 June 2022)

DfT 2020a *Carriage of Dangerous Goods: Approved Derogations and Transitional Provisions, GOV.UK Website* https://gov.uk/government/publications/the-carriage-of-dangerous-goods-approved-derogations-and-transitional-provisions/carriage-of-dangerous-goods-approved-derogations-and-transitional-provisions (accessed 28 June 2022).

DfT 2020b *Security Requirements for Moving Dangerous Goods by Road and Rail. Gov.Uk Department for Transport.* 24/09/2020 https://gov.uk/government/publications/security-requirements-for-moving-dangerous-goods-by-road-and-rail (accessed 30 June 2022)

EA(S)R 2018 *The Environmental Authorisations (Scotland) Regulations 2018* http://legislation.gov.uk/ssi/2018/219 (accessed 31 January 2023)

EA 2012 Environment Agency—How To Comply with your EPR RSR Environmental Permit—Open Sources and Receipt *Accumulation and Disposal of Radioactive Waste on Non-nuclear Sites. 2.0* (Bristol: Environment Agency) https://gov.uk/government/publications/rsr-environmental-permit-open-sources-radioactive-waste (accessed 4 March 2023)

EA 2018 *How to Comply with Your EPR RSR Environmental Permit—Sealed Sources. 2.0* (Bristol: Environment Agency) https://gov.uk/government/publications/rsr-environmental-permit-how-to-compy-sealed-sources (accessed 4 March 2023)

EA 2021 *Environmnt Agency—Non-Nuclear Radioactive Substances Regulation: Technical Guidance, GOV.UK Website.* 2/12/2021 https://gov.uk/government/collections/non-nuclear-radioactive-substances-regulation-technical-guidance (accessed 30 June 2022)

EPR 2016 *Environmental Permitting (England and Wales) Regulations* https://legislation.gov.uk/uksi/2016/1154 (accessed 31 January 2023)

HSE 2003 Security of Radiation Generators in Hospital Nuclear Medicine Departments *Radiation Protection News Issue 23 [Preprint]* (Health and Safety Executive) https://webarchive.nationalarchives.gov.uk/ukgwa/20170902124949/http://hse.gov.uk/radiation/rpnews/rpa23.htm#SECURITY (accessed 11 March 2023)

HSE 2018 L121 Work with ionising radiation Ionising Radiations Regulations 2017 *Approved Code of Practice and Guidance* 2nd edn (Norwich: Health and Safety Executive) https://hse.gov.uk/pubns/books/l121.htm (accessed 31 January 2023)

HSE, SEPA, EA, HSENI and Service, E. and H 2010 Control of radioactive substances *Ionising Radiation Protection Series No 8 [Preprint]* (Health and Safety Executive) https://hse.gov.uk/pubns/irp8.pdf (accessed 11 March 2023)

Human Tissue Act 2004 https://legislation.gov.uk/ukpga/2004/30 (accessed 11 February 2023)

Human Tissue (Scotland) Act 2006 https://legislation.gov.uk/asp/2006/4 (accessed 11 February 2023)

IAEA 2006 Storage of Radioactive Waste *Safety Guide, Standard Safety Series No. WS-G-6.1* (Vienna: International Atomic Energy Agency) https://iaea.org/publications/7441/storage-of-radioactive-waste (accessed 11 March 2023)

IAEA 2019 Security of Radioactive Material in Use and Storage and of Associated Facilities *Nuclear Security Series No. 11-G (Rev.1)* (Vienna: International Atomic Energy Agency) https://iaea.org/publications/12360/security-of-radioactive-material-in-use-and-storage-and-of-associated-facilities (accessed 11 March 2023)

IAEA and United Kingdom 2018 *Agreement between the United Kingdom of Great Britain and Northern Ireland and the International Atomic Energy Agency for the Application of Safeguards in the United Kingdom of Great Britain and Northern Ireland in Connection with the Treaty on the Non-Proliferation of Nuclear Weapons* https://assets.publishing.service.gov.uk/government/uploads/system/uploads/attachment_data/file/754735/MS_13.2018_VOA_Agreement.pdf (accessed 11 March 2023)

ICAO 2015 *Technical Instructions For The Safe Transport of Dangerous Goods by Air (Doc* 9284)

IMO 2016 *The International Maritime Dangerous Goods (IMDG) Code* (Interntional Maritime Organisation Publishing)

IRR(NI) 2017 *The Ionising Radiations Regulations (Northern Ireland)* http://legislation.gov.uk/nisr/2017/229 (accessed 31 January 2023)

IRR 2017 *The Ionising Radiations Regulations* https://legislation.gov.uk/uksi/2017/1075 (accessed 31 January 2023)

Lawson R S, Davies G, Hesslewood S R, Hinton P J and Maxwell A 2004 Delivery and collection of radioactive packages to and from UK hospital nuclear medicine departments *Nucl. Med. Commun.* **25** 1161–7

ONR 2016 *Transporting Radioactive Material—Guidance on Radiation and Contamination Monitoring Requirements, and Determining a Transport Index* (Office for Nuclear Regulation) https://onr.org.uk/transport/transport-guidance-monitoring-ti.pdf (accessed 11 March 2023)

ONR 2020 *(ONR 2023)five-steps-transport-emergency-planning*. Bootle, UK: Office for Nuclear Regulation https://www.onr.org.uk/transport/five-steps-transport-emergency-planning.docx (accessed 27 April 2024)

ONR 2021a *Dangerous Goods Safety Advisers Annual Report TD-TCA-GD*-001. 2.1 *ONR Guidance Document. 2.1* (Bootle) https://onr.org.uk/transport/resources.htm (accessed 25 December 2022)

ONR 2021b Transporting radioactive material—security guidance on the carriage of Class 7 radioactive material (TD-TCA-GD-002). 2.1 *ONR Guidance Document. 2.1* (Office for Nuclear Regulation) https://onr.org.uk/operational/other/td-tca-gd-002.pdf (accessed 18 February 2023)

ONR 2022a Emergency/contingency plans *ONR Website* (Office for Nuclear Regulation) https://onr.org.uk/transport/emergency-plans.htm (accessed 21 March 2023)

ONR 2022b Transport of Radioactive Materials—Guidance and Resources *Office for Nuclear Regulation Website* https://onr.org.uk/transport/resources.htm (accessed 30 June 2022)

REPPIR(NI) 2019 *The Radiation (Emergency Preparedness and Public Information) Regulations (Northern Ireland)* http://legislation.gov.uk/nisr/2019/185 (accessed 31 January 2023)

REPPIR 2019 The Radiation (Emergency Preparedness and Public Information) Regulations 2019 http://legislation.gov.uk/uksi/2019/703 (accessed 31 January 2023)

RSA(A)R(NI) 2011 The Radioactive Substances Act 1993 *(Amendment) Regulations (Northern Ireland)* https://legislation.gov.uk/nisr/2011/290 (accessed 31 January 2023)

RSA 1993 *Radioactive Substances Act 1993* http://legislation.gov.uk/ukpga/1993/12 (accessed 11 February 2023)

Scottish Government 2020 *Scope of Radioactive Substances Legislation in Scotland—Guidance Document* https://gov.scot/publications/scope-radioactive-substances-legislation-scotland-guidance-document/pages/1 (accessed 22 July 2022)

SEPA 2020a *Authorisation Guide for Radioactive Substances Activities. 1.2* (Stirling: Scottish Environmental Protection Agency) https://sepa.org.uk/media/371985/rs-authorisation-guide.pdf (accessed 11 February 2023)

SEPA 2020b *Radioactive Waste Advisers, SEPA Website* https://sepa.org.uk/regulations/radioactive-substances/radioactive-waste-advisers/ (accessed 6 March 2023)

SEPA 2021 *Guidance and Reports, sepa.gov.uk* https://sepa.org.uk/regulations/radioactive-substances/guidance-and-reports/ (accessed 30 June 2022)

The Dangerous Goods in Harbour Areas Regulations 2016 2016 https://legislation.gov.uk/uksi/2016/721 (accessed 11 March 2023)

The Merchant Shipping (Dangerous Goods and Marine Pollutants) Regulations 1997 https://legislation.gov.uk/uksi/1997/2367 (accessed 11 March 2023)

The Nuclear Safeguards (EU Exit) Regulations 2019 https://legislation.gov.uk/uksi/2019/196 (accessed 11 March 2023)

IOP Publishing

Medical and Dental Guidance Notes (Second Edition)
A good practice guide on all aspects of ionising radiation protection in the clinical environment:
IPEM Report 113
John Saunderson, Mohamed Metwaly, William Mairs, Philip Mayles,
Lisa Rowley and Mark Worrall

Chapter 19

Accumulation and disposal of radioactive waste

General requirements

19.1 Management of radioactive waste is controlled under the Environmental Authorisations (Scotland) Regulations (EASR) 2018 (Scotland) (EA(S)R 2018), the Radioactive Substances Act (RSA) 1993 (Northern Ireland) (RSA 1993, RSA(A)R(NI) 2011) or the Environmental Permitting Regulations (EPR) 2016 (England and Wales) (EPR 2016), (referred to hereafter as the Radioactive Substances Regulation (RSR)) through site permits/authorisations or in certain circumstances sites may operate under a relevant exemption (BEIS *et al* 2018) or general binding rule (GBR) (EA(S)R 2018).

19.2 This is enforced by the Scottish Environment Protection Agency (SEPA) in Scotland, the Northern Ireland Environment Agency (NIEA) in Northern Ireland, the Environment Agency (EA) in England, and Natural Resources Wales (NRW) in Wales. Exemptions fall under the Radioactive Substances Exemption (Northern Ireland) Order 2011 (RSE(NI) 2011) and EPR (EPR 2016), both as amended. GBRs apply in Scotland and are included in Schedule 9 of EASR (EA(S)R 2018). Joint guidance from the respective Environment Agencies for exemptions to the regulations and permits/authorisation (referred to hereafter as permits) for England, Wales and Northern Ireland are available (BEIS *et al* 2018).

19.3 If the radioactive waste is exempt from permitting but has other properties that, if it were not radioactive, would make it special waste, then the Hazardous Waste Regulations (England and Wales) (2005), the Hazardous Waste Regulations (Northern Ireland) (2005) and the Special Waste Regulations (1996) (in Scotland) must also be complied with when disposing of the waste. Requirements of the Carriage of Dangerous Goods and Use of

doi:10.1088/978-0-7503-2332-1ch19

Transportable Pressure Equipment Regulations (as amended) (CDG 2009, CDG(NI) 2010, CDG(A) 2019, CDG(A)(NI) 2019) must be fulfilled if radioactive material is transported off-site for disposal.

19.4 Waste that is classed as 'out-of-scope of regulation' is not subject to the requirements of the radioactive substances legislation (BEIS *et al* 2018, Scottish Government 2020). Waste that is 'exempt from permitting' or under GBRs is still subject to some requirements under the legislation, outlined in the relevant exemption or GBRs, but is exempt from the requirement of a permit (BEIS *et al* 2018, EA(S)R 2018) so long as the rules and conditions are complied with. Management of all radioactive waste from medical practices must be carried out in accordance with either permit conditions or a relevant exemption or GBRs.

19.5 The following types of radioactive waste may occur in medical practice:

(a) sealed and other solid sources;
(b) spent radionuclide generators, which may contain depleted uranium in their shielding;
(c) excreta from patients treated or tested with open radionuclides;
(d) unused radionuclides originally intended for diagnostic or therapeutic use;
(e) low activity liquid waste, e.g. from washing of apparatus;
(f) liquids immiscible with water, such as liquid scintillation fluids, counting residues;
(g) solid waste in the form of pipettes, syringes, cyclotron wastes (activated components, targets, foils) etc, for relatively long-lived radionuclides (e.g. ^{32}P);
(h) low activity solid waste, e.g. paper, glass, syringes, vials, gloves and PPE, waste from spills and decontamination;
(i) gases (including gaseous waste from patients);
(j) animal waste from medical research;
(k) incinerator ash;
(l) human tissue containing radioactive sources (including bloods); and
(m) contaminants with a longer half-life than the original predominant waste radionuclide.

The employer must ensure that appropriate authorised waste transfer/disposal routes exist for each type of waste. Radioactive waste should be included in the overall waste management system.

19.6 Appropriate risk assessments, including environmental impact assessments for aqueous waste disposal, should be performed, documented, and made available for inspection and review. These will contribute to the requirements for the completion of the application form needed for a permit to accumulate and dispose of radioactive waste. Template spreadsheets for impact assessment of aqueous disposals are available on request from local Environment Agency inspectors, or on the SEPA website for discharges to sewers in Scotland (SEPA 2021), and will be sufficient for the majority of nuclear medicine departments if the dose to critical groups is under 20 μSv per annum (EA *et al* 2012, SEPA 2019). This methodology uses worst-case scenario principles, and for those performing large numbers of therapeutic administrations, a more detailed assessment will likely be required to gain a more accurate result. This may include more accurate modelling of radionuclides not listed in the spreadsheets, obtaining the actual flow rate from the local sewage treatment works and river flow downstream, information from local anglers clubs, information on abstraction points, and information on special areas of conservation. Consideration should also be given to any waste that might be generated by outpatients that require further management (e.g. bandages, iodine seeds, sharps, incontinence pads etc) such as returning them to the hospital.

19.7 The vast majority of solid radioactive waste from medicinal practices (excluding spent sealed sources) is low level waste (LLW), less than 12 GBq beta/gamma (4 GBq for alpha) activity per tonne (DECC *et al* 2016). Very low level waste (VLLW) represents a limit below which it is acceptable to dispose of the waste with normal refuse, and is classed as; individual items of less than 40 kBq and containing less than 400 kBq beta/gamma per 0.1 m^3 of non-radioactive waste. For Carbon-14 (^{14}C) or Tritium (^{3}H) the limit is 400 kBq for a single item or 4000 kBq total (DECC *et al* 2016).

19.8 Disposals will need to be covered by an appropriate permit or exemption. The employer must ensure that an appropriate permit is obtained for each type of radioactive waste (solid, liquid or gaseous) or that the proposed disposal is within the terms of an exemption.

19.9 If a site holds a permit for management of radioactive waste under the relevant radioactive substances regulations, the employer must appoint a person who has gained a Certificate of Core Competence from an Assessing Body recognised by the UK Environment Agencies to act as a Radioactive Waste Adviser (RWA) with the appropriate experience in the sector, to advise the employer on radioactive waste disposal (SEPA 2020).

19.10 The employer must ensure that the conditions specified in the permit or exemption are complied with by adhering to a robust waste management system. The advice of the RWA should be sought to this effect.

19.11 Authorised routes for the transfer/disposal of each type of waste are specified or referenced in the permit/exemption/GBRs. The conditions listed will include specification of:

(a) permitted method(s) of transfer/disposal;
(b) the maximum activities and activity concentrations of the waste which may be disposed of in a particular period (day/week/month/year);
(c) the containment of the waste while awaiting disposal;
(d) the limit for accumulation on amount and volume;
(e) the maximum time which waste may be stored prior to disposal; and
(f) the records which must be kept.

19.12 Records of the type and amount of activity and the time of disposal should be kept for inspection by regulatory authorities, to show compliance with the permit/exemption/GBRs (see also paragraphs 18.39–18.51, *Stock records and control procedures*). The records should be submitted to the appropriate regulatory authorities in the prescribed format and at the frequency specified in the permit, usually on an annual basis by the 28th of February for the previous year using the EA Pollution Inventory Electronic Data Capture (PIEDC) system in England, OSPAR Radionuclides Data Collection forms to Natural Resources Wales, the Scottish Pollutant Release Inventory (SPRI) or submission separately to the RS Notification mailbox in Scotland, and the annual return by email to the NIEA.

19.13 Permits, relevant procedures, and records must be accessible to all staff working within its scope. No information regarding sealed sources should be displayed in public areas for security reasons. However, there may be a need for signage indicating the presence of radiation.

19.14 Radioactive waste disposal is only one component of the clinical waste management process. The employer should ensure that this is:

(a) legal (i.e. authorised or exempt under RSR (RSA 1993, EPR 2016, EA(S) R 2018));
(b) safe (the waste should be contained, shielded and segregated);
(c) simple (the waste system should be easily understood by all those who are to apply it); and
(d) economical (the final disposal of some radioactive waste can be very expensive, e.g. high activity sealed sources, such as tritium tubes used for generating neutrons, Am/Be neutron sources, and redundant ^{60}Co and ^{137}Cs radiotherapy sources).

19.15 It is a condition of permits issued under the relevant RSR that Best Practical Means (BPM) or Best Available Techniques (BAT) are followed to minimise all radioactive waste generated and disposed of as far as is practicable. This should be considered in the design of facilities, taking into account radioactive waste that may arise during decommissioning, for example, for cyclotron sites, but also contaminated items, such as sink traps, pipework, etc, that should be maintained through the life of the permit. In Scotland, a list of contaminated items is required, as these are no longer exempt while on the premises where they were contaminated. The list needs to identify things like sink traps, pipework, fume cupboards, extract ducting, bits of floor/bench where longer-lived radionuclides were spilt, etc. This is to facilitate decommissioning and to ensure that these items are not removed without some consideration.

Types of waste

Solid waste

19.16 Much of the radioactive waste generated in hospitals arises from the preparation and administration of radionuclides to patients or from *in vitro* tests in laboratories and may be considered solid waste. Other sources of solid waste include patient-generated waste that is not disposable via the sewer system and waste arising from decontamination procedures. Research laboratories are also likely to generate a great deal of solid waste.

19.17 Vials containing small volumes of radioactive materials that are no longer in stock may be considered solid radioactive waste and disposed of in a relevant sharps bin. The sharps bin should be labelled and, if appropriate, shielded while waste is accumulated.

19.18 Solid waste may be disposed of by any of the authorised means as follows:
- (a) collection by the local authority or their contractor as refuse for co-disposal with large amounts of non-radioactive waste at a landfill site or municipal incinerator (VLLW and some small sealed sources);
- (b) collection by a contractor as clinical waste for co-disposal with large amounts of non-radioactive waste at a landfill site or municipal incinerator (VLLW) or transfer to an authorised/permitted contractor for incineration at their facilities (LLW);
- (c) return to the manufacturer (e.g. sealed sources, spent generators); or
- (d) transfer to a permitted site for accumulation and disposal off-site.

19.19 For solid radioactive waste disposed of as VLLW or out of scope in the non-radioactive waste stream, all radioactive marking and labelling must be removed prior to disposal.

19.20 Financial provision must be made for the disposal of all High Activity Sealed Sources (HASS) during purchase. This should also be considered for non-HASS sealed sources, and good practice would include the removal of old sealed sources with the purchase of new ones. If the waste undergoes a recycling process by the waste contractors, it should be ensured that disposal meets the radiological impact assessment assumptions for dilution by co-disposal, set out in the guidance (BEIS *et al* 2018) for England, Wales, and Northern Ireland. In Scotland, metallic sources can only go directly to landfills (EA(S)R 2018) and must not pass through any kind of materials recycling facility as the metallic sources will be removed. Radioactive sealed sources should only be disposed of with advice from the RWA (and radiation protection adviser, RPA, as appropriate).

Liquid waste

19.21 Aqueous liquid waste may be disposed of via the hospital drainage system, either via sinks/sluices for waste arising in laboratories or direct to the main drains via toilets/sluices for patient excreta. Dilution of liquids with high radioactivity is recommended. Sinks/toilets/sluices designated for disposal should be clearly marked, as should the waste outlet. The designated discharge points should afford the highest dilution of the waste, e.g. those that drain directly into a sewer or main drain. There should be records of periodic contamination monitoring of the sink and surfaces associated with nearby pipes and outlets. Plans for the drainage system should be retained by the organisation.

19.22 For centres with large volumes of molecular radiotherapy inpatient administrations, consideration should be given to the use of holding tanks to allow for the decay of the radionuclide before disposal to the sewage system. The impact assessment might provide guidance for this choice.

19.23 Organic liquid waste includes ^{14}C, tritium, and other beta emitters mixed with liquid scintillation fluid. The scintillant material, which is immiscible with water, toxic, and/or flammable, should not be disposed of via the drains. The new generation of non-flammable low-toxicity biodegradable scintillants that do not penetrate plastics may be disposed of via the drains, subject to prior permission being obtained from the local sewerage undertaker. The options for disposal of scintillant wastes, where permitted on the site authorisation/permit, are:

(a) incineration of the whole vials and contents;
(b) separation of the scintillant from the vials, followed by incineration of bulky scintillant by a licensed contractor; and
(c) separation of biodegradable scintillant(s) from the vials, followed by disposal to drains, with approval from the local sewerage undertaker.

19.24 For some laboratory waste, vial crushers and splitters may be an effective means of rendering solid waste into liquid form, disposing of it in the sewer, and thereby operating within the terms of an exemption for liquid waste if there is no permit in place for the facility. Some crushers are effective on both glass and plastic vials, while those for plastic vials usually split them rather than crush them. Some can be effective in achieving separation of the broken vials from the scintillant, but some granulators can reduce plastic vials to such small particles that the large surface area absorbs the scintillant causing problems for subsequent separation. The efficacy of their use for disposal needs to be determined and documented prior to use.

Liquid discharges from hospitals via patients

19.25 When a radiopharmaceutical is administered to a patient, a percentage of the administered dose is excreted in the patient's urine and faeces and is thus discharged to the drain. The percentage varies between different radiopharmaceuticals. The fraction of activity administered to patients that are recorded as waste should be in accordance with the latest guidelines issued on the Institute of Physics and Engineering in Medicine (IPEM) website, as approved by the relevant Environment Agencies and kept under review by the Nuclear Medicine Special Interest Group (IPEM 2018). These excretion factors take into account disposals at both the administering hospital and once the patient has returned home, with the assumption that they are in the same sewage catchment area. If a large number of patients are administered outside this area, a reduced percentage can be applied in agreement with the local regulatory inspectors. Records should be kept of all local agreements and calculations.

19.26 If a patient is administered in one hospital but returns to another in a different sewage catchment area as an inpatient, a care home, or a hospice, a fraction of the administered activity may be accounted for in the receiving hospital and should be recorded by that hospital from information provided by the administering hospital (IPEM 2018).

19.27 If the receiving hospital does not have a permit, it may account for the disposal under an exemption, e.g. 10% of ^{99m}Tc administrations or all of ^{131}I administrations for thyrotoxicosis (IPEM 2018). Contaminated articles, such as incontinence pads, may be stored and disposed of as VLLW, under an exemption (BEIS *et al* 2018) or GBRs (EA(S)R 2018), or under section I of the standard conditions (EA(S)R 2018). These items may also be returned to the administering hospital. Up to 10 GBq of ^{99m}Tc and 5 GBq of all other radionuclides may be disposed of as patient excreta per annum under an exemption (BEIS *et al* 2018). In Scotland, the disposal of radioactive waste in the form of human excreta where the disposal occurs at a place other than the place of administration of the radioactive material is not classified as 'radioactive substances activity' and is therefore not 'regulated activity' under schedule 8 of EASR (EA(S)R 2018). The employer of the facility where the administration took place should keep a record of all patients being returned to sites without permits and should advise management at each site of the need to maintain suitable records of waste discharges.

19.28 If a patient is being returned to a facility that possesses a permit, both the administering and receiving facilities should check that this can be done without breaching permit conditions. This may be an issue for patients undergoing therapy with alpha-emitting radionuclides at one hospital whilst being an inpatient at a second hospital. If the receiving hospital owns a permit with an allowance to dispose of only small amounts of unspecified beta/gamma-emitting radionuclides then the disposal of an alpha emitter will not be authorised, unless it falls under a relevant exemption or GBR. In such circumstances, the RWA for each employer should discuss the situation with their local environment agency inspector.

19.29 In specific cases, users may propose, and respective environment agency officers may agree, alternative arrangements if the disposer can provide information to justify departing from the guidelines.

Methodology for calculation of excreted activity for various radiopharmaceuticals

19.30 Biokinetic data for commonly used nuclear medicine radiopharmaceutical compounds has been published by the International Committee for Radiological Protection (ICRP) in Report 128 (ICRP 2015). This includes data on radiopharmaceuticals that are or may be coming into common use in human patients, provided that acceptable and sufficient metabolic data are available (for making the absorbed dose calculations). For each compound, a biokinetic model is used to give quantitative estimates for the distribution and metabolism of the radiopharmaceutical in the body based on results available in the literature. Where appropriate, a range of pathological variations is given. Further detail on the methodology for the calculation of excreted activity is given in the IPEM guidance note (IPEM 2018)

Gaseous or airborne waste

19.31 Gaseous radioactive waste may arise from many different operations, for example:

(a) patient investigation involving either a radioactive gas or aerosol (^{133}Xe) or a radioactive substance (^{14}C Urea);
(b) research involving animals;
(c) radiochemical operations using gaseous or volatile radioactive substances, e.g. ^{3}H, ^{14}C (dioxide or monoxide), ^{125}I, ^{131}I in a fume cupboard or safety cabinet;
(d) discharges from an incinerator; or
(e) exhaust from a store for radioactive materials/radioactive waste, or fume cupboards;
(f) decay products from radionuclides, e.g. ^{226}Ra, ^{131}I and cyclotron discharges.

19.32 For some radioactive gases used in diagnostic nuclear medicine and positron emitters such as ^{13}N, ^{15}O and ^{11}C and their derivatives, quantification of their disposal can be difficult, and the advice of the RWA should be sought. They may require a suitable extractor system and measurement of any vented product using stacks (see chapter 11, *Positron emission tomography*). If the half-life of the gas and any radionuclides it contains is under 100 seconds, it is not considered radioactive material or waste under the radioactive substances regulations.

19.33 Release of radioactive emissions from opening containers, e.g. ^{131}I capsules, is exempt (IPEM 2014, BEIS *et al* 2018), however, the opening of such containers should take place in a suitable vented area, such as a fume cupboard.

19.34 All gaseous radioactive waste should normally be discharged from the building via suitable ducting in such a manner that it cannot easily re-enter other parts of the same building or enter other buildings nearby; some very short half-life gases do not require specific discharge conditions.

19.35 Filters in discharge stacks and microbiological safety cabinets should be checked periodically and replaced. Any contaminated filters should be treated as solid waste and disposed of accordingly.

Animal waste products

19.36 Animal waste products (including carcasses, dismembered if necessary) should be incinerated or macerated. The macerator should be plumbed directly into the waste system. Large animals can present particular problems, as their urine and faeces can be highly radioactive. These should be collected and disposed of as aqueous radioactive waste in drains or as solid waste by incineration. For small carcasses containing longer half-life radionuclides (e.g. ^{32}P, ^{125}I) some form of segregation and decay storage may be required.

Alpha emitters

19.37 Alpha emitting radioactive waste should not be mixed with other radionuclides. It may be preferable to accumulate alpha-emitting (e.g. ^{223}Ra) waste if permitted and justified (for the decay storage period chosen) until it is VLLW or out of scope and may be disposed of in non-radioactive waste streams.

Special waste

19.38 Some radioactive waste is also special waste and, as such, needs to conform to the requirements of the legislation outlined in paragraph 19.3. All radioactive medicinal products are prescription-only medicines (POMS) (Medicines Act 1968). Waste POMS are defined as special waste under the Hazardous/Special Waste Regulations (The Special Waste Regulations 1996, The Hazardous Waste (England and Wales) Regulations 2005, The Hazardous Waste Regulations (Northern Ireland) 2005).

19.39 The statutory authorities have indicated that, with the exception of small quantities of waste POMS, shipments of radioactive waste that is special waste will be required to be accompanied by consignment notes. These consignment notes, which meet the requirements of the Hazardous/Special Waste Regulations (The Special Waste Regulations 1996, The Hazardous Waste (England and Wales) Regulations 2005, The Hazardous Waste Regulations (Northern Ireland) 2005), are in addition to those conditions imposed by permits issued under radioactive substances legislation (RSA 1993, RSA(A)R(NI) 2011, EPR 2016, EA(S)R 2018).

Storage and accumulation of radioactive waste material

19.40 A permit may allow waste to be accumulated up to a certain activity for a certain period of time, pending disposal, subject to certain conditions. Where waste is accumulated on a short-term basis at the point of work prior to its transfer to a main store (e.g. sharps bin in an injecting room), the period of accumulation should normally begin at the point of transfer to the store. This is only acceptable where waste is held at the point of work for the shortest reasonable time and never longer than one week (EA 2012). All EASR permits allow waste storage with unspecified decay storage times, left to the authorised person to determine and justify. This may facilitate the storage of radionuclides to allow radioactive decay to reduce the activity disposed of and accumulation for final disposal in an economic manner, as VLLW or out of scope.

19.41 Accumulated waste should be disposed of as soon as is practical, once it is consigned as waste, and within 26 weeks in the case of sealed sources (BEIS *et al* 2018) for England, Wales, and Northern Ireland. A time frame is no longer specified under EASR.

19.42 Waste may be segregated into the following categories, or combinations of these, as follows:

(a) radioactive waste of a range of activities;
(b) radioactive waste of half-lives in different ranges (e.g. short-lived and long-lived);
(c) radioactive waste of different physical types, solid, sealed, aqueous, scintillant, organic, alpha emitters; and
(d) radioactive waste to be disposed of by transfer, e.g. solid waste to landfills, incineration, or return to the manufacturer (sealed source).

The categories should reflect those contained in the relevant permit/exemption.

19.43 Waste containers should not be overfilled. Normally, they should be sealed when two-thirds full or in accordance with the local hospital infection control policy. The activity of individual items (i.e. hypodermic needles) disposed of into the waste container need not be recorded for short-lived radionuclides or if the total activity of the container can be determined using a suitably calibrated monitor. If the intent is to store waste until it can be classed as VLLW or out of scope, the time to store until such levels are reached should be calculated to ensure accumulation time limits are not breached.

19.44 On removal from the store, the container(s) should be checked with a contamination monitor to ensure activity levels are not higher than expected or that there is no residual activity that would indicate the presence of radionuclides other than those expected. If this is the case, the container should be returned to the store and thereafter considered as other radioactive waste to be disposed of accordingly through an approved route. An attempt should be made to determine the radionuclide present.

19.45 Waste containers out of scope or at VLLW levels should be disposed of **promptly** within the appropriate waste stream (clinical or domestic), according to type. A record of the total activity of VLLW should be recorded to ensure a maximum of 200 MBq is disposed of annually (2000 MBq/annum for ^{3}H and ^{14}C) if operating under an exemption or GBR (BEIS *et al* 2018, EA(S) R 2018).

19.46 A record should be kept of each container placed in the store. This should record the date that the container entered the store, the activity on that date, and the date of disposal (i.e. removal from the store). The record should also describe the radionuclide(s) in each container, or at least the longest-lived radionuclide of a mixture.

19.47 A robust management system is required for the segregation and storage of radioactive waste. Radioactive waste must not be stored with flammable materials. The waste store must be secure and of a design appropriate to the hazards present. It will require surfaces that are easily decontaminated and may require shielding and an appropriate ventilation system. If putrescible waste (such as patient tissue/samples, animal carcasses, etc) is to be stored, a cold room or freezer will also be required.

19.48 In the case of human tissue, e.g. prostate removed from the body after death following ^{125}I brachytherapy treatment, the relevant Human Tissue regulations (Human Tissue Act 2004, Human Tissue (Scotland) Act 2006) need consideration. The prostate may be stored for two years following administration and returned to the family, or the tissue may be dissolved and the seeds retrieved for appropriate disposal. Hospitals administering ^{125}I seeds should include provisions within their permit for the accumulation and disposal of seeds in such a contingency.

19.49 Waste stores should have restrictive access and be clearly marked with radiation warning signs and suitable information on their contents, bearing in mind possible security issues. External waste stores separate from departments may have radiation warning signs clearly visible on entry to the store, rather than displayed externally. The waste management systems and procedures for the waste store should be clearly documented. An auditable accounting system is required for logging in, checking, and logging out the waste. Clear, auditable pathways should be maintained. Periodic checks should be performed to review the procedures, verify that waste is still present, and ensure that the facilities are in good condition. The RPA/RWA should advise management as to the suitability of their arrangements.

19.50 Contingency arrangements should be accounted for in the permit application for the accumulation of waste. These may include excretions into nappies, spills of therapeutic radiopharmaceuticals, resulting in large volumes and activities of waste.

19.51 Some authorisations/permits allow the decay storage of longer half-life radionuclides, e.g. ^{125}I, ^{35}S, ^{59}Fe, ^{89}Sr, if the employer can demonstrate a well-organised and safe system for the medium-term storage of radioactive materials. It should be noted, however, that the waste from some procedures, e.g. ^{125}I iodinations, may require storage for over two years before the waste has decayed to a level at which it can be disposed of. Appropriate risk assessments will identify the control measures required, e.g. ventilation systems and the risk of contamination by volatilisation, or penetration or breakdown of the plastic containers. The decay storage system should conform to all conditions specified in the authorisation.

19.52 Radioactive daughters of radionuclides listed on the permit do not require consideration; however, long-lived contaminants/impurities present in waste must be accounted for, e.g. ^{177m}Lu with ^{177}Lu, ^{121}Te and ^{125}I for ^{123}I, and ^{154}Eu from ^{153}Sm. The maximum activity of contaminants present can be measured or calculated from the percentage provided in the Summary of Product Characteristics (SPC) for the product and may be accumulated to decay to VLLW levels, if compliant with the permit conditions.

Records

19.53 Various computer packages are available commercially to cover the whole accounting procedure, from the ordering of radionuclides through their receipt and usage to the accumulation and disposal of waste. Their use should be considered, particularly by large NHS Employers with multiple user departments, as they may lead to the standardisation and simplification of record systems.

19.54 Records must be clear and legibly written. Certain specified records of radioactive waste disposals must be kept indefinitely, or, in Scotland, as long as necessary to ensure and demonstrate compliance with the permit. The date, radionuclide(s), type and quantities disposed of, and disposal route must be recorded on the date of disposal. A record of all stored waste (solid or liquid) must also be kept. Monthly summaries of all waste from a site should be kept by a designated person/department, particularly if there are multiple users on the site. Records should include disposals of patient excreta via the sewage system. Such monthly liquid waste disposal records will be retrospective and must be completed within 14 days of the end of the calendar month, according to the conditions of the authorisation or permit (note—this is not required under EASR (EA(S)R 2018)).

19.55 Records relating to the accumulation and disposal of radioactive waste and contaminated items for Scotland, under EASR (EA(S)R 2018), which could include monitoring of waste pipes and the environment) should be retained for the periods specified by the appropriate Authority and as specified on the permit, or exemption. See appendix 10 *Record keeping* for more detail.

19.56 The shipment of radioactive waste across frontiers is regulated by the Transfrontier Shipment of Radioactive Waste and Spent Nuclear Fuel Regulations 2008 (UK Parliament 2008) which are enforced by the relevant Environment Agencies. Authorisations must be obtained prior to all transfrontier shipments into, from, and through the United Kingdom. Further, help and advice is available at the following contact email: askshipments@environment-agency.gov.uk. SEPA automatically authorises transfrontier shipments, provided they follow BPM. Further advice can be obtained using the RS Enquires mailbox.

References

BEIS, DEFRA, Welsh Government and DAERA 2018 Scope of and Exemptions from the Radioactive Substances Legislation in England *Wales and Northern Ireland Guidance Document* https://gov.uk/government/publications/guidance-on-the-scope-of-and-exemptions-from-the-radioactive-substances-legislation-in-the-uk (accessed 11 February 2023)

CDG 2009 *The Carriage of Dangerous Goods and Use of Transportable Pressure Equipment Regulations 2009* https://legislation.gov.uk/uksi/2009/1348 (accessed 31 January 2023)

CDG(A) 2019 *The Carriage of Dangerous Goods (Amendment) Regulations* https://legislation.gov.uk/uksi/2019/598 (accessed 18 February 2023)

CDG(A)(NI) 2019 *The Carriage of Dangerous Goods (Amendment) Regulations (Northern Ireland)* https://legislation.gov.uk/nisr/2019/111 (accessed 18 February 2023)

CDG(NI) 2010 *The Carriage of Dangerous Goods and Use of Transportable Pressure Equipment Regulations (Northern Ireland) 2010* https://legislation.gov.uk/nisr/2010/160 (accessed 31 January 2023)

DECC, Scottish Government, Welsh Government and DOE 2016 *UK Strategy for the Management of Solid Low Level Waste from the Nuclear Industry* https://assets.publishing.service.gov.uk/government/uploads/system/uploads/attachment_data/file/497114/NI_LLW_Strategy_Final.pdf (accessed 4 March 2023)

EA 2012 Environment Agency—How to Comply With Your EPR RSR Environmental Permit—Open Sources and Receipt *Accumulation and Disposal of Radioactive Waste on Non-Nuclear Sites. 2.0* (Bristol: Environment Agency) https://gov.uk/government/publications/rsr-environmental-permit-open-sources-radioactive-waste (accessed 4 March 2023)

EA, SEPA, NIEA, HPA and FSA 2012 *Principles for the Assessment of Prospective Public Doses arising from Authorised Discharges of Radioactive Waste to the Environment* (Environmental Protection Agency, Scottish Environmental Protection Agency, Northern Ireland Environment Agency, Health Protection Agency, Food Standards Agency) https://assets.publishing.service.gov.uk/government/uploads/system/uploads/attachment_data/file/296390/geho1202bklh-e-e.pdf (accessed 4 March 2023)

EA(S)R 2018 *The Environmental Authorisations (Scotland) Regulations 2018* http://legislation.gov.uk/ssi/2018/219 (accessed 31 January 2023)

EPR 2016 *Environmental Permitting (England and Wales) Regulations* https://legislation.gov.uk/uksi/2016/1154 (accessed 31 January 2023)

Human Tissue Act 2004 https://legislation.gov.uk/ukpga/2004/30 (accessed 11 February 2023)

Human Tissue (Scotland) Act 2006 https://legislation.gov.uk/asp/2006/4 (accessed 11 February 2023)

ICRP 2015 Radiation Dose to Patients from Radiopharmaceuticals: A Compendium of Current Information Related to Frequently Used Substances. ICRP Publication 128 *Annals of the ICRP* ed S Mattsson, L Johansson, S. Leide Svegborn, J Liniecki, D Noßke, K Riklund, M Stabin, D Taylor, W Bolch, S Carlsson, K Eckerman, A Giussani, L Söderberg, and S Valind 44(2S) https://www.icrp.org/publication.asp?id=ICRP Publication 128 (accessed 27 April 2023)

IPEM 2014 *Report 109: Radiation Protection in Nuclear Medicine* ed M McJury and C Tonge (York: Institute of Physics and Engineering in Medicine)

IPEM 2018 *Excretion Factors: The Percentage of Administered Radioactivity Released to Sewer for Routinely Used Radiopharmaceuticals* (York: Institute of Physics and Engineering in Medicine) https://ipem.ac.uk/resources/other-resources/statements-and-notices/advice-notice-on-excretion-factors-the-percentage-of-administered-radioactivity-released-to-sewer-for-routinely-used-radiopharmaceuticals/ (accessed 4 March 2023)

Medicines Act 1968 https://legislation.gov.uk/ukpga/1968/67 (accessed 8 February 2023)

RSA 1993 *Radioactive Substances Act 1993* http://legislation.gov.uk/ukpga/1993/12 (accessed 11 February 2023)

RSA(A)R(NI) 2011 *The Radioactive Substances Act* 1993 *(Amendment) Regulations (Northern Ireland)* https://legislation.gov.uk/nisr/2011/290 (accessed 31 January 2023)

RSE(NI) 2011 *The Radioactive Substances Exemption (Northern Ireland) Order* https://legislation.gov.uk/nisr/2011/289 (accessed 4 March 2023)

Scottish Government 2020 *Radioactive Substances Legislation—Scope: Guidance, GOV.SCOT Website* https://www.gov.scot/publications/scope-radioactive-substances-legislation-scotland-guidance-document/documents/ (accessed 27 April 2024)

SEPA 2019 *Principles for the Assessment of Prospective Public Doses Arising from Authorised Discharges of Radioactive Waste to the Environment (RS-JG-016)* (Stirling: Scottish Environmental Protection Agency) https://sepa.org.uk/media/478051/rs-jg-016-principles-for-assessment-of-public-doses.pdf (accessed 4 March 2023)

SEPA 2020 *Radioactive Waste Advisers, SEPA Website* https://sepa.org.uk/regulations/radioactive-substances/radioactive-waste-advisers/ (accessed 6 March 2023)

SEPA 2021 *Guidance and Reports, Sepa.Gov.Uk* https://sepa.org.uk/regulations/radioactive-substances/guidance-and-reports/ (accessed 30 June 2022)

The Hazardous Waste (England and Wales) Regulations 2005 2005 https://legislation.gov.uk/uksi/2005/894 (accessed 8 February 2023)

The Hazardous Waste Regulations (Northern Ireland) 2005 https://legislation.gov.uk/nisr/2005/300 (accessed 8 February 2023)

The Special Waste Regulations 1996 https://legislation.gov.uk/uksi/1996/972 (accessed 4 March 2023)

UK Parliament 2008 *The Transfrontier Shipment of Radioactive Waste and Spent Fuel Regulations* https://legislation.gov.uk/uksi/2008/3087 (accessed 31 January 2023)

IOP Publishing

Medical and Dental Guidance Notes (Second Edition)

A good practice guide on all aspects of ionising radiation protection in the clinical environment: IPEM Report 113

John Saunderson, Mohamed Metwaly, William Mairs, Philip Mayles, Lisa Rowley and Mark Worrall

Chapter 20

Contingency planning and emergency procedures for radioactive substances activity

Scope

20.1 This chapter gives guidance on the plans needed to deal with accidents and other incidents related to working with radioactive substances. The purpose of any plan should be to restrict possible exposures or damage to the environment or property. The plan, or at least a reference to it, should be included in the local rules.

20.2 Detailed advice is given in chapter 7, *Radiotherapy*, regarding preparations for dealing with incidents involving beam therapy or remotely operated afterloading equipment, and in chapter 9 regarding brachytherapy sources. Advice regarding contingency plans for x-ray sources can be found in chapter 3, *Diagnostic radiology (excluding dental) and fluoroscopically guided interventions*, and chapter 5, *Dental radiography*. In relation to contingency planning for transport events, more detailed advice is provided in chapter 18, *Keeping, accounting for, and moving radioactive substances (including transport)*, as part of the broader emergency planning considerations required by transport legislation (CDG as amended refers (CDG 2009). Chapter 1, *General measures for radiation protection* paragraphs 1.82–1.87, *Contingency plans*, also contains general advice.

20.3 Contingency planning is required at a number of levels:

- Internal incidents:
  - (i) local contingency planning for incidents under the Ionising Radiations Regulations, IRR (IRR 2017, IRR(NI) 2017);

(ii) internal Radiation (Emergency Preparedness and Public Information) Regulations, REPPIR, incidents (REPPIR 2019, REPPIR(NI) 2019).

- External incidents:
  (i) involvement in REPPIR response to incidents off-site;
  (ii) chemical, biological, radiological and nuclear incident responses;
  (iii) radiation monitoring units (RMUs);
  (iv) national planning for National Incidents and Accidents Involving Radiation (NAIR) where an establishment may be a responder;
  (v) response to the national threat level.

20.4 It is important to ensure that, where the emergency services are involved in off-site emergency plans, instruction, information, training, and adequate equipment are available. Local plans should dovetail with these emergency plans, and there must be cooperation in the rehearsal of these plans. For example, for NAIR, the local plans must supplement the national plans. Plans will need to be flexible enough to cover the range of incidents that could occur and be viewed as part of a wider system of responses.

Local contingency planning

20.5 All reasonably foreseeable incidents should be identified during the radiation risk assessment required by IRR (IRR 2017, IRR(NI) 2017), in consultation with staff working in the area. Reasonably foreseeable means accidents that are less likely but realistically possible. Typical incidents which are likely to involve the need for a contingency plan include:

(a) the loss or suspected loss (or theft) of a sealed or solid source;
(b) failure of engineering controls;
(c) failure/misuse of Personal Protective Equipment (PPE);
(d) unauthorised use of a High Activity Sealed Source (HASS) (EA 2018);
(e) the loss or spillage of open radioactive substances (including excreta);
(f) spread of contamination;
(g) breakage of a sealed or solid source. This should include the procedure for a check of the integrity of the source following the incident (EA 2018);
(h) failure to retract a radioactive source, close a shutter or switch off a radiation beam by normal methods;
(i) fire or explosion, including such occurrences during transport;
(j) flood;
(k) medical emergency of a patient with radioactive material implanted or administered;
(l) faulty equipment;
(m) user error or unauthorised entry into a controlled area.

20.6 There may be unique situations that apply to an organisation or to a location, e.g., proximity to fast-moving traffic, unusual drainage arrangements, or water supply. These would need to be considered if an accident could reasonably be envisaged.

20.7 The risk assessment should determine whether the accident or cleanup could result in significantly greater than normal exposure. Below this level, normal procedures should be followed, and a contingency plan would not be enacted. Experience has shown that most incidents involving spills of open radioactive substances in hospitals or clinics do not warrant any drastic emergency action but, on assessment, require only simple remedial action by local staff, often as part of their routine procedures for the control of the spread of contamination, for example, the uncontrolled release of patient excreta.

20.8 Nevertheless, more serious incidents are foreseeable for which a contingency plan must be prepared, for example, the release of a stock vial. There should be a clear distinction between a situation requiring the enactment of a contingency plan and normal procedures for small spillages. The point at which a contingency plan is to be enacted should be clearly defined within the local rules based on risk assessment. Where a contingency plan has been enacted, records must be kept for two years (IRR 2017, IRR(NI) 2017); see paragraph 1.85.

20.9 Part of the risk assessment process should be to determine the potential doses involved during incidents so that exposures can be restricted as far as practicable in an emergency. Staff should be made aware of these potential doses to help inform actions in an incident situation and to ensure the contingency plan helps reduce those exposures. If this assessment identifies the requirement for dose sharing, the details must be simple and clear. If possible, during an incident situation, an individual should be designated as a timekeeper to remove this responsibility from those dealing with the incident. The importance of safety over the continuation of service must be made clear to all staff.

20.10 The risk assessment should also identify and justify the scale and frequency of rehearsals required based on the factors set out in L121 (HSE 2018). Where rehearsals are undertaken, the following records should be made:

(a) list of participants;
(b) details of the rehearsal;
(c) actions resulting from the rehearsal;
(d) analysis of when the rehearsal should be repeated based on the outcome.

20.11 The outcome of the risk assessment should be a simple, practical, step-by-step plan of actions to be taken in the event of an incident, produced with the involvement of the Radiation Protection Adviser (RPA). Paragraph 244 of L121 (HSE 2018) provides the suggested contents of the contingency plan.

20.12 There should be a readily available, up-to-date list of all places in the establishment where there are or may be radiation hazards, and plans should cover all hazards. This list should show the exact location of and the means of access to all rooms likely to contain amounts of radioactive substances in excess of those in column 5 of Schedule 8 of the IRR (IRR 2017, IRR(NI)

2017). A copy should be available to those that require access in each place where there is a potential hazard, and a copy should be kept by the Chief Fire Officer and by the Emergency Planning Officer, if appointed, with due consideration given to source security. Those involved in contingency plans, such as for large-scale fires, who require access to this information should be able to readily access it in the event of an incident.

20.13 Arrangements should be made with the local Fire and Rescue Service to visit the establishment on behalf of the Chief Fire Officer to obtain information about the layout of the premises, warning notices and signs, and the location and types of radiation sources, should they feel it necessary to do so. Useful information on radioactive releases during fire engulfment is given in a Health and Safety Laboratory report (Thyer *et al* 2000). The Chief Fire Officer should also be consulted about the suitability of the fire-fighting equipment.

20.14 Notices should be posted at places where foreseeable accidents may occur, and they should show:

(a) how to contact the Radiation Protection Supervisor (RPS) and RPA, or an alternative person, who should be notified immediately of any emergency;
(b) how to call the Fire and Rescue Service and medical services if required; and
(c) the location of emergency equipment, if required.

20.15 Staff working in the area must be able to quickly and accurately determine the level of risk associated with an incident. To this end, the local rules could contain a simple triage assessment to determine whether the accident is of minor or major consequence, which is readily available to those working in the area.

20.16 It is also suggested that, for high-risk areas, a flow chart of the main actions to take in an emergency be kept within the area as an aide memoire and included in the contingency plan, though this should not be considered a replacement for adequate training and rehearsal.

20.17 Equipment should be kept available for use in an emergency wherever open radioactive substances are used. This may need to include some or all of the following items:

(a) overshoes, protective clothing (including gloves, disposable aprons, and caps), and respirators or breathing apparatus, depending on the level of risk determined by prior risk assessment. Where PPE is used, it must be used in compliance with the Personal Protective Equipment at Work Regulations 1992 (PPER 1992), which include requirements on specification, training, storage, and maintenance of equipment;
(b) decontamination materials for the affected area, including absorbent material for wiping up spills;
(c) decontamination materials for people;

(d) barriers or means for cordoning off affected areas;
(e) warning notices;
(f) handling tools and receptacles for contaminated articles;
(g) portable monitoring equipment[1];
(h) personal monitoring devices such as electronic dosemeters[1] and pocket dosemeters[1];
(i) non-porous floor covering, to be used only after any liquid spill has been cleared up; and
(j) sundry items, such as adhesive tape, spare batteries for all battery-operated equipment, waste bags, labels, a torch, a notebook and pencils, a calculator, a copy of the *Radionuclide and Radiation Protection Data Handbook* (Delacroix *et al* 2002, 2022), a copy of the NAIR Handbook (if involved in the scheme) (NRPB 2002), and simple first-aid equipment.

20.18 Careful consideration should be given to the selection of monitoring equipment, and the limitations of monitors should be well understood by those using them. The NAIR technical handbook provides useful advice on instrument selection (NRPB 2002). It is suggested that instruments be accompanied by a card stating the batteries they require and the responses to the different forms of radiation that they may be used for.

20.19 An appropriate selection of emergency equipment should be kept in a clearly labelled portable container in a readily accessible place. A list of the emergency equipment should be fixed to the container, and checks should be made periodically at a frequency determined by and specified in the risk assessment and immediately after use to ensure that all items are present or replaced as necessary due to observable dysfunction or passing a use by date.

20.20 It is suggested that at least annual exercises be performed (the frequency should be determined by local risk assessment) to test the effectiveness of the arrangements and to ensure that all persons concerned know what action to take in an emergency. Training in emergency situations should form part of the competency record for staff. Mock or low-activity sources should be used during training exercises.

20.21 A contact list should be readily available with all relevant contacts. For hospitals, it might be advantageous for the switchboard to keep a copy of the radiation emergency contact list on hand in case of after-hours incidents. Rehearsals should include checks to ensure that contact lists are up-to-date and that individuals are contactable as required. Consideration should be given to all potential contacts, for example, the contact to halt waste transfers and removals from the site.

[1] Consider removing the batteries from devices if they are not used routinely, in case they leak over time and damage the device.

20.22 Where there are outside workers or shared work areas, all workers within the area must be aware of potential accidents, their severity, and contingency plans. This will involve the cooperation of employers to ensure adequate training for all involved. The cooperation of employers may also be necessary if an incident requires control of an area outside the normal working area of the employer. This should be considered during the planning process, and it should be clear who takes control of the area and in what circumstances.

Action during the enactment of the local contingency plan

20.23 The best course of action in a local contingency plan depends very much on local circumstances and the nature of the incident. If radioactive substances are present, the emergency may involve actual or potential dispersal of the activity or a high dose rate. Occasionally, both hazards may be present.

20.24 If an activity greater than that specified in Column 5 of Part 1 of Schedule 7 IRR (IRR 2017) has been dispersed, urgent actions would include:

(a) deciding on the need for fire services, medical assistance and advice from the RPA;
(b) in accordance with the situation, warning those who are nearby the accident or evacuating the area;
(c) rendering first aid to any person who may be injured;
(d) notifying the RPS and local management;
(e) dealing with the emergency;
(f) notifying the HSE, HSENI, or ONR as appropriate, and the appropriate environment agency if a permit is breached.

20.25 If a source is lost or thought to be lost or stolen:

(a) The person responsible for the security of sources (see paragraphs 18.6–18.7, *Responsibility*), as well as the RPS, should be informed without delay; the RPS should arrange for a competent person to make an immediate search for the lost source; the possibility that a lost sealed or solid source might have fallen into a gap in protective material should not be overlooked; or a small source (e.g. Iodine-125 seed) is lodged in footwear or clothing.
(b) All means by which the lost source might be moved further astray should, as far as possible, be eliminated until the search has been carried out; floors should not be swept, furniture should not be moved, waste transfers on-site and off site should be stopped as necessary, sinks should not be used, and incinerators ashes should not be disturbed. Staff must not circulate in the room or leave the area.
(c) The possibility of contamination by spilt radioactive substances should be borne in mind and If there is any reason to suspect that the lost sealed source might have become damaged, rigorous precautionary measures should be instituted as soon as the contamination is detected.

(d) Contingency plans should make it clear which external bodies should be informed and who is responsible for this notification to enable immediate notification. Consideration should be given to the appropriate internal escalation mechanisms when determining the appropriate person to report to an external body.

(e) There should be arrangements for the prompt reporting of security events and incidents to your local police, which should also include notifying your Counter Terrorism Security Advisor (Counter Terrorism Policing no date). The relevant Environment Agency must also be informed of the requirements of a site permit/registration/authorisation, or for exempt sources if the above activity levels are outlined in the exemption guidance or binding rules (BEIS *et al* 2018, EA(S)R 2018). HSE, HSENI or ONR must also be informed if the activity lost or stolen exceeds column 6 of Part 1 of Schedule 7 of the IRR (IRR 2017, IRR(NI) 2017).

(f) The prior risk assessment should identify whether a radiation detector should be present in the area to support the search for the source, e.g. a contamination monitor (Iodine-125 seeds) or sentinel node probe (SLNB) in theatres or pathology for Iodine-125 seeds in prostate specimens.

20.26 It is suggested that a reduced version of Schedule 7 of the IRR (IRR 2017, IRR(NI) 2017) be included in the contingency plans, covering only the radionuclides in use for ease of access.

20.27 In the case of solids and liquids, the dispersal of the radioactive substance should be contained as far as possible. In the case of radioactive gases or vapours, they should be dispersed as quickly as possible. The design of facilities in which the release of radioactive gases or vapours may occur should take into account potential contingency situations.

20.28 If evacuation is required and time permits, all functioning apparatus in the area should be made safe. Ventilation and all laboratory services except lighting should be switched off, and all doors and windows should be closed. However, when radioactive gas or vapour, e.g. tritiated water vapour, has been released, mechanical ventilation should be left on and, with discretion, doors and windows should be opened.

20.29 The spread of contamination, particularly on the shoes or clothing of people leaving the affected area, should be minimised. Contamination may be contained by the use of cat litter or sand, or an area may be closed off to allow decay, depending on the use of the area and the radionuclide involved. It may sometimes be preferable to wash contamination down the drain where permitted to do so.

20.30 Where possible, a change barrier should be put into place with monitoring taking place of individuals crossing the barrier. All evacuated persons who might be contaminated should be monitored. Appropriate arrangements should be made for their decontamination.

20.31 Clothing contaminated in an emergency should be removed, bagged, ladled, and left in or near the affected area in a safe manner until it can be placed in an appropriate store for decay or disposal off-site.

20.32 Contaminated parts of the body should be washed thoroughly but gently until monitoring shows that contamination will not be significantly reduced further by this method. Care is being taken not to roughen or break the skin which would allow contamination to enter the bloodstream. Any contaminated wound, however trivial, should be irrigated with water or saline solution, and care is taken to limit any spread of contamination to or from other parts of the skin. If measurable contamination persists, an assessment of the resulting radiation dose should be obtained from a Medical Physics Expert (MPE). If morbidity is likely, specialist medical advice should also be sought. Consideration may be given to cutting nails and hair where contamination persists.

20.33 Persons entering the affected area to carry out emergency procedures should wear appropriate protective clothing, which they should monitor and remove when they leave the area. Prior to entering the area, as much planning as is possible should be undertaken, and roles on this occasion should be clearly defined. In the event of an emergency involving a serious spill, those concerned with the cleaning-up procedures should, if necessary, wear properly fitting respirators or breathing apparatus appropriate to the radioactivity being handled, so that even under these conditions, air monitoring will rarely be needed.

20.34 A direct radiation hazard may arise through loss of shielding or failure of a shutter or source transport mechanism. Temporary shielding should be used as necessary, and the immediate area should be evacuated and barriers erected to restrict access. Recovery of the source should then be undertaken in accordance with pre-planned procedures that take into consideration the doses likely to be incurred. Advice on dealing with an emergency that occurs during the treatment of a patient with solid sources is given in chapter 7, *Radiotherapy*, and chapter 8, *Radiotherapy and brachytherapy equipment*.

20.35 Access to an area affected by an emergency should be restricted until radiation surveys show that the area may be reoccupied. Any radiation generator should be examined for defects before it is re-energised. Leakage tests should be carried out if it is suspected that a radioactive substance is leaking from any container.

20.36 In consultation with the RPA, the employer should review all emergencies and near misses in order to learn any appropriate lessons from the way they have been managed and, where appropriate, revise the contingency plan in light of those lessons. Such reviews should be undertaken as soon as possible following the incident. L121 Paragraph 247 (HSE 2018) details a checklist of items that should be covered in this record where a contingency plan is enacted:

(a) why the plan was used;
(b) the specific cause or causes of the accident;
(c) whether any precautionary measures failed;
(d) risk assessments are reviewed if needed;
(e) whether any general management failures were the cause or contributed to the cause of the accident;
(f) the involved people and whether any injuries or exposures occurred;
(g) the effectiveness of the contingency plan and whether it needs to be revised;
(h) measures are in place to learn from errors and incidents, including an action plan to limit the risk of recurrence.

20.37 Incident reporting mechanisms and taxonomies should facilitate the detailed recording of an enacted contingency plan. Exposures due to the enactment of the contingency plan should be noted on the personnel's dose record. Where an individual may have received an 'overexposure' as defined by the regulations (IRR regulation 2(1), e.g. an annual dose limit exceeded (IRR 2017, IRR(NI) 2017)), an immediate investigation should be carried out, and if an overexposure cannot be ruled out beyond reasonable doubt, the HSE/HSENI, employee, appointed doctor, and employer (management and occupational health) of the employee should be notified as soon as practicable. An assessment of the dose received shall also be made, as well as a full investigation into the circumstances of the exposure. Immediate investigations must be kept for at least two years, and detailed investigations should be kept until the person affected reaches or would have reached the age of 75 years and at least 30 years from the date of the investigation. The individual will need to undergo medical surveillance.

20.38 Local dose investigation levels should also be set at levels that indicate where ALARP (as low as reasonably practicable) practices may not have been followed. These will be set at much lower levels than the statutory investigation level but must also be investigated locally. Dose investigation levels should be included in the local rules.

20.39 Given the infrequency of emergencies, it is helpful to ensure that the learning from any such incidents is shared within the wider medical physics community.

The Radiation (Emergency Preparedness and Public Information) Regulations (REPPIR) 2019

20.40 REPPIR 2019 (REPPIR 2019, REPPIR(NI) 2019) defines a radiation emergency as a non-routine situation or event arising from work with ionising radiation that necessitates prompt action to mitigate the serious consequences:

(a) of a hazard resulting from that situation or event;
(b) of a perceived risk arising from such a hazard; or
(c) to any one of the following; human life, health and safety, quality of life, property, or the environment.

The ONR and HSE have published guidance on compliance with the regulations, available online (HSE and ONR 2019).

20.41 The REPPIR (REPPIR 2019, REPPIR(NI) 2019) applies to any work with ionising radiation that involves having radioactive substances on site containing more than the quantity specified for that radionuclide in Schedule 1 of the REPPIR (REPPIR 2019, REPPIR(NI) 2019). However, if the operator's risk assessment demonstrates that, in an emergency situation, the radionuclides held on-site would not cause an annual effective dose to persons off-site of greater than 1 mSv, then the regulations would not apply.

20.42 Where more than one radionuclide is involved, the Operator must calculate a quantity ratio for each individual radionuclide, and the total sum must be less than 1.

20.43 Although the REPPIR is not expected to apply in a medical setting, it is incumbent on all employers to carry out an assessment to ensure that this is the case, to retain this calculation, and to revisit it if there are any significant changes in holding levels. In most cases, this will involve a simple calculation of quantity ratios for all radionuclides held on-site. In any situation where a radionuclide is not listed in Schedule 1 of REPPIR, the Employer must perform an evaluation of potential doses to persons off-site in an emergency situation. If the effective dose is likely to exceed 1 mSv, then the regulations will apply.

20.44 These Regulations are unlikely to apply to the transportation of radioactive materials between hospitals or to any other similar transport operations, because of the exemptions provided. Any transport contingency plan prepared under IRR (IRR 2017, IRR(NI) 2017) may be helpful in preparing the emergency plans needed under REPPIR (REPPIR 2019, REPPIR(NI) 2019).

20.45 The HSE and HSENI, which have a number of crucial roles to play in relation to the regulations, enforce REPPIR for healthcare services. A number of other bodies, such as local authorities, are statutory consultees on the emergency plans, as outlined in REPPIR regulation 10(5) (REPPIR 2019, REPPIR(NI) 2019). Actions concerning other transport modes (see chapter 18, *Keeping, accounting for, and moving radioactive substances (including transport)*) do not come within the REPPIR and will be implemented and enforced by:

- The Office for Nuclear Regulation (ONR) for road transport in Great Britain and NIEA for road transport in Northern Ireland (Department of Agriculture Environment and Rural Affairs 2020, ONR 2022);
- The Civil Aviation Authority for air transport; and

- The Maritime and Coastguard Agency for transport by sea and inland waterways.

Incidents involving radioactivity which occur off-site

Incident management principles

20.46 All radiation employers must have contingency plans to deal with the medical consequences of transport accidents, lost or stolen radioactive material, or the effects of a large-scale incident elsewhere.

20.47 The Civil Contingencies Act 2004 (CCA 2004) requires that NHS bodies operating hospitals with accident and emergency services:

(a) assess the risk of an emergency occurring;
(b) maintain plans to ensure that the employer can continue to function in an emergency and to mitigate/control the effects of an emergency;
(c) have a plan for carrying out exercises. NHS guidance (NHS England and NHS Improvement 2019a) states this should include (as a minimum);
 i. communications exercise every 6 months;
 ii. tabletop exercise once a year;
 iii. live exercise every three years;
 iv. a command post exercises every three years;
(d) have a plan for the training of staff, including training on the use of RAMGENEs, where applicable;
(e) warn and inform the community of the risks from hazards or incidents that have occurred;
(f) share information and cooperate with other emergency responders.

Similar requirements are implemented in Scotland through *The Civil Contingencies Act* 2004 *(Contingency Planning) (Scotland) Regulations* 2005.

20.48 All acute hospitals with Accident and Emergency departments are required to have arrangements in place for dealing with contaminated casualties. They must have specific plans and resources for dealing with incidents involving radioactivity.

20.49 In England, all plans should be developed according to the *NHS England* Core Standards for Emergency Preparedness, Resilience, and Response (EPRR) (NHS England and NHS Improvement 2019a) and in accordance with the NHS England guidance and framework (NHS England 2019a), and all health bodies should nominate a director level accountable emergency officer. Planning for radiation incidents should be carried out in conjunction with this officer. In Northern Ireland, the *Department of Health* has the responsibility to develop policy, guidance, and advice on planning and response to emergencies (DoH 2016). In Scotland, information is published on the Ready Scotland website (Scottish Government no date). In Wales, the Wales Resilience Forum addresses these issues (Welsh Government no date).

20.50 All Emergency Departments need to ensure that they have access to expert advice on how to manage patients who may have been exposed to radioactivity. In the event of a casualty having received a high radiation dose, the first signs of which may be nausea and vomiting, immediate transfer to a designated hospital is imperative.

20.51 First responders, such as the ambulance service, should have access to a multipurpose, rugged monitor such as the RAMGENE. The following must be ensured:
(a) staff are trained in how to use the monitor, with easy-to-use instructions kept with it;
(b) there is a spare battery kept in the monitor case;
(c) there are regular checks undertaken on the function of this monitor, and the responsibility for this is clearly defined;
(d) staff are aware of any techniques particular to the monitor required (e.g. do not cover the light sensor on the RAMGENE when using it without the cap on).

20.52 There are some principles that should be applied to the response to all incidents, regardless of their cause (PHE 2018):
- Maintain comprehensive, contemporaneous, and legible records with times and dates of actions. These records should be signed. A good practice is to maintain these in a bound log book with numbered pages.
- Any corrections should be struck through and signed, but remain legible.
- All advice given or received should be recorded.
- All protective actions should be recorded.
- Where there may be any subsequent investigation, it is important to maintain the chain of evidence. This will mean any samples taken or clothes removed must be adequately tracked. Details of how to do this are provided by Public Health England (PHE 2018).

REPPIR

20.53 Health bodies with fixed nuclear sites within their area must be involved in the site operators' emergency planning and response arrangements and in the off-site emergency plan. Carriers who make a carrier's emergency plan under REPPIR (REPPIR 2019, REPPIR(NI) 2019) may also consult them. There should be liaisons between hospitals and local employers who use radioactive substances so that arrangements for dealing with contaminated casualties can be made and the hospital can ensure that access to the appropriate monitors is available for all possible contaminants.

20.54 Any emergency worker who is involved in the response to a radiation emergency must have adequate information, instruction, and training to undertake this role and regarding the risks associated with the role, and must have this training updated at regular intervals. Where it is not possible to determine who will be involved in the response, on-the-day training must be provided.

20.55 No member of staff will be required to undertake duties where they believe that they are not competent to do so, even after training. Pregnant and breastfeeding staff should not attend to the incident.

20.56 A volunteer list for out-of-hours, off-site incidents should be kept. Requests for off-site support are to the hospital, not the volunteer; therefore, any radiation exposure to the volunteer is occupational.

Radiation emergencies

20.57 Public Health England (now UKHSA) (PHE 2021) and the NHS England Core Standards for Emergency Preparedness, Resilience, and Response (NHS England and NHS Improvement 2019b), the Ready Scotland website (Scottish Government no date), the Northern Ireland Department of Health (Department of Health 2016), and Wales NHS (Wales NHS no date, Public Health Wales 2016) give the requirements for NHS bodies and ambulance services who may need to deal with casualties likely to be contaminated by radioactivity. These core standards list the equipment that is required at hospitals to deal with radiation incidents. All monitoring equipment must be maintained in working order at all times.

20.58 All health bodies should have major incident plans for radiation incidents commensurate with the level of risk. Advice should be sought from the RPA and, where appointed, the radioactive waste adviser (RWA) during the preparation of this plan. During the planning process, consideration should be given to whether it is possible to contain wastewater from the decontamination process, and the sewerage undertaker and relevant authority must be involved in this planning. The guidance states that due to the potential long-term effects of radioactive contaminants, all reasonably practicable steps should be taken to contain runoff (Water UK *et al* 2018). The core standards state decontamination facilities should have a waste bladder, and where this is used, the risk of radiation doses from the contained material should be assessed.

20.59 While UK radioactive substances regulations, RSR (RSA 1993, EPR 2016, SEPA 2021a, 2021b), provide a defence if discharges are made to avoid danger to human health, all reasonable steps must be taken to reduce pollution, and acts undertaken must be reported to the competent authority as soon as possible after the event. It will be the responsibility of the Fire and Rescue Service to contact the sewerage undertaker and regulator as soon as possible should this be required (Water UK *et al* 2018).

20.60 The RPA and RWA should be consulted if it is known or suspected that a casualty is contaminated, and as much information as possible about the incident and the radionuclide(s) involved should be sought.

20.61 In the case of radiation incidents, the advice for life-threatening injuries is that first responders should stabilise the patient before decontamination using standard precautions, and if there is no life-threatening injury, decontamination at the scene prior to treatment is highly recommended (PHE 2018). The decontamination plan should take into account the potential records required to maintain the chain of evidence.

20.62 While recent guidance on the Initial Operational Response (NHS England 2022) states that dry decontamination of individuals should be used for non-caustic chemical contamination (NHS England 2019b), the consensus is that this is not as effective as wet decontamination for radiological contamination (IPEM 2016), and this should be reflected in local plans until further scientific evidence is provided.

20.63 Emergency Department staff should wear surgical masks, face visors (to offer eye protection), surgical hats, gowns with arm protection, and overshoes, and be double-gloved when dealing with potential radiation incidents. All patients involved in a radiation incident should be assumed to be contaminated until proven not to be. Staff should be monitored for external radiation doses while dealing with contaminated casualties and should also be checked for contamination immediately afterwards. Arrangements must be in place to ensure that the doses received by all members of staff are ALARP. The REPPIR (REPPIR 2019, REPPIR(NI) 2019) do allow, under certain limited circumstances, emergency exposures in excess of the dose limits under the IRR (IRR 2017); however, this is for pre-identified employees (as per risk assessment) who have received information and training in advance of an emergency and are properly equipped (HSE and ONR 2019). The PHE (now UKHSA) has provided detailed guidance on triaging and managing patients exposed to a radiological risk in a Chemical, Biological, Radiological, and Nuclear (CBRN) incident (PHE 2018). Pregnant and breast-feeding staff should not normally be involved in treating casualties contaminated with radioactivity, and all staff should be aware of this.

20.64 Medical physics staff may assist in responding to such an incident, as defined by the local CBRN plans, which may include advice on;

- radiation screening of the patient;
- identification of the radioisotope and resultant hazard;
- providing advice on whether to isolate or provide cohort care for mass casualties;
- establishment of dirty and clean control zones;
- safety precautions, for example, no eating, drinking or smoking, decontamination, etc;
- dose estimation of casualties;
- management of radioactive waste.

20.65 Although effective and timely communication of the risks and precautions to take in such an incident is paramount, all Employers should have agreed communication plans with emergency services, which include role cards, and these should be followed.

Radiation monitoring units

20.66 Medical physics staff may also be involved in the establishment of RMUs to carry out large-scale monitoring for potential contamination of the public in a radiation emergency. The PHE (now UKHSA) has issued extensive guidance on factors for consideration during the planning and set-up of such a facility, including suggested layout, equipment, staffing, measurement methods, operational aspects, and training required (HPA 2011). Guidance has also been provided on establishing RMUs in Scotland (NHS Scotland 2017).

20.67 While medical physics staff may be asked to contribute to the running of RMUs, priority must always be given to the emergency response at hospitals.

National arrangements for incidents involving radioactivity (NAIR)

20.68 In the case of accidents and situations that are not readily foreseeable or for which formal plans cannot be made, a scheme exists to support the police, who are usually the first officials on the scene, with expert help and advice. The National Arrangements for Incidents involving Radioactivity (NAIR) (PHE 2021) are coordinated by the Radiation, Chemical, and Environmental Hazards Directorate (RCE) in England, Scotland, and Wales. A similar system is in place in Northern Ireland under RIPP (Radiation Incidents in Public Places). These national arrangements are required for orphan sources under The Ionising Radiation (Basic Safety Standards) (Miscellaneous Provisions) Regulations 2018 (IR(BSS)(MP)R 2018).

20.69 In NAIR, help is offered in two stages:

(a) Stage 1 is monitoring and advice from a radiation expert on what action to take, typically medical physics departments;

(b) Stage 2 is invoked in more serious cases and when more extensive action than can be provided by Stage 1 is needed to protect the public, particularly where there is a risk of contamination spreading.

20.70 Under NAIR, help is provided by major nuclear installations, the Ministry of Defence (MOD), and hospitals' medical physics departments nominated under the scheme. The police must start a NAIR response. Guidance on the contents of the NAIR responder kit can be found in the NAIR technical handbook (NRPB 2002).

20.71 All exposures during a NAIR incident should be planned where possible, and doses should always be kept as low as reasonably practicable.

20.72 Although NAIR response is provided voluntarily and coordinated by the UK Health Security Agency (UKHSA), establishments that provide support for NAIR should ensure that arrangements are in place to provide assistance and that contact lists are kept up-to-date. Permits under the relevant RSR (RSA

1993, EPR 2016, SEPA 2021a, 2021b) should include permission to temporarily store radioactive materials arising from a NAIR incident when requested. Establishments providing a NAIR response are under no obligation to recover sources from their premises. However, if they believe that it is safe and appropriate to do so, then the legal provision is made to ensure that this is possible. It is recommended that advice from the relevant competent authorities be sought before recovering sources.

20.73 Arrangements exist for the collection of radioactive material arising from a NAIR incident at the scene of the incident or from a location of temporary safekeeping. These are instigated by the relevant environmental regulator (EA, EANI, NRW, SEPA) in cooperation with the person(s) responding under NAIR. Further information on NAIR is available via the gov.uk website (PHE 2021)

Response to national threat level

20.74 Those that hold High Activity Sealed Sources (HASS) are required to have contingencies in place for increasing security in the event of an increase in the national threat level. The CTSA (counter terrorism security advisers) should be consulted about the precautions that may be necessary on a site-by-site basis and act as a source of information regarding the threat level.

References

BEIS, DEFRA, Welsh Government, and DAERA 2018 Scope of and Exemptions from the Radioactive Substances Legislation in England *Wales and Northern Ireland Guidance Document* https://gov.uk/government/publications/guidance-on-the-scope-of-and-exemptions-from-the-radioactive-substances-legislation-in-the-uk (accessed 11 February 2023)

CCA 2004 *Civil Contingencies Act* https://legislation.gov.uk/ukpga/2004/36 (accessed 4 March 2023)

CDG 2009 *The Carriage of Dangerous Goods and Use of Transportable Pressure Equipment Regulations 2009* https://legislation.gov.uk/uksi/2009/1348 (accessed 31 January 2023)

Counter Terrorism Policing no date *Report Suspicious Activity, Action Counters Terrorism Website* https://act.campaign.gov.uk (accessed 22 July 2022)

Delacroix D, Guerre J P, Leblanc P and Hickman C 2002 *Radionuclide and Radiation Protection Data Handbook 2002* (Nuclear Technology Publishing) (Radiation Protection Dosimetry, vol 98)2nd edn

Delacroix D, Guerre J and Leblanc P 2022 *Guide Pratique Radionucleides & Radioprotection* 3rd edn (Les Ulis, France: EDP Sciences)

Department of Agriculture Environment and Rural Affairs 2020 *Radioactive Transport, Transfrontier Shipments and Justification, Department of Agriculture Environment and Rural Affairs Website* https://daera-ni.gov.uk/articles/transport-transfrontier-justification (accessed 22 July 2022)

Department of Health 2016 *Chemical, Biological, Radiological and Nuclear, Department of Health (Northern Ireland) Website* (Belfast) https://health-ni.gov.uk/articles/chemical-biological-radiological-and-nuclear (accessed 6 March 2023)

DoH 2016 *Emergency Planning and Response, Department of Health, Northern Ireland Website* https://health-ni.gov.uk/articles/emergency-planning-and-response (accessed 22 July 2022)

EA 2018 *How to Comply with Your EPR RSR Environmental Permit—Sealed Sources. 2.0* (Bristol: Environment Agency) https://gov.uk/government/publications/rsr-environmental-permit-how-to-compy-sealed-sources (accessed 4 March 2023)

EA(S)R 2018 *The Environmental Authorisations (Scotland) Regulations 2018* http://legislation.gov.uk/ssi/2018/219 (accessed 31 January 2023)

EPR 2016 *Environmental Permitting (England and Wales) Regulations* https://legislation.gov.uk/uksi/2016/1154 (accessed 31 January 2023)

HPA *HPA-CRCE-017: Radiation Monitoring Units: Planning and Operational Guidance* 2011 (UK: Health Protection Agency) https://gov.uk/government/publications/radiation-monitoring-units-planning-and-operational-guidance (accessed 4 March 2023)

HSE 2018 *L121 Work with Ionising Radiation Ionising Radiations Regulations 2017—Approved Code of Practice and Guidance* 2nd (Norwich: Health and Safety Executive) https://hse.gov.uk/pubns/books/l121.htm (accessed 31 January 2023)

HSE and ONR 2019 L126 the radiation (emergency preparedness and public information) Regulations 2019 *Approved Code of Practice and Guidance* 2nd (London: TSO) https://onr.org.uk/documents/2020/reppir-2019-acop.pdf (accessed 31 January 2023)

IPEM 2016 *Written evidence Unpublished by the Institute of Physics and Engingeering in Medicine to the Commons Science and Technology Committee 'Science in Emergencies: Chemical, Biological, Radiological or Nucleaar Incidents' Inquiry*

IR(BSS)(MP)R 2018 *The Ionising Radiation (Basic Safety Standards) (Miscellaneous Provisions) Regulations* https://legislation.gov.uk/uksi/2018/482 (accessed 4 March 2023)

IRR 2017 *The Ionising Radiations Regulations* https://legislation.gov.uk/uksi/2017/1075 (accessed 31 January 2023)

IRR(NI) 2017 *The Ionising Radiations Regulations (Northern Ireland)* http://legislation.gov.uk/nisr/2017/229 (accessed 31 January 2023)

NHS England 2019a *Emergency Preparedness, Resilience and Response (EPRR). Guidance and Framework, NHS England Website* https://england.nhs.uk/ourwork/eprr/gf/ (accessed 22 July 2022)

NHS England 2019b *Guidance for the Initial Management of Self Presenters from Incidents Involving Hazardous Materials* 1st edn (London: NHS England) https://england.nhs.uk/publication/eprr-guidance-for-the-initial-management-of-self-presenters-from-incidents-involving-hazardous-materials/ (accessed 4 March 2023)

NHS England 2022 *Hazardous Materials (HAZMAT) and Chemical, Biological, Radiological and Nuclear (CBRN), NHS England Website* https://england.nhs.uk/ourwork/eprr/hm/ (accessed 22 July 2022)

NHS England and NHS Improvement 2019a *NHS Core Standards for Emergency Preparedness, Resilience and Response Guidance* (NHS England: NHS Improvement) https://england.nhs.uk/publication/nhs-england-core-standards-for-emergency-preparedness-resilience-and-response/ (accessed 4 March 2023)

NHS England and NHS Improvement 2019b *NHS England Core Standards for Emergency Preparedness, Resilience and Response* 5th (NHS England: NHS Improvement) https://england.nhs.uk/publication/nhs-england-core-standards-for-emergency-preparedness-resilience-and-response/ (accessed 4 March 2023)

NHS Scotland 2017 *Radiation Monitoring Unit Template Plan*

NRPB 2002 *NRPB-W7: NAIR Technical Handbook 2002 Edition Technical Handbook on the National Arrangements for Incidents Involving Radioactivity* ed N P Mccoll and P Kruse

(Chilton: National Radiological Protection Board) https://assets.publishing.service.gov.uk/government/uploads/system/uploads/attachment_data/file/347760/2002_NrpbW7.pdf (accessed 4 March 2023)

ONR 2022 *Transport of Radioactive Material, Office for Nuclear Regulation Website* https://onr.org.uk/transport/index.htm (accessed 30 June 2022)

PHE 2018 *Chemical, Biological, Radiological and Nuclear Incidents: Clinical Management and Health Protection* ed N Gent and R Milton 2nd edn https://gov.uk/government/publications/chemical-biological-radiological-and-nuclear-incidents-recognise-and-respond (accessed 4 March 2023)

PHE 2021 *Guidance: National Arrangements for Incidents Involving Radioactivity (NAIR)*, *GOV.UK Website* https://gov.uk/guidance/national-arrangements-for-incidents-involving-radioactivity-nair (accessed 22 July 2022)

PPER 1992 *The Personal Protective Equipment at Work Regulations 1992* https://legislation.gov.uk/uksi/1992/2966 (accessed 31 January 2023)

Public Health Wales 2016 *Civil Contingency and Emergency Planning Arrangement* ed D Rixon (Public Health Wales) 1st edn

REPPIR 2019 *The Radiation (Emergency Preparedness and Public Information) Regulations 2019* http://legislation.gov.uk/uksi/2019/703 (accessed 31 January 2023)

REPPIR(NI) 2019 *The Radiation (Emergency Preparedness and Public Information) Regulations (Northern Ireland)* http://legislation.gov.uk/nisr/2019/185 (accessed 31 January 2023)

RSA 1993 *Radioactive Substances Act 1993* http://legislation.gov.uk/ukpga/1993/12 (accessed 11 February 2023)

Scottish Government no date *Ready Scotland—Preparing for and Dealing with Emergencies*, *Safer Scotland Website* https://readyscotland.org/ (accessed 8 July 2022)

SEPA 2021a *Environmental Authorisations (Scotland) Regulations 2018 Standard Conditions for radioactive substances activities V2.0. 2.0.* (Stirling, UK: Scottish Environmental Protection) Agency https://www.sepa.org.uk/media/593756/standard-conditions-for-radioactive-substances-activities-v2.pdf (Accessed: 26 April 2024)

SEPA 2021b *Environmental Authorisations (Scotland) Regulations 2018 Guide to standard conditions for radioactive substances activities v2* (Stirling, UK: Scottish Environmental Protection) https://www.sepa.org.uk/media/591433/guide-to-standard-conditions-v2.pdf (accessed 24 April 2024)

The Civil Contingencies Act 2004 (Contingency Planning) (Scotland) Regulations 2005 https://legislation.gov.uk/ssi/2005/494

Thyer A, Allen J, Atkinson G and Thyer A M 2000 Release fractions for radioactive sources in fires (FS-99-19) https://webarchive.nationalarchives.gov.uk/ukgwa/20170111122518/http://www.hse.gov.uk//research/hsl/fire.htm (accessed 27 April 2024)

Wales NHS no date *Standard 4: Civil Contingency and Emergency Planning Arrangements*

Water UK, EA, NIEA, NRW, DWI, Fera, Defra and NFCC 2018 *Protocol for the Disposal of Contaminated Water and Associated Wastes at Incidents* 16th edn (London: Water UK) https://water.org.uk/wp-content/uploads/2018/11/Contaminated-Water-Disposal-Protocol.pdf (accessed 4 March 2023)

Welsh Governement no date *Emergency Preparation, Response and Recovery*, *GOV.WALES Website* https://gov.wales/emergency-preparation-response-recovery (accessed 22 July 2022)

IOP Publishing

Medical and Dental Guidance Notes (Second Edition)
A good practice guide on all aspects of ionising radiation protection in the clinical environment:
IPEM Report 113
John Saunderson, Mohamed Metwaly, William Mairs, Philip Mayles, Lisa Rowley and Mark Worrall

Appendix 1

Roles and responsibilities of the employer using ionising radiation

While every attempt has been made to include as many of the Employer's responsibilities as possible, this list is not necessarily exhaustive. It is an indication of the measures to be taken and is not a replacement for expert advice. The Regulations addressed are:

- The Ionising Radiations Regulations 2017, referred to as the IRR (IRR 2017, IRR(NI) 2017).
- The Ionising Radiation (Medical Exposure) Regulations 2017 (GB) and 2018 (NI), referred to as IR(ME)R (IR(ME)R 2017, IR(ME)R(NI) 2018).
- The Environment Permitting (England and Wales) Regulations 2016 and Environmental Authorisations (Scotland) Regulations 2018, as amended, referred to as EPR (EPR 2016, EA(S)R 2018).
- The Radioactive Substances Act 1993 (Amendment) Regulations (Northern Ireland) 2011, referred to as RSA (RSA(A)R(NI) 2011).
- The Carriage of Dangerous Goods and Use of Transportable Pressure Equipment Regulations 2009 (GB) and 2010 (NI), as amended, referred to as CDG (CDG 2009, CDG(NI) 2010).

Be aware of regulatory requirements and appoint expert advisers to help ensure essential measures are in place before commencing work with radiation.

Identify and consult one or more expert advisers that apply to the services being established, as required under the regulations or likely associated permit conditions. The advisers can support Employers as they establish services and legally must be involved in certain elements that must take place before work with ionising radiation is commenced. The experts to be sought out and appointed, as required, are as follows:

- IRR—Radiation Protection Adviser (RPA) (see appendix 3);
- IR(ME)R—Medical Physics Expert (MPE) (see appendix 15);

doi:10.1088/978-0-7503-2332-1ch21

- EPR/RSA permit holders—Radioactive Waste Adviser (RWA) (see appendix 21);
- CDG consignors or carriers—Dangerous Goods Safety Adviser (DGSA).

Establish if there is a need to notify the HSE or HSE(NI) of work with ionising radiation, register a practice or gain consent for specific practices under the graded approach. If any material changes are made to a previous application, then notify the HSE or HSE(NI).

Undertake a radiation risk assessment before starting a new activity involving work with radiation. It would be prudent to consult a Radiation Protection Adviser (RPA) at this stage. Establish suitable dose constraints for the restriction of exposure for each category of person likely to be exposed.

Provide relevant local rules in compliance with the legislation, appointing Radiation Protection Supervisors (RPS) as necessary to ensure the local rules are implemented.

Prepare contingency plans and incorporate them into local rules if the risk assessment shows that an accident is reasonably foreseeable.

Designate, as appropriate, the necessarily controlled and supervised areas (to be described in local rules) with the necessary monitoring and controls (demarcation, signs, restricted access, systems of work, written arrangements, etc) to provide adequate protection from external radiation and radioactive contamination, including washing and changing facilities as required.

Provide suitable and sufficient monitoring equipment and arrange for its maintenance and testing (with input from a Qualified Person); ensure records of tests of monitoring equipment are kept.

Ensure records of monitoring in designated areas are kept.

Obtain information from the manufacturer and installer about the proper use, testing, and maintenance of radiation equipment after its critical examination, and involve the RPA.

Ensure the necessary steps are taken to restrict exposures to ionising radiation for staff, patients, and others who may be exposed, setting investigation levels and providing written arrangements if necessary.

Demonstrate commitment to radiation protection, for example, through a written radiation safety policy, the establishment of a radiation protection committee, clear management lines, clear actions, and the involvement of senior staff.

Provide sufficient engineering controls, design features, safety features, and warning devices to restrict exposures so far as is reasonably practicable, and ensure that these are properly maintained and tested at suitable intervals.

Ensure personal protective equipment is provided, worn, stored, and maintained as appropriate after all other measures have been considered.

Ensure that all employees (including RPSs) are given appropriate radiation protection training sufficient to understand the risks and precautions needed, including female workers who may be pregnant or breastfeeding. Training should include general and specific measures, the use of personal protective equipment (PPE), contingency plans, and any specific functions required of that individual. Ensure the training is assessed as sufficient.

Co-operate with other Employers concerning the exposure of employees, as appropriate, to ensure all Employers can meet the requirements of the regulations. Of particular concern is personal dose limitation.

Designate classified persons, if necessary, and provide appropriate radiation monitoring and medical surveillance (health record); inform all persons when they are designated as classified persons.

Provide personal radiation monitoring and dosimetry records as necessary; ensure that the results of personal monitoring are kept under review (as low as reasonably practicable; ALARP); and that any unusual results are investigated.

Ensure that dose limits are not exceeded, nor is there a constraint for those who are pregnant and have notified the Employer (in writing).

Establish a mechanism for reporting, investigating, evaluating, and recording incidents (including near misses) that lead to or have the potential for unnecessary or excessive exposure to ionising radiation.

Investigate and notify the HSE or HSE(NI) about overexposures. Uphold subsequent dose limitations for staff who were overexposed.

Review procedures at regular intervals (as specified in the Employer's policies), preferably with the support of the Radiation Protection Committee/Medical Exposures Committee.

More specifically, on using radioactive substances

If the work will involve the storage, use, or disposal of radioactive materials, then the Employer must seek legal authorisation to do so from the regulator. In the UK, the relevant regulations and regulators are:

- In England, the Environmental Permitting (England and Wales) Regulations (EPR) (2016) (as amended 2018), enforced by the Environment Agency (EA).
- In Wales, the Environmental Permitting (England and Wales) Regulations (EPR) (2016) (as amended 2018), enforced by Natural Resources Wales (NRW).
- In Scotland; the Environmental Authorisations (Scotland) Regulations (2018), enforced by the Scottish Environment Protection Agency (SEPA).
- In Northern Ireland; the Radioactive Substances Act (RSA) (1993) (Amendment) Regulations (Northern Ireland) 2011, enforced by the Northern Ireland Environment Agency (NIEA).

If the quantities of radionuclides involved are small, then it may be possible to operate under an exemption from the relevant legislation. It may be useful for the Employer to consult a Radioactive Waste Adviser (RWA) to assist with this process.

Consult and appoint a suitable RWA to provide expert advice on radioactive waste management and environmental radiation protection.

Consult and appoint a suitable Dangerous Goods Safety Adviser (DGSA) to provide expert advice on the transportation (as a consignor or carrier) of radioactive materials. There is no requirement to appoint a DGSA if the workload only involves the carriage of 'excepted' packages.

If the work involves the transportation of radioactive material, either as a consignor or a carrier, then the Employer should perform a radiation risk assessment to determine whether a radiation emergency may arise. If an emergency situation is foreseeable, then the Employer must develop an emergency plan to account for potential hazards present during the transport operation. It is advisable for the employer to consult an RPA and a DGSA to assist with this process.

Ensure radioactive substances in use are sealed wherever practicable to prevent leakage. Where this is impracticable, the substance should be contained to prevent leakage as much as possible. Keep records of appropriate leakage tests in both cases.

Account for and keep records of the quantity and location (including ultimate disposal) of all radioactive substances.

Ensure radioactive substances are suitably contained, labelled and stored when not in use.

Ensure radioactive substances are suitably packaged, labelled and accompanied by appropriate documentation when in transit.

Ensure systems are in place to minimise the number of radioactive sources in use and the volume and activity of radioactive waste generated.

Ensure the conditions of any permit/authorisation to work with radioactive materials are adhered to.

Notify the HSE or HSE(NI) if a quantity of radioactive material (in excess of that identified for the particular radionuclide) is spilt or accidentally released, resulting in significant contamination, having made an immediate investigation and a report of the incident. For incidents involving the transportation of radioactive substances between sites, the Office for Nuclear Regulation (ONR) is the GB authority.

Notify immediately the relevant authority if a radioactive source is damaged, lost, or stolen (the EA in England, NRW in Wales, SEPA in Scotland, and the NIEA in Northern Ireland).

Assess whether the quantities of radionuclides on the premises could exceed those listed in Schedule 1 of The Radiation (Emergency Preparedness and Public Information) Regulations 2019 (REPPIR 2019, REPPIR(NI) 2019) and, if this is confirmed, undertake an evaluation to identify the hazards that have the potential to cause a radiation emergency. The employer must consult an RPA for advice on radiation protection issues relating to radiation emergency planning.

More specifically, on making medical exposures

Appoint one or more Medical Physics Experts (MPEs) as appropriate to the scope of the exposures planned as required under IR(ME)R.

Consult an MPE on the preparation of technical specifications for equipment and installation design.

Hold a licence in respect of each radiological installation at which radioactive substances are to be administered, for such purposes as may be specified in that licence.

Set up a framework for the entitlement of appropriately trained Referrers, IR (ME)R Practitioners, and Operators (including MPEs) in each service delivered, e.g. diagnostic radiology, diagnostic nuclear medicine, radiotherapy, and nuclear medicine therapies. These roles will also be required for non-medical imaging with medical radiological equipment.

Maintain records (available for inspection) of the training and continuing education of these Practitioners and Operators (including MPEs). Referrer training is not strictly required under IR(ME)R but it is good practice to provide appropriate training to this group to the extent an Employer has control over this.

Ensure an MPE is closely involved in every radiotherapeutic medical exposure, except those involving standardised therapeutic nuclear medicine.

Ensure an MPE is available in standardised therapeutic nuclear medicine, diagnostic nuclear medicine practices, high-dose interventional radiology, and high-dose computed tomography.

Ensure an MPE is involved as appropriate for consultation on optimisation in all other radiological practices not covered in the previous two paragraphs.

Ensure written protocols are in place for every type of standard radiological practice for each piece of equipment, and a current inventory of equipment is kept for and at each radiological installation, ensuring the amount of equipment is limited to that necessary for the proper carrying out of exposures.

Establish quality assurance (QA) programmes for written procedures and protocols.

Establish recommendations concerning referral guidelines for exposures, including radiation doses, and ensure these are available to the referrer.

Take measures to raise awareness of the effects of ionising radiation among individuals capable of childbearing or breastfeeding.

Establish local diagnostic reference levels, undertake regular reviews, and ensure corrective action is taken as necessary.

Collect dose estimates from medical exposures for radiodiagnostic and interventional procedures, taking into consideration the distribution by age and gender of the exposed population, and when so requested, provide the estimates to the Secretary of State or Department of Health.

Establish dose constraints for research programmes where no direct medical benefit for the individual is expected from the exposure and for the exposure of carers and comforters.

Ensure a clinical evaluation of the outcome of each exposure, including factors relevant to the patient's dose where appropriate, is recorded.

Ensure clinical audit is carried out as appropriate.

Implement Employer's procedures as specified in Schedule 2 of IR(ME)R and ensure these are complied with by IR(ME)R Referrers, Practitioners and Operators.

Ensure the quality assurance programme includes a study of the risk of accidental or unintended exposures in respect of radiotherapeutic practices.

Establish a system for recording analyses of events involving or potentially involving accidental or unintended exposures proportionate to the radiological risk posed.

Undertake a preliminary investigation of any known or suspected accidental or unintended exposure that has or may have resulted in a person being exposed to levels of ionising radiation significantly greater than those considered proportionate (or significantly lower than those considered proportionate in a radiotherapeutic exposure), and unless that investigation shows beyond reasonable doubt that no such exposure has occurred, immediately notify the appropriate authority. Then, undertake a detailed investigation, assess the dose received, and notify the authority of the outcome of the investigation and corrective measures. An MPE must be involved. Consider reporting equipment problems to the Medicines and Healthcare Products Regulatory Agency (MHRA), as appropriate (MHRA 2023).

More specifically, on using equipment for (medical or non-medical) exposures

Ensure that the equipment available for the range of examinations or treatments using ionising radiation is appropriate and is not used for procedures for which it is not suitable.

Implement and maintain a quality assurance programme for equipment (as defined in IR(ME)R). Ensure that the equipment is tested before first use, routinely, and following maintenance that is capable of affecting the equipment's performance.

Identify, provide, maintain, and calibrate appropriate test equipment as part of the QA programme.

Ensure that new or replacement equipment is capable of meeting the specifics set out in Regulation 16 of IR(ME)R, as appropriate. These requirements are related to the indication of the dose delivered during an exposure, the verification of key treatment parameters, and the transfer of dose information to a person's record.

References

CDG 2009 *The Carriage of Dangerous Goods and Use of Transportable Pressure Equipment Regulations 2009* https://legislation.gov.uk/uksi/2009/1348 (accessed 31 January 2023)

CDG(NI) 2010 *The Carriage of Dangerous Goods and Use of Transportable Pressure Equipment Regulations (Northern Ireland) 2010* https://legislation.gov.uk/nisr/2010/160 (accessed 31 January 2023)

EA(S)R 2018 *The Environmental Authorisations (Scotland) Regulations 2018* http://legislation.gov.uk/ssi/2018/219 (accessed 31 January 2023)

EPR 2016 *Environmental Permitting (England and Wales) Regulations* https://legislation.gov.uk/uksi/2016/1154 (accessed 31 January 2023)

IR(ME)R 2017 *The Ionising Radiation (Medical Exposure) Regulations* www.legislation.gov.uk/uksi/2017/1322 (accessed 31 January 2023)

IR(ME)R(NI) 2018 *The Ionising Radiation (Medical Exposure) Regulations (Northern Ireland)* https://legislation.gov.uk/nisr/2018/17 (accessed 31 January 2023)

IRR 2017 *The Ionising Radiations Regulations.* Great Britain https://legislation.gov.uk/uksi/2017/1075 (accessed 31 January 2023)

IRR(NI) 2017 *The Ionising Radiations Regulations (Northern Ireland)* http://legislation.gov.uk/nisr/2017/229 (accessed 31 January 2023)

MHRA 2023 *Yellow Card Reporting Site, MHRA Website* (London: Medicines & Healthcare products Regulatory Agency) https://yellowcard.mhra.gov.uk/ (accessed 6 March 2023)

REPPIR 2019 *The Radiation (Emergency Preparedness and Public Information) Regulations 2019* http://legislation.gov.uk/uksi/2019/703 (accessed 31 January 2023)

REPPIR(NI) 2019 *The Radiation (Emergency Preparedness and Public Information) Regulations (Northern Ireland)* http://legislation.gov.uk/nisr/2019/185 (accessed 31 January 2023)

RSA(A)R(NI) 2011 *The Radioactive Substances Act* 1993 *(Amendment) Regulations (Northern Ireland)* https://legislation.gov.uk/nisr/2011/290 (accessed 31 January 2023)

IOP Publishing

Medical and Dental Guidance Notes (Second Edition)
A good practice guide on all aspects of ionising radiation protection in the clinical environment:
IPEM Report 113
John Saunderson, Mohamed Metwaly, William Mairs, Philip Mayles, Lisa Rowley and Mark Worrall

Appendix 2

Ionising radiation protection and medical exposure policy

Introduction

The purpose of a radiation safety policy is to restrict, so far as is reasonably practicable, the dose received by staff and the public (including visitors) as a result of an organisation's work with ionising radiation and to ensure that medical exposures of patients and non-medical exposures using medical radiological equipment are optimised. The policy should describe the management structures, procedures, and organisational arrangements to be applied in practice and should establish means for preventing and mitigating the consequences of accidents (IAEA 2015). The Employer's general management arrangements for complying with the Ionising Radiations Regulations 2017 (IRR 2017, IRR(NI) 2017) form part of the general health and safety arrangements required by regulation 5 of the Management Regulations (MHSWR 1999, HSE 2018) and provide evidence that the Employer has discharged their responsibilities under IR(ME)R.

The policy should identify:

- Responsibilities for the management of radiation protection, including the management of staff.
- Arrangements for the protection of staff and members of the public.
- Arrangements for the protection of patients undergoing medical exposure, or non-medical exposure using medical radiological equipment.

The policy should outline the principles to guide decisions made in respect of work with ionising radiation. It will be a local decision as to how much detail the high-level policy contains. Related policies should be referenced (there may be a separate radioactive materials policy, a personal dosimetry policy, or cooperation between Employers policy, for example). Departmental-level standard operating procedures (or detailed work instructions) enacting the high-level policy should also reference the relevant documents to demonstrate the pathway/structure to the Employer.

Key stakeholders identified in the document should be involved in the drawing up of the policy, and a suitable review period should be chosen to ensure that the policy remains up to date.

An example policy is given below. It is not intended to provide an exhaustive list of contents or prescribe how the policy should be written, but rather to be used as a guide as to how it could be set out and what it should cover. It is written in the style required for a large English NHS organisation and can be adapted for other UK countries as appropriate. For smaller organisations and those who are self-employed, the management structures will differ, but the same principles will apply.

Enforcement agencies have identified a risk that such an example policy is copied directly to address a particular Employer's policy needs. This is not appropriate. Development work will be required with the guidance and support of appointed expert advisers, as appropriate. Decisions can also be made about the inclusion or otherwise of non-ionising radiation risks such as lasers and UV.

Example policy

Contents

1. Policy statement

The Trust uses ionising radiation within its facilities for medical (and non-medical) imaging and for medical treatment. This includes diagnostic and interventional X-rays, external beam radiotherapy, and the use of sealed and unsealed sources for diagnosis and treatment. The Trust will only carry out radiation exposures for medical or research purposes. It will not carry out any imaging exposures for non-medical purposes (e.g. medico-legal purposes).

This policy covers the provisions of the Ionising Radiations Regulations 2017 and the Ionising Radiation (Medical Exposures) Regulations 2017, their subsequent amendments, and associated guidance for work with ionising radiation at all sites within the organisation. It supports and augments the organisation's general health and safety policy. It also addresses the Trust's responsibilities under the Carriage of Dangerous Goods and Transportable Pressure Equipment Regulations 2009, as amended in 2011, in so far as they relate to radioactive materials, and to the Radiation (Emergency Preparedness and Public Information) Regulations 2019.

The purpose of this policy is to restrict, so far as reasonably practicable, the dose received by staff, students, outside workers, volunteers, and the public (including visitors) as a result of the Trust's work with ionising radiation and to ensure that the medical exposures of patients, research volunteers, and 'carers and comforters' are optimised. The document sets out to achieve this purpose by identifying the responsibilities of those involved and stating the principles that should guide decisions made in respect of radioactive sources, radiation-producing equipment, radiation-safety equipment, and the arrangements for working with radiation.

The keeping and usage of radioactive materials and the accumulation and disposal of radioactive waste are covered in a separate policy (Management of Radioactive Materials Policy). The 'Reporting of Injuries, Diseases, and Dangerous Occurrences Regulations (RIDDOR)' has requirements that may apply to ionising radiation exposure; this is covered in the Trust's health and safety policy.

This policy does not cover non-ionising radiation, which has its own Trust policy ('Artificial Optical Radiation Safety Policy Incorporating Management of Laser Safety').

2. Definitions and associated documents

Associated documents *(list Trust wide and departmental policies of relevance).*

Term	Meaning
Appointed doctor	Registered medical practitioner who meets criteria specified by HSE.
Approved Dosimetry Service (ADS)	A dosimetry service approved in accordance with regulation 36 of IRR17.
As low as reasonably practicable (ALARP)	A requirement of IRR17 to ensure that radiation doses are kept as low as possible while taking practical and economic factors into account.
Chief Executive	The person who has delegated responsibility from the Board of Directors for the management of

(*Continued*)

(*Continued*)

Term	Meaning
	governance arrangements within the Trust and is ultimately responsible for ensuring that the Trust meets its obligations with regard to the safe and effective delivery of services. This is delegated to responsible individuals within the Trust.
Carer and comforter	An individual knowingly and willingly incurring exposure to ionising radiation by helping, other than as part of their occupation, in the support and comfort of individuals undergoing or having undergone an exposure.
Carriage of Dangerous Goods and Transportable Pressure Equipment Regulations 2009, as amended in 2011 (CDG)	Regulations defining transport requirements for radioactive substances.
Classified person	A person delegated as such pursuant to regulation 21(1) of IRR17.
Controlled area	An area that has been designated as such in accordance with regulation 17(1) of IRR17. This will include areas where it is necessary to follow special procedures to restrict significant exposure to ionising radiation.
Dangerous Goods Safety Adviser (DGSA)	A person who is currently qualified to advise on the safe Transport of hazardous goods and compliance with CDG.
Examination	The act of being examined or the state of being examined. To look at, inspect, or scrutinise carefully or in detail. To investigate the patient's state of health.
Image Optimisation Team	A multidisciplinary team of experts working to ensure a consistent approach to optimisation across all modalities (COMARE 16th report).
Local rules	Written rules made pursuant to regulation 18(1).
Medical Physics Expert (MPE)	An individual or a group of individuals, having the knowledge, training, and

	experience to act or give advice on matters relating to radiation physics applied to exposure, whose competence in this respect is recognised by the Secretary of State.
Non-classified person	Any person who has not been classified in accordance with IRR17 regulation 21(1) (also see classified person definition).
Outside worker	Any person who is carrying out services in a controlled area or supervised area but who does not have an individual contract of employment with the Employer responsible for that area.
Patient	Means a person who is receiving medical care from the Trust.
Practitioner	A registered healthcare professional who is entitled, in accordance with the Employer's procedures, to take responsibility for an individual's exposure.
Radiation	Throughout this policy, the term radiation shall only cover ionising radiation, viz., electromagnetic or particulate radiation capable of producing ions either directly or indirectly.
Radiation (Emergency Preparedness and Public Information) Regulations 2019 (REPPIR)	Regulations that require preparation for possible radiation emergencies.
Radiation Protection Adviser (RPA)	An individual (or body) who has the specific knowledge, experience, and competence required for giving advice on the particular working conditions or circumstances for which the employer is making the appointment of an expert adviser.
Radiation Protection Supervisor (RPS)	An employee appointed under regulation 18(5) (IRR17) to provide supervision to ensure all work is carried out in line with the requirements of the Regulations.

(*Continued*)

(Continued)

Term	Meaning
Reasonably foreseeable radiation incident	An incident involving radiation that has been identified in a radiation risk assessment as foreseeable and possible. Steps should be taken to mitigate the risk of the potential incident and its impact. Reasonably foreseeable radiation 'accidents' require a contingency plan.
Regulations	The Ionising Radiation Regulations 2017 (IRR17), the Ionising Radiation (Medical Exposures) Regulations (IR(ME)R), and their subsequent amendments.
Reporting of Injuries, Diseases and Dangerous Occurrences Regulations (RIDDOR)	Regulations that put duties on employers to report certain serious workplace accidents, occupational diseases, and specified dangerous occurrences.
Significant Event	An event that triggers the enactment of a contingency plan.
Supervised area	An area that has been so designated by the Employer in accordance with Regulation 17(3) of IRR17. This will include areas where it is necessary to keep the conditions of the area under review to determine whether the area should be designated as a controlled area.
Written procedures	Written procedures for exposures established in accordance with regulation 6(1) of IR(ME)R, including those Employer's procedures set out in Schedule 2 of IR(ME)R.
Written protocols	Written protocols for exposures established in accordance with regulation 4 of IR(ME)R. Examples would include x-ray exposure factors in diagnostic radiology and standard prescriptions in radiotherapy.

3. Duties and responsibilities

3.1 Committees

3.1.1 Radiation protection committee (RPC)—see Trust H&S policy for reporting structure.

The committee will:

- Provide advice to the organisation on compliance with the Regulations.
- Identify specific problems concerning the safe use of ionising radiation.
- Recommend for ratification organisational policies (including this one) regarding the safe use of ionising radiation and radioactive materials.
- Ensure that the policy is reviewed on a regular basis (according to Trust policy) and in response to any changes in legislation, guidance, or best practice.

3.2 Groups

3.2.1 Board of directors.

The Board of Directors has overall responsibility for ensuring the safe use of ionising radiation throughout the organisation.

3.3 Individuals

3.3.1 Chief executive.

The Chief Executive has overall responsibility for compliance with the relevant ionising radiation legislation by ensuring the relevant management of risk. In particular, the Chief Executive will:

- Determine whether there is a need to notify the HSE under their notification, registration, and consent system. Any material changes are to be reapplied for.
- Ensure a management system, organisation structure, and resources that are sufficient to achieve compliance with the legislation.
- Maintain a Radiation Protection and Medical Exposures Committee to provide advice to the Trust on the state of compliance with the Regulations and to identify specific problems relevant to the use of ionising radiation.
- Ensure that all necessary steps are taken to restrict, so far as is reasonably practicable, the extent to which persons are exposed by the organisation's work with ionising radiation.
- Appoint a suitable Radiation Protection Adviser(s).
- Appoint Radiation Protection Supervisors from within the line management structure of all departments that use ionising radiation to provide an adequate level of day-to-day supervision of work with ionising radiation.
- Fulfil the role of the Employer as described in IR(ME)R.
- Appoint suitable Medical Physics Expert/Experts to aid with the application of IR(ME)R.

- Will ensure advice from the MPE and RPA is sought by the Trust during procurement of any radiation-producing equipment or equipment that could affect patient, public, or staff dose.
- Provide for cooperation between Employers to the extent necessary for each Employer to comply with the Regulations.
- Delegate the task of entitlement of all duty holders as described in the Employer's procedures.
- Appoint an external DGSA to provide advice on CDG compliance in the transport of radioactive materials (including consignments).
- Maintain a mechanism for reporting, investigating, evaluating, and recording incidents (including near misses) that result in or have the potential to result in unnecessary or excessive ionising radiation exposure.

3.3.2 Divisional managers.

Divisional Managers have responsibility for ensuring that suitable arrangements are in place in their respective divisions. Specifically that:

- In accordance with IR(ME)R 15(1b), an inventory of radiological equipment is kept up to date.
- An RPA is consulted when necessary, as outlined in Schedule 4 of IRR17, the associated Approved Code of Practice and non-binding guidance.
- An MPE is involved when necessary, as outlined in IR(ME)R Regulation 14.
- The MPE and RPA are consulted on the specification and selection of new radiation equipment.
- An RPA is consulted on the assessment of whether REPPIR applies beyond the requirement to carry out the assessment.
- To cooperate with other divisional managers to establish and support multidisciplinary IR(ME)R and 'Image Optimisation Teams'.
- Identify and cooperate with other Employers as necessary to ensure that shared facilities or designated areas are managed within the Regulations to protect staff, visitors, patients, and the public. Where other Employers work on site, there should be contractual agreement that they will follow the Regulations and cooperate with the Trust.

3.3.3 Departmental service managers.

Departmental Service Managers will, seeking the advice and support of the RPA and/or MPE when appropriate, ensure that:

- They request advice from an RPA or MPE for those matters required under the Regulations (as set out in the expert adviser appointment letter), including for procurement exercises of any radiation-producing equipment or equipment that could affect patient, public, or staff dose.
- They provide the time and resources required for those with radiation protection responsibilities to fulfil their roles satisfactorily.
- The necessary controlled and supervised areas are designated (to be described in local rules) with the necessary monitoring and controls

(demarcation, signs, restricted access, systems of work, written arrangements, etc) to provide adequate protection from external radiation and radioactive contamination, including washing and changing facilities as required.

- An adequate degree of supervision of the work with ionising radiation is maintained via the system of Radiation Protection Supervisors.
- All relevant staff (including students and visiting staff) receive information, guidance, and training in radiation protection to enable them to understand the risks and precautions needed to comply with the requirements of the Regulations, this policy, all relevant local rules, and other key work instructions. This includes raising awareness of risks and requirements with those who may be pregnant and/or breastfeeding. Training should include general and specific measures, the use of PPE, contingency plans, and any specific functions required of that individual. Ensure the training is assessed as sufficient. Records must be maintained and available for inspection.
- All equipment in use is designed, installed, and maintained so as to keep doses to staff and the public as low as reasonably practicable and patient doses as low as reasonably practicable, consistent with the intended purpose.
- Information is obtained from the manufacturer and installer about the proper use, testing, and maintenance of radiation equipment after its critical examination, which involves the RPA.
- Ensure a critical examination of any installed radiation producing equipment has been carried out following advice of the RPA and that the commissioning of equipment is performed before clinical use following the advice of an MPE.
- All relevant radiation risk assessments (including individual risk assessments, e.g. for pregnancy and breastfeeding), local rules, and other key work instructions are in place and subject to regular review.
- Suitable and sufficient monitoring equipment is provided; arrange for its maintenance and testing (with input from a Qualified Person); and ensure records of tests of monitoring equipment are kept.
- Area dose monitoring is carried out in accordance with the risk assessment, and records are maintained.
- The necessary arrangements for classification of staff are in place.
- The management of personal monitoring provision (sufficient use of, distribution of, and timely collection of), compliance of staff, analysis of results, and appropriate actions taken as required. This may require direct cooperation with the contracted Approved Dosimetry Service.
- The records of personal dosimetry for non-classified staff are kept for at least two years from the date of monitoring.
- Ensure any Employers that need to be approached as part of cooperation between Employers are identified and that the relevant Divisional Director is aware of these other Employers.

- The necessary arrangements for controlled area/equipment handover are in place.
- Adequate training in local arrangements is provided for outside workers.
- Request previous or concurrent dose records for staff who are working for another employer or who have worked for another employer in the current calendar year, where radiation dose may have been recorded.
- Any suspected incidents involving unexpected exposures of members of staff, outside workers, or members of the public are investigated and reported as appropriate.
- Any suspected (or potential/near miss) accidental, unintended or clinically significant medical exposures are investigated and reported as appropriate.
- An up-to-date record is maintained of the training of all IR(ME)R Referrers, Practitioners and Operators.
- Written procedures, including those specified in IR(ME)R Schedule 2, are in place for all medical exposures and non-medical exposures with medical radiological equipment.
- Written protocols are in place for every type of standard radiological practice for each type of equipment.
- An effective QA programme is operational for radiological equipment and ancillary equipment that may affect patient dose and image quality.
- Recommendations made during service, repair, and QA of radiation equipment and ancillary equipment are followed up appropriately.
- A quality and clinical auditing programme is in place.
- Advice is taken from the DGSA as well as the RPA to ensure all documentation required for transport of radioactive materials is maintained.
- Radiation protection resources are included within business cases as appropriate.
- Records are retained in line with regulatory requirements (and Trust or NHS requirements, given that radiation requirements are not often the only demand in these matters).

N.B. If any of the above activities are delegated, Departmental Service Managers will remain accountable for all health and safety issues in their areas of work.

3.3.4 Employees.

All staff:

- Must follow local rules (or systems of work) in radiation designated areas.
- Must follow the IR(ME)R Employer's procedures as required by role.
- Must not intentionally or recklessly misuse or interfere with radiation equipment, radioactive substances, or protective equipment.
- Must take reasonable care to ensure their own and others' doses are kept as low as practicable.
- Must use any relevant protective devices that have been provided by the Employer, return such equipment after use, and report any defects.

- Must wear any dosemeters issued to them, either by name or job role, whenever carrying out work with ionising radiation, and change and return the dosemeter promptly at the end of each monitoring period.
- Must report any incidents or near misses that could lead to an unintended dose for a member of staff, outside worker, or member of the public.
- Must report any incidents (or potential incidents/near misses) that could lead to an accidental, unintended, or clinically significant exposure of a patient, carer, or comforter.
- They must only perform tasks for which they have been trained, deemed competent, and are authorised.
- Must tell their line manager if they are working with ionising radiation (or working in a facility where they may be exposed to ionising radiation) for another Employer or, if recently appointed, have done so within that calendar year.
- Staff working in designated areas who become pregnant or are breastfeeding should notify their line manager in writing as soon as the individual discovers they are pregnant.

3.3.5 Radiation protection supervisors (RPSs).

The principal duty of the Radiation Protection Supervisor is to supervise the work with ionising radiation to ensure that the local rules are complied with and any other appropriate radiation protection measures are observed by staff. To this end, they must be closely involved in the work, in a position of sufficient authority, and given time to carry out their responsibilities.

Appropriately trained RPSs will be appointed in writing by the chair of the Radiation Protection and Medical Exposures Committee (or a person nominated by them) for each area of work with ionising radiation.

The RPS should:

- Bring to the attention of their departmental manager and the RPC any matters of concern regarding radiation safety.
- Submit an annual report to the RPC on the state of radiation safety in their area of responsibility.
- Where required, make an entry of the estimated dose in the radiation passbooks of classified outside workers.

As part of the Trust's commitment to radiation protection, they will support and appropriately resource the RPS role. A list of current RPSs is available in each area's local rules.

3.3.6 Classified persons.

Employees who are designated as classified shall comply with the requirements of that designation, namely, they must:

- Attend a medical examination by the appointed doctor.
- Wear, change, and return on time all dosemeters supplied for the assessment of their dose.
- Inform all Employers if they are classified through employment with another Employer.

3.3.7 Bank/agency staff/voluntary workers.

If staff from the Bank, staff supplied via employment agencies, or voluntary workers undertake duties within the Trust involving entry into radiation-controlled areas, they must follow local rules and comply with all radiation protection requirements, including receiving training and wearing dose monitoring badges appropriate to the area. It is likely that some of these staff will be called 'outside workers'.

3.3.8 Appointed doctor.

The Appointed Doctor will perform an annual medical review for all classified persons for the purpose of determining whether they can be certified as fit for their work with ionising radiation.

3.3.9 IR(ME)R duty holders.

All IR(ME)R duty holders must:

- Follow written procedures for the conduct of medical and non-medical exposures using medical radiological equipment.
- Apply professional judgement to any reasonable deviations from written protocols for standard radiological practices and evidence the rationale for the deviation.
- Cooperate with other IR(ME)R duty holders, in particular to participate in multidisciplinary IR(ME)R and image optimisation teams as appropriate.

Referrers

- Referrers shall supply the Practitioner with sufficient medical data (such as previous diagnostic information or medical records) relevant to the medical exposure they are requesting to enable the Practitioner to decide on whether there is a sufficient net benefit.
- All Referrers will attend the required in-house training and ensure that any referrals made are within their agreed scope of practice.

Practitioners

- Are responsible for the justification of medical exposures, taking into account all relevant factors, including those specified in IR(ME)R.
- Will undertake continuing education and training relevant to their role.
- ARSAC licence holders must ensure their licence is current and covers the scope of their practice.

Operators

- Are responsible for each and every practical aspect of a medical exposure that they carry out, including quality control checks, ensuring that doses arising from the exposure are kept as low as reasonably practicable and consistent with the intended purpose.
- They must not use any equipment on which they are not trained unless under the supervision of a qualified Operator.
- Will undertake continuing education and training relevant to their role.

3.3.10 Expert advisers.

Provide advice to the Trust as required by the Regulations in the scope outlined in the expert adviser appointment letter(s).

Radiation Protection Adviser (RPA)

- Provides general advice on request to the Trust on compliance with IRR17 with a focus specifically on matters listed in Schedule 4 of IRR17.
- Where the assessment requests it, give advice on REPPIR compliance.

Medical Physics Expert (MPE)

An MPE shall be involved in medical exposures, give advice, and contribute to those matters, as set out in Regulation 14 of IR(ME)R and as summarised in the appointment letter.

The MPE will also:

- Participate in Image Optimisation Teams.
- Consult the RPA and RWA as needed.
- Provide a dose and risk assessment for research exposures (lead sites and secondary site local reviews).

Dangerous goods safety adviser

- Provides advice to the Trust, on request, on compliance with CDG.

Clinical radiation expert

- Undertakes a review of research doses and risks.
- Must be authorised by the Health Research Authority.
- May work either within or outside of the Trust.

4. Record retention

Advice on record retention can be sought from the RPA with regard to IRR requirements.

- Records are typically kept for at least two years, e.g. area monitoring or assessment of engineering controls.
- Dosimetry records for classified staff are to be kept by the Approved Dosimetry Service (until the individual is 75 years old, but in any event for at least 30 years from when the record is made).
- Only the most current risk assessment or local rules need to be kept.

Advice on record retention can be sought from the MPE with regard to IR(ME)R requirements.

- Typically, QA records are kept for at least the lifetime of the equipment.
- Local Diagnostic Reference Levels are retained as long as they are the most recent version.
- Only the most recent IR(ME)R Employer Procedure must be retained.

Radiation protection regulations requirements may not be the only driving factor for document/record retention; therefore, Trust procedures should be consulted in conjunction with expert advice.

5. Administrative arrangements for the management of work with ionising radiation

The Departmental managers, via the relevant RPSs, will ensure that:

- A radiation risk assessment is made and documented prior to the start of any new or modified radiation practice, including changes to equipment or techniques. This applies to both permanent and temporary changes.
- Existing radiation risk assessments are reviewed regularly (at least annually) to assess their on-going suitability and sufficiency.
- All radiation risk assessments include contingency arrangements for reasonably foreseeable radiation accidents and an estimate of the dose that might be received in such circumstances.
- Regular reviews of radiation safety are carried out, and a written record is kept of all significant findings and recommendations. These reviews should be carried out at least once a year and form the basis of a report to the Radiation Protection Committee.
- Where relevant, that transport documentation is in place and is followed (management and radiation protection programme, security and contingency, audit, as appropriate).
- The Radiation Protection Adviser should be consulted on any amended or new risk assessments and, as necessary, in respect of any other queries arising from the annual review. The appropriate RPA can advise on the structure and content of a risk assessment on request.

6. Dose monitoring

6.1 Personal monitoring

Personal dosimetry will be carried out for all non-classified persons who, as determined by the risk assessment, are likely to receive an annual effective dose greater than 1 millisievert (whole body), 50 mSv (skin/extremities) or 6 mSv (lens of the eye) as a result of routine work with ionising radiation or reasonably foreseeable radiation incidents.

Specific groups of (or individual) employees identified as likely to receive effective doses greater than 6 mSv per annum, or an equivalent dose that exceeds 150 mSv (skin/extremities) or 15 mSv (lens of the eye), are designated as classified persons in line with regulation 21 of IRR17, following the advice of the RPA.

Dosemeters from an Approved Dosimetry Service will be issued for the assessment of any significant doses, i.e. 1 mSv effective dose, and doses considered significant by the RPA for the eye and extremities based on monitoring results, professional judgement, and guidance.

'Job badges' may be used in appropriate cases to demonstrate that individual monitoring is not necessary. Each individual will wear the monitor while they are doing the role (e.g. scrub nurse), and the reading on the dosemeter will indicate the worst-case scenario, which is that a single person receives all the exposure during the wear period. Job badges are only appropriate when used to confirm that doses are negligible (i.e. below the threshold of the dosemeter) and where there is no identified risk of a radiation accident in the risk assessment.

Trial monitoring for periods of a number of months may periodically be used to demonstrate that doses are in line with the risk assessment and to support the decision not to monitor individuals.

Advice on the procurement of personal dosimetry will be taken from the RPA.

Those departments working with radiation must keep records of personal dosimetry for non-classified staff for at least two years from the date of monitoring.

Classified staff records are kept by the ADS in line with the Regulations.

6.2 Investigation level

An investigation level will be set locally for each modality or specialist practice, in line with the risk assessment, to demonstrate that exposures are being restricted as far as reasonably practicable. Within diagnostic radiology and radiotherapy, these levels would not be expected to be set above 2 mSv. In nuclear medicine, the level would not be expected to be set above 3 mSv. In interventional radiology, the level would not be expected to exceed 4 mSv.

6.3 Area monitoring

Monitoring of external radiation will be carried out in accordance with RPA advice after any new radiation installation, significant increase in workload, or change in practice. In every case, passive dosimetry will be used to assess conditions during typical clinical use. This may be supplemented by active dose rate monitoring. Routine area monitoring is carried out on a schedule determined by the risk assessment.

7. Carers and comforters

Carers and comforters are addressed within the IR(ME)R procedures for the various departments. Dose constraints are contained within those procedures.

8. Equipment

8.1 Procurement

The purchasing decision making panel will ensure that equipment is selected that complies with current radiological safety criteria as given in IRR and IR(ME)R and described in the relevant codes of practice, is CE marked, has adequate specifications, and is suitable for its intended clinical use.

The MPE must be consulted and contribute to the preparation of technical specifications for equipment and installation design (along with other relevant experts, e.g. the RPA).

The Trust will appropriately resource the radiation protection of existing and new services, such as through inclusion by service managers in the appropriate business case, etc. The Radiation Protection and Medical Exposures Committee or expert advisers may at any time bring a lack of resources to the attention of the Trust.

8.2 Quality assurance and maintenance

While it is the responsibility of the installer to carry out a critical examination of an article that can affect radiation protection by the way it is installed (e.g. installed radiation producing equipment), the relevant service manager will ensure that it has been performed prior to the equipment going into first use.

A critical examination of the facility (rather than the installed equipment) will be performed by an appointed RPA. Departmental managers are to seek their advice on engineering controls and safety/warning features, e.g. room shielding/warning signs, or lights, so that this can be assessed before radiation is used in the facility (including before the supplier uses radiation in the facility while installing equipment).

Equipment will be subject to appropriate maintenance and quality assurance programmes, following advice from manufacturers or an MPE. Equipment that develops a fault will be subject to a process to ensure it is not returned to use until the fault has been rectified.

Employees will be supported in taking equipment out of use where there are concerns about safety. Clinical pressures are not a reason to reduce safety standards. Where necessary, the appropriate experts and managers will be consulted, and a decision on use will be documented.

Quality control checks of radiotherapeutic and radiographic equipment will be carried out by suitably trained Operators at installation, periodically throughout its use, and following maintenance and repair. The routine periodic quality control checks will be carried out on a schedule based on recommendations in national publications and guidance and/or upon the recommendations of the MPE.

8.3 Replacement/disposal

In consultation with the MPE, equipment nearing end of life or equipment that no longer complies with national guidance on acceptable quality will be added to the risk register and managed appropriately, e.g. consideration of higher levels of QA.

Replacement may be required due to the deterioration of optimisation.

The route of disposal shall be considered, e.g. return to the supplier, removal by the supplier of new equipment, onward sale/donation, or as waste from the Trust. If appropriate, while awaiting disposal, the equipment will be disabled (e.g. removal of the power supply).

Equipment that may be radioactive, either due to contamination with radioactive materials or activation of components, is covered by the Trust radioactive materials policy.

9. Designation of areas

The ALARP principles will apply to the design of facilities, and in any case, an annual public dose constraint of 0.3 mSv will be used. Areas where dose rates exceed these values will be designated controlled areas or supervised areas, as appropriate, following consultation with the RPA.

All designated areas must be suitably demarcated with appropriate signage.

9.1 Arrangements for entry into controlled areas

Employees (and any other person) who are not classified persons may only enter a controlled area to carry out work in accordance with the systems of work set out in the local rules.

9.2 Permits to work

Some non-clinical locations, including roofs, plant rooms, and basements, are designated controlled areas due to their proximity to radiation sources. The appropriate wording on the pertinent signs will be able to identify these. Access to these areas for essential repair work, etc is controlled by a permit-to-work system operated by the Estates Department. A permit to work must be issued to all contractors and members of the Estates Department, etc who require access to these areas and must be authorised by a member of the Estates Department and someone from the relevant clinical area. Permits are also required for work in controlled areas.

10. Personal protective equipment (PPE)

PPE is provided where risk assessment demonstrates that it is required, in consultation with the RPA. The specific PPE required must be stated within the risk assessment. PPE is chosen to provide adequate protection from radiological risks.

Where PPE is provided, employees must ensure that it is worn and stored correctly and that missing or defective equipment is reported promptly to the RPS.

PPE must undergo periodic visual and/or radiological inspection as appropriate to ensure that it is not excessively worn or damaged and is repaired or replaced as necessary.

11. Radiation incidents

Hazards with the potential to cause a radiation accident will be identified during risk assessment, along with an assessment of likelihood and impact. Actions will be taken to limit the likelihood of its occurrence and the impact.

All incidents involving exposures must be recorded on the Trust incident database in line with the ‘incident reporting and investigation policy’.

The Radiation Protection and Medical Exposures Committee will review the list of all incidents and near misses involving ionising radiation and recommend action on common themes or trends, sharing the learning with other relevant departments.

11.1 External notifications

The MPE/RPA will advise the Trust on any incidents where it is likely that notification should be made to an external body, in line with regulator and professional body guidance, which will include:

- Clinically significant (accidental or unintended) patient exposures: Care Quality Commission.
- Overexposure of staff or members of the public: HSE.
- Loss of a source: Environment Agency incident helpline—0800 807060.
- Equipment fault leading (or potentially leading) to an accidental or unintended exposure: Medicine and Healthcare Products Regulatory Authority Yellow Card scheme.

Notification will be made jointly by the Health, Safety, and Emergency Planning Lead or, if they are not available, a colleague within the Quality and Standards division and an appropriate expert adviser.

12. Research exposures

The Trust participates in many clinical trials that require exposure to ionising radiation for the purpose of diagnosis or therapy. All such trials must have a dose and risk assessment completed by an MPE and a clinical review of these risks by a Clinical Radiation Expert (CRE).

In addition, all trials must undergo a local review to ensure that the Trust is able to comply with and is in agreement with the assessment by the MPE and CRE. This will ensure that any doses delivered by the Trust are in line with the approval granted by an ethics committee. These reviews will be undertaken by qualified individuals in the departments of radiotherapy, radiology, and nuclear medicine.

Dose constraints will be set for any exposures that are not of direct benefit to trial participants. Should a dose be delivered that is much greater than a dose constraint, this should be considered for notification under IR(ME)R Reg 8(4). For any trials that involve nuclear medicine procedures, nominated staff from the nuclear medicine department will ensure that ARSAC study approval is in place and the procedures required are listed on the relevant ARSAC Practitioner and Employer licences.

13. Cooperation with other employers

This may be required due to:

- Trust employees who work with, or are exposed to, ionising radiation on other Employers' sites while undertaking Trust business or while employed by other organisations.
- Employees of other organisations that undertake work with, or are exposed to, ionising radiation on a Trust site.
- Trainees and students who are members of an educational establishment who work with, or are exposed to, ionising radiation while on a Trust site.

Service managers should identify managers and RPSs to assist with the identification and monitoring of any external Employers whose staff work with ionising radiation or who are occupationally exposed to radon in their department within the Trust.

Management oversight/effort should be related to the associated risk, with very low-level cooperation being appropriate for the least risky scenarios. Once identified, the service manager should undertake an assessment of the work involved and identify details of any information that would be required to be exchanged to ensure the safety of any staff involved. This information should be provided to the Divisional Director, who will arrange for a cooperation agreement between Employers to be in place and identify a suitable contact at the Trust and with the other employer so that information can be exchanged. The Directorate Manager will ensure that any information that is required is supplied in a timely manner.

Service managers are responsible for undertaking suitable risk assessments for all such situations where Trust staff work with ionising radiation on other Employers' sites or where other Employers' staff work with ionising radiation on Trust premises.

The service manager will ensure that any training is adequate, personal dosimeters are worn or provided if necessary, radiation passbooks are presented for classified staff and required personal protective equipment is provided before any work commences.

A summary of the matters to be considered in an agreement for cooperation between employers is given in appendix A.

13.1 Handover of controlled areas

In radiotherapy, the radiation risk assessments are sufficient to include service and maintenance activities for in-house engineering work. There are cooperation arrangements documented in the local rules. The majority of Trust radiation risk assessments will not be suitable or sufficient for non-clinical operation of x-ray imaging equipment or external engineer use of radiotherapy equipment. Where activities require it, the controlled area should be handed over by an appropriately trained individual to the visitor, using a suitable handover form. This process should be monitored via a regular audit by service leads.

13.2 Classified outside workers

Where classified persons from another Employer are carrying out duties for that Employer in a Trust controlled area, i.e. the area has not been handed over, they are Classified Outside Workers. Classified Outside Workers must provide the RPS with their radiation passbook so that the RPS can record an estimate of the dose received while working in the Trust's controlled area. Dose estimates may be determined by calculation in conjunction with the RPA, or the use of active, e.g. real-time, personal dosemeters. Both Employers will need to consider sharing risk assessments, local rules, staff training, and monitoring arrangements. If a radiation passbook is not available, then the RPS or manager in control of the designated area should not allow the classified outside worker to work.

13.3 Outside workers

Where non-classified persons from another Employer are carrying out a service for the Trust in a Trust controlled area, i.e. the area has not been handed over, they are Outside Workers. Both Employers will need to consider sharing risk assessments, local rules, staff training, and monitoring arrangements. This applies to non-classified outside workers who normally work with radiation AND workers such as prison guards, trainees, care assistants from nursing homes, etc if they are providing a service to the Trust. Advice from the RPA should be sought regarding these arrangements. If an outside worker is not able to provide current dose data and training information, then it should be considered if they may be prevented from working, based on the risk of exceeding constraints or dose limits. Seek advice from the RPA if in doubt. If general and specific radiation protection training cannot be demonstrated or records provided, then direct supervision is essential.

13.4 Sharing of dose records

Co-operation between Employers is required in respect of the sharing of dose records. On appointment (in line with the Trust induction process but in advance of the first day on site if possible), the dose records for a new member of staff who was working with radiation within the current calendar year (including agency, locum, and bank staff) must be requested from their previous Employer(s). This must be requested for the current calendar year by a suitable individual with responsibilities for personal dosimetry, e.g. the RPS, in conjunction with the line manager/service manager. If the new staff member was or is to be classified in the current calendar year, special procedures are required and the RPA should be consulted by the RPS.

For staff that have left the employment of the Trust, on request from their new Employer, the Trust will supply historic dose records as requested. This is the responsibility of the previous employing department, which had overall responsibility for the former employee's personal dosimetry.

13.5 Employees with multiple places of employment

Employees must tell their line manager if they are working with ionising radiation (or working in a facility where they may be exposed to ionising radiation) for another Employer. The Trust should also ask their staff periodically (e.g. annually at PDR).

Once aware of an employee with multiple Employers where exposure to ionising radiation may occur, collection of required information will be through the Trust's 'previous or multiple employment—radiation dose recording form' which includes employee specific information, details (including contact details) of other Employers, the type of work carried out and details of personal monitoring. Contact will then be made with the other Employer(s) by an appropriate individual with personal dosimetry responsibilities, e.g. the RPS. An arrangement for sharing dose records between employers should be established, and an appropriate strategy for managing the Employee's radiation dose developed by the RPS.

Appendix A—Matters to be considered in an agreement for cooperation between employers

- The nature of the work involved.
- The Employers involved.

- The employees involved.
- The expected levels of exposure and any risks associated with accidental or unintended exposures.
- The responsibility for dose assessment and compliance with any relevant dose limits.
- The entering of information into passbooks as necessary.
- Identifying any information relating to staff that would need to be shared by the Employers and any data protection issues; taking into account the legal requirement for the cooperation between Employers.
- How compliance with the relevant Regulations will be demonstrated by each Employer.
- Training requirements from each Employer.
- Sharing of risk assessments relevant to the work.
- Requirements for PPE and the responsibility for the provision and testing of such PPE.
- Responsibility for classification of areas and local rules.
- Entry by staff into controlled areas under the control of a different Employer.
- The effect of any work on other employees in the vicinity.
- Handover of any controlled areas or radiation equipment.
- The responsibilities for the return of any radiation equipment into service following maintenance.
- The responsibilities of each organisation to collaborate in investigating incidents or to share in a timely manner any investigation outcomes that impact on each other.
- The route for the exchange of information and contact information.

References

HSE 2018 L121 Work with ionising radiation Ionising Radiations Regulations 2017 *Approved Code of Practice and Guidance* 2nd edn (Norwich: Health and Safety Executive) https://hse.gov.uk/pubns/books/l121.htm (accessed 31 January 2023)

IAEA 2015 Safety Reports Series No. 84 Radiation Protection of Itinerant Workers *Safety Reports Series* (Vienna: International Atomic Energy Agency) https://iaea.org/publications/10788/radiation-protection-of-itinerant-workers (accessed 31 January 2023)

IRR 2017 *The Ionising Radiations Regulations* https://legislation.gov.uk/uksi/2017/1075 (accessed 31 January 2023)

IRR(NI) 2017 *The Ionising Radiations Regulations (Northern Ireland)* http://legislation.gov.uk/nisr/2017/229 (accessed 31 January 2023)

MHSWR 1999 *The Management of Health and Safety at Work Regulations* www.legislation.gov.uk/uksi/1999/3242/regulation/3 (accessed 31 January 2023)

John Saunderson, Mohamed Metwaly, William Mairs, Philip Mayles,
Lisa Rowley and Mark Worrall

Appendix 3

The role of the radiation protection adviser

The Radiation Protection Adviser (RPA) is an individual or corporate body that meets the criteria of competence specified by the HSE in Great Britain (IRR 2017), or either the HSENI or GB HSE in Northern Ireland (IRR(NI) 2017) and, for on-going consultation, is appointed in writing by an Employer. The appointment includes the scope of the advice that is required, such as advice on medical exposures in diagnostic and interventional radiology or on the transport of radioactive material.

The Ionising Radiations Regulations (IRR 2017, IRR(NI) 2017) state that a suitable RPA must be consulted as appropriate on the following matters:

1) The implementation of requirements for controlled and supervised areas (Schedule 4).
2) The prior examination of plans for installations and the acceptance into service of new or modified sources of ionising radiation in relation to any engineering controls, design features, safety features, and warning devices provided to restrict exposure to ionising radiation (i.e. the nature and extent of critical examination and results of that examination) (Schedule 4).
3) The regular calibration of equipment provided for monitoring levels of ionising radiation and the regular checking that such equipment is serviceable and correctly used (Schedule 4).
4) The periodic examination and testing of engineering controls, design features, safety features, and warning devices and regular checking of systems of work, including any written arrangements provided to restrict exposure to ionising radiation (Schedule 4).
5) Advice on the use of dose limits when averaged over periods greater than a single calendar year (Schedule 3).
6) Dosimetry for accidents (Regulation 24).
7) Installer's critical examination (Regulation 32).

doi:10.1088/978-0-7503-2332-1ch23

In addition, the Approved Code of Practice (ACOP) for IRR (HSE 2018) states that the RPA must be consulted on the following matters:

8) The radiation risk assessment (Regulation 8) and local rules (Regulation 18).
9) The designation of controlled and supervised areas (Regulation 17).
10) The conduct of investigations and analysis required by the Regulations e.g. for a suspected overexposure of a member of staff or staff exceeding an investigation level.
11) Contingency plans (Regulation 13).
12) Dose assessment and recording (Regulation 22), including dosimetry for accidents (Regulation 24).

Published HSE guidance to IRR (HSE 2018) advises that the RPA should be consulted on the following matters:

13) Selection of adequate and suitable personal protective equipment (PPE).
14) Optimisation and establishment of appropriate dose constraints.
15) Classification of workers and the suitability of dosimetry services for assessment and recording of doses received by the classified persons.
16) Outside workers.
17) Workplace and individual monitoring programmes and related personal dosimetry.
18) Appropriate radiation monitoring instrumentation.
19) Quality assurance.
20) Arrangements for the prevention of accidents and incidents.
21) Training and re-training programmes for exposed workers (information, instruction, and training).
22) Investigation and analysis of accidents and incidents and appropriate remedial actions.
23) Employment conditions for pregnant and breastfeeding workers.
24) Preparation of written procedures.
25) Suitability of radiation protection supervisor (RPS) appointments.
26) Selecting investigation levels.
27) The working lifetime of sealed sources, as required.

The HSE Criteria of Competence are detailed on their website (HSE no date). An individual awarded a Certificate of Core Competence from an HSE RPA Assessing Body or holding a Radiological Protection Level 4 National or Scottish Vocational Qualification (N/SVQ) is recognised by HSE as an RPA. RPA 2000 is currently the only assessing body that operates a competence certification scheme to assess the competence of individuals wishing to act as RPAs. Re-accreditation takes place on a five-year basis.

RPA 2000 was established by the Association of University Radiation Protection Officers (AURPO), the Institute of Physics and Engineering in Medicine (IPEM), the Society for Radiological Protection (SRP), and the Institute of Radiation Protection (IRP), which has since been incorporated into the Society for Radiological Protection. Details on making an application are available on the RPA 2000 website.

References

HSE 2018 L121 Work with ionising radiation Ionising Radiations Regulations 2017 *Approved Code of Practice and Guidance* 2nd edn (Norwich: Health and Safety Executive) https://hse.gov.uk/pubns/books/l121.htm (accessed 31 January 2023)

HSE no date *Radiation Protection Advisers, HSE Website* (Bootle: Health and Safety Executive) https://hse.gov.uk/radiation/rpnews/rpa.htm (accessed 7 December 2022)

IRR 2017 *The Ionising Radiations Regulations* https://legislation.gov.uk/uksi/2017/1075 (accessed 31 January 2023)

IRR(NI) 2017 *The Ionising Radiations Regulations (Northern Ireland)* http://legislation.gov.uk/nisr/2017/229 (accessed 31 January 2023)

IOP Publishing

Medical and Dental Guidance Notes (Second Edition)

A good practice guide on all aspects of ionising radiation protection in the clinical environment: IPEM Report 113

John Saunderson, Mohamed Metwaly, William Mairs, Philip Mayles, Lisa Rowley and Mark Worrall

Appendix 4

Examples of dose calculations for risk assessments

These examples are intended to illustrate the calculation techniques that may be used for risk assessment dose estimations and should not be used verbatim. They are not exhaustive of all possible scenarios, and the resulting doses are highly dependent on the assumptions made, which are likely to vary considerably between different places of work and equipment. In particular, the worst-case time taken to carry out individual tasks should be confirmed locally, with consideration given to the potential for additional time required in stressful situations. It is recommended that any calculations performed for risk assessments be verified by direct measurements of dose rate and/or personal dosimetry where this is possible and relevant.

The appointed, certified Radiation Protection Adviser (RPA) must be consulted on the use of such calculations in assessing radiation risk.

1. Cleaning minor spill of Tc-99m on the floor or work surface

Assumptions:

- 100 MBq of Tc-99m (approximately 0.05 ml) is spilt onto the floor or work surface from the syringe following injection or drawing up. The spill was immediately noticed and has not spread.
- For operational reasons, the spill needs to be cleaned up and cannot be left to decay.
- The member of staff is wearing appropriate PPE to prevent themselves from becoming contaminated (e.g. disposable gloves, apron/gown, etc).
- A staff member takes 5 min to clean up the spill.
- The hands, body, and eyes are at distances of 1, 30, and 50 cm from the spill, respectively.
- The dose rate from a point source (i.e., the worst case) of Tc-99m is 2.61×10^{-4} mSv h^{-1} MBq^{-1} at 30 cm (Delacroix *et al* 2002).

Dose calculation:

- Finger dose =

$$2.61 \times 10^{-4}\ \mathrm{mSv\ h^{-1}\ MBq^{-1}} \times \left(\frac{30\ \mathrm{cm}}{1\ \mathrm{cm}}\right)^2 \times 100\ \mathrm{MBq} \times \frac{5\ \mathrm{min}}{60\frac{\mathrm{min}}{\mathrm{h}}} = 1.96\ \mathrm{mSv}$$

- Whole-body dose =

$$2.61 \times 10^{-4}\ \mathrm{mSv\ h^{-1}\ MBq^{-1}} \times 100\ \mathrm{MBq} \times \frac{5\ \mathrm{min}}{60\frac{\mathrm{min}}{\mathrm{h}}} = 2.18 \times 10^{-3}\ \mathrm{mSv}$$

- Eye dose =

$$2.61 \times 10^{-4}\ \mathrm{mSv\ h^{-1}\ MBq^{-1}} \times \left(\frac{30\ \mathrm{cm}}{50\ \mathrm{cm}}\right)^2 \times 100\ \mathrm{MBq} \times \frac{5\ \mathrm{min}}{60\frac{\mathrm{min}}{\mathrm{h}}} = 7.83 \times 10^{-4}\ \mathrm{mSv}$$

2. Extremity dose to a member of staff from drawing up and injecting F-18 for PET scanning

2.1 Drawing up injection

Assumptions:

- Assume an average activity of 2 GBq in the stock vial. The vial is contained within a shielded pig in a dispensing jig.
- Fingers are at a distance of approximately 20 cm from the vial when drawing up.
- The maximum instantaneous dose rate to any part of the fingers was measured locally to be approximately 1.12×10^{-3} mSv h^{-1} MBq^{-1}. (Note that you will need to determine your own local factor for your equipment and should not use this one.) This is significantly lower than if the relevant factor from Delacroix *et al* (2002) was used due to the presence of lead shielding around the vial.
- The dose rate at the surface of a syringe in a syringe shield containing 3.8 MBq F-18 measured to be 80 μSv h^{-1}. This equates to a dose rate of 0.021 mSv h^{-1} MBq^{-1}. (Again, note that you will need to determine your own local factor for your equipment and should not use this one.)
- A radiographer takes approximately 20 s to draw up a dose (400 MBq).
- Assume that the mean activity in the syringe over these 20 s is 200 MBq.
- A syringe shield is used while drawing up. The syringe is only removed from the shield to check the activity in the dose calibrator. The distance between the syringe and fingers is maintained at 20 cm by the use of tongs. The unshielded syringe is handled for approximately 5 s.

- The dose rate at 1 m from a 10 ml glass vial of F-18 is 1.58×10^{-4} mSv h^{-1} MBq^{-1} (Delacroix *et al* 2002).
- A radiographer prepares seven injections per day, 250 days per year.

Dose calculation:

- Finger dose from stock vial (drawing up) =

$$1.12x10^{-3}\text{mSv h}^{-1}\text{MBq}^{-1} \times 2000\text{ MBq} \times \frac{20\text{ s}}{60\frac{\text{s}}{\text{min}}60\frac{\text{min}}{\text{h}}} \times 7\text{d}^{-1} \times 250\frac{\text{d}}{\text{y}} = 22\text{ mSv/annum}$$

- Finger dose from shielded syringe (drawing up) =

$$0.021\text{ mSv h}^{-1}\text{MBq}^{-1} \times 200\text{ MBq} \times \frac{20\text{ s}}{60\frac{\text{s}}{\text{min}}60\frac{\text{min}}{\text{h}}} \times 7\text{d}^{-1} \times 250\frac{\text{d}}{\text{y}} = 41\text{ mSv/annum}$$

- Finger dose (dose calibrator check) =

$$1.58 \times 10^{-4}\text{mSv h}^{-1}\text{MBq}^{-1} \times 400\text{ MBq} \times \left(\frac{100\text{ cm}}{20\text{ cm}}\right)^2 \times \frac{5\text{ s}}{60\frac{\text{s}}{\text{min}}60\frac{\text{min}}{\text{h}}} \times 7 \times 250 = 3.8\text{ mSv/annum}$$

The total maximum extremity dose equivalent from this task is 67 mSv/annum.

2.2 Extremity dose during injection

Assumptions:

- Syringe shields are used at all times.
- The dose rate at the surface of a syringe in a syringe shield containing 3.8 MBq F-18 measured to be 80 μSv h^{-1}. This equates to a dose rate of 0.021 mSv h^{-1} MBq^{-1}. (Note that, you will need to determine your own local factor for your equipment and should not use this one.)
- The patient is administered 400 MBq of F-18 in the resting bay.
- Radiographer/technologist administering the F-18 is holding the shielded syringe for 5 s.
- Radiographer/technologist administers seven injections per day, 250 days per year.

Dose calculation:

- Finger dose =

$$0.021\text{ mSv h}^{-1}\text{MBq}^{-1} \times 400\text{ MBq} \times \frac{5\text{ s}}{60\frac{\text{s}}{\text{min}}60\frac{\text{min}}{\text{h}}} \times 7\text{d}^{-1} \times 250\frac{\text{d}}{\text{y}} = 20\text{ mSv/annum}$$

- Total estimated extremity dose for drawing up and injecting = 87 mSv/ annum.

3. Whole-body dose due to proximity to patients post-injection of F-18 for PET scanning

Assumptions:

- The patient is administered 400 MBq of F-18 in the injection room.
- The radiographer/technologist administering the F-18 is within 0.5 m of the patient for 30 s immediately after the administration.
- Radiographer/technologist administers seven injections per day, 250 days per year.
- The dose rate at 1 m from the patient is 0.092 μSv h^{-1} MBq^{-1} immediately after injection (Madsen *et al* 2006).
- The patient rests for 1 h (uptake phase). In most cases, the patient will void prior to imaging, removing approximately 15% of the administered activity (Madsen *et al* 2006) and thereby decreasing the dose rate by a factor of 0.85. The dose rate is further decreased by the physical decay of F-18 which has a half-life of 110 min.
- The radiographer/technologist spends approximately 4 min with each patient after the uptake phase (collecting them from the rest bays, setting them up on the scanner, and getting them off the couch after the scan). Approximately 1 min is spent at 1 m from the patient and 3 min at 2 m.

Dose calculation:

- Whole-body dose immediately after injection =

$$0.092 \times 10^{-3}\,\text{mSv h}^{-1}\text{MBq}^{-1} \times 400\,\text{MBq} \times \left(\frac{1\,\text{m}}{0.5\,\text{m}}\right)^2 \times \frac{30\,\text{s}}{60\frac{\text{s}}{\text{min}}60\frac{\text{min}}{\text{h}}} \times 7\text{d}^{-1} \times 250\frac{\text{d}}{\text{y}} = 2.1\,\text{mSv}$$

- Dose rate at 1 m from patient post uptake phase =

$$0.092 \times 10^{-3}\,\text{mSv h}^{-1}\text{MBq}^{-1} \times 400\,\text{MBq} \times \exp\left(\frac{-0.693 \times 60\,\text{min}}{110\,\text{min}}\right) \times 0.85 = 0.021\,\text{mSv h}^{-1}$$

- Whole-body dose from proximity to patient post uptake phase =

$$0.021\,\text{mSv h}^{-1} \times \left[\left(\frac{1\,\text{min}}{60\frac{\text{min}}{\text{h}}}\right) + \left(\left(\frac{1\,\text{m}}{2\,\text{m}}\right)^2 \times \frac{3\,\text{min}}{60\frac{\text{min}}{\text{h}}}\right)\right] \times 7\text{d}^{-1} \times 250\frac{\text{d}}{\text{y}} = 1.1\,\text{mSv}$$

- Total whole-body dose = 3.2 mSv/annum

4. Whole-body/eye dose to radiographer present in linac bunker when exposure accidentally made

Assumptions:

- The radiographer was located 2 m from the couch for a 10 MV exposure of 100 monitor units (MUs).

- A patient was positioned in the primary beam on the couch with zero gantry rotation.
- A whole-body exposure of 100 MUs for a 10 MV photon beam at a scattering angle of 90° gives a scatter fraction of 3.81×10^{-4} Sv Gy^{-1} at 1 m from a human-size phantom for a target-to-phantom distance of 1 m and field size of 400 cm^2 (IPEM 2017).
- For linacs, leakage is also significant. The maximum permitted leakage at 1 m from the isocenter is specified as 0.1% of the isocenter dose rate (IPEM 2017).
- 100 MU is equivalent to 1 Gy under standard conditions.

Dose calculation:

- The radiographer would approximately receive a whole-body/eye dose of

$$1000\ \text{mGy} \times 3.81 \text{ x } 10^{-4} \times \left(\frac{1\ \text{m}}{2\ \text{m}}\right)^2 + 1000\ \text{mGy} \times 0.1\% \times \left(\frac{1\ \text{m}}{2\ \text{m}}\right)^2 = 0.1 + 0.25 = 0.35\ \text{mSv}$$

5. Manual retrieval of HDR brachytherapy source following failure of afterloader retrieval

5.1 Retrieval by turning handle on the equipment to retract the source

Assumptions:

- 10 Ci = 3.7×10^5 MBq (nominal) Ir^{192} source (activity when new).
- The dose rate at 1 m from the source is 48 mSv h^{-1} (based on the Reference Air Kerma Rate).
- Dose rate at 2 m = 12 mSv h^{-1} (by inverse square law).
- The dose rate at 0.5 m from the source is 192 mSv h^{-1} (by inverse square law).
- A member of staff takes approximately 5 s to move from the shielded position (2 m away from the patient) to the handle location (approximately 1 m from the patient). The mean dose rate during this time is assumed to be 30 mSv h^{-1} (the true average dose rate is likely to be lower, so this is a conservative assumption).
- During retraction, the distance from the source to the hands and body/eyes of staff members was assumed to be 0.5 and 1 m, respectively.
- During staff training, the longest time to retract the source was 5 s. This has been doubled for the purposes of the dose calculation to take into account the potential stress of the situation.

Dose calculation:

- Whole-body/eye dose =

$$\left[30\ \text{mSv h}^{-1} \times \frac{5\ \text{s}}{60\frac{\text{s}}{\text{min}}60\frac{\text{min}}{\text{h}}}\right] + \left[48\ \text{mSv h}^{-1} \times \frac{10\ \text{s}}{60\frac{\text{s}}{\text{min}}60\frac{\text{min}}{\text{h}}}\right] = 0.18\ \text{mSv}$$

- Hand dose =

$$\left[30\text{ mSv h}^{-1} \times \frac{5\text{ s}}{60\frac{\text{s}}{\text{min}}60\frac{\text{min}}{\text{h}}}\right] + \left[192\text{ mSv h}^{-1} \times \frac{10\text{ s}}{60\frac{\text{s}}{\text{min}}60\frac{\text{min}}{\text{h}}}\right] = 0.57\text{ mSv}$$

5.2 Retrieval by removing the applicator containing the source and placing it in the lead pot using forceps

Assumptions:

- 10 Ci = 3.7×10^5 MBq (nominal) Ir^{192} source (activity when new).
- The dose rate at 1 m from the source is 48 mSv h^{-1} (based on the Reference Air Kerma Rate).
- Dose rate at 2 m = 12 mSv h^{-1} (by inverse square law).
- Dose rate at 0.5 m = 192 mSv h^{-1} (by inverse square law).
- Dose rate at 20 cm = 1200 mSv h^{-1} (by inverse square law).
- A member of staff takes approximately 5 s to move from a shielded position 2 m away from the patient to 0.5 m away from the patient, ready to remove the applicator. The mean dose rate during this time is assumed to be 102 mSv h^{-1}.
- The source is located in the applicator. The channel in which the source is sitting will be indicated on the treatment console. The treatment radiographers will be able to tell the location of the source well enough to enable them to keep their hand at a distance of 20 cm from the source when using the forceps.
- The time taken to remove applicator may vary depending on treatment. For cylinder treatment, estimated to be 30 s (1 min worst case). During this time, the body/eyes of the radiographer are 0.5 m from the source.

Dose calculation:

- Whole-body/eye dose =

$$\left[102\text{ mSv h}^{-1} \times \frac{5\text{ s}}{60\frac{\text{s}}{\text{min}}60\frac{\text{min}}{\text{h}}}\right] + \left[192\text{ mSv h}^{-1} \times \frac{1\text{ min}}{60\frac{\text{min}}{\text{h}}}\right] = 3.3\text{ mSv}$$

Hand dose =

$$\left[102\text{ mSv h}^{-1} \times \frac{5\text{ s}}{60\frac{\text{s}}{\text{min}}60\frac{\text{min}}{\text{h}}}\right] + \left[1200\text{ mSv h}^{-1} \times \frac{1\text{ min}}{60\frac{\text{min}}{\text{h}}}\right] = 20.1\text{ mSv}$$

6. The radiologist's fingers stray into primary beam during interventional CT procedure

Assumptions:

- The dose rate in primary beam is approximately 0.1 mGy s^{-1} (based on CTDI in air measurements).
- Fingers in the beam for whole procedure (60 s).
- 1 mGy air kerma $\approx$ 1 mSv dose equivalent for these purposes.

Dose calculation:

- The total maximum dose to 1 cm^2 area of skin on fingers from one procedure is 6 mSv.

7. Staff member not wearing lead apron during fluoroscopy procedure

Assumptions:

- Patient undergoing interventional ERCP procedure.
- A staff member is standing upright at a distance of 0.5 m from the patient during the procedure.
- The x-ray tube is pointing vertically upwards for the duration of the procedure.
- Maximum scatter kerma, S_{max}, at 1 m from the patient is S_{max} = [(0.031 × kV)+2.5] μGy Gy^{-1} cm^{-2} (Sutton *et al* 2012).
- Assume 100 kV accelerating potential.
- The typical Dose Area Product (DAP) for the procedure is 10 Gy cm^2.
- 1 mGy air kerma gives approximately 1 mSv effective dose (Sutton *et al* 2012).

Dose calculation:

$$S_{max} = [(0.031*100) + 2.5] \times 10 \times \left(\frac{1\text{ m}}{0.5\text{ m}}\right)^2 = 224\mu\text{Gy}$$

- Assuming whole-body exposure, the effective dose for a member of staff is 0.22 mSv.

References

Delacroix D, Guerre J P, Leblanc P and Hickman C 2002 Radionuclide and radiation protection data handbook 2nd edition (2002) *Radiat. Prot. Dosim.* **98** 1–168

IPEM 2017 *Report 75: Design and Shielding of Radiotherapy Treatment Facilities* 2nd ed P Horton and D Easton. (Bristol: Institute of Physics)

Madsen M T, Anderson J a, Halama J R, Kleck J, Simpkin D J, Votaw J R, Wendt R E, Williams L E and Yester M V 2006 AAPM Task Group 108: PET and PET/CT shielding requirements *Med. Phys.* **33** 4–15 (accessed 31 January 2023)

Sutton D G, Martin C J, Williams J R and Peet D J 2012 *Radiation Shielding for Diagnostic Radiology* 2nd edn (London: The British Institute of Radiology)

Appendix 5

Radiation risk assessment pro-forma

This example pro-forma should be used in conjunction with the Ionising Radiations Regulations (IRR 2017, IRR(NI) 2017) and Approved Code of Practice (HSE 2018). It is only a guide and must be tailored to the local Employer's needs. The ONR, HSE, and HSENI all agree that a radiation risk assessment is suitable and sufficient if it covers all the important parts of the IRR Approved Code of Practice (ACOP) Paragraphs 70 and 71 (HSE 2018). Structuring a risk assessment under those headings would be an appropriate alternative to that set out below.

Supporting information can be linked to the risk assessment if it is felt that the detail should not sit within the document. Any specific terminology used within this appendix can be understood through a review of the general chapters on IRR (and IRMER) in this publication.

doi:10.1088/978-0-7503-2332-1ch25

Employer

Radiation risk assessment for

Possible headings	Considerations	Elements of ACOP paragraphs 70 and 71 covered
Work practice	Type of procedure (including quality control testing exposures). Location(s). Workload (and audit information unless in the audit section). Date of commencement.	70: (a).
Groups of people exposed	Permanent staff and also others considered atypical, including: • Students; • Ancillary staff; • Visitors; • Visiting staff; • External radiation specialists, e.g. service engineers; • Contractors; • Members of the public. Consider if anyone is an Outside Worker. Carers and Comforters should be considered to the extent that it is clear who is and who is not included in this category. As Carers and Comforters fall within IR(ME)R (IR(ME)R 2017, IR(ME)R(NI) 2018) under the control of medical exposures, they may be excluded from the IRR radiation risk assessment as arrangements and optimisation of their exposure should be addressed in the IR(ME)R Employer's procedures. However,	Feeds into 71: (a), (d), and (i).

Nature of source/identified routine risk	Sealed or unsealed. Energy and dose rate. Primary or secondary radiation. Likelihood of contamination arising and being spread. Estimated levels of airborne and surface contamination likely to be encountered.	70: (a), (b), (c), (h).
Control measures for routine work	Designation of an area. Design advice details, including constraints. Engineering controls, design features, safety, and warning devices. Consider leak tests for sealed sources, which may sit under audit (below). Systems of work, including controlling access where necessary. These should reflect the different groups of individuals exposed to the extent they have control over events (identified above), e.g. permanent staff, outside workers, ancillary staff, visitors, etc. Provision of appropriate/effective PPE and who must wear it. Any advice from manufacturers? This may extend to activities such as planned preventative maintenance and servicing, etc, or these may fall under a specific section (below). Handover procedures.	70: (e), (f), (g), (i), (j).71: (a), (b), (c), (d), (g), (j), (k).
Possible accident situations, the likelihood of occurrence and severity	Identify possible accident situations and estimate their likelihood and severity. Equipment or procedural failure.	70: (k), (l).

(*Continued*)

(Continued)

Possible headings	Considerations	Elements of ACOP paragraphs 70 and 71 covered
	Failure of control measures. Could a source be lost/damaged and how would this be known? If training is not suitable, what could go wrong? Could individuals knowingly cause harm?	
Control measures to prevent accidents and limit the consequences	Limit the likelihood of the accident or the consequences if the risk cannot be eliminated. This can be risk-based so that unlikely accidents with low impact can be covered briefly.	70: (e), (f), (g), (i), (j).71: (a), (b), (c), (d), (j).
Contingency plans *for reasonably foreseeable* accidents	In two stages: make it safe, and then follow it up. What is practical and available? Details on rehearsal—may be picked up in the training section.	70: (m).71: (a), (b), (c), (d), (h), (j).
Estimated or projected dose	Estimates based on current knowledge. Make use of past monitoring evidence where available (area or personal). Include the estimated projected dose from accident situations. Dose estimates to outside workers in particular roles. Are there applicable dose constraints? Are doses as low as reasonably practicable (ALARP)? Are further measures required to restrict doses, e.g. PPE? Does anyone need to be designated as an IRR Classified Person, and why (or why not)? IRR requires a realistic estimate of the dose to a member of the public using the highest dose to a representative person—see HSE IRR guidance Paragraphs 231–236 (HSE 2018).	70: (b), (d), (i).71: (c), (d), (j), (l).

Training requirements	Depending on their level of involvement, staff will require differing levels of instruction and training. Document what this is and how often they should get refresher training. Consider classified and non-classified staff and outside workers. Others may be under constant direct supervision. Training should also cover contingency plans and rehearsals.	71: (a), (i).
Maintenance and examination of control measures	Schedule and actions covering: • Engineering controls, design features, safety and warning devices, likely to include critical examination details, ongoing service and quality assurance programme to the extent that they impact employees and members of the public. • PPE quality assurance programme.	70: (e), (i).71: (a), (b), (g).
Extra precautions/special considerations	For pregnant or breastfeeding staff. Young or inexperienced staff. Temporary staff or those who don't speak English.	70: (g), (i), (j).71: (a), (b), (c), (d), (e).
Audit	What measures will be taken to ensure the content of the risk assessment is being upheld? This might include observation and auditing. This may also be an indicator of training adequacy. Personal and area monitoring programme details lead to conclusions on the effectiveness of the measures in place. Location of dosimeters and wear period. Routine or periodic (frequency).	70: (d), (i), (l).71: (a), (b), (c), (d), (f), (l), (m), (n), (p).

(*Continued*)

(Continued)

Possible headings	Considerations	Elements of ACOP paragraphs 70 and 71 covered
	What action is to be taken at what level of monitoring results? e.g. including formal investigation levels. Review of workloads. May link to the Maintenance of Examination of Control Measures section (above). Leak testing of radioactive sources.	
Summary of outstanding actions to be taken	What has yet to be closed off to ensure risks are managed? Have the right people been consulted and informed? Is there a need to share the risk assessment with anyone in particular, including other Employers to facilitate Cooperation Between Employers (see chapter 1 from paragraphs 1.127)	71: (a), (b).
Responsibilities	Who is responsible for ensuring compliance with these requirements (managers, workers, and outside workers)?	71: (a), (b), (o).
Review arrangements	Who and when?	71: (a), (b), (p).
Signatures and dates	Management, RPS, RPA, H&S advisor, etc, as stipulated in the Employer's management procedures.	

References

HSE 2018 *L121 Work with Ionising Radiation Ionising Radiations Regulations 2017—Approved Code of Practice and guidance* 2nd edn (Norwich: Health and Safety Executive) https://hse.gov.uk/pubns/books/l121.htm (accessed 31 January 2023)

IR(ME)R 2017 *The Ionising Radiation (Medical Exposure) Regulations* www.legislation.gov.uk/uksi/2017/1322 (accessed 31 January 2023)

IR(ME)R(NI) 2018 *The Ionising Radiation (Medical Exposure) Regulations (Northern Ireland)* https://legislation.gov.uk/nisr/2018/17 (accessed 31 January 2023)

IRR 2017 *The Ionising Radiations Regulations* https://legislation.gov.uk/uksi/2017/1075 (accessed 31 January 2023)

IRR(NI) 2017 *The Ionising Radiations Regulations (Northern Ireland)* http://legislation.gov.uk/nisr/2017/229 (accessed 31 January 2023)

**John Saunderson, Mohamed Metwaly, William Mairs, Philip Mayles,
Lisa Rowley and Mark Worrall**

Appendix 6

Designation of controlled and supervised areas

The flow chart below is designed to help in the process of determining whether an area should be designated as a controlled or supervised area. A suitable radiation protection adviser (RPA) must always be consulted in making this decision. Note that changes of practice (e.g. new techniques, new equipment) will generally require a reassessment of the designation of the surrounding area, in consultation with the RPA.

The Ionising Radiations Regulations (IRR 2017, IRR(NI) 2017) require the designation of:

- a 'controlled area' where an assessment has shown that either:
    1. it is necessary for any person (other than a person undergoing medical exposure) to follow special procedures (local rules) to restrict significant exposure to radiation or to prevent or limit the magnitude of radiation accidents; or
    2. if a worker is likely to receive greater than 6 mSv effective dose, greater than 15 mSv eye dose, or greater than 150 mSv skin or extremity dose in a year.
- a 'supervised area', not being a controlled area, where either:
    1. it is necessary to keep conditions under review; or
    2. any person (other than a person undergoing a medical exposure) is likely to receive greater than 1 mSv effective dose, greater than 5 mSv eye dose, or greater than 50 mSv skin or extremity dose in a year.

Paragraph 297 of the HSE Approved Code of Practice to IRR contained in L121 (HSE 2018) gives circumstances in which special procedures should always be necessary to restrict the possibility of significant exposures, and therefore a controlled area is required. Paragraphs 302 to 314 of HSE Guidance to IRR contain additional guidance on when special procedures will or will not be necessary. This guidance should be carefully considered in addition to dose and dose rates, and in particular where open radioactive sources are used.

The final line of Paragraph 297 of the Approved Code of Practice says, '*In addition, an area should be designated as a controlled area if the dose rate (averaged over a minute) exceeds 7.5 μSv per hour and employees untrained in radiation protection are likely to enter that area unless the only work with ionising radiation involves a radioactive substance dispersed in a human body and none of the conditions in (a) to (e) in Paragraph 307 apply*'. Note that Paragraph 307 does not contain any conditions, but Paragraph 297 was numbered 307 in a previous draft of L121, 2nd edition, so it is clear that this is intended to refer to cases (a) to (e) in Paragraph 297.

For some applications in healthcare, there may be accessible areas (e.g., an external wall of some modern radiotherapy linacs) where the dose rate averaged over a minute can exceed 7.5 μSv h^{-1} for very short periods (i.e. minutes, or fractions of a minute), but where the potential annual dose to any person is significantly lower than 0.3 mSv (dose constraint used for optimising exposures to members of the public—see paragraph 1.60 in chapter 1, *General measures for radiation protection*), and there are no special procedures necessary, with the doses already as low as reasonably practicable (ALARP).

In the introduction to L121 (page 2), it is explained that you may use alternative methods to those set out in the Approved Code of Practice in order to comply with the law, providing that you can prove (if necessary in court) that you have complied with the law in some other way.

Publications by the Institute of Physics in Engineering in Medicine (IPEM 2017) and the British Institute of Radiology (Sutton *et al* 2012) contain specific guidance for radiotherapy and diagnostic radiology, respectively, on the shielding and designation of controlled areas. In cases such as those described above (instantaneous dose rates exceed 7.5 μSv h^{-1} but the annual dose to any person is significantly lower than 0.3 mSv), these publications might be used in consultation with the RPA to show that further shielding of the controlled area is unjustified, doses are as low as reasonably practicable, and the law does not require the area in question to be designated. Such a justification should be carefully documented, and it is recommended that records be kept for at least two years beyond the time the area is used for radiation work.

HSE Guidance to IRR (HSE 2018) Paragraph 298 goes on to say '*The designation of a controlled area is also likely to be required when the instantaneous dose rate exceeds 100 μSv per hour even though the dose rate, when averaged over a working day, is less than 7.5 μSv per hour…*'.

This 100 μSv per hour value was consulted on before adoption in the IRR ACOP and Guidance, and Employer's should strive to achieve the provisions of Paragraph 298. In exceptional circumstances, as this paragraph is guidance rather than an Approved Code of Practice, an Employer can justify not controlling an area based on this criterion (indicated as '*' in the flow chart below). This will need to be supported with an appropriate (documented) risk assessment plus appropriate ongoing dose monitoring and review. The overriding requirement is that all exposures be ALARP.

Flow chart to aid in the designation of controlled and supervised areas

The following flow chart should only be used in conjunction with a detailed understanding and use of the Regulations, Approved Code of Practice and guidance, and in consultation with an RPA.

IDR is '*instantaneous dose rate*', defined as the dose rate averaged over one minute.

TADR is a '*time-averaged dose rate*', defined as the dose rate averaged over an eight-hour working day (taking into consideration use/workload and should be based on the worst case scenario. Assumes an occupancy factor of 100%, i.e. a person will receive all the exposure in that 8-hour day.

Note that, as well as in consideration of external dose rate, special procedures to restrict exposure may be necessary based upon:

a) consideration of internal dose from radioactive substances ingested, inhaled, absorbed through skin absorption, accidentally injected, etc. in the area; or
b) consideration of the risk of the spread of contamination to other areas and subsequent exposure of persons.

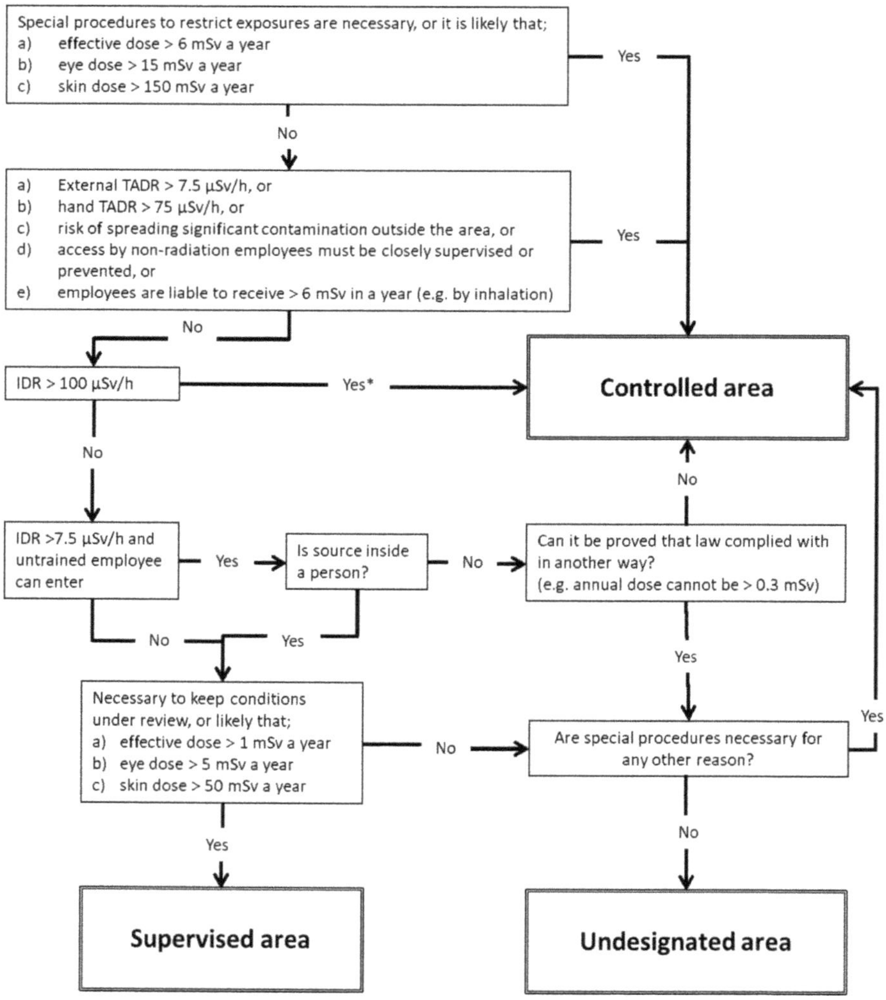

References

HSE 2018 *L121 Work with Ionising Radiation Ionising Radiations Regulations 2017—Approved Code of Practice and Guidance* 2nd edn (Norwich: Health and Safety Executive) https://hse.gov.uk/pubns/books/l121.htm (accessed 31 January 2023)

IPEM 2017 *Report 75: Design and Shielding of Radiotherapy Treatment Facilities* 2nd edn ed P Horton and D Easton (Bristol: Institute of Physics)

IRR 2017 *The Ionising Radiations Regulations* https://legislation.gov.uk/uksi/2017/1075 (accessed 31 January 2023)

IRR(NI) 2017 *The Ionising Radiations Regulations (Northern Ireland)* http://legislation.gov.uk/nisr/2017/229 (accessed 31 January 2023)

Sutton D G, Martin C J, Williams J R and Peet D J 2012 *Radiation Shielding for Diagnostic Radiology* 2nd edn (London: The British Institute of Radiology) (accessed 31 January 2023)

IOP Publishing

Medical and Dental Guidance Notes (Second Edition)
A good practice guide on all aspects of ionising radiation protection in the clinical environment:
IPEM Report 113
**John Saunderson, Mohamed Metwaly, William Mairs, Philip Mayles,
Lisa Rowley and Mark Worrall**

Appendix 7

Warning signs and notices

1. All safety signs giving health or safety information or instruction to persons at work (apart from the exceptions referred to in paragraphs 18 and 19 below) should comply with the Health and Safety (Safety Signs and Signals) Regulations 1996 (HS(SSS)R 1996), or Northern Irish Equivalent (HS(SSS)R(NI) 1996), (the Regulations) and relevant standards (BSI 2011a, 2011b, 2012, 2015a). These apply to all workplaces and activities where people are employed in the United Kingdom. However, they exclude signs used in connection with transport, which are separately covered under other regulations. Further guidance on the application of the Safety Signs and Signals Regulations can be found in HSE Guide L64, '*Safety Signs and Signals. The Health and Safety Regulations* 1996. *Guidance on Regulations*' (HSE 2015).
2. The need for safety signs should be determined as part of the risk assessment required under regulation 8 of the Ionising Radiations Regulations (IRR 2017, IRR(NI) 2017) and should be considered as a method of warning of risks within the area and of identifying the measures required to reduce risks while in the area (e.g. necessary PPE). They should not be considered a substitute for other methods of reducing risks of exposure.
3. Employers should ensure that all employees that may have access to the areas that are marked with safety signs have received suitable and sufficient training to understand the prohibitions, warnings, and mandatory requirements contained on the signs.
4. BS EN ISO 361:2015 *Basic ionizing radiation symbol* (BSI 2015a) specifies the basic symbol (see figure A7.1) to denote the actual or potential presence of ionising radiation and to identify objects, devices, or substances that emit ionising radiation. The standard does not specify any radiation levels at which the symbol is to be used.

doi:10.1088/978-0-7503-2332-1ch27

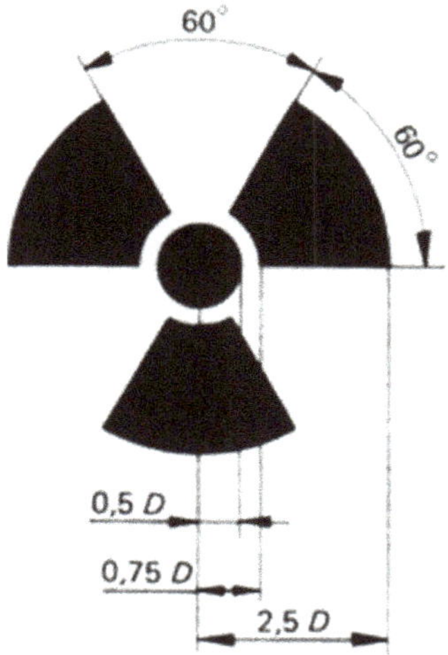

Figure A7.1. Basic ionising radiation warning symbol.

Figure A7.2. Warning sign—ionising radiation.

5. The safety signs and signals regulations (HS(SSS)R 1996, HS(SSS)R(NI) 1996), BS EN ISO 7010 *Graphical symbols: Safety colours and safety signs. Registered safety signs* (BSI 2012) and HSE Guidance (HSE 2015) specify various types of safety signs, including a warning sign indicating 'Warning; Radioactive material or ionising radiation'. This warning sign (see figure A7.2) is triangular in shape with rounded corners, a yellow background (which should take up at least 50 percent of the area of the safety sign), a black border, and the radiation symbol placed centrally on the background. The design and colour of the symbol should meet the requirements specified in the ISO 3864 series *Graphical symbols—Safety colours and safety signs* (BSI 2011a, 2011b, 2015b). Some variation in

detail is acceptable, provided the original meaning is retained and comprehension of the safety sign is maintained.

6. Other signs may be useful to include with the yellow warning sign, such as prohibition signs (a round shape with a black pictogram on white background with red boundary and diagonal line taking up at least 35% of the sign) or mandatory signs (a round shape with a white pictogram on blue with at least 50% of the background being blue). These signs should also follow the requirements of BS EN ISO 7010 (BSI 2012) and the BS ISO 3864 series (BSI 2011a, 2011b, 2015b). Examples are the 'No unauthorised access' and 'Wear eye protection' signs shown in figures A7.3 and A7.4. The previously used prohibition notice with the walking person is intended to convey 'No thoroughfare' or 'No access for pedestrians'.
7. Explanatory wording, through supplementary signs, is often useful to give further clarity to the sign and alert employees to the nature of the area (e.g. Controlled or Supervised), specific risks (e.g. x-rays or Radioactive Substances), prohibitions (e.g. No Unauthorised Access) or draw attention to special precautions (e.g. Eye protection must be worn). Consideration could also be given to adding the name and contact details of the RPS, however, these details must be reviewed and kept up-to-date. Care should be taken when choosing the wording to ensure that the conditions can be met. For example, for illuminated signs in theatres, 'Do Not Enter' is not appropriate, as people may enter under set conditions, and instead 'No Unauthorised Entry' may be more applicable.
8. Supplementary signs must meet the requirements of BS 3864–1 (BSI 2011a). They should be rectangular, with the text on a background that is either white or of the same colour as the safety sign (i.e. yellow in the case of a warning sign) and either below or to the right of the symbol it is

Figure A7.3. Prohibition sign—no unauthorised access.

Figure A7.4. Mandatory sign—Wear eye protection.

Figure A7.5. Example x-ray signboard.

Figure A7.6. Example Radioactive material signboard.

referring to. The number of words should be kept to the minimum necessary.

9. It can be beneficial to have multiple signs on the same signboard to reflect the requirements of the room, and these should meet the requirements of BS ISO 3864–1 (BSI 2011a). These combinations should be mounted on a common white background in order of priority. Examples of multiple signs with supplementary text are given in figures A7.5 – A7.7 and A7.9.
10. If a suitable sign is not available in Schedule 1 of the Regulations (HS (SSS)R 1996, HS(SSS)R(NI) 1996), it is possible for you to design your own, provided it meets the general requirements of the Regulations and

Figure A7.7. Example x-ray room warning light and signboard suitable for temporary controlled area designation (see chapter 1, paragraph 1.107).

Figure A7.8. Example of a mandatory sign designed for lead PPE.

Figure A7.9. Example of an x-ray signboard in a theatre using warning, prohibition, and mandatory signs with supplementary text.

relevant standards (BSI 2011a, 2011b, 2012, 2015b, 2015a). In general, they should be kept as simple as possible. One example of a mandatory sign that could be used for Lead PPE is shown in figure A7.8, and its use in a vertically arranged multiple sign signboard is given in figure A7.9. It should be noted that mandatory signs are actions that must be followed and should only be used if all staff will follow the action when displayed.

Alternatively, the supplementary text may explain when and to whom, the conditions apply.

11. Signboards, and the signs on them should be sufficiently large and clear to be distinctive and should be fixed in a prominent position in a well-lit and easily visible location at the access point to an area or in the immediate vicinity. Care should be taken not to place too many signs in close proximity, potentially causing 'sign blindness' where people become overwhelmed and ignore the signs.
12. A warning sign may have been attached to a piece of equipment by its manufacturer, but it is the Employer's responsibility to ensure compliance with the Regulations (HS(SSS)R 1996, HS(SSS)R(NI) 1996).
13. L121 ACOP (HSE 2018) Paragraph 126 requires suitable warning devices where sources of ionising radiation can cause significant exposure in a very short time and goes on to describe expectations for exposure indications associated with radioactive sources and generators. Illuminated warning signs are essential for radiotherapy rooms (see chapter 7, *Radiotherapy*, paragraphs 7.36 to 7.38). An example is given in figure A7.10. Guidance in L121 Paragraph 127 says that for most x-ray generators, it should be reasonably practicable to have automatic warning devices (indicating the exposure conditions). Specific advice on this can be found in the main body of the publication (chapter 1, *General measures for radiation protection*, paragraphs 1.104 to 1.115 on *Controlled areas and supervised areas* and chapter 3, *Diagnostic radiology (excluding dental) and fluoroscopically guided interventions*, paragraphs

Figure A7.10. Example of radiotherapy illuminated signs.

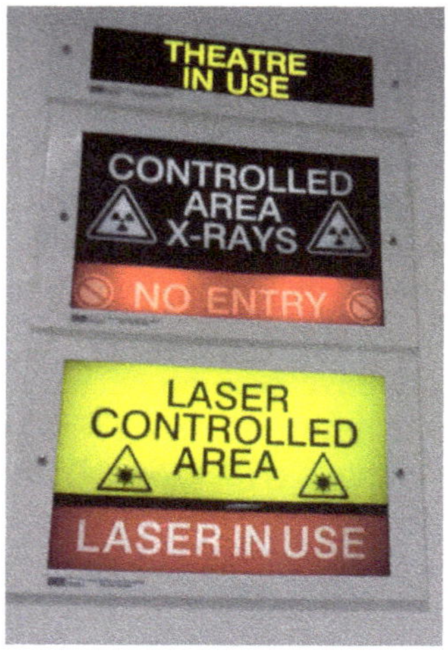

Figure A7.11. Example of contrasting x-ray and laser warning lights.

3.20 to 3.32 on Facility design). It should be noted that two illuminated signs that are likely to be confused are not to be used at the same time. One option to differentiate between similar lights (e.g., x-ray and laser-illuminated signs in theatres) would be to have one with black text on a yellow background and the other with yellow text on a black background, as shown in figure A7.11.

14. Illuminated signs must remain on for as long as the indicated conditions exist and should be bright enough to be seen in the ambient conditions without being glaring. Lights that are triggered by the radiation exposure process should be on for a sufficient duration, and, as these exposures could be short, the light source should be one that has a quick response time (e.g. LEDs but not fluorescent bulbs).
15. Signs should be made of shock and weather-resistant material suitable for the surrounding environment and should be regularly checked and maintained. This includes cleaning signs and ensuring that illuminated signs are functioning correctly.
16. Safety signs and illuminated signs should only be displayed/illuminated when the risk exists and should be removed or switched off when not required. Where safety signs are used for temporary Controlled Areas, this could be achieved by having a reversible sign in a holder or with a sign that can be covered with a slider.
17. Labels, if required, on pipework should be on a clearly visible side and may also contain sufficient supplementary wording to indicate the hazard, e.g., Before dismantling this radioactive waste pipe, contact the RPA'. The labels should be durable and can be in self-adhesive or painted form.

Figure A7.12. Label on a bottle containing radioactive solution.

18. Labels or markings on a package or container to be transported are excluded from the Regulations (HS(SSS)R 1996, HS(SSS)R(NI) 1996). This allows the use of labels and placards when transporting radioactive materials as specified in the Carriage of Dangerous Goods and Use of Transportable Pressure Equipment Regulations (CDG 2009, CDG(NI) 2010)] (and in other documents). Such labels include the radiation symbol on a white or yellow background but not the radiation warning sign.
19. This exclusion is not restricted to transport, and it is acceptable to label a bottle in a laboratory, as in figure A7.12. However, the door of the cupboard in which the bottle is kept should be labelled with the triangular radiation warning sign.

References

BSI 2011a BS ISO *3864–1:2011* Graphical Symbols *Safety Colours and Safety Signs—Design Principles for Safety Signs and Safety Marking* (London: British Standards Institute)

BSI 2011b *BS ISO* 3864–4:2011 *Graphical Symbols*—Graphical Symbols *Safety Colours and Safety Signs—Colorimetric and Photometric Properties of Safety Sign Materials* (London: British Standards Institute)

BSI 2012 *BS EN ISO 7010:2012+A7:2017 Graphical Symbols. Safety Colours and Safety Signs. Registered Safety Signs* (London: British Standards Institute)

BSI 2015a *BS EN ISO 361:2015 Basic Ionizing Radiation Symbol* (London: British Standards Institute)

BSI 2015b *BS ISO 3864-2. Graphical Symbols. Safety Colours and Safety Signs. Part 2. Design Principles for Product Safety Labels* (British Standards Institute)

CDG 2009 *The Carriage of Dangerous Goods and Use of Transportable Pressure Equipment Regulations 2009* https://legislation.gov.uk/uksi/2009/1348 (accessed 31 January 2023)

CDG(NI) 2010 *The Carriage of Dangerous Goods and Use of Transportable Pressure Equipment Regulations (Northern Ireland)* 2010 https://legislation.gov.uk/nisr/2010/160 (accessed 31 January 2023)

HSE 2015 *L64 Safety Signs and Signals. The Health and Safety Regulations 1996. Guidance on Regulations* 3rd edn (Health and Safety Executive) https://hse.gov.uk/pubns/books/l64.htm (accessed 13 February 2023)

HSE 2018 *L121 Work with Ionising Radiation Ionising Radiations Regulations 2017—Approved Code of Practice and Guidance* 2nd edn (Norwich: Health and Safety Executive) https://hse.gov.uk/pubns/books/l121.htm (accessed 31 January 2023)

HS(SSS)R 1996 *The Health and Safety (Safety Signs and Signals) Regulations* https://legislation.gov.uk/uksi/1996/341 (accessed 13 February 2023)

HS(SSS)R(NI) 1996 *Health and Safety (Safety Signs and Signals) Regulations (Northern Ireland)* 1996 https://legislation.gov.uk/nisr/1996/119 (accessed 13 February 2023)

IRR 2017 *The Ionising Radiations Regulations* https://legislation.gov.uk/uksi/2017/1075 (accessed 31 January 2023)

IRR(NI) 2017 *The Ionising Radiations Regulations (Northern Ireland)* http://legislation.gov.uk/nisr/2017/229 (accessed 31 January 2023)

IOP Publishing

Medical and Dental Guidance Notes (Second Edition)
A good practice guide on all aspects of ionising radiation protection in the clinical environment:
IPEM Report 113

**John Saunderson, Mohamed Metwaly, William Mairs, Philip Mayles,
Lisa Rowley and Mark Worrall**

Appendix 8

Identifying persons exposed to ionising radiation

This flowchart is written assuming that the management (primary control) of a facility/site (with or without a radiation source) is asking the questions to their employees.

doi:10.1088/978-0-7503-2332-1ch28

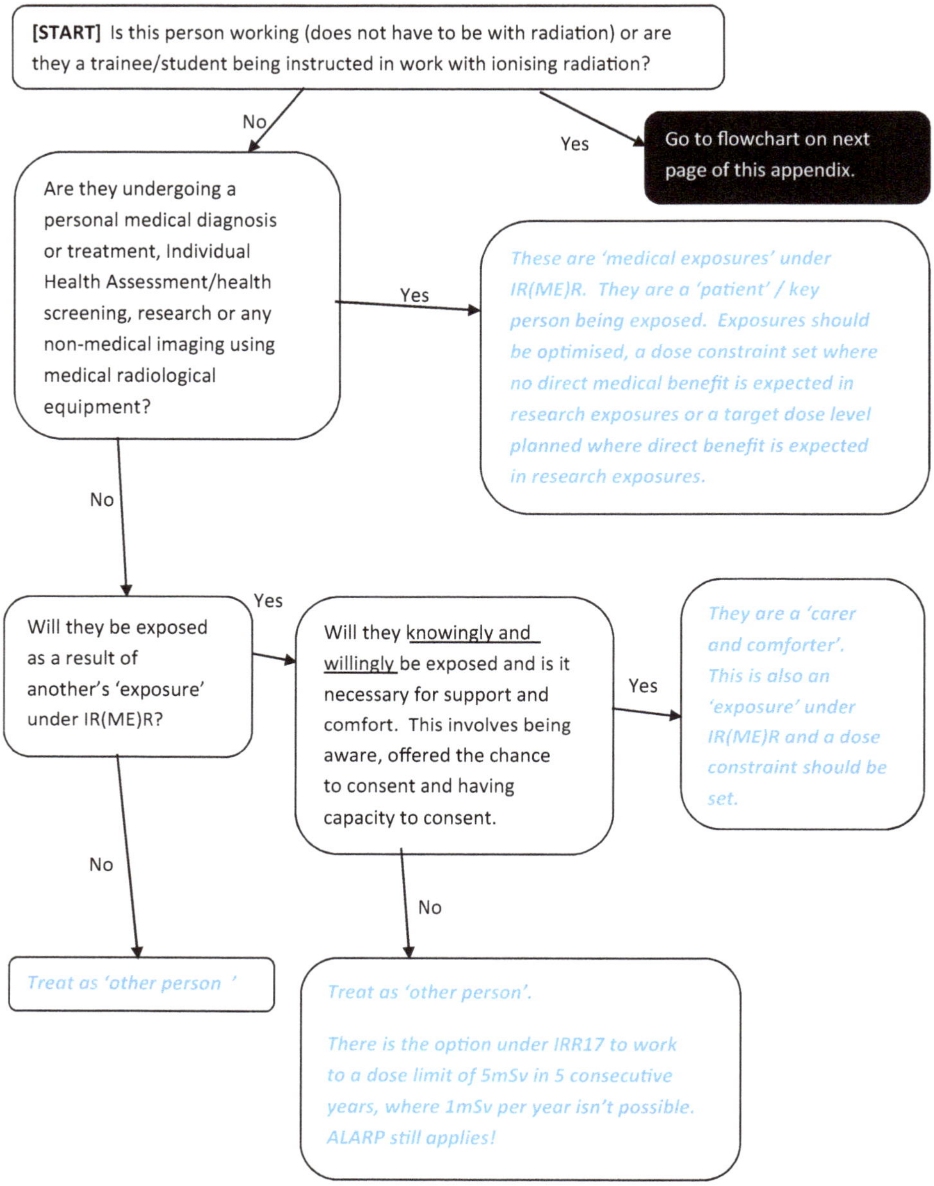
[START] Is this person working (does not have to be with radiation) or are they a trainee/student being instructed in work with ionising radiation?
No
Yes
Go to flowchart on next page of this appendix.
Are they undergoing a personal medical diagnosis or treatment, Individual Health Assessment/health screening, research or any non-medical imaging using medical radiological equipment?
Yes
These are 'medical exposures' under IR(ME)R. They are a 'patient' / key person being exposed. Exposures should be optimised, a dose constraint set where no direct medical benefit is expected in research exposures or a target dose level planned where direct benefit is expected in research exposures.
No
Will they be exposed as a result of another's 'exposure' under IR(ME)R?
Yes
Will they knowingly and willingly be exposed and is it necessary for support and comfort. This involves being aware, offered the chance to consent and having capacity to consent.
Yes
They are a 'carer and comforter'. This is also an 'exposure' under IR(ME)R and a dose constraint should be set.
No
Treat as 'other person '
No
Treat as 'other person'.
There is the option under IRR17 to work to a dose limit of 5mSv in 5 consecutive years, where 1mSv per year isn't possible. ALARP still applies!

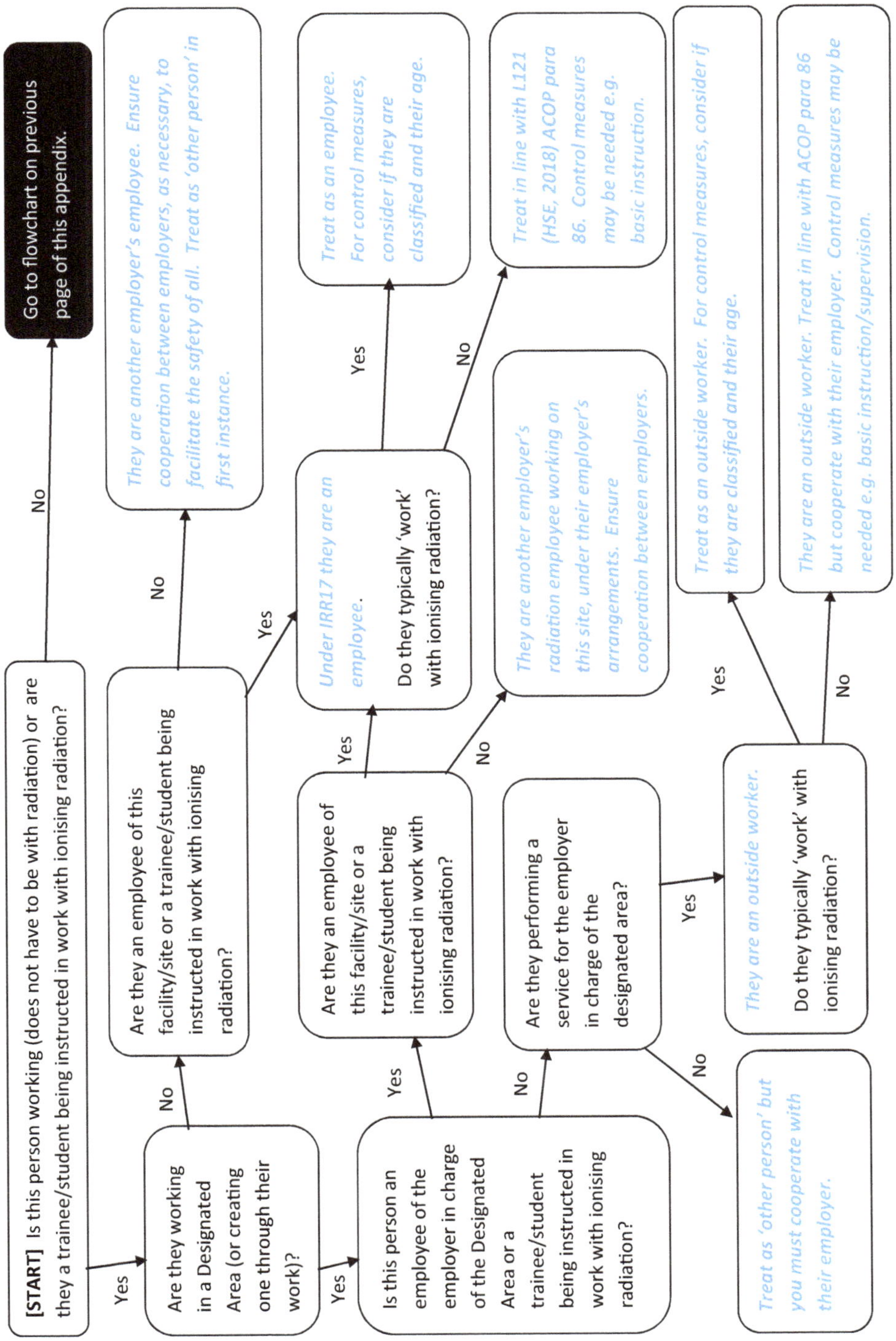
[START] Is this person working (does not have to be with radiation) or are they a trainee/student being instructed in work with ionising radiation?
No
Go to flowchart on previous page of this appendix.
Yes
Are they working in a Designated Area (or creating one through their work)?
No
Are they an employee of this facility/site or a trainee/student being instructed in work with ionising radiation?
No
They are another employer's employee. Ensure cooperation between employers, as necessary, to facilitate the safety of all. Treat as 'other person' in first instance.
Yes
Under IRR17 they are an employee.
Do they typically 'work' with ionising radiation?
Yes
Treat as an employee. For control measures, consider if they are classified and their age.
No
Treat in line with L121 (HSE, 2018) ACOP para 86. Control measures may be needed e.g. basic instruction.
Yes
Is this person an employee of the employer in charge of the Designated Area or a trainee/student being instructed in work with ionising radiation?
Yes
Are they an employee of this facility/site or a trainee/student being instructed in work with ionising radiation?
Yes
No
They are another employer's radiation employee working on this site, under their employer's arrangements. Ensure cooperation between employers.
No
Are they performing a service for the employer in charge of the designated area?
Yes
They are an outside worker.
Do they typically 'work' with ionising radiation?
Yes
Treat as an outside worker. For control measures, consider if they are classified and their age.
No
They are an outside worker. Treat in line with ACOP para 86 but cooperate with their employer. Control measures may be needed e.g. basic instruction/supervision.
No
Treat as 'other person' but you must cooperate with their employer.

John Saunderson, Mohamed Metwaly, William Mairs, Philip Mayles, Lisa Rowley and Mark Worrall

Appendix 9

Aide memoire to developing a policy on cooperation between employers (CBE)

This appendix should be helpful when drafting a cooperation between Employers policy and when managing visitors to a radiation facility (or where radon in the workplace is a concern). It does not address sending radioactive individuals to nursing homes, etc, nor the management of 'carers and comforters'. A 'designated area' may be a Controlled Area or a Supervised Area under the IRR (IRR 2017, IRR(NI) 2017).

Note 1—dosimetry specific communication requirements for non-classified outside workers

For non-classified outside workers, there should be prior communication between Employers as necessary to establish dosimetry arrangements. It will be appropriate in this situation for the Employer at the facility where the outside worker is visiting to only communicate with the Employer with whom there is a direct contract to engage the employee (rather than all Employers of that individual, as they would in line with the management of dosimetry across multiple Employers). Share:

- policies on cooperation between Employers (or at least the key points);
- details of who will be the ongoing point of contact for communication;
- employee details as appropriate;
- appropriate risk assessments to allow the main Employer to establish risks and likely exposures when their employee is working in the controlled area of the other Employer;
- the main Employer should share an estimate of the total projected annual exposure for the outside worker so the Employer of the facility that is being visited knows how to manage the dosimetry;
- dose history (if available);
- details of personal dosemeters that are currently provided.

doi:10.1088/978-0-7503-2332-1ch29

Types of personnel	Information gathering	Radiation protection management	Information sharing	Include in the contract details
Visiting employees who are not working with ionising radiation				
Visiting employees who do not enter designated areas or work with radioactive individuals and who are not working in an atmosphere of radon —no requirement for CBE under IRR17.	Not applicable (N/A).	Standard restricted access signs as for members of the public.	N/A	N/A
Visiting employees who do not enter designated areas or work with radioactive individuals but work in an atmosphere of radon >300 Bq m^{-3}. **(HSE notification required.)**	N/A	Restrict access or follow standard systems of work.	Notify other Employers of the risks, the measures taken to control them, and any relevant actions for their employees.	N/A
Visiting employees, who do not work with radioactive individuals but do enter designated areas, who are not	N/A	Control access to times when it is safe and provide basic radiation protection training to identify	If they are performing a service for the facility, they are outside workers, and a copy of the risk assessment (with details of their	N/A

working with ionising radiation and should not be exposed. Includes when areas are declassified to allow entry (e.g. cleaners, building maintenance staff, inspectors, etc)		signs, etc or ensure under supervision	role included) is to be shared with their Employer. Otherwise, N/A	
Visiting employees who may be exposed to radioactive individuals, or who enter designated areas while not working with ionising radiation, *who may be exposed* (e.g. support workers, prison guards, interpreters).	Discussion before undertaking a role or before entry is granted, as detailed in the local rules for visitors e.g. in x-ray, it is common to ask if they are pregnant or if they have been exposed in a similar role before.	Control access and/or limit exposure. Consider avoiding exposure if pregnant or if they have undertaken this role before. Explain risk, give instructions, and ensure supervision. Provide PPE/ shielding. Record details of exposure, e.g. on RIS. May consider trial monitoring periodically, depending on risk.	Provide a copy of the risk assessment and possibly a generic letter for their Employer explaining they may have had an exposure through their employment.	N/A

(*Continued*)

(Continued)

Types of personnel	Information gathering	Radiation protection management	Information sharing	Include in the contract details
Visiting employees who are working with ionising radiation				
Visiting employees who enter designated areas to work with ionising radiation and equipment/controlled area *handover forms are used.* (e.g. medical physics, equipment manufacturer, engineers).	Prior communication between Employers (contractual arrangements, communications when agreeing a time to enter a controlled area to test equipment, etc), discussion before entry to ensure satisfactory risk assessments and local rules are in place, and there are no conflicting arrangements.	Control access, i.e., meet authorised local staff to gain entry and carry out the handover procedure. Ensure any local-specific arrangements are brought to the visitor's attention.	Ensure appropriate risk assessments and local rules are in place and there are no conflicting arrangements (at least once).	Training, dosimetry, risk assessment, and local rules responsibilities. Arrangements for handover and handback.
Visiting employees who enter designated areas to work with ionising radiation (non-classified outside workers e.g. RPA/ MPE in an advisory	Prior communication between Employers, pre-work meeting checklist (first visit and suitable recurring frequency, e.g. 3, 6, or 12 monthly).	Control access, and ensure both general and specific training are in place (read local rules) or under supervision. Ensure PPE and appropriate	Share a copy of the risk assessment and local rules with their Employer, and share dosimetry arrangements and results as necessary.	Training and dosimetry responsibilities (including action on dosimetry results).

capacity, applications specialists, manufacturer reps). *See note* 1 starting above		personal dosimetry arrangements.		
Visiting employees who are classified who enter *controlled* areas to work with ionising radiation (classified outside workers). *See note* 2 *below*	Prior communication with the Employer, pre-work meeting checklist (first visit and suitable recurring frequency, e.g., each visit if infrequent or each time new dosimetry results are available). Radiation passbook.	Control access, and ensure both general and specific training is in place (read local rules) or under supervision. Ensure PPE and appropriate personal dosimetry arrangements (assessment with an Approved Dosimetry Service provided by the main Employer and estimates with dosemeters for all significant components of dose in the local controlled area). Record details in the passbook (appropriate and trained contacts).	Share a copy of the risk assessment and local rules with their Employer. Share dosimetry arrangements and results regularly. A passbook is also used to share and track details.	Training and dosimetry responsibilities (including action on dosimetry results).

(*Continued*)

(Continued)

Types of personnel	Information gathering	Radiation protection management	Information sharing	Include in the contract details
Visiting trainees				
Visiting Trainees (including Students). *See note* 3 *below* NB: Once competent, rotating trainees performing work/ 'services' may be considered outside workers.	Prior communication with the education provider. Pre-work meeting (start of rotation and any return visit or each time new dosimetry results are available).	Treat them as employees. Control access, and ensure both general and specific training (read local rules) is in place or under supervision. Ensure appropriate PPE and personal dosimetry arrangements.	Share a copy of the risk assessment and local rules with their education provider. Share dosimetry arrangements and results as necessary.	Training and dosimetry responsibilities (including action on dosimetry results).
Local employees with 'multiple Employers'				
Local employees with multiple Employers (including self-employed) where they may be exposed to radiation (particularly important where staff may be classified).	Use a cooperation between Employers form to gather information on their other employment, exposure, and monitoring. It is very important that staff inform Employers if they are classified. Contact other Employers.	The main additional concern is to ensure doses are ALARP across employment and ultimately remains below limits. If staff are classified, individual monitoring by each Employer through ADS. Decide who will report to CIDI.	Share cooperation between Employers policy, risk assessment, and monitoring arrangements. If staff are classified, share dose records for each wear period.	N/A

Decisions can then be made and agreed upon, including:

- what dosemeters will be provided at the facility being visited;
- how and when dose records will be shared;
- if and when classification of individuals will occur (the main Employer must classify employees);
- action to be taken when doses are higher than expected and when investigation levels are exceeded;
- who will report overexposures.

In situations where an outside worker arrives on site and prior communication between Employers has not been carried out, a pre-work checklist and induction process should extract sufficient information to allow decisions to be made and work to proceed if appropriate.

Note 2—dosimetry specific communication requirements for classified outside workers

For classified outside workers, there should be prior communication between Employers as necessary to establish dosimetry arrangements. It will be appropriate in this situation for the Employer at the facility where the outside worker is visiting to only communicate with the Employer with whom there is a direct contract to engage the employee (rather than all Employers of that individual, as they would in line with the management of dosimetry across multiple Employers). Share:

- policies on cooperation between Employers (or at least the key points);
- details of who will be the ongoing point of contact for communication;
- employee details as appropriate;
- appropriate risk assessments to allow the main Employer to establish risks and likely exposures when their employee is working in the controlled area of the other Employer;
- the main Employer should share an estimate of the total projected annual exposure for the outside worker so the Employer of the facility that is being visited knows how to manage the dosimetry;
- dose history; see also the radiation passbook;
- details of personal dosemeters that are currently provided.

Decisions can then be made and agreed upon, including:

- what method of dose estimation will be used in the controlled area to allow entry in the passbook? This is in addition to the main Employer's provided dosemeters. A method that gives a quick result is best, e.g. electronic personal dosemeter;
- how and when dose records will be shared; make use of the radiation passbook;
- action to be taken when doses are higher than expected and when investigation levels are exceeded;
- who will report overexposures?

In situations where a classified outside worker arrives on site and prior communication between Employers has not been carried out, a pre-work checklist and radiation passbook should provide sufficient information to allow decisions to be made and work to proceed if appropriate. Do not proceed unless there is sufficient information.

Note 3—dosimetry specific communication requirements for trainees

Although trainees are considered employees under IRR17 (IRR 2017, IRR(NI) 2017), they should be given a pre-work meeting to gather information in a similar way to outside workers, and there is still a requirement for managing dosimetry across multiple Employers and cooperation between Employers.

References

IRR 2017 *The Ionising Radiations Regulations* (https://legislation.gov.uk/uksi/2017/1075) (accessed 31 January 2023)

IRR(NI) 2017 *The Ionising Radiations Regulations (Northern Ireland)* (http://legislation.gov.uk/nisr/2017/229) (accessed 31 January 2023)

**John Saunderson, Mohamed Metwaly, William Mairs, Philip Mayles,
Lisa Rowley and Mark Worrall**

Appendix 10

Record keeping

UK regulatory requirements

The following records in **bold type** in the table below must be kept for the time stated, as required by one of the following UK regulations:

1. IRR—Current Ionising Radiations Regulations (IRR 2017, IRR(NI) 2017).
2. Previous ionising radiations regulations (IRR(SS)(NI) 1969, IRR(SS)R 1969, IRR 1999, IRR(NI) 2017) *(with regard to classified worker dose records prior to* 2018).
3. L121 Working with Radiation 2nd edition—IRR Approved Code of Practice (HSE 2018).
4. REPPIR—Current Radiation (Emergency Preparedness and Public Information) Regulations (REPPIR 2019, REPPIR(NI) 2019).
5. Previous radiation emergency regulations (REPPIR 2001, REPPIR(NI) 2001) *(with regard to classified worker dose records prior to* 2019).
6. L126 REPPIR—Approved Code of Practice. 2nd edition (HSE and ONR 2019).
7. CDG—Carriage of Dangerous Goods and Use of Transportable Pressure Equipment Regulations (CDG 2009, CDG(NI) 2010) as amended (CDG(A) 2011, CDG(A)(NI) 2011) and (CDG(A) 2019, CDG(A)(NI) 2019).
8. ONR 2021—Guidance Document: Transporting Radioactive Material —Security Guidance on the Carriage of Class 7 Radioactive Material (ONR 2021).
9. IR(ME)R—Ionising Radiation (Medical Exposure) Regulations (IR (ME)R 2017, IR(ME)R(NI) 2018).
10. MHSWR—Management of Health and Safety at Work Regulations (MHSWR 1999, MHSWR(NI) 2000).

doi:10.1088/978-0-7503-2332-1ch30

11. RSR—Radioactive Substances Regulations (RSA(A)R(NI) 2011, EPR 2016, EA(S)R 2018).
12. BEIS 2018—Scope of and Exemptions from the Radioactive Substances Legislation in England, Wales, and Northern Ireland Guidance Document (BEIS *et al* 2018).

In addition, the regulations and ACOPs require some records to be kept but do not state a retention period.

Regulators' and professional bodies' guidance

Additional guidance has been drawn from:

- ARSAC—Notes for Guidance on the Clinical Administration of Radiopharmaceuticals and Use of Sealed Radioactive Sources (ARSAC 2023).
- CQC 2020—Significant accidental and unintended exposures under IR (ME)R Guidance for employers and duty-holders. Care Quality Commission, Healthcare Inspectorate Wales, the Regulation and Quality Improvement Authority, and Healthcare Improvement Scotland. (CQC *et al* 2020).
- L121 Working with Radiation—HSE Guidance (HSE 2018).
- L126 A Guide To The Radiation (Emergency Preparedness And Public Information) Regulations (HSE and ONR 2019).
- NHSX 2021—Records Management Code of Practice 2021 (NHSX 2021) (See note below on *Radioactive records*).
- PHE 2018—Guidance: Retention, storage and disposal of mammograms and screening records (PHE 2018).
- RCR oncology records retention guidance (RCR 2021).
- SEPA 2021—Environmental Authorisations (Scotland) Regulations 2018 Standard conditions for radioactive substances activities (SEPA 2021).

Other records or documents recommended to be kept to demonstrate regulatory compliance (in plain text) should be kept up to date. The list is not necessarily exhaustive.

Where no regulatory requirements or published recommendations for a retention period have been found, we have indicated '*No defined period*'. If these records are not to be kept indefinitely then a local policy should be developed for such records, and regulators consulted if appropriate.

Note

NHSX 2021 (NHSX 2021) *appendix II: Retention schedule* includes in its table a row on '*Radioactive records*' with a 30 year minimum retention period for certain information. Note that records required by radioactive substances permits, authorisations, or certificates of registration can only be disposed of with the permission of the environment agency issuing the permit. Without such permission, these records must be retained, even if 30 years have elapsed. If records are over 30 years old, then it may be appropriate to review whether permission should be sought from the relevant regulator to dispose of the records.

Record or documentation	Minimum retention period	Regulation or reference
ALL WORK WITH IONISING RADIATION		
IRR notification, registration, and consent certificates	*No defined period*	IRR rr. 5, 6, 7
Risk assessments and their reviews	*No defined period*	IRR r. 8; MHSWR r.3(6); L121 (guidance) paras 63, 77
Systems of work for any work with ionising radiation, including local rules and contingency plans	*No defined period*	IRR r.9(2), IRR r.13, IRR r.18(1), and IRR r.19(3)
Investigation report for employee doses over the investigation level	2 years	L121 (guidance) para 188
Record of examination of engineering control, etc to restrict exposure	2 years after the subsequent test	L121 (guidance) para 203
Record of examination of respiratory protective equipment	**2 years**	**IRR r.11(2)**
Record of analysis of circumstances which triggered contingency plans	**2 years**	**IRR r.13(2)**
Noted on any relevant dose record of exposure arising from circumstances which triggered contingency plans	*No defined period*	**IRR r.13(2)**

(*Continued*)

(Continued)

Record or documentation	Minimum retention period	Regulation or reference
Equipment monitoring—general testing and maintenance work	Lifetime of installation	NHSX 2021 (Estates)
Appointment of RPA and RPS	*No defined period*	IRR rr.14(2), 18(5)
Staff training records—Records of significant training (e.g. clinical training)	75th birthday or 6 years after the staff member leaves	NHSX 2021 (Staff Records and Occupational Health)
Staff training records—Statutory and mandatory training records	10 years	NHSX 2021 (Staff Records and Occupational Health)
Staff training records—Other training records	6 years	NHSX 2021 (Staff Records and Occupational Health)
Demonstration by personal dose monitoring or other suitable measurements that doses in controlled areas are restricted	**2 years**	**IRR r.19(8)**
Area monitoring of controlled or supervised areas.	**2 years**	**IRR r.20(5)**
Testing of monitoring equipment used before use and at appropriate intervals	**2 years**	**IRR r.20(5)**
For classified person, personal dose records	**To age 75 years but at least 30 years if made after 2017, 50 years if made between 1985 and 2018, and 30 years if made between 1969 and 1985**	**IRR r.22(3); IRR 1999, IRR(NI) 2000 r.21(3), IRR 1985, IRR(NI) 1985 r.13 (3); IR(SS)R1969, IR(SS)R(NI) 1969 r.10**

For classified persons, personal dose record summary	**At least 2 years from the end of the calendar year to which the summary relates.**	**IRR rr.22(6), 22(7)**
Staff exposure monitoring information	Where the record is for identifiable employees, for at least 40 years, or in any other case, for at least 5 years.	NHSX 2021 (Staff Records and Occupational Health)
For classified persons—investigation and correction of incorrect dose record	**2 years**	**IRR r.23(4)**
Dose assessment following accident leading to more than 6 mSv effective dose, 15 mSv eye dose or 150 mSv skin or extremity dose to a person	**To age 75 years but at least 30 years if made after 2017, 50 years if made between 1985 and 2018, and 30 years if made between 1969 and 1985**	**IRR r.24(2); IRR 1999, IRR(NI) 2000 r.23(2); IRR 1985, IRR(NI) 1985 r.14 (2); IR(SS)R1969, IR(SS)R(NI) 1969 r.10**
Health records for classified employees, for overexposed employees (i.e. overdose limit), and for employees certified unfit for radiation work	**To age 75 years but at least 30 years if made after 2017, 50 years if made between 1985 and 2018, and 30 years if made between 1969 and 1985**	**IRR r.25(2); IRR 1999, IRR(NI) 2000 r.24(3); IRR 1985, IRR(NI) 1985 r.16 (2); IR(SS)R1969, IR(SS)R(NI) 1969 r.10**
Immediate investigation report of suspected overexposures (i.e. overdose limit) showing none occurred	**2 years**	**IRR r.26(2)**
Full investigation report of overexposures (i.e. overdose limit)	**To age 75 years but at least 30 years if made after 2017, and 50 years if made between 1985 and 2018**	**IRR r.26(2); IRR 1999, IRR(NI) 2000 r.25(2); IRR 1985, IRR(NI) 1985 r.29 (2)**

(Continued)

(Continued)

Record or documentation	Minimum retention period	Regulation or reference
Reasons for using Schedule 3 Part 2 dose limits (i.e. effective or eye dose < 50 mSv y^{-1} & < 100mSv/5y)	**To age 75 years but at least 30 years if made after 2017, and 50 years if made between 2000 and 2018**	**IRR r.12(2), schedule 3(15); IRR 1999, IRR(NI) 2000 r.11(2), schedule 4(15)**
Critical examination	*No defined period*	IRR r.32(2)
Report to the Justification Authority of reviewing non-medical imaging exposures using non-medical radiological equipment.	*No defined period*	JoPIIR r.21D(1)
Committees (major)—listed in Scheme of delegation or report directly to the board (including major projects)	Up to 20 years	NHSX 2021 (Corporate Governance)
Committees (minor)—not listed in the scheme of delegation	6 years	NHSX 2021 (Corporate Governance)
MEDICAL EXPOSURES		
Licence to administer radioactive substances	Current	IR(ME)R r.5(1)
Written procedures, protocols, referral guidelines, QA programme for procedures and protocols, established diagnostic reference levels, dose constraints for biomedical and medical research programmes, and for carers and comforters	*No defined period*	IR(ME)R rr.6(1), 6(4), 6(5), schedule 2; ARSAC 2022 para 5.3

Clinical protocols	20 years	NHSX 2021 (Event and Transaction Records)
Referrals not accepted	2 years as an ephemeral record	NHSX 2021 (Event and Transaction Records)
Review where DRLs consistently exceeded	*No defined period*	IR(ME)R r.6(7)
Clinical audit	5 years	IR(ME)R r.7; NHSX 2021 (Event and Transaction Records)
Study of risk of accidental or unintended medical exposures	*No defined period*	IR(ME)R r.8(2)
Incidents (serious)	20 years	NHSX 2021 (Corporate Governance)
Incidents (not serious)	10 years	NHSX 2021 (Corporate Governance)
Record analyses of events involving or potentially involving accidental or unintended medical exposures	In accordance with local procedures	IR(ME)R r.8(3)CQC 2020
Reports to HSE on patient exposures much greater than intended between 1986 and 6th February 2018	**50 years**	**IRR 1985 r.33IRR 1999 r.32**
Ethics committee approval for research exposure	*No defined period*	IR(ME)R r.11(1)

(*Continued*)

(Continued)

Record or documentation	Minimum retention period	Regulation or reference
Oncology diagnosis and treatment records	30 years, or 8 years after death	IR(ME)R r.12(2); NHSX 2021 (Care Records); RCR 2021
In treatments with radioactive substances, target dose and non-target doses.	*No defined period*	ARSAC 2022 para 5.15
Record of clinical evaluation of each medical exposure	*No defined period*	IR(ME)R r.12(9)
Collection of dose estimates for PHE	*No defined period*	IR(ME)R r.13
Appointment of MPE	*No defined period*	IR(ME)R r.14(1)
QA programme for equipment	*No defined period*	IR(ME)R r.15(1)
Non-Clinical Quality Assurance Records	12 years	NHSX 2021 (Corporate Governance)
Radiation equipment inventory	**'keep up-to-date and preserve'**	**IR(ME)R r.15(1)**
Radiation equipment performance criteria and corrective action	*No defined period*	IR(ME)R r.15(6)
Equipment maintenance logs	11 years	NHSX 2021 (Event and Transaction Records)
Inspection of equipment records	11 years	NHSX 2021 (Event and Transaction Records)

Relevant training of practitioners and operators	**'up-to-date'**	**IR(ME)R 17(4)**
Record of increased administered (radio)activity for individual or group of patient	*No defined period*	ARSAC 2022 paras 5.8, 5.9, 5.10
Instructions to breastfeeding patients administered with radioactive substances	*No defined period*	ARSAC 2022 para 7.18
Patient X-rays and scans	8 years	NHSX 2021 (Care Records)
Destruction Certificates or Electronic Metadata destruction stub or record of clinical information held on destroyed physical media	20 years	NHSX 2021 (Care Records)
Screening where no cancer/illness detected	10 years	NHSX 2021 (Event and Transaction Records)
Screening—children	10 years or 25th birthday, whichever longer	NHSX 2021 (Event and Transaction Records)
Advanced medical therapy research—master file	20 years	NHSX 2021 (Clinical Trials and Research)
Clinical trials—applications for ethical approval	5 years	NHSX 2021 (Clinical Trials and Research)
Mammography screening	(i) Normal result—9 years (ii) Cancer detected, interval cancers, interesting cases—indefinitely (iii) Clinical trial—15 years (iv) Died—9 years	PHE 2018

(Continued)

Record or documentation	Minimum retention period	Regulation or reference
RADIOACTIVE SUBSTANCES		
Records of holding, transferring and disposing of radioactive substances under an authorisation, registration, or permit	In accordance with conditions of the permit(s)/authorisations held.	RSR
Records of authorised sealed and unsealed sources, including location, transfer and disposal or discharge (Scotland)	In accordance with conditions of the permit(s) held.	EA(S)R 2018 schedule 8 para 20; SEPA 2021 schedule 1
Leak tests of sealed radioactive sources	**2 years after disposal, or until next test recorded**	**IRR r.28(3)(b); L121 para 577 (guidance)**
Quantity and location of radioactive substances	**At least 2 years from audit and at least 2 years after disposal**	**IRR r.29(1); L121 para 581 (guidance)**
Records of exempt radioactive sources held or waste accumulated (England, Northern Ireland and Wales)	At least 1 year after transfer or disposal	BEIS 2018 paras 3.11, 3.22, 3.34, 3.50, 3.63
Records of radioactive sources held under General Binding Rules or waste accumulated (Scotland)	**At least 2 years after transfer or disposal**	**EA(S)R 2018 schedule 9 para 3**
Immediate investigation report following suspicion of release, loss or theft of radioactive substance, where none occurred.	**2 years**	**IRR r.31(5)**

Full investigation report of notifiable release, loss of theft of radioactive substance	**30 years if made after 2017, or 50 years if made between 1985 and 2018**	**IRR r.31(5)IRR 1999, IRR(NI) 2000 r.30(5); IRR 1985, IRR(NI) 1985 r.31 (4)**
Record of significant events involving or potentially involving accidental or unintended public exposure to radioactivity (Scotland)	*No defined period*	EA(S)R 2018 schedule 8 para 20
HASS record keeping	*No defined period*	EPR schedule 23 para 6; HASS 2005 r.7; EU 2003 article 6; EA(S)R 2018 schedule 8 para 25
Transport documents for radioactive substances	**3 months after delivery**	**CDG 2009 r.31, CDG(NI) 2010 r.27**
For transport of radioactive substance, emergency plan (consignor and carrier)	*No defined period*	CDG schedule 2 para 4(1)CDG 2011 amendments, GB r15, NI r.10
Assessment of transport radiation emergency that occurs	**50 years**	**CDG 2019 amendment r.6**
Emergency exposure of employee from transport radiation emergency	**To age 75 years but at least 30 years**	**CDG schedule 2 para 7**
Instructions in writing to driver	*No defined period*	ADR 2021 para 1.4.2.2
Security training records for carriage of radioactive material	4 years	ONR 2021
Dangerous Goods Safety Adviser (DGSA) reports	**5 years**	**ADR 2017 para 1.8.3.3**

(*Continued*)

(Continued)

Record or documentation	Minimum retention period	Regulation or reference
REPPIR—Assessment showing source is non-dispersible	While source on-premises	L126 REPPIR guidance 2019 para 117
REPPIR—written evaluation, assessment of hazards and risks, review of assessments, emergency plans, etc	For at least the period of their validity	L126 REPPIR guidance 2019, Paragraph 648REPPIR r.23; REPPIR(NI) r.22
REPPIR—written evaluation providing the details of the hazard evaluation	While applicable and at least 3 years from writing	L126 REPPIR guidance 2019, Paragraph 174
REPPIR—written records, updates or statements of a 'material change.' (Updates or statements pertaining to the previous records)	While applicable and at least 3 years from writing(3 years)	L126 REPPIR guidance 2019, Paragraph 223
Full assessment of radiation emergency that occurs where REPPIR applies	**Emergencies after 2019 regulations are in force, to age 75 years for persons receiving an emergency exposure and at least 30 years from termination of work that gave rise to the exposure. But whether or not anyone was exposed, at least 50 years. Emergencies when 2001 regulations are in force, to an age of 75 years for persons receiving an emergency exposure and at least 50 years from termination of work that gave rise to the exposure. But whether or not anyone exposed, at least 50 years**	**REPPIR 2019 rr.17(8), 18(11); REPPIR(NI) 2019 rr.16(8), 17(11); REPPIR/REPPIR(NI) 2001 rr.13(3), 14(10);**

References

ARSAC 2023 *Notes for Guidance on the Clinical Administration of Radiopharmaceuticals and Use of Sealed Radioactive Sources* (Administration of Radioactive Substances Advisory Committee) https://gov.uk/government/publications/arsac-notes-for-guidance (accessed 25 February 2023)

BEIS, DEFRA, Welsh Government and DAERA 2018 Scope of and Exemptions from the Radioactive Substances Legislation in England *Wales and Northern Ireland Guidance Document* https://gov.uk/government/publications/guidance-on-the-scope-of-and-exemptions-from-the-radioactive-substances-legislation-in-the-uk (accessed 11 February 2023)

CDG 2009 *The Carriage of Dangerous Goods and Use of Transportable Pressure Equipment Regulations 2009* https://legislation.gov.uk/uksi/2009/1348 (accessed 31 January 2023)

CDG(A) 2011 *Carriage of Dangerous Goods and Use of Transportable Pressure Equipment (Amendment) Regulations* https://legislation.gov.uk/uksi/2011/1885 (accessed 18 February 2023)

CDG(A) 2019 *The Carriage of Dangerous Goods (Amendment) Regulations*. https://legislation.gov.uk/uksi/2019/598 (accessed 18 February 2023)

CDG(A)(NI) 2011 *Carriage of Dangerous Goods and Use of Transportable Pressure Equipment (Amendment) Regulations (Northern Ireland)* https://legislation.gov.uk/nisr/2011/365 (accessed 18 February 2023)

CDG(A)(NI) 2019 *The Carriage of Dangerous Goods (Amendment) Regulations (Northern Ireland)* (Northern Ireland) https://legislation.gov.uk/nisr/2019/111 (accessed 18 February 2023)

CDG(NI) 2010 *The Carriage of Dangerous Goods and Use of Transportable Pressure Equipment Regulations (Northern Ireland)* 2010 https://legislation.gov.uk/nisr/2010/160 (accessed 31 January 2023)

CQC, RQIA, HIS and HIW 2020 *Significant Accidental and Unintended Exposures under IR(ME)R: Guidance for Employers and Duty-Holders. Version 2* (Care Quality Commission, The Regulation and Quality Improvement Authority, Healthcare Improvement Scotland, Healthcare Improvement Wales) https://cqc.org.uk/guidance-providers/ionising-radiation/saue-criteria-making-notification (accessed 31 January 2023)

EA(S)R 2018 *The Environmental Authorisations (Scotland) Regulations 2018* http://legislation.gov.uk/ssi/2018/219 (accessed 31 January 2023)

EPR 2016 *Environmental Permitting (England and Wales) Regulations* https://legislation.gov.uk/uksi/2016/1154 (accessed 31 January 2023)

HSE 2018 *L121 Work with Ionising Radiation Ionising Radiations Regulations 2017—Approved Code of Practice and Guidance* 2nd edn (Norwich: Health and Safety Executive) https://hse.gov.uk/pubns/books/l121.htm (accessed 31 January 2023)

HSE and ONR 2019 *L126 The Radiation (Emergency Preparedness and Public Information) Regulations 2019—Approved Code of Practice and Guidance* 2nd edn (London UK: TSO) https://onr.org.uk/documents/2020/reppir-2019-acop.pdf (accessed 31 January 2023)

IR(ME)R 2017 *The Ionising Radiation (Medical Exposure) Regulations* www.legislation.gov.uk/uksi/2017/1322 (accessed 31 January 2023)

IR(ME)R(NI) 2018 *The Ionising Radiation (Medical Exposure) Regulations (Northern Ireland)* https://legislation.gov.uk/nisr/2018/17 (accessed 31 January 2023)

IRR 1999 *The Ionising Radiations Regulations* https://legislation.gov.uk/uksi/1999/3232 (accessed 31 January 2023)

IRR 2017 *The Ionising Radiations Regulations* Great Britain https://legislation.gov.uk/uksi/2017/1075 (accessed 31 January 2023)

IRR(NI) 2017 *The Ionising Radiations Regulations (Northern Ireland)* http://legislation.gov.uk/nisr/2017/229 (accessed 31 January 2023)

IRR(SS)(NI) 1969 *Ionising Radiations (Sealed Sources) Regulations (Northern Ireland)* NISR 1969/318 https://legislation.gov.uk/nisro/1969/318 (accessed 18 February 2023)

IRR(SS)R 1969 *Ionising Radiations (Sealed Sources) Regulations* https://legislation.gov.uk/uksi/1969/808 (accessed 18 February 2023)

MHSWR 1999 *The Management of Health and Safety at Work Regulations* www.legislation.gov.uk/uksi/1999/3242/regulation/3 (accessed 31 January 2023)

MHSWR(NI) 2000 *Management of Health and Safety at Work Regulations (Northern Ireland)* 2000 http://legislation.gov.uk/nisr/2000/388 (accessed 31 January 2023)

NHSX 2021 *Records Management Code of Practice* 2021 *A Guide to the Management of Health and Care Records*. AUGUST 202, *NHS*. AUGUST 202. NHSX https://nhsx.nhs.uk/information-governance/guidance/records-management-code/ (accessed 31 January 2023)

ONR 2021 Transporting Radioactive Material—Security Guidance on the Carriage of Class *7* Radioactive Material (TD-TCA-GD-*002*). *2.1 ONR Guidance Document. 2.1* (Office for Nuclear Regulation) https://onr.org.uk/operational/other/td-tca-gd-002.pdf (accessed 18 February 2023)

PHE 2018 Guidance: Retention, Storage and Disposal of Mammograms and Screening Records *GOV.UK Website* (Public Health England) https://gov.uk/government/publications/breast-screening-manage-mammograms-and-records/retention-storage-and-disposal-of-mammograms-and-screening-records (accessed 29 December 2022)

RCR 2021 Statement on the Retention of Oncology Records *Royal College of Radiologists Website* (Royal College of Radiologists) https://rcr.ac.uk/posts/statement-retention-oncology-records-0#:~:text=The RCR also recommends that,information in DICOM image format (accessed 18 February 2023)

REPPIR 2001 *Radiation (Emergency Preparedness and Public Information) Regulations* https://legislation.gov.uk/uksi/2001/2975 (accessed 18 February 2023)

REPPIR 2019 *The Radiation (Emergency Preparedness and Public Information) Regulations 2019* http://legislation.gov.uk/uksi/2019/703 (accessed 31 January 2023)

REPPIR(NI) 2001 *Radiation (Emergency Preparedness and Public Information) Regulations (Northern Ireland)* https://legislation.gov.uk/nisr/2001/436 (accessed 18 February 2023)

REPPIR(NI) 2019 *The Radiation (Emergency Preparedness and Public Information) Regulations (Northern Ireland)* http://legislation.gov.uk/nisr/2019/185 (accessed 31 January 2023)

RSA(A)R(NI) 2011 *The Radioactive Substances Act* 1993 *(Amendment) Regulations (Northern Ireland)* https://legislation.gov.uk/nisr/2011/290 (accessed 31 January 2023)

SEPA 2021 Environmental Authorisations (Scotland) Regulations *2018 Standard Conditions for Radioactive Substances Activities V2.0. 2.0* (Scottish Environmental Protection Agency) https://www.sepa.org.uk/media/593756/standard-conditions-for-radioactive-substances-activities-v2.pdf (accessed 18 February 2023)

IOP Publishing

Medical and Dental Guidance Notes (Second Edition)
A good practice guide on all aspects of ionising radiation protection in the clinical environment:
IPEM Report 113
**John Saunderson, Mohamed Metwaly, William Mairs, Philip Mayles,
Lisa Rowley and Mark Worrall**

Appendix 11

Example pre-work checklists for outside radiation workers and trainees

Appropriately trained staff are to fill out the following form before letting a new visiting employee working with ionising radiation (e.g. an outside worker or trainee) start working in a controlled area. Use in conjunction with information gathered during communication with their Employer. Also for periodic use to ensure accurate information is held in line with the management of outside workers policy.

Full name (and NI number if similar to another employee's)
Contact details
Employer name, address and contact details
Role (in the controlled area)

doi:10.1088/978-0-7503-2332-1ch31

Classified outside workers

Do they have their 'passbook'? If not, they cannot work		
Is there an entry in their passbook that states they are certified fit to work with radiation? If not, they cannot work		
What is their projected total annual radiation dose across all employment and their dose to date? Concern: could they exceed a limit (don't forget ALARP)?	*Projected annual dose (mSv)* Body Eye Foot Finger Other	*Dose to date (mSv)* Body Eye Foot Finger Other
Has this exceeded, or is this approaching, your local investigation level?		
Are they pregnant or breastfeeding?		
Have you provided them with additional personal monitoring to estimate doses in line with your local staff and any significant dose they may be exposed to, i.e. body, eyes, and extremities?		
Have they had general and specific radiation protection training? If not, can you provide sufficient training or ensure they are always under direct supervision?		
Have the risks and likely doses been identified for the worker?		
Is suitable and sufficient PPE available? See risk assessment and local rules for requirements.		
Has a copy of the risk assessment been provided so they can share it with their Employer?		
Have arrangements been made to record an estimate of the dose in the passbook on completion of the work or sooner?		

Classified trainee (must be >18 years of age)

Is there an entry in their passbook that states they are certified fit to work with radiation? If not, they cannot work		
What is their projected total annual radiation dose across all employment and their dose to date? Concern: could they exceed a limit (don't forget ALARP)	*Projected annual dose (mSv)* Body Eye Foot Finger Other	*Dose to date (mSv)* Body Eye Foot Finger Other
Has this exceeded, or is this approaching, your local investigation level?		
Are they pregnant or breastfeeding?		
Have you provided them with personal monitoring through an ADS to assess doses in line with your risk assessment and any significant doses they may be exposed to, i.e. body, eyes, and extremities?		
Have they had general and specific radiation protection training? If not, can you provide sufficient training or ensure they are always under direct supervision?		
Have the risks and likely doses been identified for the worker?		
Is suitable and sufficient PPE available? See risk assessment and local rules for requirements.		
Has a copy of the risk assessment been provided so they can share it with their Employer (or education provider)?		
Are there any additional dosimetry arrangements (not covered by the risk assessment) or dose records that should be shared with their other Employers or education providers?		

Non-classified outside workers and non-classified trainees

What age is this person? Tick a box Note that there are different dose limits for each of these age categories	>18 yrs	16–18 yrs	<16 yrs
What is their projected total annual radiation dose across all employment and their dose to date? Concern: could they exceed a limit (don't forget ALARP)	*Projected annual dose (mSv)* Body Eye Foot Finger Other	*Dose to date (mSv)* Body Eye Foot Finger Other	
Has this exceeded or is this approaching your local investigation level?			
Are they pregnant or breastfeeding?			
What are the applicable dose limits (Schedule 3 IRR17) and constraints (from the risk assessment)?			
Use the managing dosimetry across multiple Employers' policies to establish the best monitoring method and record details here.			
Have they had general and specific radiation protection training? If not, can you provide sufficient training or ensure they are always under direct supervision?			
Have the risks and likely doses been identified for the worker?			
Is suitable and sufficient PPE available? See risk assessment and local rules for requirements.			
Has a copy of the risk assessment been provided so they can share it with their Employer (or education provider)?			
Are there any additional dosimetry arrangements (not covered by the risk assessment) or dose records that should be shared with their Employer (or education provider)?			

This person has been/has not been allowed to work in the controlled area.
Sign and Date:

Appropriate person	**Outside worker**

IOP Publishing

Medical and Dental Guidance Notes (Second Edition)
A good practice guide on all aspects of ionising radiation protection in the clinical environment: IPEM Report 113
John Saunderson, Mohamed Metwaly, William Mairs, Philip Mayles, Lisa Rowley and Mark Worrall

Appendix 12

Incidents and shared learning

As well as the learning that should take place within an organisation as a result of the investigation and analysis of incidents (and near misses), it can help to share those occurrences with a wider audience so that measures can be taken to prevent a repeat elsewhere. A number of pathways relevant in the UK are detailed below. Note that these are in addition to the legal requirement to report to regulators under IRR or IRMER, for example.

Learning from radiotherapy incidents

The planning and delivery of radiotherapy treatment is complex. It is reliant on input from a team of skilled professionals using sophisticated technology to manage and process large amounts of unique data generated on a per-treatment basis and interpret it correctly for each fraction of treatment. The potential for errors across the patient pathway is high. When the opportunity for error is weighed against the reported incidence of error, radiotherapy may be seen as a safe form of treatment for cancer (HPA 2010). However, when an error does occur, the consequences can be significant for the patient.

The fundamental role of reporting and learning systems is to enhance patient safety by learning from failures in the healthcare setting (WHO 2020). It is known that most errors are not just a series of random, unconnected, one-off events; they are provoked by poor systems and often have common root causes that can be generalised and corrected. Although each error is unique, there are likely to be similarities and patterns in the sources of error that may go unnoticed if incidents are not reported and analysed.

Towards Safer Radiotherapy (TSRT) (BIR *et al* 2008), published in 2008, contains practical recommendations for the radiotherapy community aimed at improving safety and reducing errors. One of these recommendations is the national reporting of and learning from radiotherapy errors. TSRT provides definitions for

doi:10.1088/978-0-7503-2332-1ch32

the terminology to be used in discussing *radiotherapy errors* (RTE) and proposes two taxonomies for use in describing the severity and where on the pathway the error occurred. In 2016, a refinement of the pathway coding was published to reflect contemporary practice and introduce safety barrier and *causative factor* (CF) taxonomies (PHE *et al* 2016, PHE 2021). These are currently under review as part of the upcoming guidance document, Advancing Safer Radiotherapy.

A voluntary national UK incident learning system for radiotherapy was established in 2009 and is now hosted by the UK Health Security Agency (UKHSA). RTE reports are submitted through multiple routes, from England via the National Reporting and Learning System (NRLS) at NHS England, from Wales via the Once for Wales Concerns Management System (OfW), or directly to the UK Health Security Agency (UKHSA) from providers in Northern Ireland and Scotland. In England, the NRLS is planned to be replaced by the Learn from Patient Safety Events Service (LFPSE) by autumn 2023 (figure A12.1). Independent providers are also able to submit RTE data directly to the UKHSA. RTE reports are quality assured and analysed by UKHSA using frequency trend analysis based on the nationally agreed taxonomies (BIR *et al* 2008). UKHSA, in conjunction with the Patient Safety in Radiotherapy Steering Group, publishes learning from these events (UKHSA no date c) on a biennial and triannual basis (UKHSA no date b) so their probability and magnitude might be mitigated.

RTE data can also be used to inform a simple risk matrix. A study of risk, or proactive risk assessment, is a process that helps organisations understand the range of risks that they face, their capacity to control those risks, the likelihood (probability) of the risk occurring, and the potential impact thereof (Radiotherapy Board 2020).

Reporting to the voluntary radiotherapy reporting and learning system can assist in improving patient safety or reducing radiation incidents, but it does not negate the responsibility to report significant accidental and unintended exposures to the

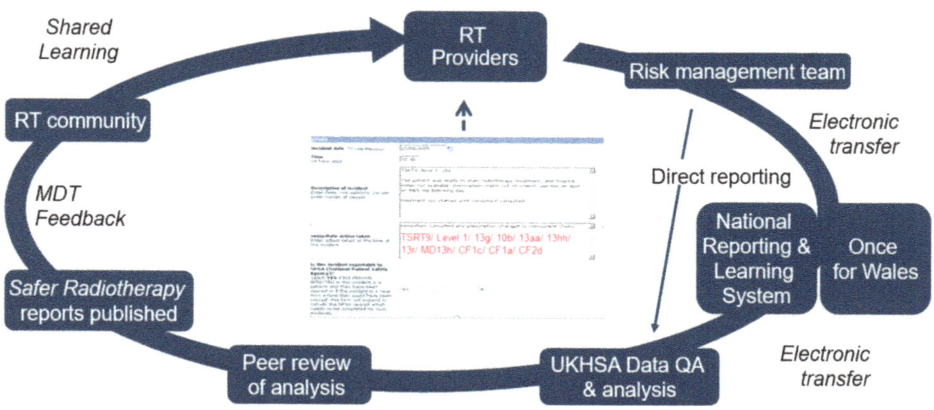

Figure A12.1. Radiotherapy error (RTE) incident learning system (Findlay *et al* 2016).

relevant enforcing authorities (IR(ME)R 2017, IR(ME)R(NI) 2018, CQC 2023). It is imperative that RTE trends continue to be reported, analysed, and monitored on a cyclical basis in order to inform ongoing safe and effective radiotherapy practice. This is especially pertinent as new techniques and technologies are implemented and new clinical radiotherapy practices are established. The radiotherapy reporting and learning system is a feedback process that supports a risk-based approach to improving patient safety (Findlay *et al* 2016) (figure A12.1). For further information, email radiotherapy@ukhsa.gov.uk.

Clinical imaging and nuclear medicine error reporting and learning system

The value of error and near-miss reporting and the associated learning is well appreciated in the UK radiotherapy community. The Clinical Imaging Board (CIB, no date) recognised the benefits of developing such a system for diagnostic imaging and nuclear medicine and commissioned a working party to take this forward. Learning from ionising radiation dose errors, adverse events, and near misses in UK clinical imaging departments (RCR *et al* 2019b) was published in June 2019. It includes a classification, contributory factor, and pathway coding system with associated guidance intended to enable organisations to code, analyse, and learn from errors. These are available from the Projects page of the Clinical Imaging Board website (CIB 2019, RCR *et al* 2019a)

The Clinical Imaging Board agreed to support the national data collection and analysis of diagnostic and nuclear medicine errors and near-miss events. This has since been expanded to include Magnetic Resonance Imaging. The work is being coordinated by the UK Health Security Agency (UKHSA no date a) with input from the professional bodies and oversight provided by the Clinical Imaging Board.

Medical device incident reporting

The Medicines and Healthcare Products Regulatory Agency (MHRA) operates a Yellow Card reporting site (MHRA 2023), which is a scheme whereby suspected side effects of medicines, vaccines, e-cigarettes, medical device incidents, and defective or falsified (fake) products can be reported to the MHRA to ensure safe and effective use. The Yellow Card Scheme is vital in helping the MHRA monitor the safety of all healthcare products in the UK to ensure they are acceptably safe for patients and those that use them. This includes medical devices. It is essential that issues are reported so that the MHRA can monitor trends and identify issues that allow actions to be taken to minimise risk and maximise benefits for patients. Their website states that an adverse incident is an event that caused, or almost caused, an injury to a patient or other person, or a wrong or delayed diagnosis and treatment of a patient.

Reports made on the MHRA Yellow Card scheme from incidents originating in the devolved UK nations will be passed to the relevant body in the relevant devolved nation (NIAIC 2021, NSS 2022), and members of the public are encouraged to use this scheme. Professionals should report a suspected problem or incident involving a medical device using their national reporting system, as outlined below:

- England and Wales (MHRA 2023)—https://yellowcard.mhra.gov.uk/
- Northern Ireland (NIAIC no date)—https://www.health-ni.gov.uk/articles/reporting-adverse-incident
- Scotland (National Services Scotland no date)—https://www.nss.nhs.scot/health-facilities/incidents-and-alerts/report-an-incident

There may also be a separate legal requirement to report an incident to a regulator, so appropriate advice should be sought on this.

OTHEA

OTHEA is a platform set up by stakeholders from a number of countries interested in radiation protection for the facilitation and dissemination of ALARA practices and, in particular, the sharing of experiences from radiological incidents (PHE *et al* no date). The site is applicable to different sectors, including the medical sector, and is supported by UKHSA, which would encourage the UK to use and contribute to the database. Incidents that are reported are anonymous and are only shared online if there are useful lessons. Reports are available for more than what might be considered a typical incident because OTHEA wishes to share learning when any event, behaviour, or anomaly occurs with the potential to cause unintended radiation exposure or a significant decrease in the existing standard of radiation protection.

To submit an incident or use the database, please see www.othea.net.

References

BIR, IPEM, NPSA, SCoR and RCR 2008 *Towards Safer Radiotherapy* (London: British Institute of Radiology, Institute of Physics and Engineering in Medicine National, Patient Safety Agency, Society and College of Radiographers, Royal College of Radiologists) https://www.rcr.ac.uk/our-services/all-our-publications/clinical-oncology-publications/towards-safer-radiotherapy/ (accessed 12 April 2024)

CIB 2019 *Clinical Imaging Board—Coding Taxonomy for Ionising Radiation Dose Errors, Adverse Events and Near Misses in UK Clinical Imaging Departments (Spreadsheet)* (UK: Clinical Imaging Board) https://rcr.ac.uk/clinical-radiology/service-delivery/clinical-imaging-board/clinical-imaging-board-projects (accessed 16 March 2023)

CIB no date Clinical Imaging Board *Royal College of Radiologists Website* https://www.rcr.ac.uk/our-services/management-service-delivery/clinical-imaging-board/ (accessed 12 April 2024)

CQC 2023 *Ionising Radiation (Medical Exposure) Regulations (IR(ME)R). CQC Website. 19/4/23* (London: Care Quality Commission) https://cqc.org.uk/guidance-providers/ionising-radiation/ionising-radiation-medical-exposure-regulations-irmer (accessed 10 May 2023)

Findlay Ù, Best H and Ottry M 2016 Improving patient safety in radiotherapy through error reporting and analysis *Radiography* **22** S3–11 https://sciencedirect.com/science/article/abs/pii/S1078817416300979 (accessed 16 March 2023)

HPA 2010 *HPA-CRCE-002—Patient Safety in Radiotherapy Steering Group Activity (November 2007–March 2010)* (Chilton: Health Protection Agency) https://webarchive.nationalarchives.

gov.uk/ukgwa/20140722062256/http://hpa.org.uk/Publications/Radiation/CRCEScientificAndTechnicalReportSeries/HPACRCE002 (accessed 14 March 2023)

IR(ME)R 2017 *The Ionising Radiation (Medical Exposure) Regulations* www.legislation.gov.uk/uksi/2017/1322 (accessed 31 January 2023)

IR(ME)R(NI) 2018 *The Ionising Radiation (Medical Exposure) Regulations (Northern Ireland)* https://legislation.gov.uk/nisr/2018/17 (accessed 31 January 2023)

MHRA 2023 Yellow Card reporting site *MHRA Website Website* (London: Medicines & Healthcare Products Regulatory Agency) https://yellowcard.mhra.gov.uk/ (accessed 6 March 2023)

National Services Scotland no date *Incidents and Alerts, National Services Scotland Website.* https://nss.nhs.scot/browse/health-facilities/incidents-and-alerts (accessed 16 March 2023)

NIAIC 2021 *Reporting Adverse Incidents and Disseminating Safety Information* (Belfast: Northern Ireland Adverse Incident Centre) https://www.health-ni.gov.uk/sites/default/files/publications/health/Guidance%20publication%20for%20reporting%20adverse%20incidents%20to%20NIAIC.pdf publication for reporting adverse incidents to NIAIC.pdf (accessed 12 April 2024)

NIAIC no date *Northern Ireland Adverse Incidents Centre (NIAIC), Department of Health (NI) Website* (Belfast) www.health-ni.gov.uk/articles/introduction-northern-ireland-adverse-incidents-centre-niaic (accessed 16 March 2023)

NSS 2022 *Report an Incident, National Services Scotland Website* (https://nss.nhs.scot/health-facilities/incidents-and-alerts/report-an-incident/) (accessed 16 March 2023)

PHE 2021 *Safer radiotherapy. Radiotherapy Error and Near Miss Reporting: The Unseen Pathway* (London: Public Health England) https://gov.uk/government/publications/development-of-learning-from-radiotherapy-errors (accessed 14 March 2023)

PHE, CEPN, IRSN, INRS, CoRPAR, SFRP and INSTN no date *OTHEA* (https://othea.net/en/) (accessed 16 March 2023)

PHE, IPEM, RCR and CoR 2016 *Development of Learning from Radiotherapy Errors. Supplementary Guidance Series* (London: Public Health England) https://gov.uk/government/publications/development-of-learning-from-radiotherapy-errors (accessed 14 March 2023)

Radiotherapy Board 2020 *Ionising Radiation (Medical Exposure) Regulations: Implications for Clinical Practice in Radiotherapy Guidance from the Radiotherapy Board* (London: Royal College of Radiologists) https://www.rcr.ac.uk/media/smmkkrsa/ionising-radiation-medical-exposure-regulations-implications-for-clinical-practice-in-radiotherapy.pdf (accessed 12 April 2024)

RCR, IPEM and SCoR 2019a *Learning from Ionising Radiation Dose Errors, Adverse Events and Near Misses in UK Clinical Imaging Departments: Working Party User Guidance* (UK: Royal College of Radiologists, Institute of Physics and Engineering in Medicince, Society & College of Radiographers) https://www.rcr.ac.uk/media/zdhdlbpt/rc5491-1.pdf (accessed 12 April 2024)

RCR, IPEM and SCoR 2019b *Learning from Ionising Radiation Dose Errors, Adverse events and Near Misses in UK Clinical Imaging Departments* (London: Royal College of Radiologists Institute of Physics and Engineering in Medicine Society and College of Radiographers) https://www.rcr.ac.uk/our-services/all-our-publications/clinical-radiology-publications/learning-from-ionising-radiation-dose-errors-adverse-events-and-near-misses-in-uk-clinical-imaging-departments-working-party-user-guidance/ (accessed 12 April 2024)

UKHSA no date a Medical Exposures Group—stakeholder work *UKHSA Radiation Protection Services Website* (Chilton: UK Health Security Agency) https://ukhsa-protectionservices.org.uk/meg/diagnosticimaging/stakeholderwork/ (accessed 16 March 2023)

UKHSA no date b Safer radiotherapy: biennial error analysis and learning *GOV.UK Websiteuk website* (UK Health Security Agency) https://gov.uk/government/publications/radiotherapy-errors-and-near-misses-data-report (accessed 16 March 2023)

UKHSA no date c Safer RT E-bulletins and newsletters , *UKHSA Radiation Protection Services Website* (UK Health Security Agency) https://ukhsa-protectionservices.org.uk/meg/radiotherapy/safer_RT/ (accessed 16 March 2023)

WHO 2020 *Patient Safety Incident Reporting and Learning Systems Technical Report and Guidance* (World Health Organization) https://apps.who.int/iris/handle/10665/334323 (accessed 14 March 2023)

IOP Publishing

Medical and Dental Guidance Notes (Second Edition)
A good practice guide on all aspects of ionising radiation protection in the clinical environment: IPEM Report 113

John Saunderson, Mohamed Metwaly, William Mairs, Philip Mayles, Lisa Rowley and Mark Worrall

Appendix 13

The life cycle of imaging systems

Figure A13.1 lays out the life cycle of imaging systems. See chapter 1 paragraphs 1.247–1.261 on critical examination. See chapter 2 paragraphs 2.88–2.94 for information on acceptance testing, commissioning, and routine testing of equipment.

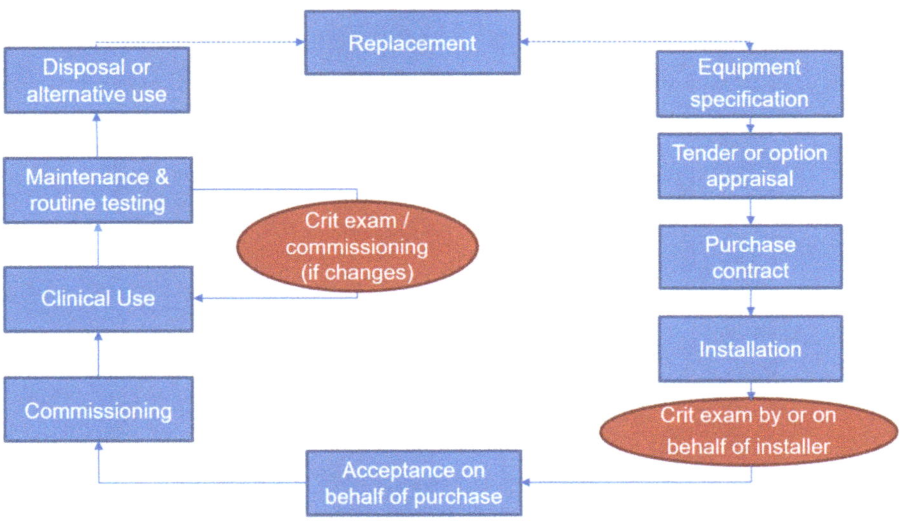

Figure A13.1. Life cycle of imaging systems, items in red fall under IRR (IRR 2017, IRR(NI) 2017).

References

IRR(NI) 2017 The Ionising Radiations Regulations (Northern Ireland) http://legislation.gov.uk/nisr/2017/229 (accessed 31 January 2023)

IRR 2017 *The Ionising Radiations Regulations* https://legislation.gov.uk/uksi/2017/1075 (accessed 31 January 2023)

Appendix 14

Issues to consider when writing IR(ME)R Employer's procedures

Ionising Radiation (Medical Exposure) Regulations 2017 as amended and Ionising Radiation (Medical Exposure) Regulations 2018 in Northern Ireland (IR(ME)R 2017, IR(ME)(A)R 2018, IR(ME)R(NI) 2018), subsequently referred to as IR(ME) R, require the Employer to establish a framework of general procedures, protocols, and quality assurance programmes under which professionals can practice. The framework of procedures must cover the matters set out in Schedule 2 of IR(ME)R as a minimum and should be included in controlled documents. It is recommended that the Employer seeks advice from experienced professional colleagues from relevant specialities when establishing the procedures (DHSC 2018a) including a recognised Medical Physics Expert (MPE) (DHSC 2018b, RPA2000 2023), and a multidisciplinary approach to writing them will be most effective.

The following guidelines can be used to help an Employer write IR(ME)R procedures that fit their local practice.

- Procedures should make it clear who is responsible for carrying out each aspect of the procedure and when. This is especially important when there are multiple entitled individuals acting as Referrers, Practitioners or Operators. There must be a record of who carries out each element.
- When the procedure requires interaction with a patient, consider and describe in the procedure how this will be enacted in situations where the patient does not speak the Operators working language, has sensory impairments, lacks capacity, is very young, or is unconscious. This is particularly important when asking the patient a question or delivering information.
- There must be contingencies for situations where these procedures cannot be enacted but the exposure must take place to contribute to lifesaving treatment, e.g., patients may be unconscious.

The procedure should reflect up-to-date professional guidance from bodies such as the National Institute for Health and Care Excellence (NICE), the Royal College of Radiologists (RCR), and other professional bodies. They should be impacted by local audit results, through the observation of staff that undertake the roles, and take into consideration relevant existing organisation-wide policies and procedures.

Note that regulation 7 of IR(ME)R says the Employer's procedures must include provisions for the carrying out of clinical audits as appropriate.

(a) Procedures to correctly identify the individual to be exposed to ionising radiation

Detailed guidance is available from the Clinical Imaging Board (CIB 2019b).

- Patient identification (ID) must take place before each exposure or before the start of an examination or procedure.
- The process must be active, e.g. ask the patient to state their full name, date of birth, and address and check this against the primary source data. This is the standard three-point check.
- It is recommended that checks of the clinical information, the site requested, and previous imaging also form part of the local checking processes. This is a six-point check.
- It is important to state who is responsible for making these checks and how they are recorded so that the member of staff can be identified at a later date from a signature, which may be electronic. This is particularly important when there are multiple Operators performing different roles.
- Consider and document the actions to take when the patient cannot actively respond to ID checks. These could be patients with dementia or sensory impairments, who lack capacity, patients who are non-English speaking, unconscious or anaesthetised, children, or, unidentified patients such as those in major accidents.
- Extra care should be taken when patients being exposed share similar or identical names, if that information is available, or may even share the same address (e.g., John Smith senior and his son John Smith junior).
- Consider how this fits in with the 'Pause and Check' process (see paragraph 2.48 in chapter 2) or the WHO checklist (WHO 2009).
- Document the steps to take if any of the ID information does not match the request.

(b) Procedures to identify individuals entitled to act as Referrer Practitioner or Operator within a specified scope of practice

- Describe the entitlement process and how this task has been delegated by the Employer.
- Individuals (or professional groups such as general practitioners or Band 7 radiographers, etc) who are entitled by the Employer to act in these roles must be clearly identified and documented.

- The training and competency required to support entitlement should be recorded. This can be included in the entitlement document or referenced. The document should make clear who holds the training records and where they are stored.
- The documentation should include the scope of practice e.g. the extent of examinations that Referrers can refer for, IR(ME)R Practitioners can justify, or the practical aspects that IR(ME)R Operators can carry out.
- Individuals performing IR(ME)R roles must be clearly identifiable for each exposure that takes place. This allows comparison with the entitlement lists and documents responsibility for aspects of the exposure that can be audited.

Referrers

Although not explicitly demanded by IR(ME)R, professionally recommended best practice is that Referrers should receive training on the practical side of completing a request form (paper or electronic) for particular organisations to which they will refer, and also how to cancel a referral if it was made in error or is no longer required (BIR *et al* 2020, Radiotherapy Board 2020). This procedure could detail the practical training requirements and means of sign-off. It could also document what good and bad practices look like and the action an Employer will take if these are not followed. A good practice could be that any part of a referral completed by another person is checked before the responsible, entitled Referrer signs it off. It is not acceptable that a Referrer pre-sign referral documents for others to complete, and electronic logins should not be misused. Such actions may constitute a breach of the IR(ME)R regulations and might be considered professional misconduct. Consider whether the local electronic referral system is capable of limiting the ability to refer for medical radiation exposures to those with entitlement, and limiting referrals to their individual or group (e.g. GPs) scope of practice.

Referrers must be registered healthcare professionals, which includes non-medical Referrers. Consider and document the process for approving the entitlement of non-medical Referrers, e.g. how do you ensure that they have sufficient competence in history taking and assessment of patients, plus the competence to understand the significance of the reported findings?

Specific clinical service further considerations:

- In breast screening, if an individual is recalled for further imaging following screening, due to meeting certain criteria, then the person doing the recall becomes a Referrer and needs to be entitled by the Employer to act in this capacity.
- For complex interventional or therapeutic procedures, the patient might not be referred by an individual but by a multidisciplinary team meeting, and procedures need to make clear which individual has the responsibility as the Referrer in this case.
- In radiotherapy, consider whether a referral for treatment would comprise all pre-treatment imaging, verification imaging, and the treatment

prescription. Also consider how repeat imaging would be requested, justified, and authorised if required, such as in the case of technical repeats where images were deemed not fit for purpose.

In all modalities, consider if a series of exposures can be referred for in one referral. For example, if a patient is on a care pathway (or trial) that requires imaging over months or years, should this require an assessment of the patient's circumstances directly before each referral, as they may have changed, or is it satisfactory that a referral is made significantly in advance of the planned exposure?

Document how requests will be cancelled and whose responsibility it is to act on the cancellation request. Consider if the IT systems can support cancellations (CQC 2018).

Practitioners

IR(ME)R Practitioners must always be registered healthcare professionals with adequate training and must be entitled by the employer. IR(ME)R Practitioners may be identified by lists of names of individuals or by staff grade, qualification, etc For example, at one hospital all radiologists may be entitled to act as IR(ME)R Practitioners for diagnostic 'general radiography' but only specific individuals or grades may be entitled to act as IR(ME)R Practitioners for CT. It is important that procedures make it clear when a member of staff is acting as an Operator and 'Authorising' under authorisation guidelines rather than as an IR(ME)R Practitioner and 'Justifying' exposures.

Specific clinical service further considerations:

- In nuclear medicine and in brachytherapy, the IR(ME)R Practitioners (for the administration of radioactive substances) are identified as those who hold valid ARASC licenses under IR(ME)R.
- In radiotherapy, justification of the pre-treatment, verification, and treatment exposures may need to be a two-stage process. Depending on the departmental procedures, if the treatment prescription is considered authorisation, and is not completed until after the treatment plan is finished, the justification of the planning exposures may need to be carried out before the patient attends for localisation imaging. A separate referral (and therefore justification) may be required for additional concomitant imaging outside of the established protocol.

Consider if radiographers are entitled to be IR(ME)R Practitioners (see paragraphs 2.27–2.44 of chapter 2 for more on this and professional guidance on the subject) (SCoR 2018).

Operators

There is a wide range of tasks that fall under 'Operator duties', so it is particularly important that the scope of practice of an Operator is well defined.

Pay close attention to exposures that take place outside conventional 'radiation departments', such as in theatres, urology, pacing, pain clinics, endoscopy, lithotripsy, etc.

Consider the scope of practice of any students and/or locums who will be acting as Operators along with the level of supervision that will be maintained. Whilst it is possible under IR(ME)R to entitle appropriately trained student radiographers and trainee assistant practitioners, the College of Radiographers advises that they are not entitled Operators (SCoR 2019). Where students are being supervised or mentored, it is the supervisor who remains responsible.

Medical physics staff will need to be entitled to act as Operators for the particular tasks undertaken by them, such as their role in undertaking quality assurance of equipment.

Image evaluation is an Operator task, and individuals must be entitled to perform this role. This includes individuals who use fluoroscopy to guide procedures and who evaluate images when radiology does not provide a report.

(c) Procedures for making inquiries about patients of childbearing potential to establish whether the individual is or may be pregnant or breastfeeding

The procedure should make it clear who from the staff is responsible for performing this and how the response and subsequent decision on whether or not to proceed are recorded. Consider:

- Will posters be displayed? e.g. each patient waiting area within radiology displays posters prompting patients to tell staff if they might be pregnant, such as that available from the Society of Radiographers (SCoR and FujiFilm 2020).
- What are the questions to ask the appropriate individuals? Is it 'Are you pregnant? and is the patient's response sufficient to satisfy the check, i.e., if they say they are not pregnant, does the procedure proceed? Or, will the question be menstrual cycle related instead? The answer to these questions leads to further points that should be considered, what is the Employer's view on contraceptive methods and other procedures such as sterilisation or hysterectomy? Questioning should be about the individual's status; therefore, it would not be appropriate to inquire about the status of a partner, for example, if they had a vasectomy.
- When should the question be asked? If the patient will be sedated for exposure, then the process should be followed prior to this.
- If a pregnancy test is ever required.
- How the response is documented. Does the patient have to sign as evidence of the check?
- The appropriate age group e.g. 12–55 year olds who may be pregnant or breastfeeding. This will be a local decision that can be based on information such as national or local pregnancy statistics. Although not an IR(ME)R issue, consider the appropriate safeguarding response if a child does say they are pregnant.
- The range of procedures for which questioning is appropriate, e.g. high dose, radionuclides, anatomical regions exposed (e.g. diaphragm to knee).

- Whether high dose examinations will be booked in the first ten days of the menstrual cycle, when conception is unlikely to have occurred, or be rebooked if, on questioning, pregnancy cannot be excluded in individuals capable of childbearing in the second half of the menstrual cycle (HPA *et al* 2009).
- In some circumstances, a clinical decision will have to be made (between the Referrer and ultimately the Practitioner) about the radiation risk associated with a potential pregnancy compared to the needs of that individual, e.g. trauma cases. The potential pregnancy is still taken into account in these situations.
- How will this process be followed regarding young individuals? Will parents be present during questioning for young patients, e.g. 12–15 years old? The CQC has reported good practice where clear and age-appropriate pre-examination information was available for children and their parents/carers which aimed to prepare them for questions that would be asked once in the x-ray room (CQC 2019).
- For patients who do not speak the same language as the Operator, a multilingual advice sheet or linguist should be available to assist in asking the pregnancy question.
- Those lacking capacity.
- What happens if the patient is pregnant? Document the steps to be taken if the patient may be or is pregnant.

Guidance to support inclusive practice in regard to pregnancy status inquiries is available (SoR 2021). See also chapter 2, paragraph 2.131, on facilitating a diverse gender spectrum.

(d) Procedures to ensure that written procedures, written protocols, and equipment quality assurance programmes are followed

This procedure refers to quality assurance and document control of the IR(ME)R framework. It should record the system of governance that is applied to the procedures required by IR (ME)R Schedule 2, standard protocols for examinations, etc, and the documentation in regard to quality assurance for equipment, which should consider the making of measurements and the recording of results.

The procedure should state:

- Who is entitled to issue the procedures and protocols?
- How frequently are the procedures reviewed to ensure they are effective, what does this entail, and who is responsible for ensuring they are completed? The review should include audits as appropriate.
- How to identify the current version.
- How the documents are disseminated among the relevant staff and how staff are trained in the procedures/protocols, as required. The period between document issues and practical implementation should be minimal.
- It is the responsibility of all staff to bring any observed failures of a procedure or protocol, or any suggestions for improvement, to the attention of the employer.

(e) Procedures for the assessment of patient dose and administered activity

Local procedures should make it clear who is responsible for recording patient doses, the parameters that should be recorded, and where these should be recorded. This is related to an individual patient's dose associated with an exposure event rather than a population assessment.

Where only exposure parameters are recorded rather than a true dose, consider that a suitably accurate estimate of patient dose may have to be calculated from these parameters at a later date. This may be when an incident has occurred. An assessment of dose will often then be required for the determination of patient risk, for recording in the patient's notes, and for reporting to the regulators.

Document if it is satisfactory to record exposure factors only when the Operator has used their professional judgement to stray away from the written protocols for standard-sized patients, such as for a large patient. The only factors that need to be recorded in those circumstances are those where there has been a change. Written protocols will, of course, contain exposure factor details for standard imaging.

Consider whether or not it is satisfactory if the only record of exposure factors is recorded on annotated patient images (such as mammograms with electronic annotation of kV and mAs, etc), which are sent to the PACS. Consider how an estimate of the dose would be determined if these images were lost.

The procedure should include details about the process and responsibility for calibrations of individual patient exposure measurement devices, where these are required, e.g. dose-area product (DAP) metres.

Examples of dose parameters to record:

- In nuclear medicine, this is typically the administered radioactive material activity. Remember to record the dose-length product (DLP) for CT imaging, such as in SPECT/CT procedures.
- In diagnostic and interventional radiology, there are a number of parameters in use depending on the modality, including the DAP (or kV and mAs if there is no DAP meter), cumulative skin dose, and DLP. Also, record the equipment used for the exposure so that machine-specific data can be used for dose estimation, e.g., tube output.
- In breast imaging, the factors used to estimate the 'mean glandular dose' include the compressed breast thickness, the x-ray tube target and filtration combination, kV, and mAs.
- In radiotherapy, the prescription and organ doses are recorded on the treatment plan. Treatment delivery is documented on the record and verification system (Oncology Management Information System) connected to the treatment machine. Careful consideration must be given to the manual recording of treatment for any equipment that does not have an automated record and verify system connected, for example, superficial treatment units.

While staff who work daily with radiation are familiar with the recording of patient doses, there must also be robust systems in place for recording doses from exposures that occur outside of traditional 'radiation' departments, e.g. theatres, pacing, endoscopy, urology, pain clinics, and dentists.

This procedure should include the processes associated with automated dose monitoring systems, where they play a role. These systems may also be documented in the procedure addressing the use and review of diagnostic reference levels (DRLs).

(f) Procedures for the use and review of such diagnostic reference levels as the Employer may have established for radio-diagnostic examinations

The procedure falls within IR(ME)R (2017) Regulations 3(a), (b), (e), and (f), which should consider:

- Which procedures DRLs will be set for and if national or local DRLs apply.
- How will the DRLs be set and who is involved, e.g., how will the MPE contribute?
- How the DRLs will be disseminated
- Who is responsible for the review, and how will they achieve that? A three-year rolling programme might be appropriate to review all DRLs unless dose management systems can provide data more frequently.
- How will the DRLs be used, e.g., are they displayed in the dispensary or x-ray room for immediate post-exposure check by the Operators, and/or will review be periodic (at what frequency) through patient dose audit?
- Facilities that may need additional radiation protection support, e.g. theatres, endoscopy, urology, pain clinics, dentists, etc.
- If the units the DRLs are displayed in correspond with room-specific dose units or if they are displayed in Gycm^2, etc, in line with national DRLs.
- The action to be taken if DRLs are consistently exceeded. What does 'consistently exceeded' mean in this situation?
- Training is required to ensure staff understand the processes, units, and use in the clinical setting.

Radiotherapy planning CT scans are not considered diagnostic scans, and therefore the use of the term DRLs is not appropriate. However, the use of dose reference levels is a useful method of demonstrating that dose optimisation has taken place (Public Health England 2019).

(g) Procedures for the use of dose constraints established by the Employer for biomedical and medical research programmes for patients or other persons where no direct medical benefit for the individual is expected from the exposure

This includes the procedures for determining whether the Practitioner or Operator is required to effect one or more of the matters set out in IR(ME)R (2017) Regulation 12(4), including criteria on how to effect those matters, and, in particular, procedures for the use of dose constraints established by the Employer for biomedical and medical research programmes falling within IR(ME)R (2017) Regulation 3(c) where no direct medical benefit for the individual is expected from the exposure.

The procedure must include a system to check that research trials that include medical radiation exposures are approved by a Research Ethics Committee (REC) and, where appropriate, the Administration of Radioactive Substances Advisory Committee (ARSAC).

Individual departments should be able to demonstrate:

- Which trials (and protocols therein) have received approval through the local research governance mechanisms? A mechanism should also be in place to inform relevant teams when a trial ends.
- The specific local IR(ME)R Practitioner who justified the research exposures on each trial.
- Where a patient in a trial will undergo some exposures that are beneficial and some that are purely for research purposes, the target doses are planned and achieved for the experimental diagnostic or therapeutic exposures for which a benefit is anticipated.
- That the dose constraints are set for individuals for whom no direct medical benefit is expected from the exposure.
- Who sets and monitors the dose constraints, and what is the response when a constraint is exceeded, e.g. is it considered an incident, and at what level might it be externally reportable to the regulators? Is it brought to the attention of the Clinical Lead and MPE? Tracking and confirmation that exposures are within the dose constraint could be part of the justification process throughout a trial.
- How referrals for research exposures are identified and the process used to check whether they meet the correct frequency of exposure within the approved protocol, etc

Further details of the required conditions for research are included in appendix 19, Guidance on medical research exposures.

(h) Procedures for the giving of information and written instructions to the patient

This procedure refers to patients undergoing treatment or diagnosis with radioactive substances, as referred to in IR(ME)R (2017) Regulation 12(6).

The procedure should make it clear who is responsible for giving this information and written instructions to the patient (or to their representative if the patient lacks capacity) and when they should be provided.

Consideration should be given to what instructions to give those patients being treated with long-lived isotopes or implantable brachytherapy devices regarding what action should be taken in the event of their death. Consider including a link from this procedure to the carer and comforter procedures, if relevant.

(i) Procedures providing that wherever practicable and prior to an exposure taking place, the individual to be exposed or their representative is provided with adequate information relating to the benefits and risks associated with the radiation dose from the exposure

Consider who will provide information, to whom, and when it is not appropriate.

In radiotherapy, the provision of information on the benefits and risks of all exposures during the radiotherapy treatment pathway, including planning and verification of imaging exposures, is usually best given to the patient by the clinical oncologist at the time the patient is invited to consent to treatment. The overall benefit and risk of the treatment, rather than those of individual exposures, are required. It is not necessary to repeat this information before each individual exposure or treatment episode.

In radio-diagnostic exposures, this information can be shared in the form of posters displayed in departments and patient information leaflets (DHSC 2018a, CIB 2019a).

Any implications for Carers and Comforters should be included.

Again, consideration will need to be given to patients whose language is different from the local language (i.e. English or Welsh, etc), those with sensory impairments, or those who lack capacity.

Sufficient time is required for the patient to both read and reflect on the benefits and risks (including the risk of not undergoing the procedure).

Consider the level of exposure/risk that might require a more detailed discussion and active documentation of consent.

(j) Procedures for the carrying out and recording of an evaluation for each exposure, including, where appropriate, factors relevant to patient dose

The procedure should describe what needs to be recorded in the clinical evaluation, by whom, and where it will be recorded. This is usually considered to be the clinical report of the imaging or treatment received. A written evaluation is currently considered the best practice to comply with this requirement.

Individuals undertaking image evaluation are IR(ME)R Operators and must be adequately trained for the role of providing comprehensive reports. This includes those using fluoroscopic imaging to guide procedures.

The evaluation report should answer the diagnostic question posed in the referral, where relevant. For example, where a patient is referred with clinical information stating that 'after a fall, the patient has pain in the wrist and an x-ray is requested to assess if it is broken', the evaluation should state that 'the wrist is/is not broken'. The evaluation may go on to document any other pathology of concern if it is seen, irrespective of whether this was of interest in the initial referral. This level of detail is a local decision, as IR(ME)R does not mandate what should be in a clinical evaluation or where and how it should be recorded.

Thought should be given to how evaluation can be completed in less straightforward situations, such as in the case of:

- intra-operative exposures using radiographic, fluoroscopic, or radioisotope guidance, e.g. Pain clinics, pacing, endoscopy, or sentinel node biopsies. It may be sufficient to document that the outcome of the radiation procedure was a 'success' or 'fail';
- dual-exposure situations, e.g. Fluoroscopic guidance for molecular radionuclide therapy;
- procedures are undertaken outside traditional 'radiation departments';
- a clinical evaluation performed by a third party. Auditing is an important part of this process.

Planning images in radiotherapy are used to derive the treatment plan. Online or offline image reviews similarly comprise the clinical evaluation for verification exposures. As such, no separate evaluation report is required.

(k) Procedures to ensure that the probability and magnitude of accidental or unintended exposure to individuals from radiological practices are reduced so far as reasonably practicable

This will be a list of good practice with adherence to the IR(ME)R (2017) framework and associated regulations. It is an opportunity to consider and record a clinical risk assessment with associated actions. Elements that may be addressed include:

- Is there an incident reporting procedure, and are all staff trained in it? Does it include identification of who is responsible for managing incidents until any actions are completed, who is responsible for analysis or review of incidents to look for trends, and does this include a review of near misses?
- The dissemination of lessons learned from incidents.
- A system to ensure equipment is properly maintained and regular QA performed.
- That equipment faults are reported and actions arising from faults or QA are resolved in a timely fashion.
- That all relevant staff have been trained in IR(ME)R procedures, the use of the equipment, and for the Practitioner and Operator responsibilities they are required to undertake.

Significant accidental and unintended exposures are defined in guidance from the Care Quality Commission (CQC 2023).

(l) Procedures to ensure that the Referrer, the Practitioner, and the individual exposed or their representative are informed of the occurrence of any relevant clinically significant unintended or accidental exposure, and of the outcome of the analysis of this exposure

The procedure should reference professional guidance (BIR *et al* 2020, Radiotherapy Board 2020) regarding the definition of ‘clinically significant’.

Duty of Candour (including any local action levels) should be considered within this procedure (see chapter 2, paragraphs 2.183–2.188, *Duty of candour*).

The procedure for managing incidents will need to identify who is responsible for informing these parties.

A desirable timeline should be included.

(m) Procedures to be observed in the case of non-medical imaging exposures

This should cover the local examples of what constitutes a ‘non-medical imaging exposure’ (examples are medico-legal exposure, occupational screening, or migration screening), their particular referral routes, and any specific requirements to allow justification of these exposures. Also, consider who is entitled to justify these exposures.

Include the action to be taken on the evaluation of these exposures, i.e. is there a care or treatment pathway that the individuals can access?

(n) Procedures to establish appropriate dose constraints and guidance for the exposure of carers and comforters

Carers and Comforters ‘knowingly and willingly’ incur an exposure in the support and comfort of individuals undergoing/having undergone an exposure. Consider:

- What procedures will preclude carers and comforters as a rule, e.g. it is acceptable to have support in general radiography and CT but not for radiotherapy?
- In what circumstances is someone a carer and comforter, or should they be considered a member of the public or an employee of another organisation, for example, a Nursing Home carer?
- What dose and risk are associated with the exposures of carers and comforters in standard scenarios? These risks can be used as the basis for justification. Calculation methods should be stated or referenced.
- Who has the responsibility to explain these risks, and in what format will it be delivered, e.g., a leaflet, a brief discussion, or as part of a more formal consent process?
- Who has/has not had the capacity to consent?

A Practitioner must justify the exposure of a carer and comforter separately to the patient. Consider who is entitled to justify these exposures, how they will be recorded, and how this will work in practice; e.g., when a radiographer realises a patient needs support, what is the process for seeking justification? A suitable solution could be that a radiographer or assistant practitioner is entitled to authorise standard exposure situations under authorisation guidelines. Non-standard situations should be individually justified by the IR (ME)R Practitioner (rather than under authorisation guidelines) and may need input from an MPE. Non-standard examples are:

- When is it not possible to exclude a pregnant individual or a child from being a carer or comforter? Ideally, they would be excluded, but exceptional situations may be justified individually by a Practitioner with a suitable dose constraint set.
- When a carer or comforter may be required, as an exception, during fluoroscopy procedures where the general anaesthetic is not used, in order to facilitate a paediatric patient, etc.

Guidance for the exposure of carers and comforters must be established to keep doses as low as reasonably practicable (ALARP) within the dose constraint.

- The procedure should require the pregnancy question to be asked and detail how this would be carried out. The risk to the foetus should be appropriately communicated to the mother.
- Appropriate protection should be provided. Consider what this protection might consist of, e.g. a 0.25 mm lead apron in radiology. Also, consider any risks associated with wearing a lead apron alongside the radiation risks.
- What actions must be taken in the case of radioactive patients to restrict others' exposure, e.g. minimise close contact time or reduce the risk of contamination?
- A record of the carer or comforter should be kept. The reason for this would be to allow audits to identify if individuals were repeatedly exposed in this way over time. It is not appropriate to document the carer/comforter's name in the patient's record on the Radiology Information System (RIS) (although you could say 'patient's father supported'), so a separate system will be required.

The regulations and guidance (DHSC 2018a) encourage a flexible approach to setting dose constraints for carers and comforters, although there must be evidence that they are not exceeded. This could be done through spot checks of individual cases of exposure that can be compared to the predicted doses. Routinely providing personal dosimetry to carers and comforters to measure the dose received in standard situations is unlikely to be necessary. For more complex procedures (e.g. nuclear medicine therapies), the dose constraint may need to be set on a case-by-case basis. Consideration should be given to whether personal dosimetry should be provided to measure the dose received. No matter what constraint is adopted, the exposure must also be optimised.

An example dose constraint statement that might be appropriate in a diagnostic radiology procedure is:

> *The organisation's appointed MPE(s) advise that it is not expected that a carer and comforter would receive more than* 0.05 *mSv per individual exposure in diagnostic radiology. If an individual is routinely acting as a carer and comforter*, e.g. *>3 times for CT and more than monthly in the*

last year for other modalities, please involve an MPE who will advise specifically on the suitability of continued use of this carer and comforter. The MPE will be able to advise on the likely dose involved, which will help with the justification process (the IR(ME)R Practitioner role) for future exposures, and the MPE will be able to help set a suitable dose constraint taking into account the flexible approach allowed in the regulations and DHSC guidance. It is anticipated that the MPE will aim to use a constraint of 1 *mSv per year for an individual carer and comforter; however, there is scope to increase this based on the benefits gained by the support offered.*

References

BIR, RCR, IPEM, SCoR and PHE 2020 *IR(ME)R Implications for Clinical Practice in Diagnostic Imaging, Interventional Radiology and Diagnostic Nuclear Medicine* https://www.rcr.ac.uk/our-services/all-our-publications/clinical-radiology-publications/ir-me-r-implications-for-clinical-practice-in-diagnostic-imaging-interventional-radiology-and-diagnostic-nuclear-medicine/ (accessed 12 April 2024)

CIB 2019a *New patient information posters on the benefits and risks of imaging RCR Website* (Clinical Imaging Board) https://rcr.ac.uk/posts/new-patient-information-posters-benefits-and-risks-imaging (accessed 5 December 2022)

CIB 2019b *Patient Identification: Guidance and Advice* ed N Strickland, S Web and M Tooley (UK: Clinical Imaging Board) https://www.sor.org/getmedia/113515a5-c1bc-4740-bd11-70c27fdd2166/cib-patient-identification-guidance-and-advice.pdf (accessed 12 April 2024)

CQC 2018 *Care Quality Commission IR(ME)R Annual Report 2017/18* (Newcastle upon Tyne: Care Quality Commission) https://cqc.org.uk/sites/default/files/20181115-IRMER-annual-report-2017-18-FINAL.pdf (accessed 31 January 2023)

CQC 2019 *Findings From CQC's IR(ME)R Inspection Programme of Specialist Paediatric Radiology Services* (Newcastle upon Tyne: Care Quality Commission) https://cqc.org.uk/sites/default/files/20190708_irmer_paediatric_radiology_inspection_programme_report.pdf (accessed 20 February 2023)

CQC 2023 *Ionising Radiation (Medical Exposure) Regulations (IR(ME)R). CQC Website. 19/4/23* (London: Care Quality Commission) https://cqc.org.uk/guidance-providers/ionising-radiation/ionising-radiation-medical-exposure-regulations-irmer (accessed 10 May 2023)

DHSC 2018a Guidance to the ionising radiation (medical exposure) regulations 2017 *Gov.UK Website* (Department of Health and Social Care) https://gov.uk/government/publications/ionising-radiation-medical-exposure-regulations-2017-guidance (accessed 31 January 2023)

DHSC 2018b *Medical Physics Experts Recognition Scheme* (Department of Health and Social Care) https://gov.uk/government/publications/medical-physics-experts-recognition-scheme (accessed 31 January 2023)

HPA, RCR and CoR 2009 *RCE-9 Protection of Pregnant Patients during Diagnostic Medical Exposures to Ionising Radiation* (Health Protection Agency, Royal College of Radiologists, and College of Radiographers) https://assets.publishing.service.gov.uk/media/5a7d9efa40f0b635051d048c/RCE-9_for_web.pdf (accessed 12 April 2024)

IR(ME)R 2017 *The Ionising Radiation (Medical Exposure) Regulations* www.legislation.gov.uk/uksi/2017/1322 (accessed 31 January 2023)

IR(ME)(A)R 2018 *The Ionising Radiation (Medical Exposure) (Amendment) Regulations* https://legislation.gov.uk/uksi/2018/121 (accessed 31 January 2023)

IR(ME)R(NI) 2018 *The Ionising Radiation (Medical Exposure) Regulations (Northern Ireland)* https://legislation.gov.uk/nisr/2018/17 (accessed 31 January 2023)

Public Health England 2019 *National Dose Reference Levels for Radiotherapy Planning CT Scans, GOV.UK Website* https://gov.uk/government/publications/diagnostic-radiology-national-diagnostic-reference-levels-ndrls/ndrl (accessed 18 January 2022)

Radiotherapy Board 2020 *Ionising Radiation (Medical Exposure) Regulations: Implications for Clinical Practice in Radiotherapy Guidance from the Radiotherapy Board* (London: Royal College of Radiologists) https://www.rcr.ac.uk/media/smmkkrsa/ionising-radiation-medical-exposure-regulations-implications-for-clinical-practice-in-radiotherapy.pdf (accessed 12 April 2024)

RPA2000 2023 *List of Certificate Holders, RPA2000 Website* http://rpa2000.org.uk/list-of-certificate-holders/ (accessed 31 August 2021)

SCoR 2018 *The Diagnostic Radiographer as the Entitled IR(ME)R Practitioner* (Society and College of Radiographers) https://www.sor.org/learning-advice/professional-body-guidance-and-publications/documents-and-publications/policy-guidance-document-library/the-diagnostic-radiographer-as-the-entitled-ir(me)(ME) (accessed 12 April 2024)

SCoR 2019 *Student Radiographers & Trainee Assistant Practitioners as 'Operators' under IR(ME)R 2017 (2018 in Northern Ireland)* (UK: Society and College of Radiographers) https://sor.org/learning-advice/professional-body-guidance-and-publications/documents-and-publications/policy-guidance-document-library/student-radiographers-trainee-assistant-practition (accessed 20 February 2023)

SCoR and FujiFilm 2020 *Pregnancy and Radiation Poster for Patients* (London: Society of Radiographers) https://sor.org/learning-advice/professional-body-guidance-and-publications/documents-and-publications/posters/pregnancy-and-radiation-poster-for-patients (accessed 10 March 2023)

SoR 2021 *Inclusive Pregnancy Status Guidelines for Ionising Radiation: Diagnostic and Therapeutic Exposures* (London: Society of Radiographers) https://sor.org/learning-advice/professional-body-guidance-and-publications/documents-and-publications/policy-guidance-document-library/inclusive-pregnancy-status-guidelines-for-ionising (accessed 8 February 2023)

WHO 2009 *Surgical Safety Checklist* (World Health Organization) https://who.int/teams/integrated-health-services/patient-safety/research/safe-surgery/tool-and-resources (accessed 18 February 2023)

Appendix 15

Role of the medical physics expert

The role of the Medical Physics Expert (MPE) is described in Regulation 14 (IR(ME)R 2017, IR(ME)R(NI) 2018). It can be broadly categorised into four areas: (1) equipment management; (2) patient dosimetry; (3) optimisation; and (4) radiation protection advice. The scope of practice of the MPE is determined by the clinical speciality they are affiliated with. This will vary from individual to individual as MPEs respond to the needs of their Employer and advances in clinical practice. For example, an individual may be considered an MPE in radiotherapy, or more specifically, photon beam radiotherapy; in nuclear medicine, or more specifically, PET imaging; in diagnostic radiology, or more specifically, mammography; in dental imaging as a whole; and so on.

The Department of Health and Social Care (DHSC) released a statement (DHSC 2018) on the means through which it recognises MPEs. DHSC has appointed RPA 2000 as the Assessing body that administers the recognition scheme (RPA2000 no date). Individuals wishing to be recognised as MPEs can apply to RPA2000 for assessment against the standards of competence for MPEs, outlined in the reference above and detailed in the syllabus (DHSC *et al* 2019). If successful, the individual is issued a certificate of competence and is added to the Medical Physics Expert Certificate Holders list (RPA2000 2023). In a hospital setting, certified MPEs are likely to be HCPC-registered clinical scientists with additional medical physics knowledge and skills relevant to the clinical speciality.

This appendix provides illustrations of what the role of the MPE might look like in the principal clinical specialities. It should be recognised that some of the duties and responsibilities identified here may not necessarily be the sole preserve of the MPE. However, some input, guidance, or advice from an MPE will be necessary on all medical exposure issues that might impinge on equipment management, radiation dose measurement, optimisation, or radiation safety.

Another useful reference is the EC guidelines on MPE (Evans *et al* 2014).

doi:10.1088/978-0-7503-2332-1ch35

1 Medical physics expert in diagnostic radiology

The MPE in diagnostic radiology will be actively involved in, or available for consultation on, all matters concerning diagnostic and interventional radiology exposures, including research applications. Their involvement will be especially warranted where doses are known to be high, e.g. computed tomography (CT) and interventional radiology, for the optimisation of doses for high-risk groups such as infants, for dose constraints in health screening, and for risk assessment in research proposals. The various matters are listed here.

1.1 Role in equipment management:
- (a) selection of equipment, including purchase and specification of radiology equipment;
- (b) commissioning of new equipment and communication with applications specialists;
- (c) Quality Assurance (QA), advice on Quality Control (QC), communication and review;
- (d) image quality evaluation and outcome performance indicators;
- (e) communication with other employees and with maintenance engineers on practical aspects;
- (f) advice on suspension of existing equipment;
- (g) imaging equipment replacement policy review.

1.2 Role in dosimetry:
- (a) systems for dose calibration and quantification;
- (b) patient dose monitoring programme and review, including dose optimisation strategy;
- (c) establishment and review of diagnostic reference levels;
- (d) dose and risk assessment for research applications;
- (e) dose constraints in research;
- (f) patient dose assessment for radiation incidents.

1.3 Role in optimisation:
- (a) communication with the Practitioner and Operators;
- (b) generic risk assessment, especially for new equipment or techniques;
- (c) audit and development of audits for medical exposures;
- (d) role in multidisciplinary medical audit and review;
- (e) operator functions as identified for the MPE in the Employer's procedures.

1.4 Role in radiation protection matters concerning medical exposures:
- (a) advice on the implementation of the Employer's procedures;
- (b) training;
- (c) radiation protection of comforters and carers;
- (d) investigation of radiation incidents;
- (e) communication role during an inspection by the relevant statutory authorities;
- (f) communication with other Employers;
- (g) clarifying overlaps with other radiation protection and exposure legislation.

2 Medical physics expert in nuclear medicine

The MPE in nuclear medicine will be actively involved in, or available for consultation on, all matters concerning diagnostic nuclear medicine exposures, including research applications and standard nuclear medicine therapies. It is essential that the MPE be closely involved in all non-standard nuclear medicine therapy administrations and be available for the following activities, either personally or by directing local medical physics support. A joint working group has published recommendations for minimum adequate safe levels of MPE and broader medical physics support for nuclear medicine (IPEM *et al* 2021).

2.1 Role in equipment management:

- (a) selection of equipment to include purchase and specification of all relevant nuclear medicine equipment;
- (b) commissioning and acceptance testing of new equipment and communication with applications specialists, where applicable;
- (c) continuous monitoring of all QA aspects of the service, including calibration of radionuclide dose calibrators;
- (d) equipment management and determination of QC procedures on equipment;
- (e) advice on the suitability of equipment for specific tasks (e.g. image quality evaluation) and the need for maintenance, repair, suspension or replacement;
- (f) communication with other Employees and with maintenance engineers on practical aspects;
- (g) involvement in the provision and assessment of dedicated nuclear medicine software.

2.2 Role in patient dosimetry:

- (a) systems for dose calibration and quantification;
- (b) advice on dosimetry for radionuclide therapy (also see items 3.3 (d) and (e));
- (c) patient dose monitoring programme and review, including dose optimisation strategy;
- (d) establishment and review of diagnostic reference levels;
- (e) advice regarding dosimetry for pregnant and breastfeeding patients;
- (f) dose and risk assessment for research applications;
- (g) dose constraints for research exposures.

2.3 Role in optimisation:

- (a) communication with the Practitioner and Operators;
- (b) agreement on protocols for all therapeutic administrations;
- (c) advice on and involvement in all scientific and technical aspects of diagnostic procedures;
- (d) introduction and validation of new procedures and protocols;
- (e) Operator functions as identified for the MPE in the Employer's procedures;
- (f) involvement in multidisciplinary clinical and external audits (e.g. by regulators).

2.4 Other radiation protection matters concerning medical exposures:
- (a) advice on the implementation of the Employer's procedures;
- (b) training;
- (c) radiation protection of comforters and carers to include the application of dose constraints;
- (d) advice to patients leaving the hospital and to their comforters and carers;
- (e) investigation of radiation incidents;
- (f) presence at all therapeutic administrations that are non-standard or being undertaken for the first time;
- (g) involvement in all applications for Employer licences under IR(ME)R;
- (h) communication role during an inspection by the relevant statutory authorities;
- (i) communication with other Employers;
- (j) clarifying overlaps with other radiation protection and exposure legislation;
- (k) involvement in the design of all nuclear medicine facilities in collaboration with the Radiation Protection Adviser (RPA);
- (l) liaise with Radiation Waste Adviser (RWA) and Dangerous Goods Safety Adviser (DGSA) as appropriate.

3 Medical physics expert in radiotherapy

The MPE in radiotherapy must be closely involved in and have responsibility, where indicated, for the following.

3.1 Role in equipment management:
- (a) specification and selection of all radiotherapy and brachytherapy equipment;
- (b) oversight of the installation and responsibility for all acceptance testing and commissioning;
- (c) responsibility for the QC programme;
- (d) responsibility for technical management of clinical computer software and computer systems;
- (e) decommissioning of radiotherapy or brachytherapy equipment, in cooperation with the RPA and RWA as appropriate.

3.2 Role in patient dosimetry:
- (a) responsibility for the definitive calibration of radiotherapy equipment and dosemeters (see appendix A of IPEM Report 81, Physics Aspects of Quality Control in Radiotherapy (IPEM 2018));
- (b) responsibility for all aspects of calibration and testing of secondary standard dosemeters;
- (c) consultation on, and responsibility for, the suitability and accuracy of the methods used to calculate dose distributions in radiotherapy and in brachytherapy procedures, and in particular the optimisation of complex treatment plans;
- (d) responsibility for the patient dosimetry programme;

(e) responsibility for other routine dosimetry programmes;
(f) investigation of any radiation incidents, including dose reconstruction;
(g) dose and risk assessment for research applications involving radiotherapy or brachytherapy;
(h) target doses for research exposures involving radiotherapy or brachytherapy.

3.3 Role in optimisation:
(a) communication with the Practitioner and Operators;
(b) consultation on the suitability of treatment techniques, with involvement at a level commensurate with the responsibility for the dosimetry and accuracy of treatment;
(c) consultation on all strategic planning issues which involve possible changes to the radiotherapy and brachytherapy service, including decisions on therapy equipment, treatment modalities and techniques, and changes to the design and layout of the building where radiation protection considerations may apply;
(d) available to give advice on the dosimetry of radionuclide therapy and, in cooperation with the RPA, on the radiation safety of these procedures;
(e) being present at all radionuclide therapy administrations not covered by a protocol (see also items 2.2(b) and 2.4(f)).

3.4 Other radiation protection matters concerning medical exposures:
(a) investigation of radiation incidents;
(b) advice on the implementation of the Employer's procedures;
(c) training;
(d) involvement in all applications for Employer licences under IR(ME)R for brachytherapy;
(e) communication role during inspections by the relevant statutory authorities;
(f) communication with other Employers;
(g) clarifying overlaps with other radiation protection and exposure legislation;
(h) cooperation with the RPA over treatment room design;
(i) advice and involvement as appropriate, in cooperation with the RPA, regarding all aspects of the safety of radiotherapy and brachytherapy equipment;
(j) collaboration with the RPA concerning the critical examination of all new and upgraded equipment.

References

DHSC 2018 *Medical Physics Experts Recognition Scheme* (Department of Health and Social Care) https://gov.uk/government/publications/medical-physics-experts-recognition-scheme (accessed 31 January 2023)

DHSC, SGHSCD and DHNI 2019 *Syllabus for Medical Physics Experts* (Department of Health & Social Care, Department of Health Northern Ireland, RPA2000, Scottish Government Health & Social Care Directorates) http://rpa2000.org.uk/wp-content/uploads/2019/05/DHSC-Syllabus-for-MPE.pdf (accessed 4 March 2023)

Evans S, Guerra A, Malone J and Bunton R 2014 *European Commission Radiation Protection N° 174 European Guidelines on Medical Physics Expert Annex* 2 *Medical Physics Expert Staffing Levels in Europe* https://efomp.org/uploads/rp_174_full.pdf (accessed 6 February 2023)

IPEM 2018 *Report 81: Physics Aspects of Quality Control in Radiotherapy. IPEM Report 81* ed I Patel, S Weston, L A Palmer, W P M Mayles, P Whittard, R Clements, A Reilly and T J Jordan (York: Institute of Physics and Engineering in Medicine) 2nd edn

IPEM, ARSAC, BNMS and BIR 2021 *IPEM Policy Statement Medical Physics Expert Support for Nuclear Medicine* ed F McKiddie, A Fletcher, C Kalirai, D McGowan, L Fraser, N Parkar, K Adamson and P Julyan. (York: Institute of Physics and Engineering in Medicine, Administration of Radioactive Substances Advisory Committee, British Nuclear Medicine Society, British Institute of Radiology) https://ipem.ac.uk/media/lbblkxyn/mpe-support-for-nuclear-medicine.pdf (accessed 4 March 2023)

IR(ME)R 2017 *The Ionising Radiation (Medical Exposure) Regulations* www.legislation.gov.uk/uksi/2017/1322 (accessed 31 January 2023)

IR(ME)R(NI) 2018 *The Ionising Radiation (Medical Exposure) Regulations (Northern Ireland)* https://legislation.gov.uk/nisr/2018/17 (accessed 31 January 2023)

RPA2000 2023 *List of Certificate Holders, RPA2000 Website* http://rpa2000.org.uk/list-of-certificate-holders/ (accessed 31 August 2021)

RPA2000 no date *RPA2000 Certifying Competence in Radiation Professional Professionals, RPA2000 Website* http://rpa2000.org.uk/ (accessed 4 March 2023)

IOP Publishing

Medical and Dental Guidance Notes (Second Edition)
A good practice guide on all aspects of ionising radiation protection in the clinical environment:
IPEM Report 113
John Saunderson, Mohamed Metwaly, William Mairs, Philip Mayles, Lisa Rowley and Mark Worrall

Appendix 16

Risk communication principles

The communication of radiation benefits and risks to a patient need not be difficult. There are some general principles of communication, psychology, and even behavioural economics that can help. Specific guidance is also available from the Society and College of Radiographers (SCoR 2019) and the Clinical Imaging Board (CIB 2019).

Preparing for communication

Possibly the most important element of communication is trust. Healthcare professionals have an inherent advantage in this regard (Skinner and Clemence 2017), but it can easily be lost if the information they provide is seen as unreliable.

Make sure that communications are consistent. A standard template or set of statements can be useful in this respect. Staff should know what benefit and risk information is provided to patients by their organisation, whether this is through posters in radiology waiting areas, leaflets (see appendix 17, *Basic x-ray benefit and risk information*, for example), or invitation letters.

A set of common questions about radiation exposure and their answers is also a good resource to have. This could be a simple list or a more extensive 'message map' (Covello 2002). It can include examples related to local practice such as:

- Why do you need to ask if I am pregnant? You didn't ask last time.
    - *We only ask about pregnancy for certain types of x-ray procedures. If there is no way that the baby could be harmed by the x-rays, for example, because the x-rays are far away from the baby, then there is no risk, so we don't need to ask.*
- Why are you not giving me a lead apron/shield to wear?
    - *Sometimes a lead shield will have no direct benefit and may, in some cases, prevent us from seeing everything we need to see. So not wearing a lead apron/shield may prevent having to repeat an x-ray scan.*

Writing about risk

Zero is not like any other number. Any change from a baseline of zero is given more emphasis than a change from a non-zero baseline, e.g. the difference between zero and ten is perceived as bigger than the difference between one and fifteen (Shampanier *et al* 2007). In this context, an explanation of ubiquitous background radiation can be reassuring to patients who may otherwise believe that their only radiation exposure is their medical exposure. In written material, the use of graphical decision aids, such as an image matrix, to show additional risks on top of the baseline risks, e.g. for cataracts or cancer, can also help in this respect (NHS National Prescibing Centre 2011).

Speaking with patients worried about their exposure

If you are having a conversation with a patient who is worried about radiation exposure that they have already had, crisis communication techniques may help. Although these are generally aimed at large-scale events, they are focussed on compassionate and concise communication, which can also work on an individual level. Some of the communication models include:

- **CCO (Compassion, Conviction, Optimism)**
  Show that you care before you show what you know; reassure the patient and provide a positive way to move ahead. The way forward might be to ask a question, such as, '*Is there something you would like me to do?*' This also engages the patient in the conversation (Covello and Milligan 2010).
- **27/9/3 (27 words, 9 seconds, 3 messages), Primacy/Recency, 1N = 3P**
  When people are upset or worried, they won't be able to take in lots of information. Keep your statements short. If you are going to explain something, put the detail in the middle and make the main points the first (primacy) and last (recency) things you say (Milligan and Covello 2012). Negative statements have three times the weight of positive ones, so don't dwell on negligible risks or harms. (Covello and Milligan 2010, Milligan and Covello 2012, Porter 2014).

Whichever approach seems more appropriate, you should always make sure that you are addressing the concerns of the patient or their relative. Do not assume a level of understanding of radiation risk. Ask them what they are worried about. The '*explore before you explain*' approach used in teaching can also work here.

Never say that you understand how someone feels; instead, acknowledge and summarise what they are saying. For example, '*I can see that…*', '*It sounds like…*'. You can name the feeling and check with the person to see if it is right.

References

CIB 2019 New patient information posters on the benefits and risks of imaging *RCR Website* (Clinical Imaging Board) rcr.ac.uk/posts/new-patient-information-posters-benefits-and-risks-imaging (accessed 5 December 2022)

Covello V T 2002 Message mapping, risk and crisis communication *Invited Paper Presented at the World Health Organization Conf. on Bio-terrorism and Risk Communication (Geneva)* http://rcfp.pbworks.com/f/MessageMapping.pdf (accessed 4 March 2023)

Covello V T and Milligan P A 2010 Radiological risk and emergency communications (Health Physics Society Mid Year Meeting, Albuquerque, NM) *nrc.gov website* (New York: Nuclear Regulatory Commission) https://www.nrc.gov/docs/ML1002/ML100290788.pdf (accessed 12 April 2024)

Milligan P A and Covello V T 2012 Radiological Risk Communication—Message Mapping for Effective Radiological Risk Communications for Nuclear Power Plant Incidents (IRPA 13) *nrc.gov website* (Washington DC: Nuclear Regulatory Commission) https://www.irpa.net/members/1815%20tue%20dochart%20milligan%20TS4b.6.pdf (accessed 12 April 2024)

NHS National Prescribing Centre 2011 *Patient Decision Aids—Directory* (UK Web Archive)

Porter J 2014 https://jrmyprtr.com/27-9-3-message-grid/ Message grid: 27/9/3–27 words, nine seconds, three messages, Jeremy Porter Communication Website (accessed 2 August 2023)

SCoR 2019 *Communicating Radiation Benefit and Risk Information to Individuals Under the Ionising Radiation (Medical Exposure) Regulations (IR(ME)R)* sor.org/learning-advice/professional-body-guidance-and-publications/documents-and-publications/archive-documents/communicating-radiation-benefit-and-risk-infor-(1) (accessed 4 March 2023)

Shampanier K, Mazar N and Ariely D 2007 Zero as a special price: the true value of free products *Market. Sci.* **26** 742–57

Skinner G and Clemence M 2017 *Politicians Remain the Least Trusted Profession in Britain* (IPSOS) ipsos.com/ipsos-mori/en-uk/politicians-remain-least-trusted-profession-britain (accessed 4 March 2023)

Appendix 17

Basic x-ray benefit and risk information

This appendix provides an example of information used in an x-ray department patient information leaflet. See also guidance from the Society and College of Radiographers (SCoR 2019) on benefit and risk communication and the Society and College of Radiographers (SCoR 2019) information posters.

X-rays and other types of radiation

X-rays are a type of radiation. Radiation can also come from radioactive substances. There are lots of natural sources of radiation in the world; for example, certain types of rocks, water, and some of our food, including bananas, nuts, and potatoes. Radiation also comes from space, but usually, the atmosphere protects us, so this radiation mostly affects people who live in high mountains or travel in airplanes.

When we want to compare radiation doses, we use the quantity millisievert (abbreviated as mSv). In the UK, the average amount of radiation that people are exposed to each year is a little over 2 mSv. That means that over a lifetime, people could receive as much as 200 mSv of radiation exposure. The exact amount depends on where you live, what you do, and how far you travel by airplane.

Why and how do we use x-rays?

X-rays can be the best way to find out what is wrong with someone or whether a condition has changed. If your healthcare professional (such as a doctor) is sending you for an x-ray or scan, you can ask:

- what information do you hope to get from it?
- is it really necessary?
- is there an alternative that doesn't use radiation that might be suitable instead?

You should mention if you have had any other x-rays or scans recently. The information from these might be useful too.

doi:10.1088/978-0-7503-2332-1ch37

Different types of machines use x-rays in different ways.

- Radiography, which used to be called 'plain film', is the most common. This is what is used to look for broken bones—in your chest and lungs or your teeth. Radiographs are very quick—the x-rays are turned on for less than 1 s.
- Fluoroscopy, or screening, uses x-rays to produce moving pictures. These pictures might be used to guide a catheter through your blood vessels or look at the way special dye moves through your body. The screening takes between a few minutes and a few hours, depending on how complicated the procedure is.
- CT scans use x-rays to provide very detailed pictures of the inside of the body. Slice-by-slice analysis or 3-D picture reconstruction are both options for these. Sometimes injections or drinks are used to make the pictures clearer. CT is quite quick. The x-rays are on for less than a minute, but you may need to spend time preparing before your scan.

X-rays and the law

Medical exposure to radiation is strictly controlled. In the UK, the Ionising Radiation (Medical Exposure) Regulations, also known as IR(ME)R, set out the requirements for diagnosis and treatment. Each request for an x-ray or scan is assessed by a registered healthcare professional with expertise in radiation imaging who makes sure that it is the most appropriate test for you. They may recommend a different test if they think it would be more helpful. Whatever decision is made, your x-ray or scan will be done with the least amount of radiation that is necessary to produce the correct images.

There are five steps in the process:

- Request/Referral: Your doctor discusses your symptoms with you and decides to ask for an x-ray or scan to help with diagnosing the problem.
- Justification: A registered healthcare professional with an understanding of radiation benefits and risks (a radiologist or a radiographer) looks at the information that your referrer has sent and decides whether that is the best way to diagnose your condition.
- X-ray: A radiologist or radiographer takes the x-ray.
- Evaluation: A radiologist or radiographer studies the x-ray images. They look for anything that shouldn't be there or any changes that might be the cause of your symptoms.
- Report: The radiologist or radiographer writes a report about your x-rays. This is then sent to your referrer.

Who are the people involved?

Referrer: This is your doctor (or sometimes a nurse, physiotherapist, or other healthcare professional) who has asked for the x-ray.

Radiologist: A doctor who specialises in x-rays and other types of medical images.

Radiographer: A healthcare professional who specialises in taking x-rays and other types of medical images.

Radiation Protection Adviser and Medical Physics Expert: A scientist who advises the hospital on the safest ways to use x-ray equipment.

You: It is important that you tell your healthcare professional everything about your condition. Remind them if you have had a similar x-ray or scan in the past. Let the staff member who is performing the x-ray know if you are or could be pregnant. Feel free to talk to any of the healthcare professionals involved, especially if something is happening that you were not expecting. For example, if you thought the x-ray was going to be on your left foot, but you have been asked to remove your right shoe.

Why do we have to think about x-ray safety?

Some x-ray radiation passes through the body to the detector. That's how the x-ray pictures are made. Other radiation can be absorbed in the tissues exposed, and sometimes this causes changes to the cells in our body. In almost all cases, these changes are repaired. Very rarely, the repair doesn't work properly, and the cell starts to misbehave. Over a long period of time (many years, or even decades), this misbehaving cell can lead to cancer. It is completely random. Although the risks are very small, we have no way of knowing when a repair might go wrong, so to reduce the chances even further, we only use x-rays when they are necessary and use the lowest amount of radiation dose possible to get the most benefit from the exam or procedure.

Because the changes caused by radiation take so long to show up, older people have an even smaller risk than younger people. That is why radiographers will ask patients if they might be pregnant and why extra care is taken when using x-rays on children. Please alert your radiographer if you think you are or might be pregnant.

References

SCoR 2019 *Communicating Radiation Benefit and Risk Information to Individuals Under the Ionising Radiation (Medical Exposure) Regulations (IR(ME)R)* https://sor.org/learning-advice/professional-body-guidance-and-publications/documents-and-publications/archive-documents/communicating-radiation-benefit-and-risk-infor-(1) (accessed 4 March 2023)

IOP Publishing

Medical and Dental Guidance Notes (Second Edition)
A good practice guide on all aspects of ionising radiation protection in the clinical environment: IPEM Report 113
John Saunderson, Mohamed Metwaly, William Mairs, Philip Mayles, Lisa Rowley and Mark Worrall

Appendix 18

Example IR(ME)R audit: procedures and document control

Answer the questions below and compare the outcomes to the standards expected (local decision).

Employer's procedures	Is there a procedure? Y/N or N/A	Date procedure last reviewed	Date of next review	Have staff read the procedure
Correct identification of the individual to be exposed				
Identity of individuals entitled to act as Referrer, Practitioner or Operator				
Enquiries of individuals of childbearing potential to establish whether the individual is or may be pregnant or breastfeeding				
Quality assurance programme in respect of written procedures, written protocols, and equipment is followed				

(*Continued*)

doi:10.1088/978-0-7503-2332-1ch38

(*Continued*)

Employer's procedures	Is there a procedure? Y/N or N/A	Date procedure last reviewed	Date of next review	Have staff read the procedure
Assessment of patient dose and administered activity				
Establishment, use of and review of Diagnostic Reference Levels				
For research programmes, determining whether the Practitioner or Operator is required to effect one or more of the matters (voluntary participation, risk of exposure, dose constraint, etc).				
Provision of information and written instructions for patients undergoing treatment/diagnosis with radioactive substances (consent where an individual lacks capacity, restriction of exposure, risks)				
Provision of adequate information before the exposure, relating to the benefits and risks associated with the radiation dose from the exposure				
Carrying out and recording of an evaluation for each exposure, including, where appropriate, factors relevant to patient dose				
The probability and magnitude of accidental or unintended exposure are reduced so far as reasonably practicable				
Ensure that the Referrer, the Practitioner, and the individual exposed or their				

representative are informed of the occurrence of any relevant clinically significant accidental or unintended exposure, and of the outcome of the analysis of this exposure	
Non-medical imaging exposures	
Establish appropriate dose constraints and guidance for the exposure of carers and comforters	
Provision for the carrying out of clinical audits as appropriate	

This is an example audit that can be expanded locally to cover whatever scope is desired. It can include Employer's procedures beyond those specified in IR(ME)R Schedule 2 and review other necessary documents, such as exposure protocols, for example. The audit should ensure the various documents are in place and check that staff are following what is expected; apply that principle when using this example audit. Learning should be fed back to the wider team.

Document control process and appropriate sign-off by, or on behalf of, 'the Employer'

Are the Employer's procedures part of a Quality Management System? Yes ☐ No ☐

How are the procedures managed/controlled and is this process documented?

....

....

....

....

....

....

....

What is the path of traceability to the Employer (e.g. to Chief Executive, Managing Director, or Board)? ....

....

....

....

....

....

How are changes communicated to duty holders/staff?
……………………………………………………………………………………
……………………………………………………………………………………
……………………………………………………………………………………
……………………………………………………………………………………
……………………………………………………………………………………
……………………………………………………………………………………
How are changes implemented?
……………………………………………………………………………………
……………………………………………………………………………………
……………………………………………………………………………………
……………………………………………………………………………………
……………………………………………………………………………………
……………………………………………………………………………………

Review of requests

How are requests received:	Electronic	☐
	Paper	☐
How are patients identified?		
Name	☐	
Gender	☐	
Address	☐	
Date of Birth	☐	
Hospital identification number	☐	

Other……………………………………………………………………………
……………………………………………………………………………………
……………………………………………………………………………………
Number of requests audited (minimum 20): …………………
How are Referrers identified?
……………………………………………………………………………………
……………………………………………………………………………………
……………………………………………………………………………………
……………………………………………………………………………………
……………………………………………………………………………………
……………………………………………………………………………………
Are requests accepted from non-medical Referrers? Yes ☐ No ☐
From which staff groups:
……………………………………………………………………………………
……………………………………………………………………………………
……………………………………………………………………………………

....

....

Are all the Referrers in the sample entitled to this task by the Employer?

....

....

....

....

....

....

Do requests contain the information required? Yes ☐ No ☐

....

....

....

....

....

....

Are the referrals appropriate? Yes ☐ No ☐

....

....

....

....

....

....

Do the requests contain suitable information regarding breastfeeding/pregnancy status?

Yes ☐ No ☐

....

....

....

....

....

....

Justification and authorisation

How is justification performed?

....

....

....

....

....

....

Who can perform justification via protocol i.e. Authorisation?

....

....

....
....
....
....
Are there suitable protocols in place? Yes ☐ No ☐
Are they up to date? Yes ☐ No ☐
Audit requests against protocol (minimum audit 20). No:
....
Name of procedure/protocol:
....
Is there a suitable authorisation signature? Yes ☐ No ☐
Has the individual performing the justification by protocol been suitably entitled?
Yes ☐ No ☐
What training is in place for IR(ME)R Operators undertaking Authorisation?
....
....
....
....
....
....

Duty holders

Is there appropriate (and current) entitlement documentation? Yes ☐ No ☐
How are duty holders identified, e.g. by name or role?
....
....
....
....
....
....
Is there a description of the duties to be undertaken by the duty holders?
Yes ☐ No ☐
....
....
....
....
....
....
....
....
....
Are individual scopes of practice reviewed to ensure they are still appropriate?
....
....

....
....
....
....

Follow the patient pathway from start to finish for a number of patients, checking 'sign-off' of tasks against entitlement. Document any irregularities:

....
....
....
....
....
....

Training

Ask to see a staff member's training records.
How are the training records kept (e.g. in a folder or on a document management system such as Q-Pulse, etc)? ....

....
....
....
....
....

Have all staff received basic training in legislation? Yes ☐ No ☐

....
....
....
....
....

Have the staff been trained to perform their duties appropriately? Yes ☐ No ☐

....
....
....
....
....

Have staff members been signed off as competent to perform these duties? Yes ☐ No ☐

....
....
....
....
....
....

How are staff members signed off as competent (e.g. observational for a given number, passing a test, etc)?

....

....

....

....

How frequently is refresher training provided?

....

....

....

....

How is it flagged if a staff member requires training or re-training?

....

....

....

....

....

Doses and optimisation

How are patient radiation doses recorded (e.g. monitor units, DAP, administered activity, etc)?

....

....

....

....

....

....

What dose audits are performed to ensure that national DRLs are met for diagnostic procedures (ARSAC levels for Nuclear Medicine)?

....

....

....

....

Is there DRL review for interventional radiological procedures?

....

How frequently is a dose audit performed?

....

How have local DRLs been established?

....

....

....

Are up to date DRLs, ARSAC figures, or research doses displayed in the department?..........................

....

....
....
For radiotherapy and nuclear medicine therapeutic doses, does the given dose match the prescribed dose? Are the doses delivered within a set tolerance given by the prescriber or by an agreed protocol with the Practitioner?
....
....
....
Has a set of images been reviewed within the past 12 months to identify areas where practice could improve?
....
....
....
What optimisation work has been performed in the past 12 months?
....
....
....
What is the process for managing the dose for carers and comforters?
....
....
....

Patient information

How are patients provided with information regarding the examination/therapy (e.g. patient information leaflet)?
....
....
....
....
How are patients informed that they will be exposed to ionising radiation?
....
....
....
....
How are patients informed of the benefit versus the risk of the exposure?
....
How are patients provided with information regarding pregnancy/breastfeeding?
....
....
....
....
How are patients administered radioactive substances provided with radiation precautions?......
....

....
....
....
Are the precautions provided suitable? Yes ☐ No ☐
....
....
....
....

Incidents

How many accidental or unintended exposures have there been in the past 12 months? Have there been any themes identified?
....
....
....
....
Is a system in place to identify themes?
....
....
....
....
How and where have these been recorded?
....
....
....
....
How many have been externally reportable?
....
....
....
....
....
What learning outcomes have there been, and what changes have been implemented? ....
....
....
....
....
How and when is information fed back to the relevant parties i.e. Referrer, Practitioner, Operator, patient?
....
....
....
....

Have there been any recorded near misses? Yes ☐ No ☐
What is the system for recording near misses?
....
....
....
....
....
What system is in place to allow learning from near misses?
....
....
....
....
What is the most common type of incident/near miss?
....
....
....
....
....
Have incidents and near misses been used to inform a study of the risk of accidental or unintended exposures?
....
....
....
....

Clinical evaluation

How many unreported exams/procedures are there?
....
What is the average length of time it takes to complete a report?
....
Do radiographers and clinical technologists perform reporting? Yes ☐ No ☐
Do any other groups perform reporting? Yes ☐ No ☐
....
....
....
....
What training and support is in place for non-radiologists performing reporting duties?
....
....
....
....

Clinical audit

Is there recent evidence of clinical audit within the IR(ME)R definition? ....

IOP Publishing

Medical and Dental Guidance Notes (Second Edition)

A good practice guide on all aspects of ionising radiation protection in the clinical environment: IPEM Report 113

John Saunderson, Mohamed Metwaly, William Mairs, Philip Mayles, Lisa Rowley and Mark Worrall

Appendix 19

Guidance on medical research exposures

1 Regulatory aspects

1.1 Employers who expose human subjects to ionising radiation for the purpose of medical research have statutory duties under IR(ME)R (IR(ME)R 2017, IR (ME)R(NI) 2018).

1.2 Before a medical research exposure can take place, the Employer needs to ensure that the following conditions have been met:

(a) The research study has been approved by a Research Ethics Committee (REC).

(b) In the case of the administration of radioactive substances, the research study has been approved by the Administration of Radioactive Substances Advisory Committee (ARSAC). Note that the IR(ME)R Practitioner and Employer need a licence to administer radioactive substances (see chapter 2, paragraphs 2.64–2.70).

(c) The research exposure has been justified by an IR(ME)R Practitioner entitled by the Employer.

(d) Individuals participate voluntarily in the research study; IR(ME)R 17 Regulation 12(4)(a).

(e) The participant has been informed in advance about the benefits and risks of the exposure.

(f) Dose constraints have been set and will be used appropriately for participants for whom no direct medical benefit is expected.

(g) Individual target dose levels have been planned for participants who undergo experimental diagnostic or therapeutic exposures from which they are expected to derive a benefit.

doi:10.1088/978-0-7503-2332-1ch39

1.3 The Employer is required to have a procedure for dealing with medical research exposures. The typical content of such a procedure would include processes in place for checking that the conditions above have been met; which persons are responsible for undertaking the checks; how research exposures will be justified and by whom; how dose constraints or dose targets are to be set and by whom; and how dose constraints or dose targets are to be used.

2 Research approval

2.1 In England and Wales, the Health Research Authority (HRA) manages the various approvals required for research studies taking place in the NHS, including REC review. The process is called HRA approval or Health Care Research Wales (HCRW) approval. The HRA also works closely with other countries in the UK to provide a UK-wide system. Further information is available on the HRA website (HRA, no date).

2.2 Applications for REC review and other regulatory approvals are made through the Integrated Research Approval System (IRAS) (HRA 2023) by the research sponsor in conjunction with the Chief Investigator (CI) at the lead site on behalf of all the participating sites. This means that the Employer undertaking the research exposure relies on the research sponsor to obtain a sound ethical opinion.

2.3 Part B section 3 of the online IRAS application form deals with exposures to ionising radiation. This section contains declarations about the use of radioactive materials and exposures to other types of radiation, a radiation dose and risk statement from the lead Medical Physics Expert (MPE), and a clinical risk/benefit analysis from the lead Clinical Radiation Expert (CRE). Completion of Part B—section 3 in the IRAS application form is commonly referred to as a lead IR(ME)R review (see section 3 below). When the IRAS application form is submitted, the IRAS system automatically generates the Preliminary Research Assessment (PRA) form, which is submitted to ARSAC in advance of any site-specific applications for ARSAC research licences.

2.4 Guidance for applicants on the use of ionising radiation in research is available through a variety of means. The HRA provides extensive information, including regulatory requirements, an eLearning module, and help on how to complete the IRAS form, on the HRA websites (HRA 2018, 2021, 2022).

2.5 The Radiation Assurance Process is a UK-wide process to clarify the information regarding radiation exposures in study documentation at an early stage in the regulatory approval pathway (prior to REC or ARSAC review). Further information, including generic risk statements to aid those completing documentation, can be found on their website (HRA *et al* 2020a, 2020b).

2.6 The REC reviews the study documentation in order to satisfy itself that the benefits of carrying out the research outweigh the risks to the participants and that the latter are being adequately informed about the risks. Note that the

Employer carrying out the research exposures depends on the research sponsor to provide a Patient Information Sheet (PIS) with adequate information about radiation risks.

2.7 Research sponsors expect study protocols to undergo several amendments during the duration of the study. Amendments can be substantial or non-substantial and may require formal approval by the original review bodies. If the changes affect the research exposures, the relevant sections of the IRAS form need to be revised and submitted with the amendment application.

2.8 Each participating site has local research approval processes. These focus on capacity and capability for undertaking the research study. However, as the legal responsibility for research exposures remains with the Employer carrying out the exposures, the latter must put in place a process for ensuring that IR(ME)R can be complied with whilst taking part in the research study. This process is commonly referred to as local IR(ME)R review (see section 4 below).

3 Lead IR(ME)R reviews

3.1 The lead Medical Physics Expert (MPE) and Clinical Radiation Expert (CRE) carry out a review of the study protocol, PIS, and other associated documentation in order to assess the radiation exposures within the study. Where the research study involves different types of exposures to ionising radiation, e.g. CT scanning and nuclear medicine imaging, the lead MPE and CRE may work with other colleagues so that collectively they have the appropriate expertise to provide the lead MPE and CRE statements. Guidance for MPEs and CREs on how to conduct reviews can be found on the HRA website (HRA 2020b).

3.2 In general, any exposure to ionising radiation required by the research study protocol, irrespective of whether it takes place as part of routine clinical care or not, is a research exposure.

3.3 The lead CRE and MPE establish, together with the CI and their research team, the number and nature of research exposures and how they map onto the standard of care. The lead MPE provides typical radiation doses for each type of research exposure. Note that these are not dose constraints. These pieces of information are captured in sub-sections A (radioactive materials) and B (other radiation) of Part B, section 3.

3.4 The lead MPE provides a statement about the total protocol dose and associated radiation risks and describes how the dose and risks compare to routine clinical care. The lead CRE provides a clinical evaluation of these risks against the benefits of the knowledge gained by carrying out the research.

3.5 The lead MPE and CRE review the PIS to confirm that adequate information on radiation risks is provided to the participants.

3.6 The HRA has published generic lead MPE and CRE statements for Part B section 3 and radiation risk statements for the PIS, which cover typical scenarios and can therefore be used for the majority of cases. They can be found on the HRA website (HRA 2020a)

4 Local IR(ME)R reviews

4.1 Research approval in the UK is centralised. This means that some aspects of the approval process, such as ethical approval and approval of the radiation risk information given to the participants, are outside the control of the Employer.

4.2 By carrying out a local IR(ME)R review, the Employer can assure themselves that they can meet the requirements of IR(ME)R for research exposures while the research study takes place. The Employer should identify who carries out local IR(ME)R reviews in the IR(ME)R research procedure.

4.3 In the first instance, the reviewers should confirm that ethical approval has been granted. They should then satisfy themselves that the ethical opinion is sound. Examples of scenarios where ethical approval may be unsound are:

(1) research exposures have been omitted from the IRAS application so that the REC has not been informed about the associated risks and therefore has not been able to consider them in its decision;

(2) the radiation risks associated with the research exposures have been incorrectly calculated.

Local MPE reviewers should be aware that procedure doses and, therefore, the total protocol dose will often differ significantly from local practice. This is to be expected, particularly where values are based on national DRLs or local practice at the lead site. The organisation's IR(ME)R research procedure should define how local dose constraints and dose targets are set based on these values to ensure optimisation in practice.

4.4 Justification of the research exposures on an individual level must be performed by an IR(ME)R Practitioner entitled by the Employer. The local reviewer should determine how the research exposures will be justified locally, by whom, and under which criteria. A possible approach is that the local CRE agrees to act as the IR(ME)R Practitioner for the research exposures, using the justification provided by the lead CRE in the IRAS form to support their decision for each participant.

The local reviewers should check that a mechanism is in place for obtaining participant consent.

4.6 The information on radiation risks provided to the participant in the PIS is outside the control of the Employer. The local reviewers should check that the information in the PIS is adequate and contact the sponsor if there are any concerns. Common problems are: the radiation risk information is given in terms of radiation dose (IR(ME)R 2017 requires information about risks); there is no information provided about the radiation risks.

4.7 For all research, either a dose constraint or a target dose level should be used. As part of the local review, it should be determined whether a dose constraint or target levels of dose are most appropriate for the study. The Employer's procedures should define how this should be calculated.

4.8 The conclusion of the local IR(ME)R review may be that the Employer will not be able to comply with the requirements of IR(ME)R whilst carrying out the research exposures. The Employer needs to decide what action to take in the eventuality of non-compliances. Possible options are: (1) contact the study sponsor and ask them to submit an amendment to the REC correcting the problem before any research exposures take place; (2) contact the study sponsor and ask them to correct the problem the next time an amendment is submitted to the REC; (3) decline to take part in the research study. The Employer's solution (s) to IR(ME)R non-compliances should be described in the organisation's IR (ME)R research procedure.

5 Using dose constraints or target levels of dose

5.1 It is expected that a dose constraint will be used for all research studies involving standard radiological procedures where no direct medical benefit is expected. A local dose constraint should be set in terms of an individual effective dose or equivalent dose over a defined time period, such as the study period or per calendar year for open-ended studies. The dose constraint should be set with consideration of the lead MPE's assessment of the total protocol dose and with sufficient flexibility to allow for variations due to local practice and patient population. The dose constraint is likely to be set for standard-sized subjects who are representative of the study population. It is not practical to set a dose constraint for each participant.

5.2 An alternative approach is to set dose constraints for individual research exposures using dose indicators such as dose-area-product for x-rays, dose-length-product for CT and activity for nuclear medicine and PET. If the research exposures are going to be carried out using standard clinical protocols, the local DRLs could be adopted for this purpose. It would then be possible to carry out regular audits of the dose indicators for the research exposures, calculate average values of the dose indicators, and compare these to the dose constraints for individual research exposures. Finally, the number of exposures undertaken can be compared against the study schedule.

5.3 Target dose levels should be planned for participants in experimental diagnostic or therapeutic exposures where direct benefit to participants is expected. This may need to be assessed on an individual participant basis or may be based on standardised protocols, for example, absorbed dose to the tumour with due consideration of absorbed doses to an organ at risk for radiotherapy.

References

HRA 2018 Ionising Radiation—Defining Research Exposures *IRAS Website* https://myresearchproject.org.uk/help/hlpradiation.aspx (accessed 4 March 2023)

HRA 2020a *Generic Ionising Radiation Risk Statements* (London: Health Research Authority) 4th edn https://hra-decisiontools.org.uk/consent/docs/Generic-ionising-radiation-risk-statements_v4_October2020.pdf (accessed 4 March 2023)

HRA 2020b *MPE and CRE Review Procedures.* 6.3.4, *IRAS Website.* 6.3.4 https://myresearchproject.org.uk/help/hlpradiationmpecre.aspx (accessed 4 March 2023)

HRA 2021 *Collated Question-Specific Guidance for Project Filter. IRAS Website.* 14th edn https://myresearchproject.org.uk/help/hlpcollatedqsg-sieve.aspx (accessed 4 March 2023)

HRA 2022 *Radiation Assurance. IRAS Website* https://myresearchproject.org.uk/help/hlpradiationassurance.aspx (accessed 4 March 2023)

HRA 2023 IRAS—Integrated Research Application System. 6.3.4 *IRAS Website. 6.3.4* https://myresearchproject.org.uk/ (accessed 4 March 2023)

HRA no date Health research authority *HRA Website* www.hra.nhs.uk (accessed 4 March 2023)

HRA, HCRW, HSC and NRS 2020a *Clinical Radiation Expert (CRE) Review Procedure* (Health Research Authority, Health and Social Care (Northern Ireland), Health and Care Research Wales, NHS Research Scotland) https://myresearchproject.org.uk/Help/Help%20Documents/CRE_Review_Procedure_FINAL_v3_0_21_October_2020.pdf (accessed 12 April 2024)

HRA, HCRW, HSC and NRS 2020b *Medical Physics Expert (MPE) Review Procedure Contents* (Health Research Authority, Health and Social Care (Northern Ireland), Health and Care Research Wales, NHS Research Scotland) https://myresearchproject.org.uk/Help/Help%20Documents/Medical_Physics_Expert_Review_Procedure_FINAL_v3_0_21_October_2020.pdf (accessed 12 April 2024)

IR(ME)R 2017 *The Ionising Radiation (Medical Exposure) Regulations* www.legislation.gov.uk/uksi/2017/1322 (accessed 31 January 2023)

IR(ME)R(NI) 2018 *The Ionising Radiation (Medical Exposure) Regulations (Northern Ireland)* https://legislation.gov.uk/nisr/2018/17 (accessed 31 January 2023)

IOP Publishing

Medical and Dental Guidance Notes (Second Edition)
A good practice guide on all aspects of ionising radiation protection in the clinical environment:
IPEM Report 113
John Saunderson, Mohamed Metwaly, William Mairs, Philip Mayles, Lisa Rowley and Mark Worrall

Appendix 20

Dosimetry in radiomolecular therapy

Introduction

A20.1 The methodology most commonly used for internal dosimetry in humans in contemporary literature is the MIRD (Medical Internal Radiation Dose) formalism, developed by the Society of Nuclear Medicine and Molecular Imaging (SNMMI). The MIRD committee has published guidelines, pamphlets, dose reports, and software (SNMMI no date).

A20.2 The other established methodology developed by the ICRP is very similar to the approach used in the MIRD methodology (ICRP 2015a, 2015b). The calculation of the absorbed dose is fundamentally similar in both systems. The MIRD pamphlet No. 21 (Bolch *et al* 2009) provides clarification on the different values and nomenclatures that have been used in the field of radiopharmaceutical dosimetry. A similar resource and methodology for dosimetry are provided on the Radiation Dose Assessment Resource (RADAR) website (RADAR no date).

A20.3 Traditional dose calculation algorithms have calculated dosimetry for standard phantoms—human models with known geometries. More recently, with the development of Monte Carlo and dose point kernel techniques, patient-specific dosimetry is beginning to be studied and employed (Bardiès and Strigari 2022).

Dosimetry procedure

A20.4 The dosimetry procedure should be specified in a protocol that takes into consideration the following:

1) Organ-based or voxel-based dose estimation.

doi:10.1088/978-0-7503-2332-1ch40

   a) The MIRD system has been extended to allow both dosimetry at the organ level (based on standard phantoms) and also at the voxel level, which allows for dose estimates of tumours within organs and also doses to patient-specific organs in terms of size and shape. A decision should be made as to which form of calculation is to be used. This will depend on the type of imaging available and the availability of appropriate software.
2) Type of imaging. Although whole-body imaging is routinely used, it is essential for at least one, if not more, SPECT/CT scan to be included in the imaging protocol, depending on resources and camera time available. For voxel-based dosimetry, multiple SPECT scans are the only technique.
3) Quantitative imaging of therapy isotopes and camera calibration.
   a) Care should be taken to develop a calibration procedure that is able to provide a quantitative assessment of activity in the organs and tumours. This should consider the quantification of both whole-body and SPECT imaging and follow the imaging protocol to be used for patients.
   b) Calibration is dependent on the choice of calibration phantoms to allow an assessment of partial volume effects by considering both large (organ-sized) and small (tumour-sized) objects, or voxel quantification.
   c) Dead time correction may be required in imaging high activities in patients shortly after administration. There should be a measurement of its effects and a method of correction for each radionuclide to be imaged if this is to be attempted.
   d) Reconstruction parameters, including attenuation correction, scatter correction, resolution modelling, and post-filtering will affect quantification. An assessment of the most appropriate reconstruction should be undertaken as part of the calibration procedure.
4) Choice of time points.
   a) For imaging-based dosimetry, the time-integrated activity in source regions or voxels must be determined from serial quantitative imaging. The temporal distribution and number of imaging points are important for defining the values of the measured time-activity curves.
   b) The number of samples required is related to the number of compartments, or separate exponential terms, in the source region time–activity function, whereas their optimal timing is related to the clearance rate for each exponential term. Rapidly clearing activity, for example, will necessitate frequent and early sampling, whereas slowly clearing activity requires fewer and more widely dispersed measurements.
   c) The time point for imaging should be chosen to represent a sampling of the expected clearance rates, and optimisation can be difficult

given the limitations of the number of occasions on which access to a gamma camera is available.

5) Imaging protocol.
   a) Imaging protocols should be developed that take into account the need for repeated quantitative imaging using planar and SPECT imaging. The protocol should be consistent with the calibration procedure. The number and type of images will be dependent on resources and camera time available but should take into account the need to cover sufficient time points in order to sample the kinetics of elimination in a patient.
   b) Included in this should be a method of recording and reproducing patient position for each scan, and ensuring all imaging procedures are recorded and according to protocol, including bed positions, timings and camera radii.
6) Registration of images.
   a) Before regions-of-interest (ROIs) or volumes-of-interest (VOIs) can be defined, the registration of sequential images should be considered. If the image series can be accurately registered to a reference scan from the sequence, a single volume of interest defined on the reference scan can be applied to all the images in the time series. If the serial images cannot be accurately registered, organ and tumour volumes should be redefined on each of the serial scans to generate the appropriate time–activity curves. The variety of registration algorithms, from simple rigid registration to multiple deformable registrations can affect the results of VOI activity considerably (Lu *et al* 2015). Centres should determine the availability of software and an assessment of errors introduced by different techniques before deciding on the implementation of a final technique.
   b) If voxel-based dosimetry is to be used, then all images need to be co-registered as closely as possible as the effects of misregistration are more critical.
7) VOI definition.
   a) The approach to the determination of ROI/VOIs must be consistent with the approach used in the calibration procedure. In general, if reasonable quality CT scans are available, then the most direct definition is to outline organs using the CT scan. There are advantages and disadvantages to performing this by hand or a commercial system developed to segment CT images.
   b) The partial volume effect must be considered, and this must have been approached in the calibration procedure so that a range of volumes have been calibrated to cover a range of tumour/organ sizes. Alternative methods that propose using larger ROIs to compensate for partial volumes may be easier to operate. Still, they may not adequately correct for activity from the background or from other close organs.

8) Time-activity curve fitting.

The counts measured in each registered organ (or voxel) will produce the time activity curve needed to calculate cumulated activity.

a) As the number of time points is limited, great care must be taken to fit the data in a meaningful fashion. The choice of fitting to trapezoidal, monoexponential, biexponential, or multiexponential fits will often give varying results depending on the number of data points and the spread of timing. The choice of fit will partly be given by the understanding of the kinetics it is representing. Non-instantaneous uptake may need to be modelled by trapezoidal estimation of the uptake phase, and then exponential fitting of the elimination phase. In particular, care should be taken with the fitting of the final elimination phase, the assumption of physical decay after the final data point may lead to higher dose estimation.
b) The goodness of fit will also vary with the choice of model and the number of data points. Great care should be taken to ensure fits are meaningful and to determine the errors associated with each fit. This will, in turn, influence the accuracy of the dose calculated.
c) Many programmes can produce curve fits. One should be chosen that allows flexibility in the choice of model and analysis of the errors of the fit.

9) S factors and phantom models.

a) The calculation of dose from cumulated activity relies on pre-calculated factors (S factors or dose factors) relating the deposition of energy from source radionuclides in a source tumour or organ to the target body organs, (In the case of voxels this is purely energy self-deposition) The use of S factors is the only form of dosimetry without access to Monte Carlo patient-specific modelling.
b) S factors are based on model patient phantoms. The original MIRD model phantom was based on a stylised mathematical representation of patient organs and used for many years in MIRDOSE 3 and Olinda1 software systems. These phantoms of organ masses have been modified by recommendations in ICRP Publication 89 (ICRP 2002) and have been generated to update the S factors used in Olinda2.
c) Through development, there is now a family of phantoms, designated as a Newborn, a One-Year-Old, a Five-Year-Old, a Ten-Year-Old, a Fifteen-Year-Old and an Adult (Male and female). The user must ensure that the most appropriate phantom for the patient is selected.
d) Tumour models are only provided by sphere and ellipse models at present. Although it is hard to extrapolate, it is recommended that a tumour size is selected that reflects the mass of the tumour.

10) Organ mass adjustment.

a) All the dose factors used are based on the standardised phantoms. In order to reflect the individual patient dose, these dose factors should be adjusted to reflect the true organ size of each patient. The variation of organ dose with size is obviously of fundamental importance when considering individual dosimetry rather than an average group dose, and mass adjustment should be performed.

b) Masses are best determined from a contemporary, high-quality CT or MR scan where organs can be segmented at the best resolution. Organ mass determination from lower resolution diagnostic scans or SPECT scans will result in larger masses and, therefore, an underestimation of dose.

11) External dose rate measurements.
   a) The use of external dose rate measurements as an addition to imaging data for whole-body dose estimates is beneficial and easy to perform, as it can be performed at additional time points between images.
   b) In addition, these measurements should be made for radiation protection purposes (see chapter 16, *Patients leaving the hospital after administration of radioactive substances)*, for the prediction of discharge, and for generation restriction advice.
   c) Measurements should be taken in a standardised fashion in fixed or repeatable geometries, and anterior and posterior measurements should be undertaken to generate geometric means. Ceiling-mounted or wall-mounted detectors have advantages if specialised facilities exist.
   d) Background measurements should also be made at each measurement point.
   e) The choice of instrumentation used should consider the sensitivity, range, stability, repeatability, and accuracy of such devices for the radionuclides to be measured. Pure beta whole-body measurements require particular care.

12) Blood sampling.
   a) Samples of blood taken at similar time points related to the kinetics are often used to determine the dose to the bone marrow.
   b) Blood samples should be clearly labelled according to local practices, especially at sample time. Counting samples and standards should be prepared according to the needs of the counting device used.
   c) Calibration of the sample counting device for each radionuclide used must be performed before starting any counting work, and a check should be made that the sample activities counted are below the dead-time characteristics of the detectors used.
   d) Counting times should be sufficient to allow accurate statistics of counts.

13) Error assessment.
   a) Currently, the uncertainty associated with internal dosimetry calculations is very large compared to that of external therapy calculations. Expected errors in external beam doses are less than 5%, while the most accurate internal dosimetry estimates one can currently achieve have errors of at least 20%.

   b) There is now a consensus that doses should be reported with an estimation of the error in the final calculated value. The sequence of activities involved in the dosimetry calculation (activity measurement, quantitative imaging, VOI definition, time–activity curve (TAC) analysis, dose factor adjustment, etc, all contribute towards the final uncertainty of the reported doses. Reported doses are best given with an uncertainty budget showing the final uncertainty of the result.
   c) The procedures for generating a combined uncertainty estimation are best outlined in the Guide to the Expression of Uncertainty in Measurement (JCGM 2008).
14) Biological Effective Dose.
   a) In molecular radiotherapy (MRT), patients may experience different radiopharmaceutical clearance rates. As a result, the dose rate to the critical organ is higher in patients with rapid clearance and may cause unexpected toxicity compared to patients with slow clearance. In order to account for the biological impact of different dose rates, radiobiological modelling is beginning to be applied to the analysis of radionuclide therapy patient data in order to derive the Biologically Effective Dose (BED).
   b) The generalised BED derivation has been based on the Medical Internal Radionuclide Dose Committee (MIRD) schema, assuming multiple source organs following exponential effective clearance of the radionuclide. The extended BED formalism is applied to red marrow dosimetry as well as kidney dosimetry, considering the cortex and the medulla separately, since both those organs are common dose-limiting in radionuclide therapy. The MIRD-based BED formalism is expected to be useful for patient-specific adjustments of activity and to facilitate the investigation of dose-toxicity correlations with respect to dose rate and tissue repair mechanisms.
15) Reporting of dose.
   a) With so many differing practices in the calculation of doses, a recommendation on the reporting of doses has been made by the European Association of Nuclear Medicine (EANM) Dosimetry Committee. EANM recommendation on Good Dosimetry Reporting (Lassmann *et al* 2011) lays out the relevant information to include, including uncertainty analysis, when providing a dose report. This is of particular use when comparing multicentre results where techniques and uncertainties may be different. This practice also brings MRT into better agreement with the established practices of external beam treatments.
16) Data recording
   a) The collection of high-quality data plays a vital role in the research and development of therapies. Clinical NHS sites in England are required to provide NHS England with monthly reports on external beam radiotherapy (EBRT) and brachytherapy treatments for the National Radiotherapy Dataset (RTDS); this resource is

acknowledged to have the potential to help plan services at a local and national level and evaluate future innovations. For MRT, there is currently no record in the UK of the number of treatments delivered, the number of centres offering MRT treatment, or the details or outcomes of the treatments themselves. However, to best assess practice and help improve the therapeutic efficacy of MRT, prospective MRT data collection should also be a core activity for clinical sites. Data collected will consist of details of treatment, dosimetry, biomarkers, and treatment outcome and should be retained in a national database. To date; data for Radium-233 therapies is now to be stored on the SACT (Systemic anticancer therapy) dataset, and data for selective internal radiotherapy (SIRT) treatments is stored in a national SIRT database.

17) Clinical trials.

Clinical trials involving MRT often have a dosimetry arm. The preparation of a clear and comprehensive dosimetry protocol that be used across many centres helps establish standardised and repeatable techniques.

Resources for dosimetry

A20.5 Because of the complexities involved in nuclear medicine dosimetry, as well as the array of methods and literature, there are several organisations and web resources which provide overviews and resources in the field.

a) **Medical Internal Radiation Dose (MIRD) Committee**

The Medical Internal Radiation Dose (MIRD) Committee of the SNMMI (SNMMI no date) pursues its mission to develop standard methods, models, and mathematical schema for assessing internal radiation dose from administered radiopharmaceuticals. The MIRD Committee continues to develop basic tools, methods, and guidance for the effective application of dosimetry techniques in clinical practice. Included is guidance for quantitative imaging, voxel dosimetry, published dose factors, and calculations on specific radionuclides and agents.

b) **Radiation Dose Assessment Resource (RADAR)**

The RADAR website (RADAR no date) seeks to provide information on dose assessment models and methods for the 21st century. The goals are to bring together the various resources that exist in the areas of internal and external dose assessment, integrate them into a single system, and put them in your hands as quickly and efficiently as possible. This includes dose factors and phantom information, modelling tools, decay data, etc.

c) **OLINDA/EXM 2.0: Dosimetry modelling code**

Originally developed in 2004, OLINDA/EXM (Organ Level Internal Dose Assessment/Exponential Modelling) calculates radiation doses to different organs of the body from systemically administered radiopharmaceuticals and performs regression analysis on user-supplied biokinetic data'. BEDOLINDA/EXM uses the RADAR method of dose calculation (Stabin and Siegel 2003) and the dose conversion factors as supplied on the RADAR website. Version 2 calculates doses to 29 different organs in SI or traditional units for 10 whole-body phantoms and 5 specific organ phantoms (peritoneal cavity, prostate gland, head and brain, kidney, and spheres). OLINDA superseded the MIRDOSE 3.0 and 3.1 codes, which were widely used in the radiopharmaceutical industry and research community for internal dose calculations for radiopharmaceuticals (Stabin *et al* 2005). OLINDA/EXM version 1 was granted an FDA 510(k)—K03396 by the US Food and Drug Administration as a software device. Appropriate approvals for version 2 are being obtained. The code is now only commercially available from Hermes Medical (Hermes Medical Solutions 2022).

d) **IAEA Radiotracer Biodistribution Template (RaBiT)**

In an effort to standardise data organisation and reporting, a freely available data template has been developed to support summary reporting of tracer biodistribution within a patient, as a sequel to the EANM recommendation on Dosimetry Reporting. The template can be used as a common point of reference for dosimetry data. This template does not contain dosimetry calculations. However, the intention is that the basic information recorded here may be used for further organ-level dosimetry calculations (IAEA 2016b).

e) **EANM Dosimetry Committee**.

The EANM dosimetry committee has published practical guidelines to aid dosimetry practice and reporting (EANM 2022).

These include guidelines for pre-therapeutic dosimetry in benign thyroid disease, blood and bone marrow dosimetry in differentiated thyroid cancer therapy, bone marrow dosimetry and good practice of clinical dosimetry reporting.

The EANM also runs training courses in dosimetry through the ESNM in Vienna.

f) **Internal Dosimetry User Group**

The Internal Dosimetry Users Group (IDUG) is a UK group set up to share ideas and best practice for radionuclide dosimetry (IDUG no date). It has developed a strong training role within the UK, providing practical workshops in dosimetry techniques through BNMS and biennial meetings with the BIR. It also undertakes annual surveys of activity in MRT and dosimetry. It provides a quarterly meeting for members to discuss practical issues of dosimetry, share best practice and promote the application of dosimetry in more centres. Practical Simple guidance is being published through BNMS.

g) **BNMS Molecular Radiotherapy Group** (BNMS no date).

The BNMS has undertaken to promote cohesion between the many groups involved in molecular radiotherapy, including CERT (CRUK/ECMC), CTRad, IDUG, the HPA, UKRG, the NPL, IPEM, and the RCR.

A20.6 Other resources include:

a) IAEA—Radiation protection of patients (IAEA no date).
b) ICRP publications (ICRP 2015a, 2015b).
c) National Council on Radiation Protection and Measurements (NCRP) (NCRP no date).
d) IAEA human health campus website (IAEA 2016a).
e) Euramet MRTDosimetry, Project Number: 15HLT06—This European-funded project is a collaboration between European metrology institutions and clinical partners. It aimed to develop standard methods for the measurement of the absorbed dose received by patients, which will form a valuable basis for the implementation of standard dosimetry methods in molecular radiotherapy clinics and support more effective targeted therapy and treatments (EURAMET 2019).
f) The current project is building on the work of the previous EMRP project HLT11 MetroMRT (EURAMET 2016) in developing a robust measurement protocol and providing standardised methods, tobjects,ects and resources to support molecular radiotherapy clinics in setting up and validating dosimetry, thereby improving patient outcomes.

References

Bardiès M and Strigari L 2022 Dose point-kernels for radionuclide dosimetry *Monte Carlo Calculations in Nuclear Medicine* 2nd edn (Bristol: IOP Publishing) pp 4–20

BNMS no date *The BNMS Molecular Radiotherapy Group, BNMS Website* (British Nuclear medicine Society) https://bnms.org.uk/page/The-BNMS-Molecular-Radiotherapy-Group (accessed 7 September 2022)

Bolch W E, Eckerman K F, Sgouros G and Thomas S R 2009 MIRD pamphlet No. 21: a generalized schema for radiopharmaceutical dosimetry—standardization of nomenclature *J. Nucl. Med.* **50** 477–84

EANM 2022 Dosimetry: guidelines *EANM Website* (European Association of Nuclear Medicine) https://eanm.org/publications/guidelines/dosimetry/ (accessed 7 September 2022)

EURAMET 2016 *Metrology for Molecular Radiotherapy (HLT11)* ed A Robinson https://euramet.org/research-innovation/search-research-projects/details/project/metrology-for-molecular-radiotherapy (accessed 8 March 2023)

EURAMET 2019 *Metrology for Clinical Implementation of Dosimetry In Molecular Radiotherapy (15HLT06)* ed A Robinson https://euramet.org/research-innovation/search-research-projects/details/project/metrology-for-clinical-implementation-of-dosimetry-in-molecular-radiotherapy (accessed 8 March 2023)

Hermes Medical Solutions 2022 HybridViewer Dosimetry™ & OLINDA/EXM® 2.0 *Hermes website* https://hermesmedical.com/our-software/dosimetry/olindaexm/ (accessed 8 March 2023)

IAEA 2016a Human health campus *IAEA Website* (Vienna: International Atomic Energy Agency) https://humanhealth.iaea.org/ (accessed 7 September 2022)

IAEA 2016b IAEA Radiotracer Biodistribution Template (RaBiT) *IAEA Website* (Vienna: International Atomic Energy Agency) https://humanhealth.iaea.org/HHW/MedicalPhysics/NuclearMedicine/InternalDosimetry/iaeaBioDistributionTemplate/ (accessed 7 September 2022)

IAEA no date Radiation protection of patients *IAEA Website* (Vienna: International Atomic Energy Agency) https://iaea.org/resources/rpop (accessed 7 September 2022)

ICRP 2002 Basic anatomical and physiological data for use in radiological protection reference values. ICRP Publication 89 *Ann. ICRP* **32** https://icrp.org/publication.asp?id=ICRP%20Publication%2089 Publication 89 (accessed 27 April 2024)

ICRP 2015a Radiation dose to patients from radiopharmaceuticals: a compendium of current information related to frequently used substances. ICRP Publication 128 *Ann. ICRP* ed S Mattsson *et al*Vol 44 https://www.icrp.org/publication.asp?id=ICRP%20Publication%20128 (accessed 27 April 2024)

ICRP 2015b Occupational intakes of radionuclides: part 1 *Ann. ICRP* **44** Publication 130 https://www.icrp.org/publication.asp?id=ICRP%20Publication%20130 (accessed 27 April 2024)

IDUG no date Welcome to Internal Dosimetry User Group *IDUG Website* (Internal Dosimetry User Group) https://idug.org.uk (accessed 8 March 2023)

JCGM 2008 *Evaluation of Measurement Data—Guide to the Expression of Uncertainty in Measurement. JCGM* 100:2008, *Joint Committee for Guides in Metrology*. https://bipm.org/documents/20126/2071204/JCGM_100_2008_E.pdf (accessed 27 April 2024)

Lassmann M, Chiesa C, Flux G and Bardiès M 2011 EANM Dosimetry Committee guidance document : good practice of clinical dosimetry reporting *Eur. J. Nucl. Med. Mol. Imaging* **38** 192–200

Lu W, Wang J and Zhang H H 2015 Computerized PET/CT image analysis in the evaluation of tumour response to therapy *Br. J. Radiol.* **88** 20140625

NCRP no date *Reports Shop, NCRP Website* (Bethesda, MD: National Council on Radiation Protection & Measurements) https://ncrponline.org/product-category/reports (accessed 8 March 2023)

RADAR no date *Welcome to RADAR—The RAdiation Dose Assessment Resource, RADAR Website* https://doseinfo-radar.com/ (accessed 7 September 2022)

SNMMI no date *Committee on Medical Internal Radiation Dose (MIRD), SNMMI Website* https://www.snmmi.org/AboutSNMMI/CommitteeContent.aspx?ItemNumber=12475 (accessed 27 April 2024)

Stabin M G and Siegel J A 2003 Physical models and dose factors for use in internal dose assessment *Health Phys.* **85** 294–310

Stabin M G, Sparks R B and Crowe E 2005 OLINDA/EXM: the second-generation personal computer software for internal dose assessment in nuclear medicine *J. Nucl. Med.* **46** 1023–7 https://jnm.snmjournals.org/content/jnumed/46/6/1023.full.pdf (accessed 8 March 2023)

IOP Publishing

Medical and Dental Guidance Notes (Second Edition)
A good practice guide on all aspects of ionising radiation protection in the clinical environment:
IPEM Report 113
John Saunderson, Mohamed Metwaly, William Mairs, Philip Mayles, Lisa Rowley and Mark Worrall

Appendix 21

The role of the radioactive waste adviser

A21.1 The Radioactive Waste Adviser (RWA) is an individual that meets the criteria of competence specified by the environment agencies (SEPA, EA, NIEA and NRW) and is suitable to advise the employer. Any employer with a permit under the Environmental Permitting Regulations 2016 (EPR 2016) or an authorisation under the Radioactive Substances Act 1993 (RSA 1993, RSA(A)R(NI) 2011) to accumulate and dispose of waste or a permit under the Environmental Authorisations (Scotland) Regulations 2018 (EA(S)R 2018) to manage waste must appoint an RWA (SEPA 2020). The need to appoint an RWA is a requirement of radioactive substances legislation, which derives from Article 68 of the Euratom Basic Safety Standards Directive—Directive 2013/59 Euratom (BSSD) (European Union 2014).

A21.2 The environment agencies have published a syllabus for RWAs on the SEPA website (SEPA *et al* 2018). An individual can be awarded a Certificate of Core Competence from an approved RWA Assessing Body. RPA2000 (RPA2000 2019) is currently the only assessing body that operates a competence certification scheme to assess the competence of individuals wishing to act as RWAs. Re-accreditation takes place on a five-year basis.

A21.3 Radioactive Waste Advisers must be appointed in writing by the Employer detailing the scope of the advice they are required to give. The role of the RWA is to advise on compliance—responsibility for ensuring compliance remains with the permit holder. The environment agencies have issued guidance on radioactive waste advisers that includes the roles and responsibilities of both permit holders and RWAs and the adequacy of the RWA in the role (SEPA *et al* 2020).

doi:10.1088/978-0-7503-2332-1ch41

A21.4 The RWA must be consulted (at a minimum) on the following matters (SEPA *et al* 2020):

a) achieving and maintaining an optimal level of protection of the environment and members of the public;
b) accepting into service adequate equipment and procedures for measuring and assessing the exposure of members of the public and radioactive contamination of the environment;
c) checking the effectiveness and maintenance of equipment as described in the point above and ensuring the regular calibration of measuring instruments.

A21.5 To achieve this, the RWA may advise on the following (SEPA *et al* 2020):

d) optimisation and establishment of appropriate dose constraints;
e) plans for new installations and the acceptance into service of new or modified radiation sources in relation to any engineering controls, design features, safety features and warning devices relevant to radiation protection;
f) appropriate radiation monitoring instrumentation;
g) quality assurance;
h) environmental monitoring programme;
i) arrangements for radioactive waste management;
j) arrangements for the prevention of accidents and incidents;
k) preparedness and response in emergency exposure situations;
l) investigation and analysis of accidents and incidents and appropriate remedial actions;
m) preparation of appropriate documentation, such as written procedures.

A21.6 While RWAs may be involved day-to-day in the processes, it is important that there be some level of audit included in ensuring compliance. The frequency of audits should be determined on the basis of risk assessment.

References

EA(S)R 2018 *The Environmental Authorisations (Scotland) Regulations 2018* http://legislation.gov.uk/ssi/2018/219 (accessed 31 January 2023)

EPR 2016 *Environmental Permitting (England and Wales) Regulations* https://legislation.gov.uk/uksi/2016/1154 (accessed 31 January 2023)

European Union 2014 *Council Directive 2013/59/Euratom of 5 December* 2013 *Laying Down Basic Safety Standards for Protection Against the Dangers Arising from Exposure to Ionising Radiation…* https://eur-lex.europa.eu/legal-content/EN/TXT/?uri=OJ:L:2014:013:TOC (accessed 31 January 2023)

RPA2000 2019 *RWA Certification Scheme, RPA2000 Website* http://rpa2000.org.uk/rwa-certification-scheme/ (accessed 6 March 2023)

RSA 1993 *Radioactive Substances Act 1993* http://legislation.gov.uk/ukpga/1993/12 (accessed 11 February 2023)

RSA(A)R(NI) 2011 *The Radioactive Substances Act* 1993 *(Amendment) Regulations (Northern Ireland)* https://legislation.gov.uk/nisr/2011/290 (accessed 31 January 2023)

SEPA 2020 *Radioactive Waste Advisers, SEPA Website* https://sepa.org.uk/regulations/radio-active-substances/radioactive-waste-advisers/ (accessed 6 March 2023)

SEPA, EA, NRW and NIEA 2018 *Radioactive Waste Adviser Syllabus* 2nd edn (Environment Agency, Natural Resources Wales, Northern Ireland Environment Agency, Scottish Environment Protection Agency) https://sepa.org.uk/media/36075/rwa-syllabus.pdf (accessed 6 March 2023)

SEPA, NRW, EANI and EA 2020 *Environment Agencies' Statement on Radioactive Waste Advisers (RWA-S-1 Version 2.1)* (SEPA) https://sepa.org.uk/media/520368/radioactive-waste-advisers-statement.pdf (accessed 25 February 2023)

Appendix 22

Authorities and organisations of interest

Regulators

In April 2021, the UK Government published *An overview of the UK regulatory framework for all aspects of radiological and civil nuclear safety across the UK* (HM Government 2021). Tables (3) to (8) of Annex B of that document outline the legislative frameworks for ionising radiation and includes a complete list of which Government Departments and Regulators are responsible for which regulations. The Regulators listed below are those of most relevance to medical and dental practices.

Ionising Radiations Regulations (IRR) and Radiation (Emergency Preparedness and Public Information) Regulations (REPPIR)	
Health & Safety Executive (HSE) http://www.hse.gov.uk/radiation/ionising For notifiable incidents email irrnot@hse.gov.uk	Regulatory authority for IRR 2017 ***in Great Britain*** (except transport—see ONR)
Health and Safety Executive for Northern Ireland (HSENI) https://www.hseni.gov.uk/topics/radiation For notifiable incidents email mail@hseni.gov.uk	Regulatory authority for IRR(NI) 2017 ***in Northern Ireland***

doi:10.1088/978-0-7503-2332-1ch42

(Continued)

Ionising Radiation (Medical Exposure Regulation (IRMER)	
Care Quality Commission (CQC) https://www.cqc.org.uk/guidance-providers/ionising-radiation/ionising-radiation-medical-exposure-regulations-irmer For reporting notifiable incidents see https://www.cqc.org.uk/guidance-providers/ionising-radiation/reporting-irmer-incidents	Enforcing authority for IRMER 2017 ***in England***
Regulation and Quality Improvement Authority (RQIA) https://www.rqia.org.uk/guidance/guidance-for-service-providers/ionising-radiation-(medical-exposure)-regulations includes guidance and forms for notification of incidents to be sent to registration@rqia.org.uk.	Enforcing authority for the IRMER(NI) 2018 ***in Northern Ireland***
Healthcare Improvement Scotland https://www.healthcareimprovementscotland.org/our_work/inspecting_and_regulating_care/ionising_radiation_regulation.aspx includes guidance and forms for notification of incidents to be sent to his.irmer@nhs.scot.	Enforcing authority for IRMER 2017 ***in Scotland***
Health Inspectorate Wales https://www.hiw.org.uk/notifying-irmer-incidents includes guidance and forms for notification of incidents to be sent to HIW.IRMERIncidents@gov.wales.	Enforcing authority for IRMER 2017 ***in Wales***
Radioactive Substances Regulations, including transport	
Environment Agency (EA) https://www.gov.uk/government/collections/radioactive-substances-regulation-for-non-nuclear-sites For notifiable incidents call 0800 80 70 60	Regulatory authority for the EPR 2016 ***in England***
Northern Ireland—Northern Ireland Environment Agency (NIEA) https://www.daera-ni.gov.uk/articles/radiation-overview	Regulatory authority for the RSA 1993 and CDG(NI) 2010 ***in Northern Ireland***
Scottish Environmental Protection Agency (SEPA) https://www.sepa.org.uk/regulations/radioactive-substances/ and https://www.daera-ni.gov.uk/articles/transport-transfrontier-justification For notifiable incidents call 0800 80 70 60	Regulatory authority for EA(S)R 2018 ***in Scotland***

Natural Resources Wales (NRW) http://naturalresourceswales.gov.uk/permits-and-permissions/non-nuclear-radioactive-substance-sites For notifiable incidents, call 0300 065 3000	Regulatory authority for EPR 2016 ***in Wales***
Office for Nuclear Regulation (ONR) http://www.onr.org.uk/transport Notification of transport incidents—see https://www.onr.org.uk/transport/incident-notifications.htm and https://www.onr.org.uk/notify-onr.htm	Regulatory authority for the transport of radioactive materials ***in Great Britain***, both CDC 2009 and IRR 2017 regulations.

Other UK authorities

Administration of Radioactive Substances Advisory Committee (ARSAC) http://www.arsac.org.uk/ *(forwards to gov.uk page)*	Issue licences under IRMER on behalf of health ministers in the ***UK***.
Committee on Medical Aspects of Radiation in the Environment (COMARE) https://www.gov.uk/government/groups/committee-on-medical-aspects-of-radiation-in-the-environment-comare	Advises on the health effects of natural and man-made radiation, both ionising and non-ionising.
Health and Care Research Wales https://healthandcareresearchwales.org	Works to promote research into diseases, treatments, services, and outcomes in ***Wales***.
Health Facilities Scotland www.hfs.scot.nhs.uk	Radiological equipment in ***Scotland*** and for radiation incidents; the Incident Reporting and Investigation Centre at the same address.
Health Research Authority (HRA) https://www.hra.nhs.uk/about-us/committees-and-services/technical-assurances/radiation-assurance	Provides guidance on the approval of research involving ionising radiation. Provides a ***UK-wide*** Research Ethics Service.
Medicines and Healthcare Products Regulatory Agency (MHRA) http://www.mhra.gov.uk/Safetyinformation/Safetywarningsalertsandrecalls https://www.gov.uk/report-problem-medicine-medical-device https://yellowcard.mhra.gov.uk	Responsible for ensuring that medicines and medical devices work and are acceptably safe.

(*Continued*)

National Physical Laboratory (NPL) https://www.npl.co.uk/ionising-radiation	The source of primary calibrations for radiation metrology in the ***UK***.
National Services Scotland, Incident Reporting and Investigation Centre (IRIC) https://www.nss.nhs.scot/browse/health-facilities/incidents-and-alerts	Reporting of adverse incidents and near misses involving medical devices, social care, estates, and facility equipment used in ***Scotland***'s health and social care services.
NHS Research Scotland https://www.nhsresearchscotland.org.uk	Promotes and supports clinical and translational research in ***Scotland***.
Northern Ireland Adverse Incidents Centre (NIAIC) www.health-ni.gov.uk/articles/introduction-northern-ireland-adverse-incidents-centre-niaic	***Northern Ireland*** Centre for the voluntary reporting and investigation of adverse incidents, including medical devices.
UK Health Security Agency https://www.gov.uk/topic/health-protection/radiation https://www.ukhsa-protectionservices.org.uk	Undertakes research to advance knowledge about protection from the risks of these radiations; provides laboratory and technical services; runs training courses; provides expert information; and has a significant advisory role ***in the UK***.

Some UK professional organisations

AHCS—Academy for Healthcare Science https://www.ahcs.ac.uk	**ACS**—Association of Clinical Scientists https://assclinsci.org
AURPO—Association of University Radiation Protection Officers http://www.aurpo.org.uk	**AXREM**—Association of X-ray Engineering Manufacturers https://www.axrem.org.uk
BDA—British Dental Association https://bda.org	**BIR**—British Institute of Radiology https://www.bir.org.uk
BNCS—British Nuclear Cardiology Society https://www.bncs.org.uk	**BNMS**—British Nuclear Medicine Society https://www.bnms.org.uk
BVA—British Veterinary Association https://www.bva.co.uk	**CoR / SoR**—College of Radiographers / Society of Radiographers https://www.collegeofradiographers.ac.uk, https://www.sor.org
CIB—Clinical Imaging Board https://www.rcr.ac.uk/clinical-radiology/service-delivery/clinical-imaging-board-cib	**CLEAPSS**—Consortium of Local Education Authorities for the Provision of Science Services https://science.cleapss.org.uk
CTRad—NCRI Clinical and Translational Radiotherapy Research Working Group https://www.ncri.org.uk/groups/radiotherapy-group	**IPEM**—Institute of Physics and Engineering in Medicine https://www.ipem.ac.uk
RAMTUC—Radioactive material Transport Users Committee https://ramtuc.org.uk	**RCR**—Royal College of Radiologists https://www.rcr.ac.uk
SCIN—Scottish Clinical Imaging Network https://www.scin.scot.nhs.uk/	**SRP**—Society for Radiological Protection https://srp-uk.org
UKMPG—UK Mammography Physics Group https://www.ukmpg.org.uk	**UKRG**—UK Radiopharmacy Group https://ukrg-meetings.org.uk

Some international organisations, with useful information available online

AAPM—American Association of Physicists in Medicine https://aapm.org	**BSI**—British Standards Institute https://www.bsigroup.com/en-IE/standards
EANM—European Association of Nuclear Medicine https://www.eanm.org	**EFOMP**—European Federation of Organisations for Medical Physics https://www.efomp.org
IAEA—International Atomic Energy Agency https://www.iaea.org	**ICRP**—International Commission on Radiological Protection https://www.icrp.org
ICRU—International Commission on Radiation Units and Measurements https://www.icru.org	**IRPA**—International Radiation Protection Association https://www.irpa.net
NCRP—National Council on Radiation Protection and Measurement (USA) https://ncrponline.org	**SNMMI**—Society of Nuclear Medicine & Molecular Imaging (USA) https://www.snmmi.org

Reference

HM Government 2021 Guidance: how we regulate radiological and civil nuclear safety in the UK—an overview of the UK regulatory framework for all aspects of radiological and civil nuclear safety across the UK https://gov.uk/government/publications/how-we-regulate-radiologicaland-civil-nuclear-safety-in-the-uk (accessed 23 March 2023)

IOP Publishing

Medical and Dental Guidance Notes (Second Edition)
A good practice guide on all aspects of ionising radiation protection in the clinical environment: IPEM Report 113
John Saunderson, Mohamed Metwaly, William Mairs, Philip Mayles, Lisa Rowley and Mark Worrall

Appendix 23

Acts and regulations in the United Kingdom

All UK legislation is available to download from www.legislation.gov.uk.

In April 2021, the UK Government published *An overview of the UK regulatory framework for all aspects of radiological and civil nuclear safety across the UK* (HM Government 2021).

Tables (3) to (8) of Annex B of that document outline the legislative frameworks for (3) *occupational exposures*, (4) *nuclear safety*, (5) *public exposures and environmental protection*, (6) *medical and non-medical exposures*, (7) *transport of radioactive materials*, and (8) *emergency preparedness and response*. These tables describe the relevant acts of parliament, regulations, guidance, government departments, regulators, and UK nations to which these apply.

In addition to the legislation contained in tables (4) to (8) of the HM Government guidance, the following legislation is also of interest.

Legislation	Application	Regulators
Justification of Practices Involving Ionising Radiation Regulations 2004	UK	DEFRA
Medical Devices Regulations 2002	UK	MHRA
Health and Safety (Safety Signs and Signals) Regulations 1996	GB	HSE
Health and Safety (Safety Signs and Signals) Regulations (Northern Ireland) 1996	NI	HSENI

doi:10.1088/978-0-7503-2332-1ch43

Other sources

A briefer summary and links on '*Regulatory controls for radiation protection in the UK*' from February 2018 can be found at BEIS, HSE and ONR (2018).

Specific advice on regulations for Northern Ireland and for Scotland is also available at:

- https://www.netregs.org.uk/legislation/northern-ireland-environmental-legislation/current-legislation/radioactive-substances
- https://www.netregs.org.uk/legislation/scotland-environmental-legislation/current-legislation/radioactive-substances

References

BEIS, HSE and ONR 2018 *Guidance: regulatory controls for radiation protection in the UK. Exemption, notification, registering and licensing of radioactive substances and radiation generators in the UK GOV.UK Website* (Department for Business, Energy & Industrial Strategy, Health and Safety Executive, and Office for Nuclear Regulation) https://gov.uk/guidance/regulatory-controls-for-radiation-protection-in-the-uk (accessed 29 December 2022)

HM Government 2021 *Guidance: how we regulate radiological and civil nuclear safety in the UK—an overview of the UK regulatory framework for all aspects of radiological and civil nuclear safety across the UK* https://gov.uk/government/publications/how-we-regulate-radiological-and-civil-nuclear-safety-in-the-uk (accessed 23 March 2023)

Appendix 24

List of acronyms and abbreviations

The following abbreviations and acronyms are used in these guidance notes and other radiation protection publications.

A	
AC	Attenuation correction (e.g. *for SPECT or PET imaging)*
ACoP	Approved Code of Practice to IRR (*or other regulations such as REPPIR where indicated*)
ADN	European Agreement concerning the International Carriage of Dangerous Goods by Inland Waterways
ADR	European Agreement concerning the International Carriage of Dangerous Goods by Road
ADS	Approved dosimetry service
AEC	Automatic exposure control
ALARA	As low as reasonably achievable (*ALARP in non-UK publications*)
ALARP	As low as reasonably practicable (*ALARA in UK legislation*)
mA	Milliampere (*x-ray tube current*)
AOR	Control of Artificial Optical Radiation at Work Regulations (SI 2010/2987, SR 2010/180)
APR	Anatomically programmed radiography
ARSAC	Administration of Radioactive Substances Advisory Committee
AXREM	UK trade association representing the interests of suppliers of diagnostic medical imaging, radiotherapy, healthcare IT and care equipment in the UK
A&E	Accidents and emergency
B	
BAT	Best Available Techniques
BED	Biologically (or biological) effective dose
BEIS	Department for Business, Energy & Industrial Strategy
BIR	British Institute of Radiology
BNCS	British Nuclear Cardiology Society

BNMS	British Nuclear Medicine Society
BPM	Best Practical Means
BSI	British Standards Institute
BVA	British Veterinary Association

C

CBCT	Cone beam computed tomography
CBE	Co-operation between employers
CBRN	Chemical, biological, radiological and nuclear
CDG	Carriage of Dangerous Goods and Use of Transportable Pressure Equipment Regulations (SI 2009/1348, SI 2011/185, SR 2010/160, SR 2011/365 SI 2019/598, and SR 2019/111)
CEMFAW	Control of Electromagnetic Fields at Work Regulations (SI 2016/588, SR 2016/266)
CEN	European Committee for Standardisation
CENELEC	European Committee for Electrotechnical Standardisation
CERT	The CRUK, ECMC, and UK Radiopharmacy Group Taskforce
CF	Causative factor
CI	Chief investigator
CIB	Clinical Imaging Board
CIDI	Central Index of Dose Information
CLEAPSS	Consortium of Local Education Authorities for the Provision of Science Services
COMARE	Committee on Medical Aspects of Radiation in the Environment
CoR	College of Radiographers
cps	Counts per second
CQC	Care Quality Commission (*England*)
CRCE	Centre for Radiation, Chemical and Environmental Hazards (*from* 2021 *RCE*)
CRE	Clinical radiation expert
CritEx	Critical examination (*of radiation safety features and warning devices*)
CRUK	Cancer Research UK
CSAUE	Clinically significant accidental or unintended exposure
CT	Computed tomography
CTDI	CT dose index
$CTDI_{vol}$	Volume CT dose index
CTRad	The NCRI Clinical and Translational Radiotherapy Research Working Group
CTSA	Counter Terrorism Security Adviser
CTSIM	CT simulator (*radiotherapy planning*)

D

DAERA	Department of Agriculture, Environment and Rural Affairs (*Northern Ireland*)
DAP	Dose-area product (*dose is usually absorbed dose to air*)
DDI	Detector dose index
DECC	Department of Energy & Climate Change
DEFRA	Department for Environment Food & Rural Affairs
DEXA	Dual-energy x-ray absorptiometry
DGN	Dangerous goods note

DGSA	Dangerous goods safety adviser
DHSC	Department of Health and Social Care
DICOM	Digital Imaging and Communications in Medicine
DLP	Dose-length product
DMS	Dose management system
DoH	Department of Health (*may refer to either the Northern Ireland DoH or, prior to* 2018, *to the health function of the DHSC).*
DR	Diagnostic radiology
DRL	Diagnostic reference level
DWI	Drinking Water Inspectorate
DXA	Dual-energy x-ray absorptiometry

E

EA	Environment Agency (*England*)
EANM	European Association of Nuclear Medicine
EASR	Environmental Authorisations (Scotland) Regulations 2018 SSI 2018/219
EC	European Commission
ECMC	Experimental Cancer Medicine Centres
ED	Emergency Department
EMF	Electromagnetic field
En	England
Environment agencies	The Environment Agency (EA), the Northern Ireland Environment Agency (NIEA), Natural Resource Wales (NRW), and the Scottish Environmental Protection Agency (SEPA)
EPD	Electronic personal dosimeter
EPR	Environmental Permitting Regulations 2016
EPRR	Emergency preparedness, resilience and response
eq	Equivalent
EPR	Environmental Permitting (England and Wales) Regulations 2016 (as amended by the Environmental Permitting (England and Wales) (Amendment) Regulations 2018 No.110
ESC	European Society of Cardiology
ESR	European Society of Radiology
EU	European Union
E&W	England and Wales

F

FDD	(*X-ray*) focus-to-detector distance
FDG	Fluorodeoxyglucose (*used for many PET scans*)
FERA	Food and Environmental Research Agency
FFD	(*X-ray*) focus-to-film distance
FoV	Field of view
FRCR	Fellow of the Royal College of Radiologists
FSD	(*X-ray*) focus-to-surface (or skin) distance

G

GB	Great Britain (i.e. *England, Scotland and Wales*)
GBq	Gigabecquerels (1 000 000 000 Bq)
GBR	General binding rules

GFR	Glomerular filtration rate
GMC	General Medical Council
GMP	Good manufacturing practice (*a minimum standard that a medicines manufacturer must meet in their production processes*)
Gy	Gray (*unit of absorbed ionising radiation dose*)
H	
HASS	High-activity sealed source
HCl	Hydrogen chloride (*hydrochloric acid when in solution*)
HCRM	High consequence radioactive material
HCRW	Health and Care Research Wales
HCRW	Health Care Research Wales
HEPA	High efficiency particulate air (*as in 'HEPA filter'*)
HIS	Healthcare Improvement Scotland
HIW	Health Inspectorate Wales
$H_p(10)$	Personal dose equivalent at 10 mm depth ($H_p(0.07)$ *is used for skin and* $H_p(3)$ *for lens of eye*)
HPA	Health Protection Agency (*radiation section now part of UKHSA*), or Hospital Physicists Association (*now IPEM*)
HPLC	High-performance liquid chromatography
HRA	Health Research Authority
HSC	Health and Social Care (Northern Ireland)
HSE	Health and Safety Executive (Great Britain)
HSENI	Health and Safety Executive for Northern Ireland
HSWA	Health and Safety at Work, etc Act 1974
HVAC	Heating, ventilation, and air conditioning
HVL	Half value layer (*a measure of x-ray beam penetrating power*)
H&S	Health and safety
I	
IAEA	International Atomic Energy Agency
ICRP	International Commission on Radiological Protection
ICRU	International Commission on Radiation Units and Measurements
IDR	Instantaneous dose rate (*For IRR, defined as averaged over* 1 *min*)
IDUG	Internal Dosimetry Users Group (UK)
IEC	International Electrotechnical Commission
IFU	Instructions for use
IGA	Information Governance Alliance
IHA	Individual health assessment
IMRT	Intensity-modulated radiation therapy
INPO	Institute of Nuclear Power Operations
IORT	Intra-Operative Radiotherapy
IPEM	Institute of Physics and Engineering in Medicine
IPEMB	Institution of Physics and Engineering in Medicine and Biology (*now IPEM*)
IPL	Intense pulsed light
IPSM	Institute of Physical Sciences in Medicine (*now IPEM*)
IRAS	Integrated Research Approval System
IRMER or IR (ME)R	Ionising Radiation (Medical Exposure) Regulations (SI 2017/1322, SR 2018/17, and SI 2018/121)

IRPA	International Radiation Protection Association
IRR	Ionising Radiations Regulations (currently SI 2017/1075, SR 2017/229, and SI 2018/482; previously SI 1969/808, SR 1969/318, SI 1985/1333, SI 1999/3232, SR 2000/375)
IRR17	Ionising Radiations Regulations SI 2017/1075 and Ionising Radiations Regulations (Northern Ireland) SR 2017/229
ISO	International Organisation for Standardisation
IV	Intravenous(ly)

J

JoPIIRR	Justification of Practices Involving Ionising Radiation Regulations (SI 2004/1769, SI 2018/430)

K

KAP	Kerma-area product (*usually air kerma*)
keV	Kiloelectron-volt
kV	Kilovolt (*x-ray tube potential/voltage*)

L

L121	Work with ionising radiation: Ionising Radiations Regulations 2017: Approved Code of Practice and guidance L121 (Second edition), HSE 2018
L126	A Guide to the Radiation (Emergency Preparedness and Public Information) Regulations 2001, HSE 2002
LBD	Light beam diaphragm
LFPS	Learn from Patient Safety Events
LLW	Low level (*radioactive*) waste
LOLER	Lifting Operations and Lifting Equipment Regulations (SI 1998/2308, SR 1999/304)

M

m	Metre
mAs	Milliampere-second (*tube current-exposure time product for x-ray tubes*)
MBq	Megabecquerels
MDGN	Medical & Dental Guidance Notes A good practice guide on all aspects of ionising radiation protection in the clinical environment. First edition, IPEM 2002. This publication is the second edition.
MDD	Medical Devices Directive
MDR	Medical Devices Regulations
MEC	Medical Exposures Committee
MeV	Megaelectron-volt
MDT	Multidisciplinary Team
MHOR	Manual Handling Operations Regulations (SI 1992/2793, SR 1992/532)
MHRA	Medicines and Healthcare products Regulatory Agency
MIRD	Medical Internal Radiation Dose
MOD	Ministry of Defence
MPE	Medical Physics Expert
MR	Magnetic resonance
MRI	Magnetic resonance imaging
MRSE	Magnetic Resonance Safety Expert

MRT	Molecular radiotherapy
MSLA	Minimum school leaving age
mSv	Millisieverts
MUGA	Multigated acquisition

N

N/A	Not applicable
NaCTSO	National Counter Terrorism Security Office
NAIR	National arrangements for Incidents involving radioactivity
NCRI	National Cancer Research Institute
NCRP	National Council on Radiation Protection and Measurement (*USA*)
NEMA	National Electrical Manufacturers' Association (*USA*)
NET	Neuroendocrine tumours
NFCC	National Fire Chiefs Council
NHS	National Health Service
NHSBSP	NHS Breast Screening Programme
NI	Northern Ireland
NICE	National Institute for Health and Care Excellence
NIEA	Northern Ireland Environment Agency
NIHP	National Institute for Health Protection (*became UKHSA*)
NM	Nuclear medicine
NPL	National Physical Laboratory
NRPB	National Radiological Protection Board (*now part of UKHSA*)
NRLS	National Reporting and Learning System
NRS	National Research Scotland
NRW	Natural Resources Wales

O

OfW	Once for Wales Concerns Management System
OLINDA/EXM	Organ Level INternal Dose Assessment/EXponential Modelling
ONR	Office for Nuclear Regulation
OSL(D)	Optically stimulated luminescence (dosimeter)
OTIF	Intergovernmental Organisation for International Carriage by Rail

P

PACS	Picture archiving and communication system
Pb	Lead
Pb eq	Lead equivalent
PET	Positron emission tomography
PGD	Patient group direction
PHE	Public Health England (*radiation section now part of UKHSA*)
PIS	Patient Information Sheet
PMMA	Poly(methyl methacrylate) (*Perspex, acrylic glass, Lucite, Plexiglass acrylic*)
PPE	Personal protective equipment
PRA	Preliminary Research Assessment form
PRRT	Peptide receptor radionuclide therapy
PSD	Patient specific direction
PSP	Photostimulable phosphor
PTCOG	Particle Therapy Co-operative Group

PUWER	Provision and Use of Work Equipment Regulations
Q	
QA	Quality assurance
QC	Quality control
RADAR	RAdiation Dose Assessment Resource
RAMP	Radiation Protection Computer Code Analysis and Maintenance program (*USA*)
R	
RCE	Radiation, Chemical and Environmental Hazards Directorate of UKHSA (*formerly CRCE*)
RCR	Royal College of Radiologists
REC	Research Ethics Committee
REPPIR	The Radiation (Emergency Preparedness and Public Information) Regulations (SI 2019/703, SR 2019/185)
RF	Radiofrequency
RID	International Carriage of Dangerous Goods by Rail
RIPP	Radiation Incidents in Public Places (*NI*)
RIS	Radiology information system
RMU	Radiation monitoring units
ROI	Region of interest
RP	Radiation protection
RPA	Radiation Protection Adviser
RPA2000	Body certifying competence of radiation protection professionals include MPE, RPA, and RWA schemes (www.rpa2000.org.uk)
RPS	Radiation Protection Supervisor
RSA	Radioactive Substance Act 1993
RSC	Radiation Safety Committee
RSR	Radioactive Substances Regulations; (*for England and Wales—SI* 2016/1154, *SI* 2018/428; *for Northern Ireland—Radioactive Substances Act* 1993, *SR* 2003/208, *SR* 2011/289, *SR* 2011/290, *SI* 2005/2686 *SR* 2018/116;*for Scotland*—2018 *SSI* 2018/219)
RT	Radiotherapy
RTDS	National Radiotherapy Dataset (UK)
RTE	Radiotherapy errors
rTSH	Recombinant thyroid-stimulating hormone
RVS	Record and verify system
RWL	Recommended working life
S	
SASG	Specialty and Associated Specialty Grade (*doctors*)
SAUE	Significant accidental or unintended exposure
Sc	Scotland
SCIN	Scottish Clinical Imaging Network
SCoR	Society & College of Radiographers
SEPA	Scottish Environmental Protection Agency
SI	Statutory Instrument
SIG	Special interest group
SIRT	Selective internal radiotherapy
SmPC	Summary of Product Characteristics

SNMMI	Society of Nuclear Medicine & Molecular Imaging (*USA*)
SOP	Standard operating procedure
SPC	Summary of Product Characteristics
SPECT	Single-photon emission computed tomography
SPR	Scan projection radiograph (*also known as CT 'topogram' or 'scout view'*)
SR	Statutory Rules of Northern Ireland
SRP	Society for Radiological Protection
SSI	Scottish Statutory Instrument
SUV	Standardised uptake value (*for nuclear medicine administrations*)
T	
TAC	Time–activity curve
TADR	Time averaged dose rate (*usually averaged over an* 8 *h working day*)
TBq	Terabecquerels (1 000 000 000 000 Bq)
TI	Transport index
TLA	Three letter abbreviation
TLD	Thermoluminescent dosimeter
TMJ	Temporomandibular joint
TSO	The Stationary Office
TSRT	Towards Safer Radiotherapy
U	
UK	United Kingdom (i.e. *England, Northern Ireland, Scotland and Wales*)
UKCA	UK Conformity Assessed
UKHSA	UK Health Security Agency (*includes former NRPB/HPA/PHE radiation sections*)
UKRG	UK Radiopharmacy Group
UN	United Nations
V	
VLLW	Very low level (*radioactive*) waste
VOI	Volume of interest
W	
Wa	Wales
WHO	World Health Organisation

www.ingramcontent.com/pod-product-compliance
Ingram Content Group UK Ltd.
Pitfield, Milton Keynes, MK11 3LW, UK
UKHW060416100426
469818UK00002B/2